# Advanced Structured Materials

## Volume 11

*Series Editors*

Andreas Öchsner
Lucas F. M. da Silva
Holm Altenbach

For further volumes:
http://www.springer.com/series/8611

P. M. Visakh · Sabu Thomas
Arup K. Chandra · Aji. P. Mathew
Editors

# Advances in Elastomers I

## Blends and Interpenetrating Networks

Springer

*Editors*
P. M. Visakh
Centre for Nanoscience
and Nanotechnology
Mahatma Gandhi University
Kottayam, Kerala
India

Sabu Thomas
Kottayam, Kerala
India

Arup K. Chandra
Apollo Tyres Ltd
Gujarat
India

Aji. P. Mathew
Department of Applied Physics
and Mechanical Engineering
Lulea University of Technology
Luleå
Sweden

ISSN 1869-8433 ISSN 1869-8441 (electronic)
ISBN 978-3-642-20924-6 ISBN 978-3-642-20925-3 (eBook)
DOI 10.1007/978-3-642-20925-3
Springer Heidelberg New York Dordrecht London

Library of Congress Control Number: 2013934535



Printed on acid-free paper

Springer is part of Springer Science+Business Media (www.springer.com)

# Preface

The book, "Advances in Elastomers-I: Their Blends and Interpenetrating Networks" summarizes many of the recent technical accomplishments in the area of elastomer-based blends and interpenetrating networks (IPNs). As the title indicates, the book emphasizes the various aspects of preparation, structure, morphology, processing, properties, and applications of elastomer-based blends and IPNs, in a systematic and comprehensive manner. Recent advances in the development and characterization of multicomponent polymer blends and IPNs based on elastomers are discussed in detail. It is important to mention that till date, there are limited books published on the recent advances in the synthesis, morphology, structure, properties, and applications of elastomer-based blends and IPNs. The various chapters in the book discuss general purpose elastomers, special purpose elastomers, compounding, vulcanization, processing of elastomers, rubber/rubber blends: micro and nanostructured, rubber/thermoplastic blends: micro and nanostructured, rubber/thermoset blends: micro and nanostructured, interphase modification and compatibilization of rubber-based blends, elastomer-based IPNs, micro and nanofillers in rubber-based blends, magnetorheological elastomers and applications: ionizing radiation processing of elastomers and electro-elastic continuum models for electrostrictive elastomers.

In this sense, the content of the book is unique. It covers an up-to-date record on the major findings and observations in the field of elastomer-based blends and IPNs. The book is intended to serve as a "one stop" reference resource for important research accomplishments in this area. The chapters in the book have been contributed by prominent researchers from the industry, academia, and government/private research laboratories across the globe. This book will be a valuable reference source for university and college faculties, professionals, post-doctoral research fellows, senior graduate students, polymer technologists, and researchers from R&D laboratories working in the area of elastomer-based blends and IPNs. The first chapter, "Advances in Elastomers: Their Blends and Interpenetrating Networks", gives an overview of the advances in elastomers on state-of-art new challenges and opportunities. This chapter is essential for beginners in these fields as it provides a thorough basic understanding of the elastomer-based blends and IPNs. The following chapter on general purpose elastomers discusses the structure, chemistry, physics, and performance. The first part of this chapter gives an introduction to thermodynamics, kinetics, structure, and polymerization. Focus is made on specific elastomer structure, properties, and thermoplastic elastomers. Later in this chapter, several topics such as elastomer-filler compositions, elastomer blends, interpenetrating elastomer blends, formulation, compounding, and shape-memory polymers are presented. Finally, the future trends are discussed. In the third chapter, special purpose-based elastomers are explained with different types of elastomers used as special purpose applications. This chapter explains different elastomers such as butyl rubber nitrile rubber, silicone rubber, fluoro rubber, polyurethane rubber, polysulfide rubber, and phlorosulfonated polyethylene rubber. The synthesis, structure–property relationship, compounding, processing, and applications of each special purpose-based elastomers are discussed in detail.

A survey on compounding and vulcanization is done in the fourth chapter. The authors deal with compounding, different types of compounding equipments, processes, and compounding ingredients. Each of the topics, single-step processes, batch process compounding ingredients and their function, different types of vulcanizing agents and vulcanizing systems, and double networking of elastomers is detailed. Finally, details are added on the advantages and disadvantages of the compounding and vulcanization process. The fifth chapter is on processing of elastomers and comprises different subtopics. The first section contains the different processing methods used for the preparation of elastomer-based blends and composites such as latex blending, two-roll mixing, extrusion blending, and other processes and the second section deals with the different types of compounding equipment, processes such as single-step processes, batch process, and other processes and advantages and disadvantages of each process. The last section of this chapter discusses the fabrication of products such as compression molding, dipping, coating, and other types of products.

The following chapter is on Rubber/Rubber Blends: Micro and Nanostructured mainly focusing on different studies on rubber/rubber blends such as recent developments, different manufacturing methods, and different characterization methods. This chapter explains various topics such as rubber/rubber-based micro blends, rubber/rubber-based nano blends, recent studies on rubber/rubber-based micro blends, recent studies on rubber/rubber-based nano blends, different manufacturing methods of rubber/rubber-based micro blends, different manufacturing methods of rubber/rubber-based nano blends, characterization methods of rubber/rubber based micro blends, and characterization methods of rubber/rubber-based nano blends. Finally, the application of these rubber/rubber blends is detailed. The next chapter discusses the Rubber/Thermoplastic Blends: Micro and Nanostructured. The authors discussed the different rubber blends of thermoplastic with micro and nanoscale, and their recent studies, manufacturing methods, and characterization methods are discussed in detail. Finally, the application of these rubbers/thermoplastic blends is explained. The eighth chapter describes Rubber-Thermoset Blends: Micro and Nanostructured. The first part of this chapter gives an introduction, morphology, and various types of rubber-thermoset blends. The second part discusses rubber-epoxy-based blends with the subtopics of epoxy toughening by liquid rubber, chemistry, thermodynamic consideration, liquid rubbers other than CTBN, saturated liquid rubbers, toughening by preformed particles and some miscellaneous studies with rubber-epoxy blends; the last part of this chapter discusses rubber toughening of cyanate esters.

The chapter on Interphase Modification and Compatibilization of Rubber-Based Blends analyzes the different topics on interphase modification and compatibilization. The current trends in development of grafts/block copolymers, crosslinking etc., are given. The other main topics discussed are compatibilizing agents, surface modifiers, plasticisers, properties of interphase modified/compatibilised elastomeric blends, interphase characterisation, and finally the future prospects. The tenth chapter focuses on elastomer-based IPNs; this is an important chapter in the book, where the basics related to IPNs, such as introduction, recent studies, different manufacturing methods, characterization methods, phase morphology, properties, and finally applications of IPNs are discussed in detail. The chapter on Micro and Nanofillers in Rubber-Based Blends discusses with an introduction to micro fillers, nano fillers and their recent studies, and finally compatiblization of rubber–rubber blends by the addition of fillers and application. The following chapter discusses the applications of elastomers, mainly focussing on magnetorheological elastomers and applications. Here, the topics discussed after the introduction are, fabrication of isotropic and anistropic MR, rheological properties, sensing properties of MR elastomers, and other applications. The thirteenth chapter explains ionizing radiation processing of elastomers, with topics such as the basics of radiation chemistry, definitions, units, sources of ionizing radiation as applied to processing of elastomers, and primary and secondary chemical reactions induced by radiation in elastomers. Finally, radiation-induced crosslinking—the most important application, selected cases and applications: radiation induced grafting, degradation of tyres, etc., are discussed. The last chapter discusses Electro-Elastic Continuum Models for Electrostrictive Elastomers. Along with an Introduction, the manufacturing process, chemical structure, physical properties, thermal properties, mechanical properties, crystallinity, and applications are explained. In the last part of this chapter the authors correlate between structure properties and applications.

Finally, the editors wish to express their sincere gratitude to all the contributors of this book, who provided excellent support for the successful completion of this venture. We are grateful to them for the commitment and the sincerity they showed in contributing to this book. Without their enthusiasm and support the compilation of this book series could not have been possible. We thank all the reviewers who spent their valuable time to make critical comments on each chapter. We also thank the publisher Springer for recognizing the demand for such a book, and for realizing the increasing importance of the area of Blends and Interpenetrating Network.

P. M. Visakh
Sabu Thomas
Arup K. Chandra
Aji. P. Mathew

# Contents

# Advances in Elastomers: Their Blends and Interpenetrating Networks-State of Art, New Challenges and Opportunities

**P. Deepalekshmi, P. M. Visakh, Aji. P. Mathew, Arup K. Chandra and Sabu Thomas**

**Abstract** Elastomers are becoming an inevitable part of day to day life. The materials based on elastomers have tremendous applications in almost all areas of life. The present chapter deals with a brief account on various types of elastomers, elastomeric based blends and interpenetrating networks (IPNs). Various classes of elastomers, and blends are addressed by giving importance to the interfacial compatibility of different phases. Topics such as immiscible rubber blends, rubber/thermoplastic blends (micro and nano structured), rubber-thermoset blends (micro and nano structured), interphase modification and compatibilization, Interpenetrating Polymer Networks (IPNs), micro and nanofillers in rubbers have been very briefly discussed. Finally the applications, new challenges and opportunities of these elastomeric based blends and IPNs are also discussed.

## 1 Introduction

Elastomers are an important class of polymers having randomly distributed chains connected by cross links in their molecular structure. Irregularity of the chains due to the difference in geometrical arrangement prevents them from having crystalline

P. Deepalekshmi · P. M. Visakh (✉) · S. Thomas
Centre for Nanoscience and Nanotechnology, Mahatma Gandhi University, Kottayam, Kerala 686560, India
e-mail: visagam143@gmail.com

P. Deepalekshmi · P. M. Visakh · S. Thomas
School of Chemical Sciences, Mahatma Gandhi University, Kottayam, Kerala 686560, India

P. M. Visakh · Aji. P. Mathew
Department of Applied Physics and Mechanical Engineering Division of Wood and Bionanocomposites Luleå, University of Technology, 97187 Luleå, Sweden

A. K. Chandra
Global RM and R&D, Apollo Tyres Ltd., Limda, Vadodara, Gujarat 391760, India

P. M. Visakh et al. (eds.), *Advances in Elastomers I*,
Advanced Structured Materials 11, DOI: 10.1007/978-3-642-20925-3_1,

nature. Among the two isomers, the cis configuration gives best elastomer properties even though the double bond restrains motion of the bonded carbon atoms. The trans configuration contributes to a regular planar zig–zag conformation that is crystallizable. A fully saturated hydrocarbon polymer can be elastomeric if substituents are in atactic configuration, or if it is a random copolymer where the segments cannot co-crystallize. The elasticity of this class of polymers is due to the random coil conformation of the macromolecular chains. This is because of the high thermodynamic stability of random coil compared to the fully extended chain. The assumption of a random coil conformation allows prediction of structural characteristics including end-to-end distance, radius of gyration, contour length, persistence length and characteristic ratio [1].

## 1.1 Classification and Compounding of Elastomers

Based on the applications, elastomers can be classified into two categories general purpose and special purpose elastomers. General purpose elastomers are applicable for a wide range of aspects as the name indicates. Natural rubber (NR) which is chemically cis-1,4-polyisoprene is the most significant among the general purpose rubbers. It is obtained from *Hevea brasiliensis* tree, and often used after compounding with vulcanizing agents, antioxidants and fillers. NR has high abrasion or wear resistance, electrical resistance, chemical resistance to acids, alkalies and alcohols and damping or shock absorbing properties and applied quite extensively in manufacturing large truck tyres, off-the-road giant tyres and aircraft tyres. Styrene-butadiene rubber (SBR) is a synthetic copolymer made of styrene and butadiene useful for many purpose. Its chemical resistance is similar to that of NR, however, it exhibits an excellent abrasion resistance than polybutadiene and natural rubber which makes it a suitable material for automobile tyres [1, 2].

Ethylene propylene diene monomer (EPDM) is another synthetic rubber having outstanding heat, ozone and weather resistance due to their stable, saturated polymer backbone structure. It is non-polar in nature and has good electrical resistivity, as well as resistance to polar solvents, such as acids, alkalies, phosphate esters, many ketones and alcohols. It is mainly used as a standard lining material for steam hoses, automotive weather-stripping and seals, radiator, electrical insulation and roofing membrane. The chlorinated rubber, polychloroprene (CR) was one of the first commercially successful synthetic rubbers. This rubber is known by the trade name Duprene or Neoprene. CR is commonly used as hose covers (resistant to oil and ozone), insulating CPU sockets, in bearings and seals for construction application and in automotive industry (waterproof seat covers). The compound is designed to have excellent low temperature, flex, ozone, and weather resistance [3].

Thermoplastic elastomers are another class of elastomers which can also be accommodated under general purpose category. Styrenic block copolymers (SBCs) are the largest volume and lowest priced category of thermoplastic elastomers. Their major applications include footwear, adhesives and sealants. SBCs can also be compounded to produce materials that enhance grip, feel, and appearance in applications. Polyolefin thermoplastic elastomers (TPOs) are an important part of the TPEs, which consist of polyolefin semi-crystalline thermoplastic and amorphous elastomeric components. TPOs are co-continuous phase system with the hard phase providing the strength and the soft phase providing the flexibility. In addition to the two most important processing methods injection moulding and extrusion, TPOs are also fabricated by calendaring, thermoforming, negative thermoforming and blow molding. TPOs ingredients generally include ethylene-propylene random copolymer (EPM), isotactic polypropylene (iPP), and other fillers and additives.

Special purpose elastomers have specific applications in various fields. For instance the poly (isobutylene-co-isoprene) rubber generally known as butyl rubber (IIR) has many applications requiring an airtight rubber. This synthetic rubber is a copolymer of isobutylene and isoprene monomer units and has a very low permeability and good electrical and shock dampening properties. The major applications of butyl rubber are in tyre inner tubes and hoses. Halogenation of IIR generates halogenated butyl rubber (XIIR) having improved air impermeability and cure reactivity required for tyre inner liner applications and to generate adhesion between the tyre inner liners and its body. The two kinds of halo butyl rubber are chloro butyl rubber (CIIR) and bromo butyl rubber (BIIR) [2–4]. Nitrile rubber with acrylonitrile content of 31–35 % can withstand a temperature range of −40 to 107 °C. It is considered to be the major oil, fuel, and heat resistant elastomer. It is useful in non-latex gloves, in dehusking rollers, where they give much longer life and greater dehusking capacity than any other polymers. Silicone rubber is well known as an electrical and thermal insulator since it contains strong silicon oxygen bonds. It has good weatherability and can maintain useful properties over a wide range of temperatures [5]. Fluoroelastomers also have excellent resistance to fuel, oil, acid and chemicals and good resistance to gas penetration and radiation and flame retardancy. Polyurethane and polysulfide rubbers are other special functional elastomers of which the latter have high resistance to petroleum solvents, organic solvents (esters, ketones etc.), aromatic fuels, oils, greases, sunlight, ozone and UV rays. Hence they are used as static seals for aircraft, building and marine industries where no other material serves the purpose.

During the processing of rubber, the raw polymer can be softened either by mechanical work termed mastication or by chemicals known as peptisers. Under processing conditions various rubber chemicals, fillers and other additives can be added and mixed into the rubber to form the rubber compound. The batch mixers like two roll mills and internal mixers allow high shear mixing. Internal mixers overcome the slowness of mill mixing by ensuring rapid mixing and large output. The internal mixers used in the industry vary in rotor, throat, chamber and floating weight design. In addition to this calenders and extruders are also used for compounding the rubber.

Vulcanization or crosslinking is the formation of three dimensional elastic networks along the polymer backbone or macromolecules to decrease its plasticity, cold and hot flow and also to improve its end-use properties like strength, elasticity and stability [6]. Materials initiating the formation of crosslinks between polymer chains are known as vulcanizing agents (e.g. Sulphur). The properties that result from a vulcanizate depend on the number and type of the crosslinks formed. Activators like zinc oxide, increases the rate of vulcanization to more than threefold. In chloroprene rubbers (CR), a combination of zinc and magnesium oxide is often used even though it can be vulcanized in the presence of zinc oxide alone. This is because magnesium oxide gives enough scorch resistance [7]. The cure behavior of peroxide vulcanization and sulphur vulcanization is different and leads to different properties in the cured compounds. The obtained rubber product can be called as mixed vulcanizate. If mixed vulcanization is just defined as vulcanization which leads to both sulphidic and C–C crosslinks, then the use of sulphur as co-agent in peroxide vulcanization, can be considered as mixed vulcanization [8, 9].

## *1.2 Blends*

Polymers are often mixed together to achieve superior properties which arise from the individual systems. Blending of two or more elastomers is carried out for several purposes, such as improving the physical and mechanical properties of the first elastomer, obtaining good processing characteristics of rubber compound and/or decreasing the compound cost [10]. Since the macromolecular chains are very larger in size the entropy of mixing should be negligible and this is the reason why most of the blends are immiscible in nature. Depending on the interactions between the individual polymers the blends can be miscible and partially miscible as well. Elastomers can be compounded themselves as well as can be mixed with thermoplastics and thermosets. A brief discussion on the different types of blend systems is given in the following subsections.

### 1.2.1 Immiscible Rubber Blends

The phases of an immiscible blend can be co-continuous, or one phase will be dispersed within a continuous matrix of the other. The former morphology is favored by equal concentrations and equal viscosities. Most blends consist of discrete particles in a continuous phase, with the latter usually the lower viscosity component, is present at a sufficient concentration. During mechanical mixing, domains of the lower viscosity material deform and encapsulate the higher viscosity phase, to produce a "globular" morphology. However, for an immiscible blend the morphology is never at equilibrium (which would correspond to macroscopic phase separation). Upon improving the compatibility of the blend components, smaller and/or more interconnected phases forms and both of these can potentially improve the properties. This can be achieved through the use of compatibilizing agents or chemical modification of the components. Compatibilizers are surfactants that modify the interfacial tension to decrease the dispersed particle size.

### 1.2.2 Rubber/Thermoplastic Blends: Micro and Nano Structured

A rubber—thermoplastic polymer blend can be obtained either when rubber-rich mixtures form soft thermoplastic elastomers, or when plastic-rich blends produce rubber toughened thermoplastic. Rubber toughened thermoplastics with flexible and high impact properties can be used as economical alternatives for ordinary plastic materials. Thermoplastic elastomers have replaced conventional rubber in a variety of applications including appliance, automotive, medical, engineering, etc. They are made by copolymerization and by blending thermoplastics with rubber component. In most cases, thermoplastic elastomers are block copolymers consisting of soft and mobile 'rubbery' blocks with a low glass transition temperature (Tg), and rigid or hard 'glassy' blocks with a high melting temperature (Tm) and/or high Tg. Rubber/thermoplastic based nano-blends can be a new class of nanocomposites, which are particle-filled polymers for which at least one dimension of the dispersed particles, is in the nanometer range [11]. Thermoplastic elastomer nanocomposites (TPE nanocomposites) based on PA6/NBR/Cloisite 30B were fabricated through a direct melt mixing process in an internal mixer by Mahallati [12]. Excellent improvement in mechanical properties like tensile strength, elongation at break, and modulus was observed on incorporation of the nanofillers in the rubber/thermoplastic based nano blends. Manufacturing methods of rubber/thermoplastic based micro blends includes injection molding, compression molding, extrusion, blow processing, calendaring etc.

### 1.2.3 Rubber-Thermoset Blends: Micro and Nano Structured

Thermosets can be toughened by rubber particles with mainly two different morphologies which are either the use of 'core–shell rubber particles' or initially 'miscible reactive rubbers'. Emulsion polymerization is the route to the preparation of core–shell rubber particles having alternate rubbery and glassy layers. One of the most important classes of thermosetting polymers is the epoxy resins as their applications can cover a wider spectrum. They are extensively used in various fields of coating, high performance adhesives and engineering applications. Cured epoxy polymers are characterized by high chemical and corrosion resistance simultaneously having good mechanical and thermal properties. Epoxy resins are reactive monomers, which are commonly cured with amine to form thermosetting polymers. Jansen et al. [13] reported the preparation of thermoset rubbery epoxy particles as novel toughening modifiers for glassy epoxy resins. They used two types of liquid epoxy resin with an aromatic backbone; diglycidyl ether of bisphenol A (DGEBA) and diglycidyl ether of bisphenol F (DGEBF). Brittle thermoset materials can be toughened successfully by blending with proper liquid rubbers in small amounts or by incorporating preformed rubber particles directly.

Boyonton and Lee [14] applied a synergistic combination of liquid carboxyl-terminated butadiene acrylonitrile rubber and solid rubber particles of different sizes, the latter obtained from recycled automobile tyres. They found no significant improvement in the fracture toughness of the composite when solid rubber alone was used. However, when the combination with liquid rubber modifier was employed, they got higher fracture toughness for the hybrid epoxies than that of those toughened with liquid rubber alone. This synergistic effect is explained in terms of crack deflection and localized shear yielding. Furthermore, they observed a slight improvement in the fracture toughness as the size of the solid rubber particles increased. Although the combination of both reactive liquid rubber and solid rubber particles as toughening agents had been practiced previously, the usage of solid rubber particles obtained from the recycled rubber tyres provides relatively low cost and created higher-value products for recycled solid rubber.

### 1.2.4 Interphase Modification and Compatibilization

Elastomers are generally immiscible with each other and their blends undergo phase separation with poor adhesion between the main matrix and dispersed phase. The properties of such blends are often poorer than the weight average property of individual components. In fact, the main objectives of blending are to combine the performance characteristics of two or more polymers, to develop high performance products. These can be accomplished by compatibilizing the blend, either by adding a third component, called compatibilizer, or by enhancing the interaction of the two component polymers, chemically or mechanically. The role of compatibilizer is in reducing the interfacial energy and improving the adhesion between two or more polymer phases, achieving finer dispersion during mixing and stabilizing the fine dispersion against agglomeration during processing and throughout the service life. Since most blended polymers are immiscible, in many cases additional compatibilization process is required to obtain maximum synergy. The compatibilization process could be reactive or non-reactive. There are several excellent techniques for the compatibilization of polymer blends. However, compatibilization of rubber based blends is more difficult because of crowd and complex compound matrix [15].

Fillers can also compatibilize blends to a greater extent. Most commonly used fillers in blends include carbon black, silica, aluminum trihydrate, aluminum oxide, clay, calcium carbonate, talc, mica, zinc oxide, magnesium oxide, calcium, magnesium silicate etc. In the rubber industry, carbon black is consumed more than three times with respect to other fillers and is one of the well-known ingredients of a tyre [16]. Carbon black (size of 10–30 nm) and precipitated silica (size of 30–100 nm) still remain the conventional fillers due to difficulty in dispersing them at nanometer level. Silica with the general formula $SiO_2$ or $SiO_2{\cdot}xH_2O$ is a naturally occurring mineral such as sand, quartz, quartzite, perlite, tripoli and diatomaceous [17–19]. They can be amorphous or crystalline forms. Its white color is a significant advantage in competition with carbon blacks. Silica with a specific surface area in the range of 125–250 $m^2/g$ is designated as "reinforcing", while products with a specific surface are in the range of 35–100 $m^2/g$ are "semi-reinforcing". Clay platelets bound together by Van der Waals forces have better compatibilizing action in polymer blends. Isomeric substitution (for e.g. tetrahedral $Si^{+4}$ by $Al^{+3}$ or octahedral $Al^{+3}$ by $Mg^{+2}$ or $Fe^{+2}$) within the layers generates negative charges that are counterbalanced by alkali and alkaline earth cations (typically $Na^{+1}$ or $Ca^{+2}$) situated inside the galleries (gaps between the layers). Fillers can also change one or more of optical properties and color, improve surface characteristics and dimensional stability, change thermal, magnetic and electrical properties, improve mechanical properties, durability, and rheology, chemical reactivity and biodegradability [6, 20].

## *1.3 Interpenetrating Polymer Networks*

Interpenetrating polymer networks (IPNs) have been studied extensively since their advent in the 1960s. In the beginning, the concept of a double network was introduced by Tobolsky and coworkers. Double-network rubber refers to an elastomer that has been crosslinked twice, the second time in a deformed state. These result in materials with unusual and enhanced properties which have been termed ''double network elastomers". The deformation employed may be uniaxial tension, biaxial tension, torsional, bending etc. Interpenetrating polymer networks (IPNs) are defined as a combination of two or more polymers in network form. IPNs come next to the class of polymer blends, blocks and grafts. IPNs involve the polymerization of one monomer in the immediate presence of the other, and the crosslinking of one or both the polymers. In most cases IPNs provide intimate mixing of phases, smaller domain sizes, better mechanical properties and damping properties than the corresponding blends, which are physical mixtures of two polymers. The mechanism of phase separation and morphology of simultaneous and sequential IPNs, are different and in most cases simultaneous IPNs leads to better phase mixing, especially when gelation point is reached at the same time.

IPNs have different applications and are the commercially successful form of polymer blends, probably owing to the crosslinked structure that provides better thermal stability, mechanical properties, chemical resistance etc. Though IPNs are traditionally used as damping materials, impact resistant materials, adhesives, etc., more recently they are used in combination with nanoparticles to produce nano-composites for use as responsive hydrogels, medical implants, porous scaffolds or catalyst supports. The number of publications in this research area has shown a regular increase and is about 400 per year, currently. The interesting and unique properties of IPNs emerge when the deliberately introduced crosslinks outnumber the accidently introduced grafts in the polymerization stage. However, some amount of graft is usually present in all IPNs and contributes to the IPN performance in a favorable manner. In IPNs the two separate glass transitions (Tg) for each polymer, with or without inward shift can occur or one broad Tg intermediate to the Tgs of the individual polymers or one sharp Tg intermediate to the Tgs of the component polymers can be obtained, depending on the extent of compatibility or phase mixing.

Kong and Narine studied about the sequential IPNs of polyurethane (PU) using canola oil based polyol and polymethyl methacrylate (PMMA) and compared it with those castor oil based PU/PMMA IPNs [21]. Also the domain sizes found in interpenetrating networks are smaller than corresponding blends, which supports this concept of forced miscibility due to crosslinking. Most IPNs and related materials show phase separation. Dongyan et al. [22] reported the synthesis and applications of castor oil based PU/PMMA simultaneous IPNs containing $BaTiO_3$ fibers. The synthesis resulted in graft IPNs and the TEM analysis showed that the domain sizes were in nanometric scale. The mechanical studies of the nanocomposites showed a change from elastomeric behavior to brittle nature, due to the addition of $BaTiO_3$. Copper nanoparticles with 10–20 nm diameter in polyvinyl alcohol (PVA)/polyacrylamide (PAAm) IPNs resulted in hydrogels which can be tuned for drug release or tissue engineering applications [23]. The study showed that the complexation of PVA and PAAm with $Cu^{2+}$ resulted in low aggregation of Cu nanoparticles and thereby controlling and stabilizing the dispersion and distribution of nanoparticles in the polymer network. There are many products like optically smooth surfaces, toughened plastics, adhesives, damping materials, ion exchange resins, impact modifiers etc. based on IPNs or related materials [24].

## 2 Conclusion

Elastomers, both synthetic and natural are very important class of polymers having great commercial importance. In general they could be classified as general purpose or special purpose based on their functional properties and end-use applications. New polymerization techniques such as controlled radical polymerizations, atom transfer radical polymerizations etc. will be very useful to synthesize new class of elastomers having controlled architecture and functional properties. The various types of elastomer blends and interpenetrating networks are gaining tremendous importance in recent years. Both blends and IPNs provide broad spectrum of property profile for a given application. They have very attractive cost/performance ratio on a commercial point of view. However, the morphology control of both blends and IPNs for the manipulation of ultimate properties is still a challenging task.

# References

1. Elias, H.G.: An Introduction to Polymer Science, p. 170. VCH, Weinheim (1997)
2. Jurkowska, B., Olkhov, Y.A., Jurkowski, B.: Thermomechanical study of butyl rubber mastication during compounding. J. Appl. Polym. Sci **68**, 2159–2167 (1998)
3. Hofmann, W.: Rubber Technology Handbook, 2nd edn, p. 611. Hanser Publishers, Munich (1989)
4. Qu, L., Huang, G., Wu, J., Tang, Z.: Damping mechanism of chlorobutyl rubber and phenolic resin vulcanized blends. J. Mater. Sci **42**, 7256–7262 (2007)
5. Silicone Electronic Materials Research Center: Silicone Fields of Application and Technological Trends, p. 5. Shin-Etsu Chemical Co., Ltd., Tokyo (2003)
6. Bhowmick, A.K.: Current Topics in Elastomers Research. CRC press, Baco Raton (2008)
7. Morton, M.: Rubber Technology, 3rd edn. Van Nostrand Reinhold, New York (1987)
8. van der Burg, T.H.: Kunststof en rubber. **46** (3), 26 (1993)
9. Van Bevervoorde-Meilof, E.W.E.: Improving mechanical properties of EPDM rubber by mixed vulcanisation. PhD, University of Twente, Netherlands (1998)
10. Rosato, D.: Plastics Processing Data Handbook. Springer, London (1997)
11. Alexandre, M., Dubois, P.: Polymer-layered silicate nanocomposites: preparation, properties and uses of a new class of materials. Mat. Sci. Eng **12**, 1–63 (2000)
12. Mahallati, P., Arefazar, A., Naderi, G.: Thermoplastic elastomer nanocomposites based on PA6/NBR. Int. Polym. Proc **25**, 132–138 (2010)
13. Jansen, B.J.P., Tamminga, K.Y., Meijer, H.E.H., Lemstra, P.J.: Preparation of thermoset rubbery epoxy particles as novel toughening modifiers for glassy epoxy resins. Polymer **40**, 5601–5607 (1999)
14. Boynton, M.J., Lee, A.: Fracture of an epoxy polymer containing recycled elastomeric particles. J. Appl. Polym. Sci **66**, 271–277 (1997)
15. Wang, W., Wu, Q., Qu, B.: Polym Eng Sci **43**, 1798–805 (2003)
16. Savran, H.Ö.: "Elastomer Teknolojisi I", Kauçuk Derneði Yayýnlarý, (2001)
17. Xanthos, M.: Functional Fillers for Plastics. Wiley-VCH, Weinheim (2005)
18. Gachter, R., Müller, H.: Plastics Additives Handbook, 4th edn, pp. 525–561. Hanser/Gardner Publications, Munich (1993)
19. Vansant, E.F., Van Der Voort, P., Vrancken, K.C.: Characterization and chemical modification of the silica surface. Elsevier Science, Amsterdam (1995)
20. Thomas, S., Stephen, R.: Rubber nanocomposites: preparation, properties and applications. Wiley, Singapore (2010)
21. Kong, X., Narine, S.S.: Biomacromolecules, **9** (8), 2221 (2008)
22. Dongyan, T., Hong, L., Weimin, C.: Synthesis and application studies of castor oil PU/PMMA IPNs with $BaTiO_3$ fiber nanocomposites. Ferroelectrics **265**(1), 259–264 (2002)
23. Lei, Z., Yang, Q., Wu, S., Song, X.J.: Appl. Polym. Sci. **111**, 3150 (2009)
24. Zhan, K., You, H., Liu, W., Lu, J., Lu, P., Dong, J.: React. Func. Polym **71**, 756 (2011)

# General Purpose Elastomers: Structure, Chemistry, Physics and Performance

**Robert A. Shanks and Ing Kong**

**Abstract** Elastomers are unique to polymers and exhibit extraordinary reversible extension with low hysteresis and minimal permanent set. They are the ideal polymers relieved of molecular interactions, crystallinity and chain rigidity constraints. The common elastomers have characteristic low modulus, though with poor abrasion and chemical resistance. Theoretical concepts have been established for their thermodynamics and kinetics and this knowledge has been applied to extending their properties by design of chemical and molecular structures, or by modification by control of crosslinking, blending or additions of fillers. This chapter reviews elastomer theory and the demanding range of properties expected. Natural rubber is the starting material for introduction of chemistries that introduce damping, abrasion resistance and higher modulus through copolymerization and interacting functional groups. Heteroatoms such as fluorine, silicon, oxygen and nitrogen are shown to extend properties and give chemical resistance. Thermoplastic elastomers move beyond typical cured systems due to formation of two-phase block copolymers. Finally modification by filler and blended systems is considered, followed by introduction to shape memory materials and a brief comment on the future trends. The unique and diverse properties and performance of elastomers continues to be a fascinating field for science and application.

## Abbreviations

| | |
|---|---|
| $\Delta H$ | Enthalpy change |
| $\Delta S$ | Entropy change |
| $\Delta \gamma$ | Wetting surface tension |
| A | Helmholtz free energy |
| COPA | Polyamide/elastomer block copolymer |
| COPE | Polyether ester/elastomer block copolymer |
| CR | Polychloroprene |

R. A. Shanks (✉) · I. Kong
Applied Sciences, RMIT University, GPO Box 2476Melbourne VIC 3001, Australia
e-mail: robert.shanks@rmit.edu.au

P. M. Visakh et al. (eds.), *Advances in Elastomers I*,
Advanced Structured Materials 11, DOI: 10.1007/978-3-642-20925-3_2,

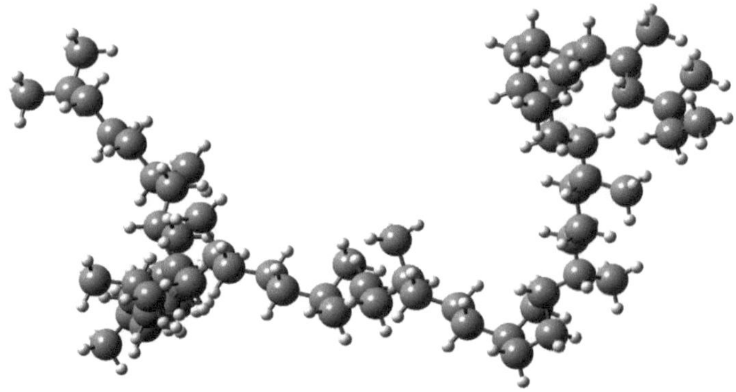

**Fig. 2** A poly(cis-1,4-isoprene) random coil (with 15 monomer units)

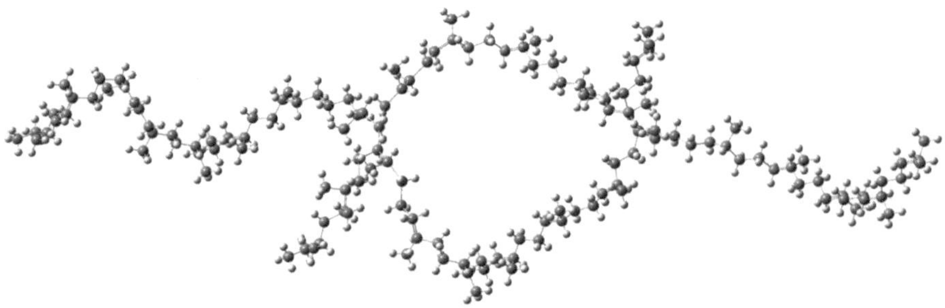

**Fig. 3** Two crosslinked 15 mers of poly(cis-1,4-isoprene)

poly(cis-1,4-isoprene linked by crosslinks that inhibit relative translational motion thereby contributing strain reversibility. This is unfortunate for studying elastomers because it means that they are insoluble, so that solution properties cannot be measured.

Solution properties contribute much to the understanding of macromolecules by virtual of hydrodynamic volume, molar mass determination, solvent swelling, theta solvent phenomenon, mean distance between chain ends, viscosity and light scattering. Solution properties of elastomers can be evaluated prior to crosslinking to characterise all of the single molecule behaviour. Crosslinking can then be performed to give the minimum crosslinks required for continuity throughout the elastomer mass that is the crosslinks must at least provide a percolation network throughout. In a solvent the elastomer will be crosslinked to or beyond the gel point. Crosslinking to a percolation network will give reversible strain. Crosslinking beyond a percolation network will decrease elastomeric performance until it is completely absent in a highly crosslinked network. Solvent swelling is a convenient method to characterise an elastomer since swelling using solvation force is analogous to straining using a physical force. Both swelling and straining are used to determine crosslink density as the molar mass between crosslinks.

The behaviour of elastomers thus far briefly introduced is why they can be described as perfect polymers. The chapter will explain the thermodynamics, structure, and polymerisation methods of these perfect molecules. The concepts will be stretched to include the less than perfect polymers that possess suitable elastomeric properties for many applications, since those perfect polymers will lack in additional properties required of elastomers in specific or harsh environments. Examples with be introduced that will include formulation and applications of commodity and advanced elastomers including blends and interpenetrating networks.

## 2 Thermodynamics

Application of thermodynamics to elastomers requires that the molecules rapidly equilibrate and comply with Hooke's law over the entire range of strain being considered. Such an elastomer is called an ideal elastomer. An ideal elastomer immediately causes recollection of an ideal gas. Thermodynamics of gases required consideration of deviation from idea behaviour. Gases deviate from ideality at high compression, due to contact of atomic radii limiting compression and due to non-zero intermolecular interactions. Elastomers deviate similarly at high extensions due to molecular chain approaching full extension, crosslinks becoming taut and crystallisation of elongated molecular segments. At low extensions initial uncoil of macromolecules is impeded by entanglements, some of which are permanent while others can be freed by uncoiling of loops. At high compressions, elastomers reach a limit of compressibility due to impinging molecules or a fully occupied volume [3].

Elastomer behaviour arises from cooperative segmental motions that depend upon a free volume into which segments can move. Segmental molecular motions involve cooperative pairs of bond rotations, translational motion does not occur due to the crosslinks. Translational motion occurring with creep corresponds to a permanent set or non-reversible component of creep. Recoverable creep is not elastomeric behaviour because it is time dependant, that is kinetic, while elastic behaviour is immediate and reversible that is thermodynamic.

The free energy for volume change of a gas under constant pressure is the Gibbs free energy ($G$):

$$\mathrm{d}G = -\mathrm{d}H - T\mathrm{d}S, \text{ while the work done on the system, } W = P\mathrm{d}V + f\mathrm{d}L \quad (1)$$

The free energy for length change of an elastomer under constant volume ($P\Delta V = 0$, assuming Poisson ratio $= 0.5$) is the Helmholtz free energy ($A$):

$$\mathrm{d}A = \mathrm{d}U - T\mathrm{d}S \text{ while the work done on the system, } W = F\mathrm{d}L \quad (2)$$

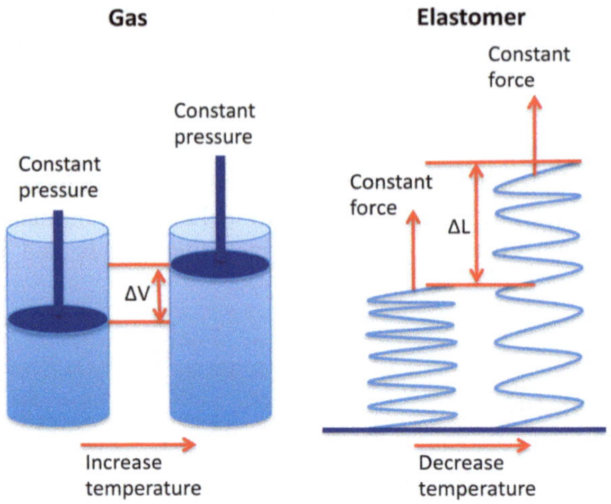


**Fig. 4** Temperature volume and length relationships for a gas and elastomer

Where $H$ = enthalpy, $U$ = internal energy, $S$ = entropy, $T$ = temperature, $V$ = volume, $L$ = length, $P$ = pressure, $F$ = force. Entropy change with length is obtained from the Maxwell equation [4]:

$$\left(\frac{dS}{dL}\right)_T = \left(\frac{-dF}{dT}\right)_L \tag{3}$$

The entropy of a gas is the disorder that is increased by expansion since there are more states that the gas molecules can occupy, while as a gas is compressed the number of states becomes limited by lack of available free volume and the entropy is decreased. The entropy of an elastomer is the disorder that is increased as the elastomer is contracted since the number of possible conformations of each macromolecule approaches infinity, while as an elastomer is elongated the macromolecules become linear and the number of conformations approaches one, a completely linear macromolecule and the entropy is at a minimum. While the thermodynamics of gases leads to an engine driven by pressure and volume change ($P$d$V$), the thermodynamics of elastomers leads to an engine driven by force and length change ($F$d$L$).

The thermodynamics of gases predicts that as a gas is heated under constant volume its pressure will increase. A consequence of the thermodynamics of elastomers is the analogous prediction that if a stretched elastomer is heated its length will decrease. Alternatively is a compressed gas is allowed to expand its temperature will decrease. The same is true for a stretched elastomer that is allowed to contract its temperature will decrease. If a gas is compressed its temperature will increase, while if an elastomer is stretched its temperature will increase (Fig. 4) [5].

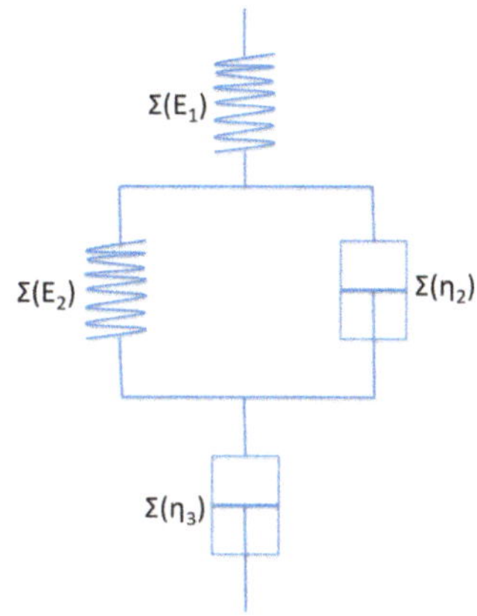


**Fig. 5** Elastic, viscoelastic and viscous component models

Like most gasses, most elastomers are not ideal. Non-ideal or real elastomers do not behave completely reversibly. Real elastomers display some time dependence, where elastomeric behaviour is accompanied by creep (viscoelastic and flow) upon stretching and viscoelastic recovery and permanent set on contraction (Fig. 5). Elastomers display hysteresis upon stretching or relaxation. This additional real behaviour produces hysteresis in the stress–strain curves, that is the elongation and contraction curves do not coincide. The area between the elongation and contraction curves is the energy lost during each cycle. The energy loss is manifest as heat.

Hysteresis is demonstrated by a cyclic stress–strain experiment. Hysteresis is the area between the increasing and decreasing stress–strain curves. The solid lines in Fig. 6 illustrate the low hysteresis of lightly crosslinked natural rubber. Repetitive cyclic stress shows that subsequent stress–strain curves differ from the first curve. This phenomenon is known as the Mullins effect. During the first straining some entanglements can be removed thus decreasing slightly the stress required for subsequent cycles. This can be described as strain softening, in contrast to strain hardening observed with a polyurethane elastomer. The time response curves in Fig. 6 show the response of NR to six repeated cycles. The data are replotted as stress–strain curves and the dotted curves show the second and subsequent cycles. The response was reversible after the first cycle such that the dotted curves are exactly superimposed. If the NR is allowed to relax a similar different first cycle can again be observed. This phenomenon has been called strain conditioning.

A series of six repetitive stress–strain cycles was imposed upon SBR and the results are shown in Fig. 7. The time-based chart illustrates the overall stress and strain data. The stress–strain chart illustrates the higher hysteresis of SBR compared with NR, especially for the first cycle. The stress–strain curves after the first cycle were almost exactly superimposed though not precise and with greater hysteresis than the results for NR. The difference of the first cycle to subsequent cycles, the Mullins effect, was greater with SBR than with NR. SBR is harder and

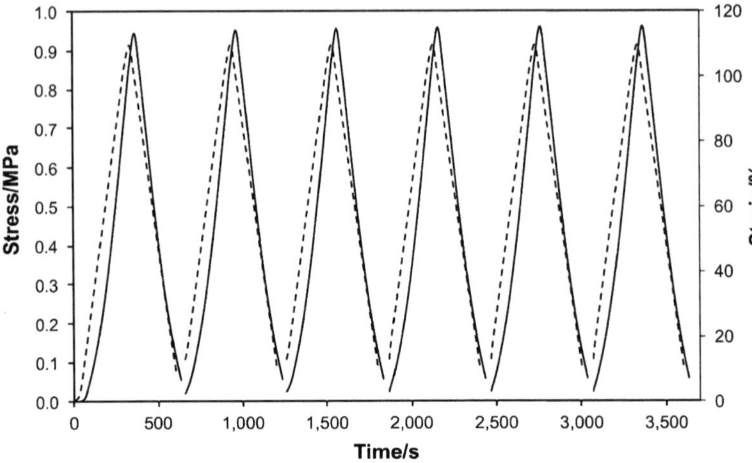


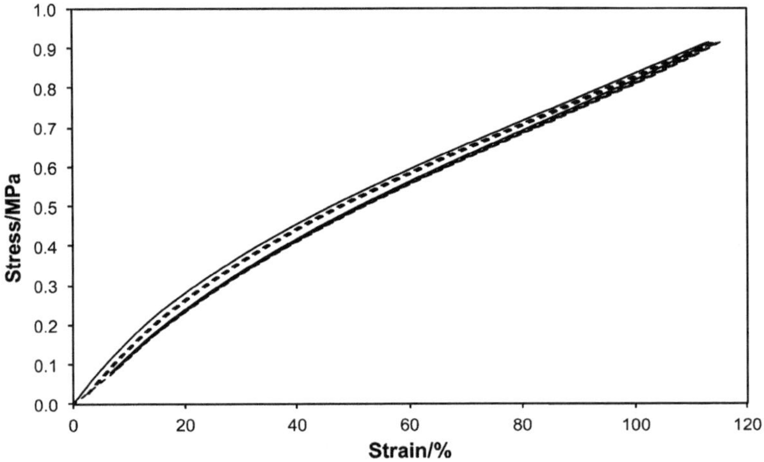


**Fig. 6** Stress–strain time (*upper*) and hysteresis (*lower*) curves for poly(cis-1,4-isoprene) elastomer

more wear resistant than NR, with a consequential increased hysteresis. The energy lost by hysteresis during cyclic stress is dissipated as heat, so that high hysteresis elastomers have a disadvantageous heat build-up. This is an advantage when stress damping is required, but not when high heat build-up will cause failure of the elastomer. Damping behaviour of elastomers can be enhanced by inclusion of fillers. Typical fillers are carbon black, silica and talc.

Elastomer hysteresis is where kinetics impinges upon thermodynamics. Kinetics is time dependant response, which with macromolecules is dependent upon temperature.

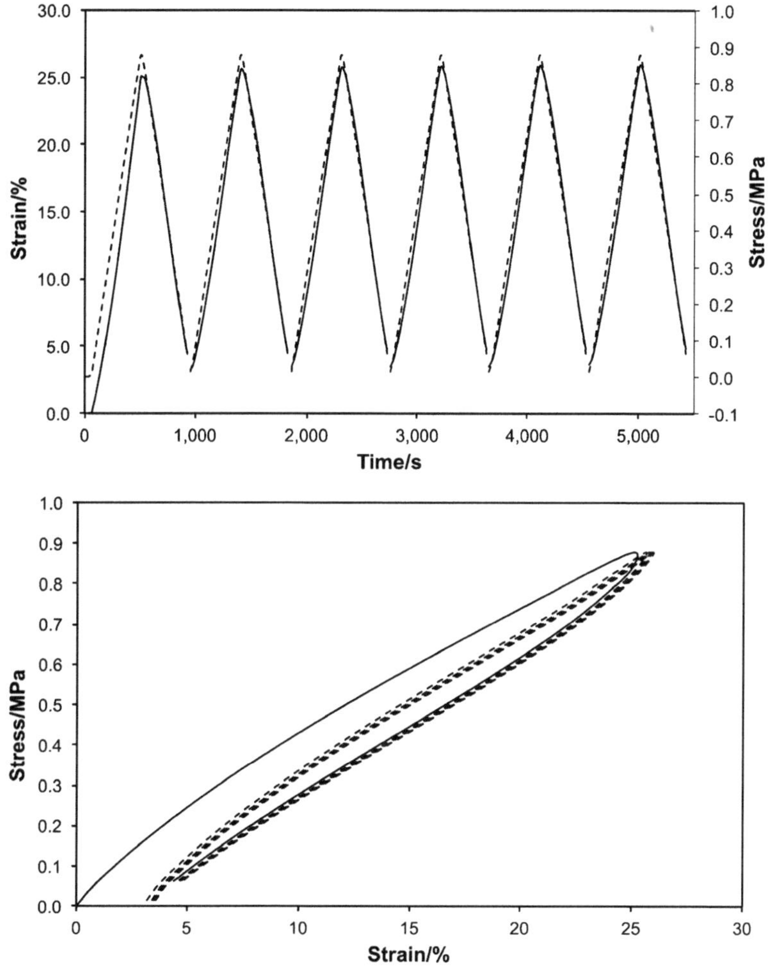


**Fig. 7** Stress–strain time (*upper*) and hysteresis (*lower*) curves for a poly(styrene-co-butadiene) elastomer

# 3 Kinetics

Elastomer kinetics is a consequence of non-ideal behaviour that is shown in the previous section results in hysteresis and heat production. Another consequence of kinetics is the glass transition of amorphous polymers or amorphous regions within semi-crystalline polymers. The glass transition occurs over a temperature range defined as the glass transition temperature (Tg). When the temperature of a polymer is below Tg there is insufficient thermal energy to overcome the activation energy for segmental motions and the polymer will be glassy. When the temperature of a polymer is above Tg segmental motions will occur and the polymer will exhibit elastomer behaviour. When the temperature is within the glass transition range the polymer will be predominately viscoelastic causing damping of forces with consequential conversion to heat. To be an effective elastomer the polymer should be well above its Tg to minimise viscoelastic behaviour. Elastomeric performance at ambient temperatures (20–25 °C) effectively requires Tg in the range −70 to −20 °C for an effective instantaneous reversible response to occur.

Creep and recovery tests were performed on NR and SBR to demonstrate that creep is almost non-existant in crosslinked elastomers, and that recovery is rapid and practically complete. After an initial instantaneous elastic strain, creep of natural rubber is low due to a network of crosslinking, while recovery was instantaneous to a strain of 1 % followed by almost complete recovery with low permanent set (Fig. 8). Similar analysis of SBR creep exhibited a slightly slower elastic response followed by greater viscoelastic creep. Recovery of SBR was rapid, though with some permanent set (Fig. 9).

Measurement of stress–strain response after a large pre-strain has been discussed under the hysteresis curves of Figs. 6 and 7, while another method is to measure the modulated stress response of an elastomer on a highly strained elastomer. Such an analysis showed that the loss modulus was independent of an underlying shear strain over most of the engineering strain range for unfilled and carbon black filled rubber compunds. These results are useful for understanding energy dissipation by strained elastomers [6].

## 4 Structure

The molecular structure of an elastomer consists of random coils connected by crosslinks. Irregularity is essential to prevent crystallinity. Irregularity can be due to geometric isomers where the cis configuration gives best elastomer properties even though the double bond restrains motion of the bonded carbon atoms. The trans configuration contributs to a regular planar zig–zag conformation that is crystallisable. A fully saturated hydrocarbon polymer can be elastomeric if substituents are in atactic configuration, or if it is a random copolymer where the segments cannot co-crystallise.

A statistical description of polymer elasticity is based upon a random distribution of chain links forming a random coil conformation of the macromolecules. A random coil is more thermodynamically stable compared with the fully extended chain because there are an infinite number of random coils, while there is only one fully extended chain. Random coil statistics is self-avoiding in that the model considers the excluded volume of the molecule. The assumption of a random coil conformation allows prediction of structural characteristics including end-to-end distance, radius of gyration, contour length, persistence length, characteristic ratio [7].

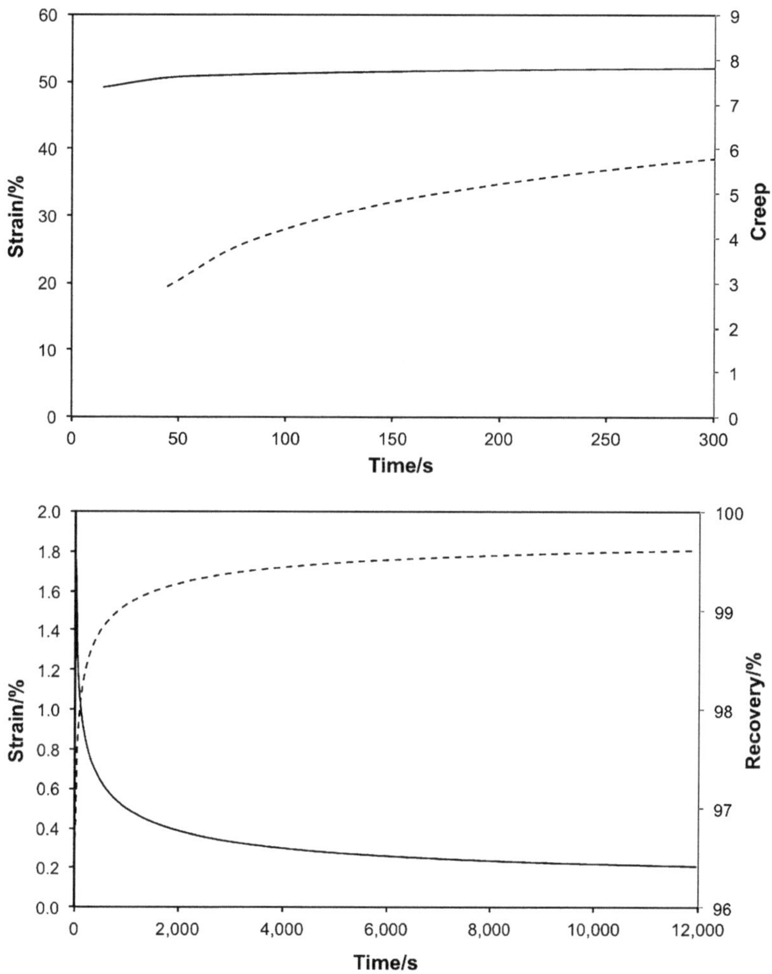


**Fig. 8** Creep strain (*upper, creep dotted*) and recovery strain (*lower, recovery dotted*) response for a poly(cis-1,4-isoprene) elastomer under a tensile creep stress of 0.5 MPa followed by recovery with a minimum restraining force

Elastomer deformational reversibility requires molecular crosslinking beyond gelation so that relative molecular positions are fixed during conformational changes. Gelation occurs when crosslinked molecules extend throughout the system forming a percolation network [8]. Crosslinking covalent bonds will prevent relative translational motions that will be irreversible. Ionic crosslinks can be introduced using covalent metal ions with a polymer with anionic charges along the chain. Alternatively, physical crosslinks will be suitable for reversibility. Physical crosslinks can be due to a phase separated block copolymer structure where a disperse phase will be glassy (high Tg) or crystallise.

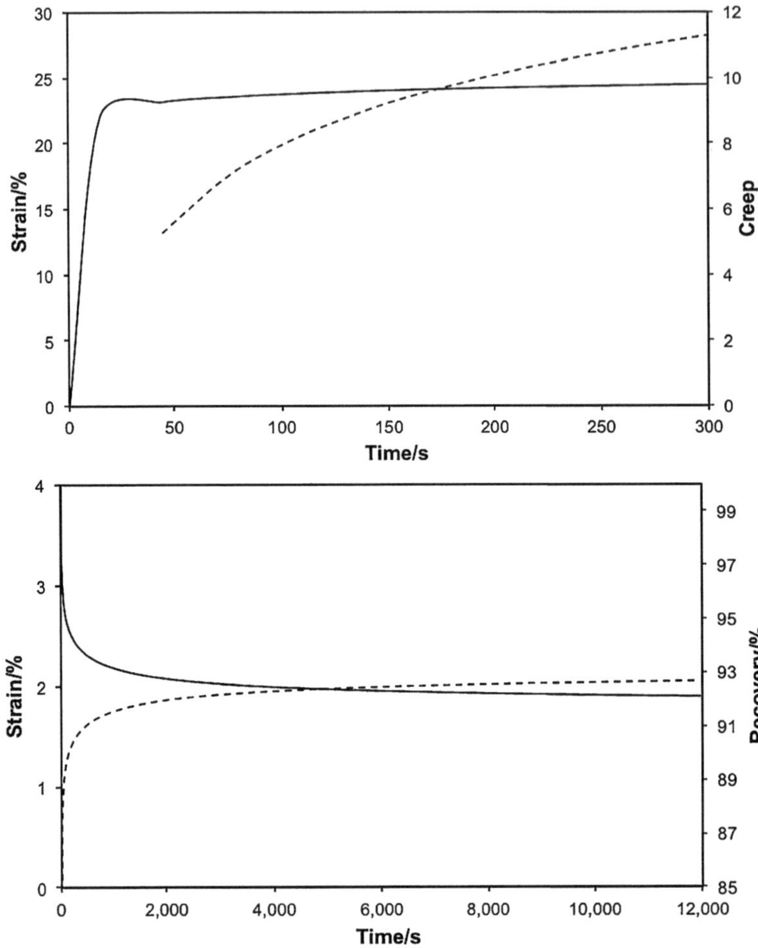


**Fig. 9** Creep strain (*upper, creep dotted*) and recovery strain (*lower, recovery dotted*) response for a poly(butadiene-co-styrene) elastomer under a tensile creep stress of 1.0 MPa followed by a minimum restraining force

# 5 Polymerisation

Elastomers are mainly synthesised using chain growth polymerisation because high molar mass is a feature of this mechanism and high molar mass is required for high chain extension. Elastomers synthesised by the mechanism with radical initiation. Are the hydrocarbon types such as polybutadien, poly (butadiene-*co*-styrene), poly(butadiene-*co*-acrylonitrile) and the fluorocarbon elastomers. Polyisobutylene is initiated using cations. Block copolymers poly(stryene-*b*-butadiene-*b*-styrene) is initiated by anions. Poly(ethyene-*co*-propylene) and its diene terpolymers are synthesis by initiation with co-ordination metal catalysts.

Elastomers synthesised by a step growth mechanism include polyurethanes, polysulfides and the polysiloxanes or silicone elastomers. Each of the main chemical structure classes of elastomer are treated in the next section.

# 6 Specific Elastomer Structure and Properties

## *6.1 Aliphatic and Aromatic Hydrocarbon Elastomers*

Natural rubber (NR) is an elastomer with a basic monomer of cis-1,4-isoprene. It is made by processing the sap of the rubber tree (i.e., *Hevea brasiliensis*) with steam, and compounding it with vulcanizing agents, antioxidants, and fillers. Natural rubber is widely used for applications requiring abrasion or wear resistance, electric resistance and damping or shock absorbing properties such as large truck tyres, off-the-road giant tyres and aircraft tyres. It is chemically resistant to acids, alkalis and alcohol. However, it does not do well with oxidizing chemicals, atmospheric oxygen, ozone, oils, petroleum, benzene, and ketones.

Styrene-butadiene rubber (SBR) is synthetic rubber copolymer consisting of styrene and butadiene. Its chemical resistance is similar to that of natural rubber, however, it exhibits an excellent abrasion resistance than polybutadiene and natural rubber that makes styrene-butadiene rubber a suitable materials for automobile tires.

Butyl rubber (IIR) is a copolymer of isobutylene and isoprene as basic monomer units. This synthetic rubber has a very low permeability rate making it a great seal under vacuum. It also has good electrical, shock dampening properties. It is used in many applications requiring an airtight rubber. The major applications of butyl rubber are tire inner tubes and hoses.

Ethylene propylene diene monomer (EPDM) is a synthetic rubber consisting of ethylene and propylene. It has outstanding heat, ozone and weather resistance due to their stable, saturated polymer backbone structure. As non-polar elastomers, EPDM has good electrical resistivity, as well as resistance to polar solvents, such as water, acids, alkalies, phosphate esters and many ketones and alcohols (Fig. 10). It is mainly used as a standard lining material for steam hoses, automotive weather-stripping and seals, radiator, electrical insulation and roofing membrane. Properties of various hydrocarbon elastomers are shown in Table 1.

## *6.2 Halogen and Nitrile Substituted Elastomers*

Polychloroprene (CR) was one of the first commercially successful synthetic rubbers with an annual consumption of about 3,00,000 tons worldwide (excluding former Soviet Union and PR of China). It is a chlorinated rubber material, which

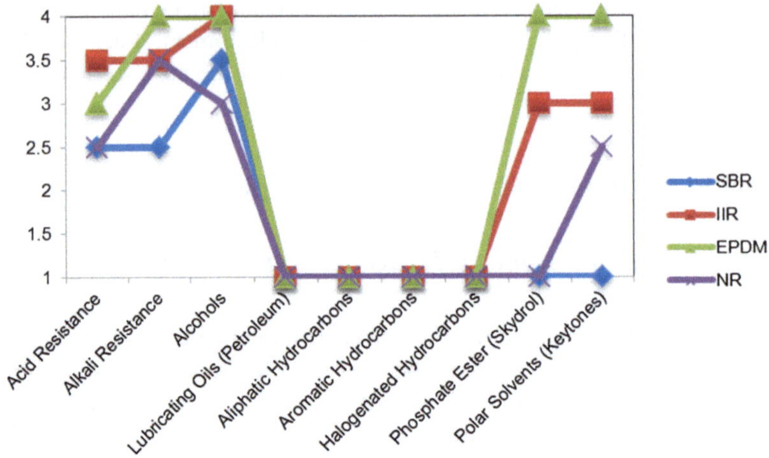


**Fig. 10** Schematic representation of hydrocarbon elastomer properties (*1* Poor; *2* Fair; *3* Good; *4* Excellent)

**Table 1** Relative comparison of the main hydrocarbon elastomers (*1* Poor; *2* Fair; *3* Good; *4* Excellent)

| Elastomers | SBR | IIR | EPDM | NR |
|---|---|---|---|---|
| Economy (Cost) | 1 | 3 | 1 | 2 |
| Tensile strength | 1.5 | 2 | 1.5 | 1 |
| Resilience/rebound | 2 | 4 | 2 | 1 |
| Compression set | 1.5 | 2.5 | 1.5 | 2 |
| Adhesion to metals | 1 | 2 | 2.5 | 1 |
| Abrasion resistance | 1.5 | 2.5 | 1.5 | 1 |
| Tear resistance | 2 | 2 | 1.5 | 1.5 |
| Weather resistance | 3 | 1.5 | 1 | 3.5 |
| Ozone resistance | 4 | 1.5 | 1 | 4 |
| Dynamic properties | 2.5 | 3 | 1.5 | 1 |
| Electrical properties | 2 | 2 | 2 | 2 |
| Water swell resistance | 1.5 | 1.5 | 1 | 1 |
| Steam resistance | 4 | 2 | 1 | 4 |
| Flame resistance | 4 | 4 | 4 | 4 |
| Gas impermeability | 2.5 | 1 | 2 | 3 |

was developed in 1932 by Carothers, Collins, and co-workers using emulsion polymerization techniques. In the same year DuPont began marketing the polymer first under the trade name Duprene and since 1938 as Neoprene. From the beginning up to 1960s, chloroprene was produced by acetylene process, which required expensive feedstock. The modern chloroprene process is a safer and more economical process. Chloroprene is produced in three steps from readily available butadiene: (1) chlorination, (2) isomerisation and (3) dehydrochlorination.

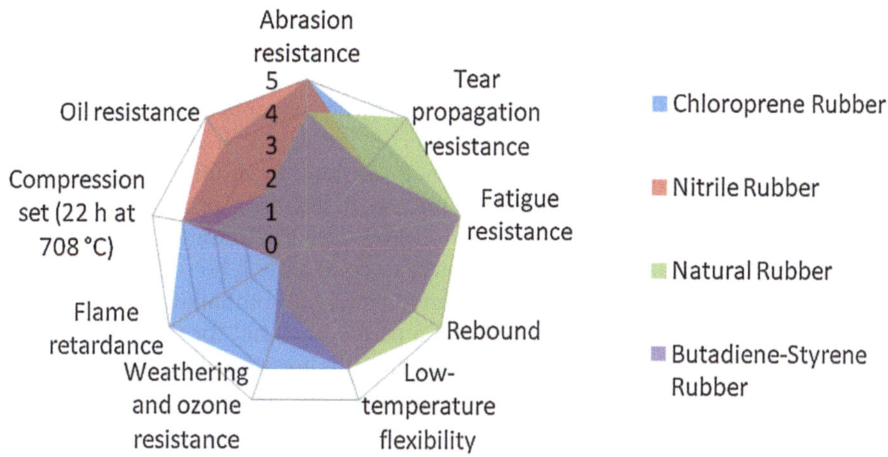


**Fig. 11** Property comparison of nitrile and chloroprene elastomers (*1* Unsatisfactory; *2* Fair; *3* Good; *4* Very Good; *5* Excellent. The ratings are compound composition dependent, hence all optimum values may not be obtained simultaneously). *Source* Reference [9]

Originally developed as an oil-resistant substitute for natural rubber, CR has a good resistance towards various organic chemicals including mineral oils, gasoline, and some aromatic or halogenated solvents. It also has good aging resistance, high ozone and weather resistance. In contrast to the majority of other rubber types, CR shows a surprisingly higher level of resistance to microorganisms, such as fungi and bacteria. Moreover, it has low flammability and outstanding resistance to damage caused by flexing and twisting, an elevated toughness (Fig. 11). Therefore, its combination of unique properties makes it well-suited for many applications throughout industry. CR is commonly used as hose covers, where the hoses resistant to oil and ozone are required. In construction application, it has been used for many years as the elastomer of choice for bearings in machinery and bridges, bellows and seals. The compound is designed to have excellent low temperature, flex, ozone, and weather resistance. In the automotive industry, it serves as gaskets, seals, boots, air springs, and power transmission belts, molded and extruded goods and cellular products adhesives and sealants. Some other applications are insulating CPU sockets, to make waterproof automotive seat covers, rollers for the printing and textile industry.

Acrylonitrile butadiene rubber (NBR), also known as nitrile rubber, is a synthetic rubber produced from a copolymer of butadiene and acrylonitrile. It was invented by the chemists Eduard Tschunkur, Erich Konrad and Helmut Kleiner and a patent for this new oil-repellent rubber was awarded on April 26, 1930. The original name was Buna N and later changed to Perbunan in 1938.

The physical and chemical properties of nitrile rubber vary depending on the composition of acrylonitrile, which typically ranges from 15 to 50 %. With increasing acrylonitrile content the rubber shows higher strength, greater resistance to swelling by hydrocarbon oils, and lower permeability to gases. However, the higher glass transition temperature makes the rubber becomes less flexible at lower temperatures.

The most common nitrile rubber with acrylonitrile content of 31–35 % withstands a temperature range of −40 to 107 °C. Nitrile rubber is considered to be the major oil, fuel, and heat resistant elastomer in the world. It is also resistant to aliphatic hydrocarbons. However, it is not resistant to aromatic hydrocarbon, ketones, ester and strong oxidizing chemicals. It has poor resistance to ozone, sunlight and weathering. Nitrile rubber is widely used in automotive industry as automotive seals and gaskets, which subject to contact with hot oils. It is used as automotive water handling applications and in fuel and oil handling hose too. In healthcare industry, its resilience makes it a perfect material for non-latex gloves. Other applications of nitrile rubber include the rolls for spreading ink in printing and hoses for oil products, as an adhesive and pigment binder.

## *6.3 Sulfide Elastomers*

Polysulfide (PSR) is a class of chemical compounds containing chain of sulfur atoms. It was first found and patented by Joseph C. Patrick and Nathan Mnookin by accident when they were trying to invent inexpensive antifreeze. They named it Thiokol and the manufacturing began in 1929.

PSR exhibits excellent chemical resistance towards oils and greases and they have very good dielectric properties. Other unique properties of PSR including dimensional stability, flexibility, low moisture vapour transmission, low gas transmission and weatherability. These properties make them particularly useful in a variety of sealant applications.

The current types of PSR are Thiokol FA, Thiokol ST and Thiokol LP. Thiokol FA, the workhorse for specialty rollers, is a tradename for polysulfide co-polymer formed from di-2-chloroethyl formal and ethylene dichloride. With its excellent solvent resistant, it is well suited for roller applications requiring resistance to ketones, highly aromatic solvents, and some chlorinated solvents. Thiokol FA is a good choice of paint spray can gaskets and hose tubes too. Thiokol ST, a branched polysulfide formed from di-2-chloroethyl formal with about 2 % 1,2,3-trichloropropane as trifunctional branching units, designed for mechanical goods. Comparison of properties of Thiokol FA and Thiokol ST is shown in Table 2. Thiokol LP is a trade name for a range of liquid polysulfide rubbers obtained by cleavage of Thiokol FA and Thiokol ST, which is one of the most widely used mercaptan-terminated polymers. It has been used as the base polymer in sealants since the early 1950s. With more than 50 years of field experience, liquid PSR have been proven successful. The outstanding resistance of PSR to petroleum products has made them the standard sealant for virtually all aircraft integral fuel tanks and bodies. It is also used in insulating glass window sealants and construction sealants.

**Table 2** Comparison of mechanical properties of two polysulfide elastomers

| Physical properties | Thiokol FA | Thiokol ST |
|---|---|---|
| Shore A | 65–70 | 65–70 |
| Modulus at 100 %, MPa | 5.1 | 3.7 |
| Tensile strength, MPa | 8.3 | 8.3 |
| Elongation, % | 380 | 220 |
| Compression set 22 h at 70 °C | 100 | 20 |
| Gehman low temperature, 8 °C | −45 | −50 |

*Source* Reference [10]

## *6.4 Polyurethane Elastomers*

Polyurethanes (PUs) are the most well-known polymers used to make foams. The raw materials for preparing PU are polyisocyanates, polyols, diamines, catalysts, additives and blocking agents. Generally, they are formed by the reaction of a diisocyanate and a diol with either ester or ether backbone in the presence of catalysts. Ether-based PUs are used to produce flexible and rigid foams while ester-based PUs are used to produce elastomers, flexible foams and coatings.

The structure of PUs is highly influenced by intermolecular forces, such as hydrogen bonding, polarizability, van de waals forces, stiffness of the chain and crosslinking. There may be crystalline regions between flexible chains. The polymers exhibit low corrosion resistance to strong acids and alkalis, and to organic solvent. Flexible foams are used for many domestic applications, such as sofas, cushions, carpet backs and car seats, while rigid foams are used as thermal insulation materials for transportation of cryogenic fluid, and frozen food products. Some others applications of PUs are shoe soles in shoe company, dashboards, bumper covers, moldings and fenders in automobile industry.

A cyclic stress–strain response for a polyether-urethane is illustrated in Fig. 12. The time chart for the data shows the regular changes of strain with stress of six cycles. The stress–strain chart illustrates high hysteresis exhibited by the polyurthane, due to its polar structure providing intermolecular interactions that cause deviation from ideal elastomer response. The second and subsequent cycles differ from the first cycle the same as noted for NR and SBR except that strain hardening has occurred with the PUs and that the repetitive cycles are not so exactly overlayed. As the strain is increased and the random coils are elongated, intermolecular interactions become more favourably and thus the stress required for any particular strain has increased. There has been a memory of the increased interactions after the first cycle, though they remain constant for subsequent cycles.

Cyclic stress response of a PU series has been resolved into elastic and inelastic contributions. The first stress cycle difference was increase with reduced hard and soft phase separation. In subsequent cycles a Mullins effect was attributed to separation of chain segments from the dispersed hard phase to increase coupling with the soft continuous phase [11].

**Table 4** Physical and mechanical properties of elastomers

| Common names | Density ($\rho$/kg·m$^{-3}$) | Elastic or young's modulus ($E$/GPa) | Ultimate tensile strength ($\sigma_{UTS}$/MPa) | Yield tensile strength ($\sigma_{YS}$/MPa) | Elongation at break ($Z$/%) | Shore A or D hardness | Minimum operating temperature range | Maximum operating temperature range | Glass transition temperature ($T_g$/°C) | Specific heat capacity ($c_p$/J·Kg$^{-1}$·K$^{-1}$) | Thermal conductivity ($k$/W·m$^{-1}$K$^{-1}$) | Referactive index ($n_D$/nil) | Relative electric permitivity (@1 MHz) ($\varepsilon\rho$/nil) | Electrical Resistivity ($\rho$/ohm·cm) |
|---|---|---|---|---|---|---|---|---|---|---|---|---|---|---|
| cis-1,4-polyisoprene | 920–1037 | 3.3–5.9 | 29 | 17.1–31.7 | 660–850 | A30-95 | −56 | 82 | −73 | 1,830 | 0.15 | 1.519–1.52 | n.a. | 2.6 |
| trans-1,4-polyisoprene | n.a. | n.a. | n.a. | n.a. | n.a. | n.a. | n.a. | n.a. | −58 | n.a. | n.a. | 1.509 | n.a. | n.a. |
| 1,2-polybutadiene | 910 | 2.1–10.3 | n.a. | 13.8–17.2 | 450 | A45–80 | −100 | 95 | n.a. | n.a. | n.a. | 1.5 | 2.5 | 1,E + 15 |
| cis-1,4-polybutadiene | n.a. | n.a. | n.a. | n.a. | n.a. | n.a. | n.a. | −102 | n.a. | 1,854 | n.a. | 1.52 | n.a. | n.a. |
| trans-1,4-polybutadiene | n.a. | n.a. | n.a. | n.a. | n.a. | n.a. | n.a. | −58 | n.a. | 2,402 | n.a. | 1.518 | n.a. | n.a. |
| Butadiene styrene or styrene butadiene rubber | 940 | 2.1–10.3 | 21 | 12.4–20.7 | 450–500 | A30–90 | −60 | 120 | n.a. | n.a. | n.a. | 1.53 | 2.4 | 1,E + 14 |
| Butyl rubber | 917 | 0.3–3.4 | 17 | n.a. | 700–950 | A30–100 | −45 | 150 | −75 to 67 | 1,950 | 0.13–0.23 | 1.5081 | n.a. | n.a. |
| cis-1,4-polychloroprene | 1230–1250 | 0.7–20.1 | n.a. | 3.4–24.1 | 100–800 | A30–95 | −43 | 107 | −20 | 2,170 | 0.192 | 1.554–1.558 | 2.0–6.3 | 1,E + 11 |
| trans-1,4-polychloroprene | n.a. | n.a. | n.a. | n.a | n.a. | n.a. | n.a. | −40 | n.a. | n.a. | n.a. | n.a. | n.a. | n.a. |
| Polysulfide butadiene rubber | 1340 | n.a. | 4.83–8.63 | n.a | 100–400 | A65–70 | −54 | 100–400 | −55 | n.a. | n.a. | 1.6–1.7 | 1.3 | 1,E + 08 |
| Acrylonitrile butadiene rubber | 1000 | n.a. | 21 | n.a | 510 | A30-90 | −40 | 121 | n.a. | n.a. | n.a. | 1.52 | 2.5 | 1,E + 15 |
| Ethylene propylene diene rubber | 850 | n.a. | 21 | n.a | 100–300 | A30–90 | −51 | 150 | n.a. | n.a. | 2.22 | 1.474 | 2.5 | n.a. |
| Epichlorohydrin rubber | 1270 | n.a. | 17 | n.a | 400 | A60–90 | −46 | 121 | n.a. | n.a. | n.a. | n.a. | n.a. | n.a. |
| Polyurethanes | 1050–1250 | n.a. | 29–49 | n.a | 10–21 | n.a. | n.a. | n.a. | n.a. | 1,800 | 0.21 | n.a. | n.a. | 1,E + 12 |
| Silicone rubber | n.a. | n.a. | 65 | n.a | n.a. | D60A | n.a. | n.a. | n.a. | n.a. | n.a. | n.a. | n.a. | n.a. |

*Source* References [12, 13]

## 6.5 Fluorocarbon Elastomers

Fluorocarbon elastomers (FKM) are synthetic copolymers derived by replacing some or all of the hydrogen atoms in hydrocarbons by fluorine atoms. They are typically used when other elastomers fail in hostile environments. The two main important properties of this class of elastomers are chemical resistance and heat resistance. The extraordinary chemical and heat resistance of FKM can be attributed to the high bond energy of the C–F bond and the higher bond energy of the C–C due to the presence of the most electronegative fluorine. Chemical resistance of FKM versus others elastomers are illustrated in a radar chart in Fig. 13.

Fluorocarbon elastomers are used widely in the harsh environment in automotive and aerospace industry (Fig. 14). In automotive industry, these materials have been designed into a number of items, such as shaft seals, valve stem seals, O-rings, engine head gaskets, filter casing gaskets, diaphragms for fuel pumps, water pump gaskets, fuel hoses, seals for exhaust gas and pollution control equipment, bellows for turbo-charger, lubricating circuits, etc. Aerospace and military uses are for shaft seals, hydraulic hoses, O-rings, electrical connectors, gaskets for firewalls, traps for hot engine lubricants, fuel tanks, heat-shrinkable tubing for wire insulation, etc. chemical and petrochemical plants utilize O-rings, expansion joints, diaphragms, blow-out preventers, valve seats, gaskets, hoses, safety clothes and gloves, stack and duct coatings, and tank lining.

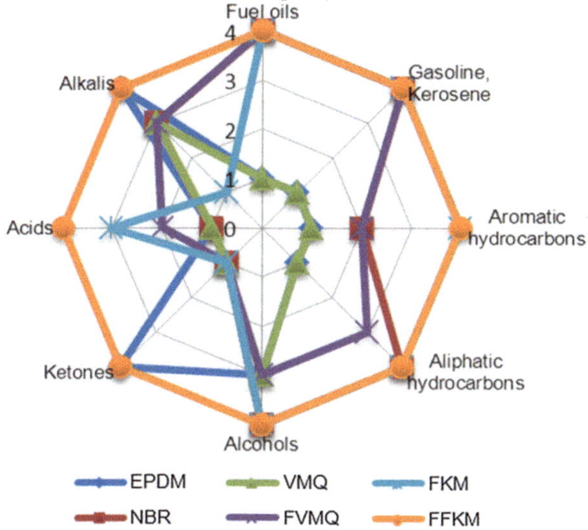


**Fig. 13** Comparison of chemical resistance of elastomers with fluorocarbon elastomers. *Source* Reference [15]

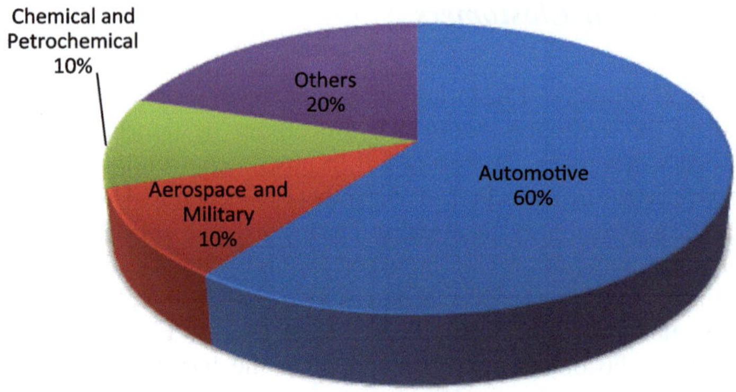


**Fig. 14** Comparative market segments of fluorocarbon elastomers showing specialty automotive dominance. *Source* Reference [16]

## *6.6 Polysiloxane Elastomers*

Polysiloxanes are mixed inorganic–organic polymers, which are well known under the common name of silicones rubber (SR). Instead of the organic carbon chain skeleton, these materials are based on a silicon and oxygen backbone (Si–O). Berzelius discovered the silicon, the basis of silicones in 1824 from the reduction of silicon tetrafluoride with potassium. In 1940s, F.S. Kipping, the father of silicone science, achieved extensive synthesis of silicone compounds and coined the name 'silicones'.

The most common siloxane is polydimethylsiloxane (PDMS). It can be prepared from the hydrolysis of dimethyl dichlorosilane that yields a silanol. The resultant silanol immediately undergo polycondensation. The mechanism of hydrolysis and the composition of silane mixture control the size of the polysiloxane molecules. Generally, silicones have an excellent resistance to heat and cold, ranging from −60 to +250 °C. They show hydrophobic properties which enable them to act as release agents for tacky substances. Silicones are also resistant to ozone and radiation. The low modulus of elasticity and a high mechanical dissipation factor make them an excellent medium for shock and sound absorption. In the cosmetic area, silicones are widely used in many skin-care products, hair products, shaving products and personal lubricants because of their high compatibility with skin, their water-repellent effect and the pleasant touch. In the automotive industry, silicones are used as a lubricant for the brake components, gaskets in automotive engines and for airbags as coatings and sealants. Transparent tubes processed from silicones are used for the food industry and medical applications and cable insulation for application at high temperature.

# 7 Thermoplastic Elastomer Structure and Properties

Thermoplastic elastomers (TPEs) are unique synthetic compounds that combine some of the properties of rubber with the processing advantages of thermoplastics. Generally, they can be categorized into two groups: multi-block copolymers and blends. The first group is copolymers consist of soft elastomers and hard thermoplastic blocks, such as styrenic block copolymers (SBCs), polyamide—elastomer block copolymers (COPAs), polyether ester—elastomer block copolymers (COPEs) and polyurethane—elastomer block copolymers (TPUs). TPE blends can be divided into polyolefin blends (TPOs) and dynamically vulcanized blends (TPVs). The world TPEs demand by types of TPE and by regions in year 2004, 2009 and 2014 are shown in Fig. 15.

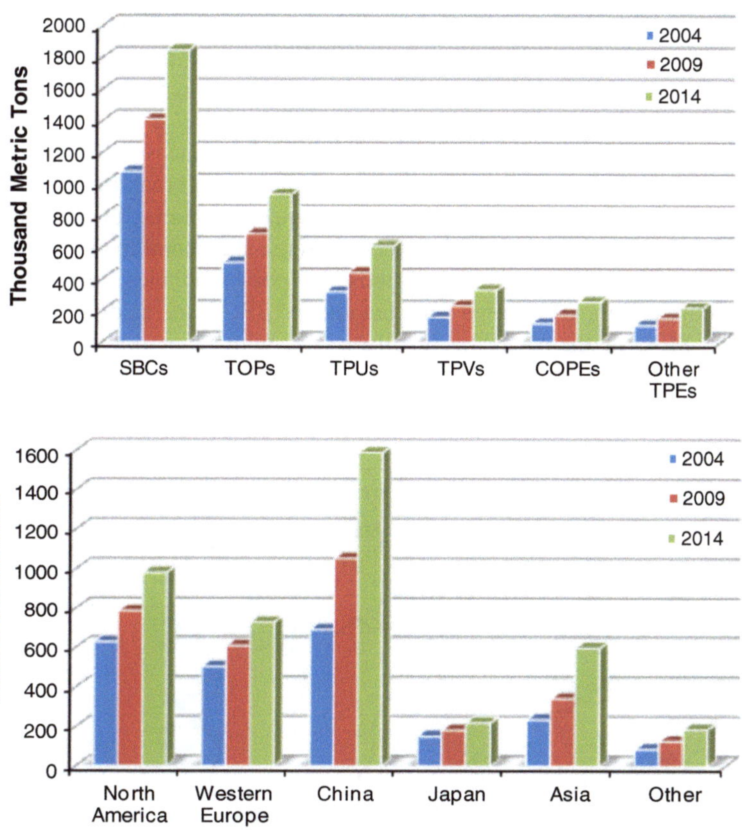


**Fig. 15** World thermoplastic elastomers demand in 2004, 2009 and 2014 by types of TPEs and by regions. *Source* Reference [17]

## *7.1 Styrenic Block Copolymer*

Styrenic block copolymers (SBCs) are the largest volume and lowest priced category of thermoplastic elastomers. SBCs are based on simple molecules (A-B-A type) that consist of at least three blocks, namely two hard polystyrene end blocks and one soft, elastomeric midblock. The midblock is typically a polydiene, either polybutadiene or polyisoprene, resulting in the well-known family of styrene–butadiene–styrene (SBS) and styrene–isoprene-styrene (SIS). Other SBCs which have been commercially successful include ethylene—butylene (SEBS), ethylene—propylene (SEPS), polyisobutylene (SIBS) and ethylene—ethylene—propylene (SEEPS). The structure of SBCs is shown schematically in Fig. 16.

It is essential that the hard and soft blocks are immiscible, so that, on a microscopic scale, the polystyrene blocks form separate domains. These domains are attached to the ends of elastomeric chains and form multifunctional junction points, thereby providing physical cross links to the rubber. When they are heated, the polystyrene domains soften and the SBCs become processable as thermoplastics. When solidified, SBCs exhibit good elastomeric properties as the polystyrene domains reform and strength returns.

Tensile strength of SBCs is much higher than those measured on unreinforced vulcanized rubbers. Most of the SBCs have elongation at break ranges over 800 % and resilience is comparable to that of vulcanized rubbers. SBCs exhibit non-Newtonian flow behavior because of their extreme segmental incompatibility. The melt viscosity is much higher than those of polybutadienes, polyisoprene and random copolymers of styrene and butadiene. SBCs are seldom used as pure materials. Most of them can be readily mixed with other polymers, oil, fillers, resins, colorants and processing aids to meet the required physical and mechanical properties [17].

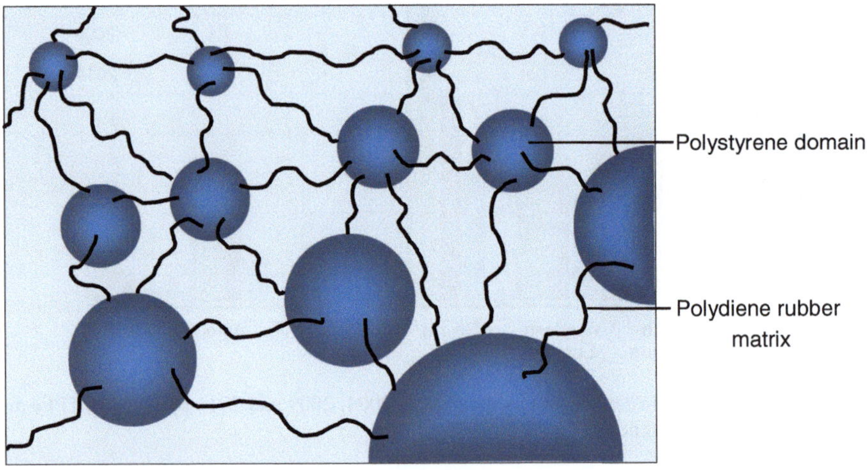


**Fig. 16** Schematic of a styrene-diene-styrene block copolymer

The major applications for SBCs are footwear and adhesives and sealants. They are also used in modifying the performance of asphalt for roofing and roads, particularly under extreme weather conditions. SBS block copolymers are among the most commonly modifiers for this application [18]. Recently, it is proven that SEBS acts as a better modifier than SBS in improving the asphalts rutting resistance due to its double bond saturation which makes the SEBS more rigid than SBS [19]. SBCs can also be compounded to produce materials that enhance grip, feel, and appearance in applications such as toys, automotive, personal hygiene and packaging.

## *7.2 Thermoplastic Elastomers Based on Polyamide*

Thermoplastic polyamide elastomers (COPAs) belong to the group of segmented block copolymer. They consist of multiblock copolymer structure with repeating hard and soft segments. The hard segments are polyamides which serve as virtual cross-links reducing chain slippage and the viscous flow of copolymer, whilst the soft segments are either polyethers or polyesters which contribute to the flexibility and extensibility of elastomers. Both segments are connected by amide linkages. The important members of COPAs are polyesteramides (PEAs), polyetheresteramides (PEEAs), polycarbonateesteramides (PCEAs) and polyether-block-amides (PE-b-As).

The properties of COPAs may vary according to such factors as the proportion of the hard and soft segments in the copolymer, chemical composition, molecular weight distribution, the method of preparation, and the thermal history of the sample that affects the degree of phase separation and domain formation. The hard segment controls the degree of crystallinity, crystalline melting point and mechanical strength while the soft segments determines the thermal oxidative stability, hydrolytic stability, chemical resistance and low temperature flexibility.

Most of the COPAs exhibit higher resistance to elevated temperature than any other commercial TPEs. They are also higher resistant to long-term dry heat aging without adding any heat stabilizers. Abrasion resistance of COPAs is comparable to that of TPUs [20]. COPAs have excellent abrasion resistance, which is comparable to that of TPUs. The hardness is in the range from Shore 80A to Shore 70D by varying the content of hard and soft segments [21]. The good insulation properties of COPAs make them suitable for low voltage applications and for jacketing. Other application areas for this material include conveyor and drive belt, footwear such as ski boots and sport shoes, automotive applications, electronics, hot melt adhesives, powder coatings for metals and impact modifiers for engineering thermoplastics [22].

Polyester amides constitute a peculiar of biodegradable family, due to the presence of both ester and amide groups that guaranties degradability. These biodegradable polymers are receiving great attention and are currently being developed for a great number of biomedical applications such as controlled drug delivery systems, hydrogels, tissue engineering, and other uses [23, 24] .

## 7.3 Thermoplastic Polyether Ester Elastomers

The polyether ester elastomers or copolyesters (COPEs) consist of sequence of hard and soft segments. The high melting blocks (hard segments) are formed by the crystalline polyester segments which are capable of crystallization and the rubbery soft segments are formed by the amorphous polyether segments with a relatively low glass transition temperature. At useful service temperature, the polyester blocks form crystalline domains embedded in rubbery polyether continuous phase. These crystalline domains act as physical cross-links. At elevated temperatures, the crystallites break down to yield a polymer melt, thus facilitating the thermoplastic processing.

COPEs are considered engineering thermoplastic elastomers because of their unusual combination of strength, elasticity and dynamic properties. They have a wide useful temperature range between the glass transition temperature (around −50 °C) and melting point (around 200 °C) [25]. These materials are elastic but their recoverable elasticity is limited to low strains. They have excellent dynamic performance and show resistance to creep. COPEs can be used in electrical applications for voltages 600 V and less. COPEs are resistant to oils, aliphatics and aromatic hydrocarbons, alcohols, ketones, esters and hydraulic fluids [26].

These materials, because of their high modulus and stiffness, have been used to replace some convention rubbers, PVC and other plastics in many applications. Uses of COPEs are reported in fuel tanks, quiet running gear wheels, hydraulic hoses, tubing, seals, gaskets, flexible couplings, wire and cable jacketing.

## 7.4 Polyolefin-Based Thermoplastic Elastomers

Polyolefin thermoplastic elastomers (TPOs) are an important part of the TPEs, which consist of polyolefin semi-crystalline thermoplastic and amorphous elastomeric components. TPOs are co-continuous phase system with the hard phase providing the strength and the soft phase providing the flexibility. The two most important processing methods of TPOs are injection moulding and extrusion. Others processing methods include calendaring, thermoforming, negative thermoforming and blow molding. TPOs ingredients generally include ethylene-propylene random copolymer (EPM), isotactic polypropylene (iPP), and the addition of other fillers and additives.

TPOs share with all TPEs the fundamental characteristics of having elastomeric properties, yet they process like a thermoplastic material. They are available in the hardness range from 60 Shore A to 70 Shore D. Their flexural modulus can range from 1,000 to 2,50,000 psi (6.9–1,725 MPa). TPOs containing high amounts of elastomers are quite rubbery with high elongation at break values while TPOs containing high amounts of polyolefin undergo a yiled at low elongation and there is little recovery after drawing. Most TPOs are resistant to ozone, unaffected by water or aqueous solutions of chemicals. TPOs are excellent electrical insulating materials with the dielectric strength of 500 V/mil and the volume resistivity at 23 °C and 50 % RH is 1.6 × 1016 [27].

Hard TPOs compound is often used for injection molded automotive interior or exterior fascia. Soft TPOs compound can be extruded into a sheet and thermoformed for large automotive part such as interior skin [28].

## *7.5 Thermoplastic Elastomers Prepared by Dynamic Vulcanization*

Dynamic vulcanization is a widely used method to prepare thermoplastic elastomers comprising partially or fully cross-linked elastomer particles in melt-processable thermoplastic matrix. The thermoplastic elastomers prepared by this method are referred to thermoplastic vulcanizates (TPVs). The difference between the TPOs and TPVs is that both the elastomeric and polyolefin phases in TPOs are co-continuous phases while in the TPVs the polyolefin phase is continuous and surrounds the cross-linked and discontinuous elastomeric phase. A two-dimensional representation of TPOs and TPVs are shown in Figure 17.

Thermoplastic vulcanizates can be processed using processing and fabrication techniques common to thermoplastics, such as extrusion, injection molding, compression molding, blow molding, thermoforming, calendaring and extrusion foaming [29]. Commercialized TPVs are commonly based on dynamically vulcanized EPDM rubber blend with a polyolefin resin. However, it is possible to make TPVs with a variety of thermoplastics and elastomers. Some that have been used are diene rubbers and polyolefins, butyl and halobutyl rubbers and polyolefin

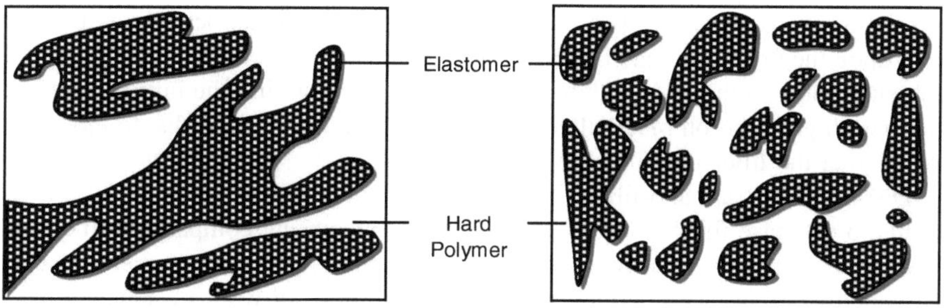


**Fig. 17** Development of morphology in a thermoplastic vulcanizates from a co-continuous phase to dispersed phase

resins, polyacrylate rubber and polyolefins [30] and butadiene-acrylonitrile rubber and polyvinyl chloride [31].

The ratio of both elastomeric and thermoplastic components in the systems is crucial for the main physical properties (Young modulus, modulus at 100 % elongation, etc.) and its influence on final properties (tensile strength, elongation at break, tear strength and resilience, etc.). There are three decisive characteristics in order to obtain TPVs with good balance of properties: the wetting surface tensions ($\Delta\gamma$) of the elastomeric and thermoplastic components, the fraction of crystallinity ($W_C$) of the thermoplastic and the critical entanglement spacing ($N_C$) of the elastomer macromolecules [32]. A decrease of both tensile strength and elongation at break has been found with an increase in $\Delta\gamma$ and $N_C$ and a decrease in $W_C$ [33].

At the beginning, TPVs were used to substitute existing applications of elastomers now they are also opening new fields of application due to their processing potential. As the second largest group of soft TPEs after styrenic-based block copolymers, TPVs play an important role in under-the-hood applications in automotive industry as vacuum tubing, body plugs, seals, air conditioning hose cover, emission tubing and fuel line hose cover. They are used widely in electrical applications as wire and cables insulation and jacketing, connectors and terminal ends too. Furthermore, TPVs can also be used mechanical rubber goods applications such as convoluted bellows, mount, bumpers, housings, oil-well injection lines, etc.

## 8 Elastomer–Filler Compositions

Elastomer formulations required for applications requiring extended performance include fillers, the most common being carbon black. Fillers such as silica, kaolin, talc and calcium carbonate are often used. Recently organomodified clays, nanoparticulate synthetic inorganic compounds and carbon nanotubes have been introduced. Filler elastomers exhibit changed beyond those expected from volume fraction theories and much has been discussed about filler distribution and interactions. Fractal nanostructure was used to interpret heterogeneous deformation where filler aggregates vary with increases in strain [34]. SBR and NR subjected to large pre-strains of 100 % showed storage and loss modulus behaviour independent upon the pre-strain when filled with 25 pph carbon black while at higher filler content to 50 pph the loss modulus increased with pre-strain. The influence of pre-strain on higher carbon black filled elastomers was interpreted as due to molecular slippage at the filler interface [35].

Dispersibility of nano-fillers in elastomers is dependent upon surface treatments to decrease filler–filler interactions while increasing compatibility between filler and elastomer. Relative interaction between filler and elastomer has been estimated from surface energies and polarity and derived work of adhesion [36].

Presence of filler aggregates and formation of filler agglomerates are determinants strain and time dependant properties. Agglomerates can be disrupted by increasing strain, though they may reform over time. Elastomer entrapped within filler agglomerates increases the effective filler volume fraction since the entrapped elastomer is immobilised and does not contribute to the elastic response until the filler agglomerate is disrupted at larger deformations. Constraint of elastomer response by filler has been described as due to a tightly bound interphase attributed to a hydrodynamic effect that increases the effective filler volume fraction [37].

Changes of filled-elastomer response with strain and over time between strains have been long observed described as Payne and Mullins effects. Much attention to the interpretation of these effects has recently occurred because of advances in instrumentation and the interest in elastomer nanocomposites. Agglomeration of filler particles has been described as the rupture and re-birth of glassy interfiller bridges allowing formation of rigid clusters for finite times. Plastic deformation due to the clusters is superimposed upon the elastic response [38].

Nanofiller–elastomer composites exhibit increased thermal stability attributed to binding of elastomer molecules to the nanofiller and the tortuosity of the path for diffusion of volatile degradation products. Polyhedral oligomeric silsequioxanes (POSS) are single molecule nanofillers that have received much recent attention, for example they have been used as part of the polyol components in polyurethane elastomers. Thermogravimetry showed that POSS-polyurethane hybrids have increased thermal stability due to an influence of POSS on ther degradation mechanism [39]. Thermoplastic elasomers based on poly(ethylene-*co*-vinyl acetate) have been used to form nanocomposies with organo-modified layered double hydroxides (LDH) where the LDH exfoliated upon dispersion. The mechanical properties where increased while thermal degradation resistance and fire retardance improved [40].

Nano-silica particles are now included in many elastomeric products, polysiloxane being the main examples, however silca has been found to enhance hydrocarbon elastomers. Silica was included in solution polymerisaed poly(butadiene-*co*-styrene) and produced a non-linear stress relaxation due to the juxtaposition of an elastomer network and a silica network that both contributed to the non-linearity [41]. Nano-fillers such as silica consist of indispersible aggregates that flocculae into larger agglomerates existing as fractal arrays within the elastomer. The agglomerates can be disrupted by elastomer strain and reform upon relaxation. The agglomerates may contain occulded elastomer reulting in an artificially high apparent filler volume fraction, a phenomenon known as the Payne effect. Up to 25 %·w/w fumed silica was dispersed in poly(dimethyl siloxane) that were found by NMR to have a two phase morphology where the silica formed percolated network structures, dependent upon surface functionalisation of the silica. Microscopic network changes occurred with storage time of the poly(dimethyl siloxane)–silica composites [42].

## 9 Elastomer Blends

Polymer blends are classified are incompatible, compatible, miscible. Polymer blends suitable for application must be compatible though they may be miscible often miscibility is not desired because properties are averaged. SBS and thermoplastic PU based upon polyether or polyester provided transparent blends due to narrow dispersed phase particle size and similar refractive index of the blended elastomers. The SBS–TPU blends showed improved thermal resistance, mechanical properties and damping at high frequency [43]. Membranes prepared from blends of nitrile rubber and epoxidized natural rubber has been prepared and their morphology, miscibility, mechanical and viscoelastic properties have been measured. Blend composition influenced homogeneity, microdomain structure and viscoelasticity [44].

Blends were prepared from natural rubber with carboxylated poly(butadiene-*co*-styrene) rubbert and composition and temperature property dependences were investigated. Two glass transitions were found and other data were consistent with a uniform two-phase system where morphology and viscoelasticity were controlled by composition [45].

## 10 Interpenetrating Elastomer Blends

Interpenetrating polymer/elastomer bnetworks (IPN) can be formed by simultaneous and sequential polymerisation or crosslinking to give a co-continuous network of chemically separate polymers (Fig. 18). The uncrosslinked, blended elastomer polymers or prepolymers must be initially miscible to allow the interpenetrating network to form at the molecular level. As the separate crosslinking reactions proceed, the polymers entropically phase separate forming the co-continuous networks. Both polymer could be elastomers, or one an elastomer and the other a polymer to enhance toughness. For example, an acrylate-modified polyurethane was interpenetrated with an unsaturated polyester resin to form a gradient IPN.

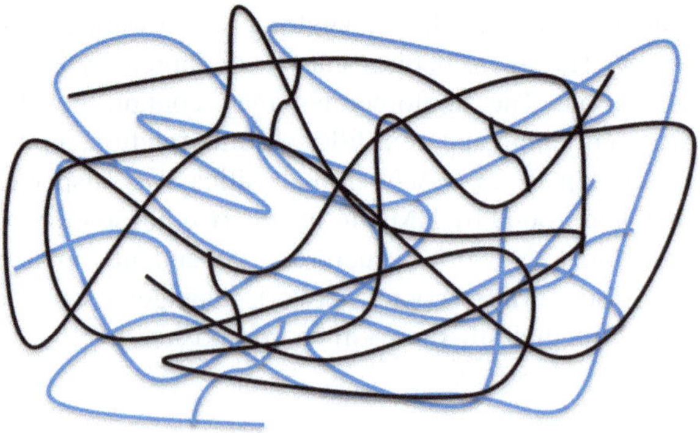

**Fig. 18** Schematic structure for an interpenetrating polymer blend

The interpenetrated domains were found to be of nanometre dimension, with glass transitions linked such that the IPNs varied from elastomeric to brittle [46].

Blends of a poly(oxypropylene)-type epoxy resin and poly(butadiene-*b*-styrene) were prepared and simultaneously cured to form a network with three glass transitions, due to two expected phases from the poly(butadiene-*b*-styrene) and a thrid epoxy phase [47]. Interphases in a polybutadiene-natural rubber blend, a poly(methyl methacrylate)–poly(vinyl acetate) structured latex film, a poly(epichlorohydrin)–poly(vinyl acetate) bilayer film, and polystyrene–polyurethane and poly(ethyl methacrylate)–polyurethane interpenetrating polymer networks were investigated using modulated temperature differential scanning calorimetry (mT-DSC). The mT-DSC results identified glass transitions of the individual components and of interphases [48].

## 11 Formulation and Compounding

Elastomer formulations include additional components such as crosslinking or curing agents, initiators and accelerators for curing, stabilisers such as antioxidants, processing lubricants, plasticisers or extenders and fillers such as carbon black and silica [49]. Elastomer based adhesives are applied as contact adhesives for example. The initial adhesion requires addition of a tacifier with rosin and phenol–formaldehyde as examples. A brominated isobutylene-co-p-methylstyrene rubber tacified with a combined hydrocarbon resin and maleated hydrocarbon resin tackifier has been found to lead to better performance during bonding and de-bonding measured using a peal test and dependent on the bulk viscoelastic properties [50].

## 12 Shape-Memory Polymers

Shape-memory polymers are elastomers in the temperature regime where they are deformed from their equilibrium shape, however, at ambient temperatures they must retain there deformed shape. At ambient temperatures they must be below the glass transition of their elastomeric continuous phase so that they retain their deformed shape for long times until the memory is activated by heating. The glass transition temperature should be conveniently above any likely ambient temperatures, though typically low enough so that elastomeric behaviour can be activated using moderate heat such as hot water, hot air from a heated air gun such as a hair dryer and other portable heating devices. When heated into the elastomeric regime shape memory materials should have true elastomeric behaviour that is they should have rapid and reversible deformations without creep, which is irreversible and with minimal viscoelasticity since immediate return to original shape is required [51, 52]. Polyurethanes are common shape-memory materials because they exhibit elastomeric properties over temperature ranges controlled by structure and

characteristic hydrogen bonding provides restraint on conformational change and molecular motions. The shape-memory response may be electrically activated in lieu of application of a mechanical force [53]. Interaction parameters between hard and soft phases and differences in glass transition temperatures proved stronger interactions in polyester–polyurethane than in polyether polyurethane. Transparency of polyester–polyurethane was due to smaller dispersed hard phase particles giving higher physical crosslink density and strain hardening [54].

## 13 Future Trends

A trend with new polymers is toward specialty applications where advanced and unique properties are required. The requirements are met through improved control of molecular structure, copolymerization and formulation of existing polymer types. Elastomers follow this trend with new polymerization techniques and catalysts to control tacticity, comonomer composition, molar mass and molar mass distribution. Formulation innovation has come from polymer blends that are compatible, though not miscible when averging of properties would occur. Recent publication frequencies show that nanocomposites are the elastomers with most rapid development. Elastomers nanocomposites have always been significant in that carbon blacks and silicas used in the traditional rubber industry are nanoparticulate materials. Enhnced elastomeric properties and energy damping are prime areas of new developments. In addition, resistant elastomers are in demand for chemical resistance, thermal resistance, radiation resistance, wear/abrasion resistance and weathering resistance.

A second trend is towards biomaterials, or materials derived from renewable resources. The source of monomers for elastomer synthesis has been mentioned in this review. Natural rubber has always been a biomaterial and it would be rational to continue to innovate with its use and enhancement through compounding. Natural rubber crop design is another area with potential. The rubber derived from plantations in different locations and hence climates or microclimates, as well as weather trends and seasonal variations for collection cause variation in the rubber. New plant breeds or genetic modification is likely to yield rubber with enhanced properties and of greater initial purity. These plantation improvement processes have been underway throughout the history of natural rubber, however biotechnology has recently advanced rapidly.

## 14 Conclusion

This chapter has described the characterisitc properties of an elastomer and that such materials must be polymeric. Elastomers are unencumbered polymers preferably of high molar mass so that random coils will form and the coils can be extended towards an entropically unstable linear conformation. Elastomeric polymers have contributed significantly to understanding of the behaviour of macromolecules. Structural refinments are required in practice, firstly some threshold of crosslinking prevent flow thus restricting molecules to reversible uncoiling. Further modification includes chemical structures to provide resistance to harsh environments, functionality to delay some elastic response to enhance toughness and energy damping. Other mechanically active compnents include fillers and blended polymers that may be other elastomers, miscible or compatibile polymers. Elastomers find application in a huge array of products where their role may be obvious or they may contribute an important but concealed function. Natural rubber was among the first polymers used and vital to even current economies. New blends and nanofillers now refine properties. Diversified elastomers classes such as fluorocarbon polymers, polyurethanes and polyacrylates now contribute unique property combinations to specialised and diversified high technology products. The trend towards biomaterials signals a return to importance of natural rubber, however, there is opportunity for development of new hybrid bioelastomers.

## References


1. Flory, P.J.: Principles of Polymer Chemistry, pp. 432–494. Cornell University Press, Ithaca (1953)
2. Elias, H.-G.: An Introduction to Polymer Science, pp. 333–354. VCH Publishers, Weinheim (1997)
3. Sperling, L.H.: Introduction to Physical Polymer Science, 4th edn, pp. 349–506. Wiley-Interscience, Hoboken (2006)
4. Ullman, R.: Rubber elasticity. In: Cahn, R.W., Haasen, P., Kramer, E.J. (eds.) Materials Science and Technology, pp. 357–388. Weiheim, VCH (1993)
5. Elias, H.-G.: Macromolecules, Volume 3, Physical Structure and Properties. Wiley VCH, Weinheim, p. 65 (2008)
6. Suphadon, N., Thomas, A.G., Busfield, J.J.C.: The viscoelastic behaviour of rubber under a small simple shear oscillation superimposed on a large pure shear. Polym. Testing **29**, 440–444 (2010)
7. Elias, H.-G.: An Introduction to Polymer Science, p. 170. VCH, Weinheim (1997)
8. Sperling, L. H.: Introduction to Polymer Sciences, 4th edn. Wiley-Interscience, p. 427 (2006)
9. Robert, C.K.: Chapter 1. In: Rudiger, M., Hans, M. (eds.) Handbook of Specialty Elastomers, CRC Press, USA, pp. 1–37 (2008)
10. Robert, C.K.: Chapter 11. In: Stephen, K.F., Robert, C.K. (eds.) Handbook of Specialty Elastomers, CRC Press, USA, pp. 371–385 (2008)
11. Buckley, C.P., Prisacariu, C., Martin, C.: Elasticity and inelasticity of thermoplastic polyurethane elastomers: Sensitivity to chemical and physical structure. Polymer **51**, 3213–3224 (2010)
12. Francois, C.: Materials Handbook, 2nd edn, pp. 691–750. Springer, London (2008). (Chapter 11)
13. Brandrup, J., Immergut, E.H.: Polymer Handbook, 2nd edn. Wiley-Interscience, Canada (1975)

14. Sonnenschein, M.F., Guillaudeu, S.J., Landes, B.G., Wendt, B.L.: Comparison of adipate and succinate polyesters in thermoplastic polyurethanes. Polymer **51**, 3685–3692 (2010)
15. Robert, C.K.: Chapter 4. In: Pascal, F. (ed.) Handbook of Specialty Elastomers, CRC Press, Raton, pp. 134–154 (2008)
16. John, S.: Chapter 2. In: Arcella, V., Ferro, R. (ed.) Modern Fluoropolymers, Wiley, West Sussex, pp. 71–90 (1997)
17. Jiri, G.D.: Handbook of Thermoplastic Elastomer, pp. 161–177. William Andrew Publishing, New York (2007). (Chapter 5)
18. Giovanni, P., Antonio, M., Dario, B., Simona, S.: Effect of composition on the properties of SEBS modified asphalts. Eur. Polym. J. **42**, 1113–1121 (2006)
19. Elseifi, M.A., Flintsch, G.W., Al-Qadi, I.L.: Quantitative effect of elastomeric modifcation on binder performance at intermediate and high temperatures. J. Mater. Civ. Eng. **15**, 32–40 (2003)
20. Geoffrey, H., Hans, R.K., Roderic, P.Q.: Thermoplastic Elastomers, pp. 217–246. Carl Hanser Verlag, Munich (2004). (Chapter 9)
21. Jiri, G.D.: Handbook of Thermoplastic Elastomer, pp. 235–347. William Andrew Publishing, New York (2007). (Chapter 10)
22. Brydson, J.A.: Thermoplastic elastomers-properties and applications. Rapra Rev. Rep. **7**, 4–18 (1995)
23. Alfonso, R.-G., Lourdes, F., Jordi, P.: Degradable poly(ester amide)s for biomedical applications. Polymers **3**, 65–99 (2011)
24. Hao, C., Paulina, S.H., Daniel, J.S., Nathaniel, V., Abigail, K.R.L.J., Seung-Woo, C., Anne, Y., Robert, L., Daniel, G.A.: A novel family of biodegradable poly(ester amide) elastomers. Adv. Healthc. Mater. **23**, 95–100 (2011)
25. White, J.R., De, S.K.: Rubber Technologist's Handbook, pp. 87–130. Rapra Technology Limited, Exeter (2001). (Chapter 4)
26. Jiri, G.D.: Handbook of Thermoplastic Elastomer, pp. 249–264. William Andrew Publishing, New York (2007). (Chapter 11)
27. Jiri, G.D.: Handbook of Thermoplastic Elastomer, pp. 191–199. William Andrew Publishing, New York (2007). (Chapter 7)
28. Hemphill, Jim: White Paper Dow Specialty Elastomers for Thermoplastic Polyolefin. Dow Chemical Company, Michigan (2009)
29. Jiri, G.D.: Handbook of Thermoplastic Elastomer, pp. 179–190. William Andrew Publishing, New York (2007). (Chapter 6)
30. Soares, B.G., Santos, D.M., Sirqueira, A.S.: A novel thermoplastic elastomer based on dynamically vulcanized polypropylene/acrylic rubber blends. eXPRESS Polym. Lett. **2**, 602–613 (2008)
31. Passador, F.R., Rodolfo, A., Pessan, L.A. : Dynamic vulcanization of PVC/NBR blends. In: Proceedings of the Polymer Processing Society 24th Annual Meeting, 2008
32. Ivan, G.: Dynamic vulcanization an accessible way to thermoplastic elastomers. Iran. J. Polym Sci. Technol. **2**, 3–11 (1993)
33. Coran, A.Y., Patel, R., Williams, D.: Rubber-thermoplastic compositions Part V. Selecting polymers for thermoplastic vulcanizates. Rubber Chem. Technol. **55**, 116–136 (1982)
34. Lame, O.: Does fractal nanostructure of filled rubber lead to fractal deformations? In situ measurements of strain heterogeneities by AFM. Macromolecules **43**(13), 5881–5887 (2010). doi:10.1021/ma100697v
35. Suphadon, N., Thomas, A.G., Busfield, J.J.C.: The viscoelastic behavior of rubber under a complex loading. II. The effect large strains and the incorporation of carbon black. J. Appl. Polym. Sci. **117**, 1290–1297 (2010)
36. Stockelhuber, K.W., Das, A., Jurk, R., Heinrich, G.: Contribution of physico-chemical properties of interfaces on dispersibility, adhesion and flocculation of filler particles in rubber. Polymer **51**, 1954–1963 (2010)

37. Sosson, F., Belec, L., Chailan, J.-F., Carriere, P., Crespy, A.: Highlight of a compensation effect between filler morphology and loading on dynamic properties of filled rubbers. J. Appl. Polym. Sci. **117**, 2715–2723 (2010)
38. Merabia, S., Sotta, P., Long, D.R.: Unique plastic and recovery behavior of nanofilled elastomers and thermoplastic elastomers (Payne and Mullins effects). J. Polym. Sci., Part B: Polym. Phys. **48**, 1495–1508 (2010)
39. Lewicki, J.P., Pielichowski, K., De La Croix, P.T., Janowski, B., Todd, D., Liggat, J.J.: Thermal degradation studies of polyurethane/POSS nanohybrid elastomers. Polym. Degrad. Stab. **95**, 1099–1105 (2010)
40. Kuila, T., Srivastava, S.K., Bhowmick, A.K.: Rubber/LDH nanocomposites by solution blending. J. Appl. Polym. Sci. **111**, 635–641 (2009)
41. Sun, J., Li, H., Song, Y., Zheng, Q., He, L., Yu, J.: Nonlinear stress relaxation of silica filled solution-polymerized styrene-butadiene rubber compounds. J. Appl. Polym. Sci. **112**, 3569–3574 (2009)
42. Serbescu, A., Saalwachter, K.: Particle-induced network formation in linear PDMS filled with silica. Polymer **50**, 5434–5442 (2009)
43. Wu, J.-H., Li, C.-H., Wu, Y.-T., Leu, M.-T., Tsai, Y.: Tsai, Thermal resistance and dynamic damping properties of poly (styrene-butadiene-styrene)/thermoplastic polyurethane composites elastomer material. Compos. Sci. Technol. **70**, 1258–1264 (2010)
44. Mathai, A.E., Thomas, S.: Morphology, mechanical and viscoelastic properties of nitrile rubber/Epoxidized natural rubber blends. J. Appl. Polym. Sci. **97**, 1561–1573 (2005)
45. Stephen, R., Raju, K.V.S.N., Nair, S.V., Varghese, S., Oommen, Z., Thomas, S.: Mechanical and viscoelastic behavior of natural rubber and carboxylated styrene-butadiene rubber latex blends. J. Appl. Polym. Sci. **88**, 2639–2648 (2003)
46. Tang, D., Qin, C., Cai, W., Zhao, L.: Preparation, morphology, and mechanical properties of modified-PU/UPR graft-IPN nanocomposites with BaTiO3 fiber. Mater. Chem. Phys. **82**, 73–77 (2003)
47. Guo, Q., Figueiredo, P., Thomann, R., Gronski, W.: Phase behavior, morphology and interfacial structure in thermoset/thermoplastic elastomer blends of poly(propylene glycol)-type epoxy resin and polystyrene-b-polybutadiene. Polymer **42**, 10101–10110 (2001)
48. Song, M., Hourston, D.J., Reading, M., Pollock, H.M., Hammiche, A.: Modulated differential scanning calorimetry: Analysis of interphases in multi-component polymer materials. J. Therm. Anal. Calorim. **56**, 991–1004 (1999)
49. Morton, M.: Elastomers, Synthetic. In: Kroschwitz, J. (ed.) Kirk-Othmer: Concise Encyclopedia of Chemical Technology, 4th edn, pp. 677–679. Wiley-Interscience, New York (1999)
50. Kumar, K.D., Tsou, A.H., Bhowmick, A.K.: Interplay between bulk viscoelasticity and surface energy in autohesive tack of rubber-tackifier blends. J. Polym. Sci., Part B: Polym. Phys. **48**, 972–982 (2010)
51. Meng, Q., Hu, J.: A review of shape memory polymer composites and blends. Compos. A Appl. Sci. Manuf. **40**, 1661–1672 (2009)
52. Ratna, D., Karger-Kocsis, J.: Recent advances in shape memory polymers and composites: a review. J. Mater. Sci. **43**, 254–269 (2008)
53. Liu, Y., Lv, H., Lan, X., Leng, J., Du, S.: Review of electro-active shape-memory polymer composite. Compos. Sci. Technol. **69**, 2064–2068 (2009)
54. Bagdi, K., Molnár, K., Sajo, I., Pukánszky, B.: Specific interactions, structure and properties in segmented polyurethane elastomers. eXPRESS Polym. Lett. **5**, 417–427 (2011)

# Special Purpose Elastomers: Synthesis, Structure-Property Relationship, Compounding, Processing and Applications

**Deepalekshmi Ponnamma, Cintil Jose Chirayil, Kishor Kumar Sadasivuni, Lakshmipriya Somasekharan, Srinivasarao Yaragalla, Jiji Abraham and Sabu Thomas**

**Abstract** Elastomers are notable as very special class of polymers due to their multifunctional applications. The superior mechanical properties, high flexibility, resilience and good viscoelastic behaviour make this class applicable in a wide range of technology and industry. Depending on the various properties and general applications elastomers are classified in to a number of categories. This particular chapter deals with a very important class of special purpose elastomers. The synthesis, structure, different properties, mode of vulcanization, processing and applications of most of the synthetic elastomers are discussed. Apart from providing a basic understanding about the materials, this chapter can facilitate wide information about the technical details and industrial importance of this class of rubbers.

## 1 Introduction

Among the various types of polymers, elastomer class has special significance due to their amazing properties like high resilience, high elasticity, viscoelastic properties, mechanical strength and frictional resistance. Their wider availability and low cost make them highly significant in technological applications. Their damping and energy absorption characteristics, resilience, long service life, ability to seal against moisture, heat, and pressure, non-toxic properties, moldability, and variable stiffness are other distinguishable attributes. Most of the elastomers

D. Ponnamma (✉) · C. Jose Chirayil · L. Somasekharan · S. Yaragalla · J. Abraham
S. Thomas
School of Chemical Sciences, Mahatma Gandhi University,
Kottayam 686 560 Kerala, India
e-mail: lekshmi.deepa09@gmail.com

K. K. Sadasivuni
LIMATB Laboratory, Université de Bretagne Sud,
rue St Maudé 56100 Lorient, France

P. M. Visakh et al. (eds.), *Advances in Elastomers I*,
Advanced Structured Materials 11, DOI: 10.1007/978-3-642-20925-3_3,

belong to the ‘Rubbers’. But there are elastomers showing properties of thermoplastics (thermoplastic elastomers) as well as thermosets. The flexibility of the elastomers are due to their segmental mobility and vulcanization makes cross links among the rubber chains for the recovery. Vulcanised rubbers are widely used in industries especially in tyre applications.

Even though elastomers have tremendous applications, their end use always depends on the selection of a particular matrix. There are several classifications of elastomers, of which the most significant is based on its functional properties. Depending on the end use applications, elastomers are classified into general purpose and special purpose. As the name indicates, the general purpose rubbers include elastomers having general applications. Natural rubber, styrene butadiene rubber (SBR), polybutadiene and polyisoprene rubber, ethylene propylene rubber, isobutylene based rubbers, etc. are known as general purpose elastomers. Those rubbers having specific applications belong to the special purpose category. All such elastomers are synthetic in nature. Butyl rubber, fluoro rubbers, sulphide rubbers, polyurethanes etc. are some of the examples. The specific properties of such elastomers are air impermeability and resistance to heat, aging and ozone in the case of butyl rubber, oil resistance for nitrile rubber, temperature stability for silicone rubber and so on. Comparing to butyl rubber (IIR), its derivatives chlorobutyl and bromobutyl rubber have better heat aging resistance. The main applications of IIR include electric wire, cable, adhesive, construction materials, sealing materials, inner tube, bladder etc. Chloro butyl and bromo butyl rubber are specifically used in inner tubes, inner liners, automotive parts, electric wires and industrial goods.

Special purpose elastomers are again classified into high volume specialty elastomers such as EPDM and polychloroprene and also low volume specialty elastomers such as silicone rubber, polyurethanes and fluoro elastomers. A detailed study about the special purpose elastomers, their mode of synthesis, curing behavior, end use applications etc. has been given in this chapter. Effort has been taken to address all the important factors in manufacturing and applying specialty elastomers in technology and industry. The main types and their significance are narrated hereafter.

# 2 Types of Rubbers

## *2.1 Butyl Rubber*

Poly (isobutene-co-isoprene) rubber generally known as butyl rubber typically contains about 98 % polyisobutylene and 2 % isoprene polymer chains distributed in a random way. The structure is shown in Fig. 1a. It has several unique properties; the most important being the superior gas impermeability and mechanical damping properties. It is the best sealant of all the known rubbers. The major application of butyl rubber in the lining of tyres (low permeability) comes from the

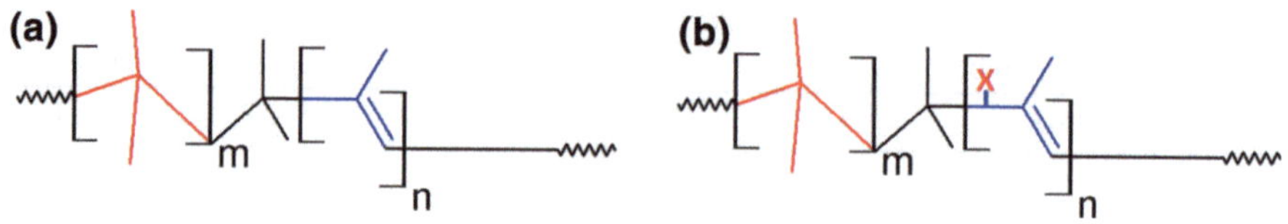


**Fig. 1** Chemical structure of (**a**) butyl rubber (m = 98 %, n = 2 %) (**b**) Halo (X) butyl rubber

low levels of unsaturation between long poly isobutylene segments. It was formerly used to make the tyre inner tubes, but now incorporating this rubber as inner liners has led to tubeless tyres.

Butyl rubber is produced via a polymerisation process taking place at −100 °C and thus the process is rather complex [1, 2]. Since the reaction is highly exothermic, control of temperature to lower level (−90 to −100 °C) is necessary particularly to synthesize high molecular weight IIR. In the most common method of polymerization, methyl chloride and boiling liquid ethylene are used as the reaction diluents which can maintain the temperature. Finally the slurry of fine particles of butyl rubber (dispersed in methyl chloride) is formed in the reactor and methyl chloride and unreacted monomers are flashed and stripped overhead by adding steam and hot water, and thus recycled. Zinc or calcium stearate compound is the slurry aid and this along with the antioxidant is introduced into the hot water/polymer slurry to stabilize the polymer and to prevent agglomeration. After screening the polymer from the hot water slurry, it is dried in a series of extrusion dewatering and drying steps [2] to obtain IIR in the final form. Butyl rubber can be produced in bulk by polymerizing in alkane solutions as well.

The allylic halide functionality of the elastomer responds to standard cure accelerators in an anomalous manner, and it is susceptible to isomerization and elimination reactions at elevated temperatures that can detract from end-use performance [3]. So researchers developed halo butyl rubber to make butyl rubber more compatible with other polymers by using substitution of halo group.

### 2.1.1 Halo Butyl Rubber

Halogenations of poly(isobutylene-co-isoprene) to generate halogenated butyl rubber (XIIR) (Fig. 1b) yields an elastomer that has the air impermeability required for tyre inner liner applications and the cure reactivity required to generate adhesion between the tyre inner liner and its body. There are two kinds of halo butyl rubber; chloro butyl rubber, CIIR and bromo butyl rubber, BIIR. While superior sulfur vulcanization kinetics are known to originate from the exo-allylic bromide structure of bromo poly(isobutylene-co-isoprene), the cure chemistry of this commercially important elastomer is typical to understand [4]. The low level of residual unsaturation in halobutyl rubber XIIR), coupled with the insolubility of crosslinked networks, complicates the structural determination of its vulcanizates [5].

**Table 1** Recipe of butyl rubber compounds (*phr* parts per hundred rubbers) [8]

| Ingredients | Butyl rubber | Zinc oxide | Stearic acid | Sulphur | MBTS | TMTD | TMQ |
|---|---|---|---|---|---|---|---|
| phr | 100.0 | 5.0 | 2.0 | 1.5 | 0.5 | 1.0 | 1.0 |

Processing of IIR, CIIR and BIIR by mastication changes the molecular weight distribution of chains between junctions of the polymer network, the coefficient of polydispersity, the share of low-temperature and high-temperature blocks, and their glass transition temperatures. These changes in the structure of rubber depend on both the presence of sulphur and the kind of sulphur (polymeric or mineral) being used during mixing. This suggests that the sequence of rubber compounding and the rubber recipe influence not only the dispersion and the distribution of ingredients (which is well known) but also the structure of the rubber matrix [6]. IIR, CIIR and BIIR must be compounded and chemically cross-linked to yield useful, durable end use products similar to all other rubbers. The vulcanization kinetics or cure-curve is monitored from the evolution of crosslinks, using an Oscillatory Disk Rheometer (ODR). The instrument measures the torque or the shear modulus, as a function of time when the vulcanizate cures, where it is assumed that the modulus is proportional to the evolving concentration of cross-link's. Different grades of this elastomer can be developed according to the processing and property needs, and a range of molecular weights, unsaturation, and cure rates are commercially available. Based on the specificity of applications, the end use attributes and the processing equipment, the right grade can be selected. The selection and ratios of the proper fillers, processing aids, stabilizers, and curatives also play critical roles in compound processing and the end product behaviour [7]. Elemental sulphur and organic accelerators are widely used to cross-link poly (isobutylene-co-isoprene). The low level of unsaturation requires aggressive accelerators such as thiuram or thiocarbamates. The usually used curing agents are given in Table 1.

As illustrated in Fig. 2, the vulcanization reaction starts at the isoprene site with the polysulfidic cross links attached at the allylic positions and displaces the allylic hydrogen. This is the general mechanism involved in the accelerated sulfur vulcanization of butyl rubber. The number of sulfur atoms per cross-link varies between one and four or more. Both the cure rate and cure state (modulus) increase at higher unsaturation rates. The basic steps in the vulcanization reactions are similar in the case of all rubbers even though some important differences remain [9]. The whole vulcanization reaction can be divided into three sub-categories:

1. The scorch delay or induction region, where the accelerator chemistry for the reactions leading to the formation of an active-sulfurating agent is involved.
2. Cure reaction which involves reactions leading to the formation of crosslinks; and
3. Post-crosslinking chemistry involving the reactions that lead to crosslink shortening and crosslink degradation [10] and here the modulus can increase, decrease or remain constant depending upon the specific vulcanization system.

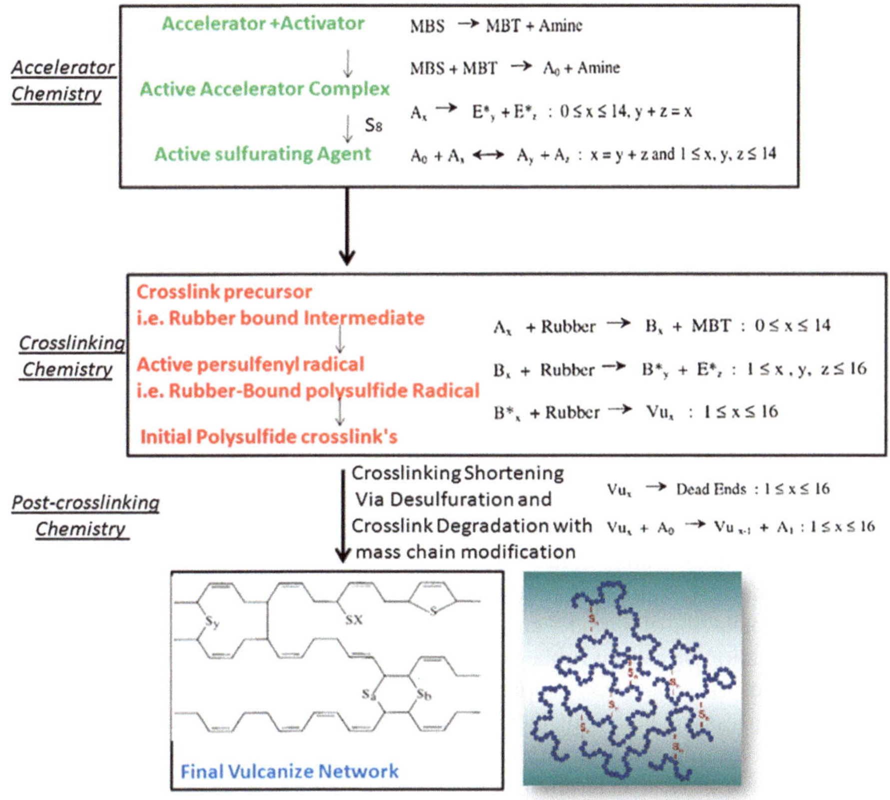


**Fig. 2** General mechanism for accelerated sulfur vulcanization for butyl rubber. Mechanism adopted from that of Morrison and Porter [9]

These three vulcanization reaction steps are evidenced from the three main regions of the cure curve [10]. Vulcanization begins with the formation of zinc-free and/or zinc-complexed accelerator species. MBTS formed from the autocatalytic dissociation of MBS act as the precursor in the formation of these accelerator polysulphides. Accelerator polysulphides are subsequently formed by the direct reaction of lower length accelerator polysulphides with sulphur. It has been proposed that these sulfuration reactions proceed by the insertion of sulphur as S8 molecules or, alternatively, that the reactions are sequential, occurring via the insertion of single sulphur atoms [11]. The exact chemical structure of the $S_x$ species and its concentration for $x < 8$ has not yet been detected. In the absence of zinc, the sulfuration reactions proceed via a radical mechanism whereas in the presence of zinc it proceeds via a polar mechanism. In zinc activated systems sulfuration reactions include a combination of radical and polar mechanisms.

In the case of halobutyl rubber, the halogen atom present in the allylic position causes easier cross-linking than allylic hydrogen alone does as halogen is a better leaving group in nucleophilic substitution reactions. Zinc oxide is the most common cross-linking agent in this kind of rubbers since it forms very stable carbon–carbon bonds by alkylation reaction through dehydrohalogenation, with zinc chloride by product. In addition to zinc oxide elemental sulfur and organic accelerators are also used to cross-link butyl rubbers. Aggressive accelerators such as thiuram or thiocarbamates are employed whenever low level of unsaturation is required. As already discussed, vulcanization proceeds at the isoprene site with the polysulfidic cross links attached at the allylic positions by displacing the allylic hydrogen. It is important to note that the sulphur cross-links formed are less stable at high temperature. The resin cure systems (commonly using alkyl phenol–formaldehyde derivatives) can eliminate this drawback since it provides carbon–carbon cross-links and thus more stable compounds.

It is generally accepted that the unique cure properties of BIIR originate from the reactivity of allylic bromides [12–15]. However, comparatively little is known about the intrinsic reactivity of the particular allylic bromide functionality found within BIIR, especially at vulcanization temperatures. This is a concern to the technologists, as degradation reactions operate concurrently with the vulcanization process with undesirable consequences as shown in Fig. 3. To facilitate the structural characterization of products arising from BIIR instability, researchers studied the behavior of a model compound and similar approach was used to study the vulcanization of natural rubber [9, 14–16] and poly(butadiene) [16, 17] and the halogenations of isobutylene–isoprene rubber and the ZnO cure of XIIR, in which complications associated with thermal instability were first highlighted [16–18]. The allylic bromide functionality within BIIR undergoes substitution with amine nucleophiles without acceleration of concurrent HBr elimination [5].

Chloro-butyl rubber can be crosslinked upon heating, the reaction being facilitated by the addition of ZnO, $ZnCl_2$, or zinc dimethyl dithiocarbamate ($Zn_2(dmtc)_4$) [4, 19]. The most widely accepted theory, as postulated by Baldwin et al. [6] and supported by Kuntz et al. [18] and Vukov [4] postulates a cationic reaction in which $ZnCl_2$, formed in situ, is responsible for crosslinking as shown in Fig. 3. Stearic acid is frequently added to commercial formulations. It is generally accepted that reactions leading to the formation of $ZnCl_2$ precede crosslinking, the induction period prior to crosslinking being ascribed to the need to form $ZnCl_2$ that is considered to act as a catalyst for crosslinking via a cationic mechanism. In ZnO formulations the $ZnCl_2$ formed may initially results from reaction of ZnO with HCl formed on thermal dissociation of the allylic chlorine [4, 6, 19]. The cationic mechanism as originally proposed by Baldwin et al. [6] also showed the $ZnCl_2$ initiated cationic reaction to lead to crosslinks without the participation of a conjugated diene butyl precursor. They reported that good co-vulcanization of CIIR with unsaturated rubbers can be effected, although care must be taken in the selection of curatives. Though sulphur and conventional accelerators were included in the formulations, the ability to link CIIR to highly unsaturated rubbers may point to dienes playing a role in the crosslinking process. $ZnCl_2$ and conjugated

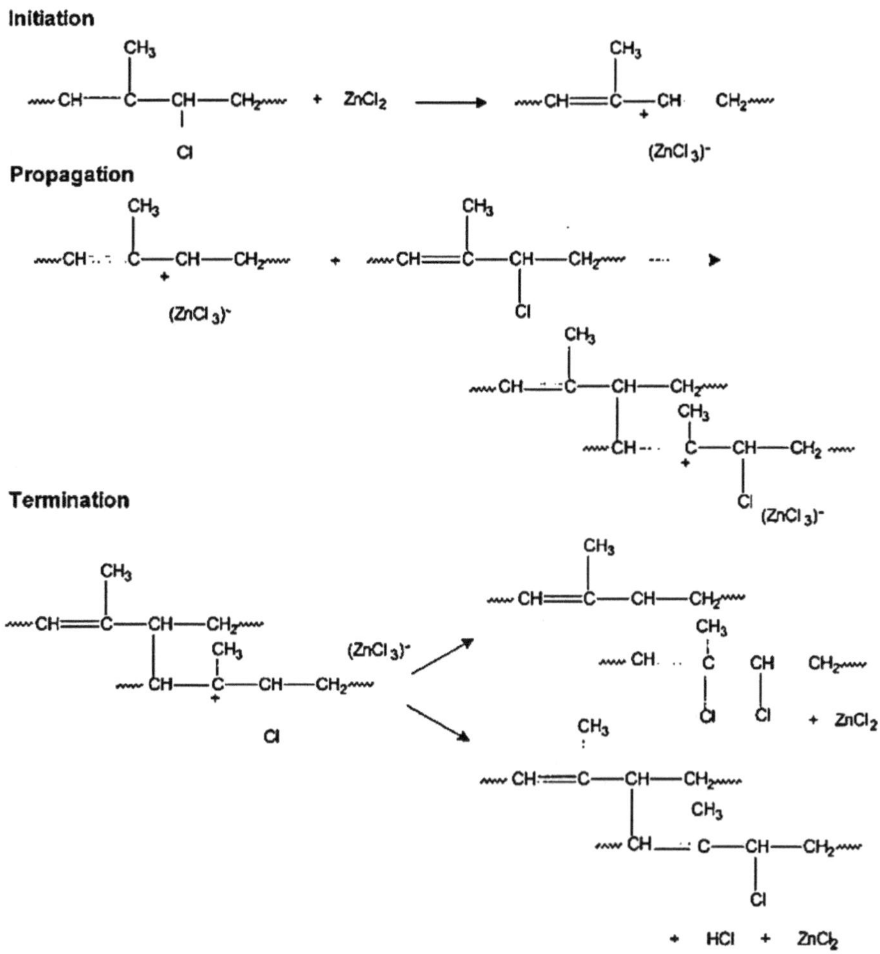


**Fig. 3** General mechanism for accelerated sulfur vulcanization for chloro butyl rubber [19]

diene butyl are products of the same reaction, dehydrohalogenation, and separation of their roles in the crosslinking process is not easy.

When the derivatives of butyl rubbers i.e. bromobutyl and chlorobutyl rubbers are considered, the former cures at a faster rate than the latter and has better adhesion to highly unsaturated rubbers. Due to this reason, the volume growth rate of bromobutyl has exceeded that of chlorobutyl in recent decades especially in tyre industry. However, utmost care should be taken while processing the halobutyls as there is a chance for premature dehydrohalogenation. To prevent this problem of premature dehydrohalogenation, stabilizers such as calcium stearate alone (for chlorobutyl) or compounded with epoxidized soybean oil (for bromobutyl) etc. are used.

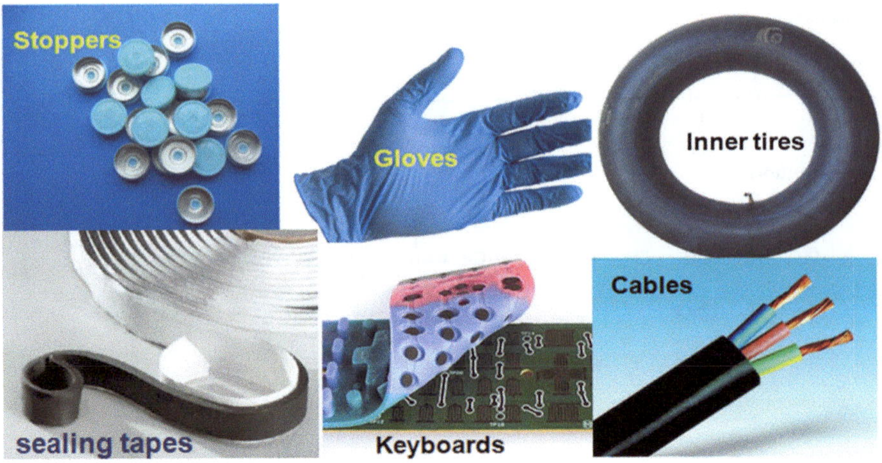


**Fig. 4** General applications of Poly (isobutylene-co-isoprene) and Halo Poly (isobutylene-co-isoprene rubbers)

### 2.1.2 Properties and Applications

One of the main important applications of butyl rubber based compounds is in the fabrication of tyre inner tubes. This is attributed to its excellent impermeability/air retention and good flex properties. Another important property of this rubber is that it is blood compatible. This feature makes it applicable in biomedical fields such as in fabricating artificial joint materials. This rubber is not toxic and so it is widely used in all chewing gum. IIR and XIIR is used in a wide range of applications, such as inner tubes, tyre liners, curing bladders, air springs, drug cap sealants, gas pipe coating and gaskets, military attires, air conditioner hoses, cable insulations, jacketing, roof membranes and sporting goods [20–22]. Butyl Rubbers are also useful as sorbent materials for the removal of oil and polycyclic aromatic hydrocarbons from seawater [22–24].

The derivative rubbers such as chloro and bromo butyls also have great significance in various fields of technology and industry. The main advantage of such halo butyl rubber is in improving the property of air impermeability. The major contribution of such rubbers is in the development of tyre inner liners. Halobutyls help in the development of more durable tubeless tyres with the air retaining inner liner chemically bonded to the body of the tyre. In short the unique combination of properties of butyl rubbers such as excellent impermeability, good flexing, good weather ability, co-vulcanization with highly unsaturated rubbers (in the case of halobutyl) makes it a preferred material for industry. General Applications of poly (isobutylene-co-isoprene) and halo poly (isobutylene-co-isoprene) are shown in Fig. 4.

## 2.2 Nitrile Rubber

Nitrile-butadiene rubber (NBR) commonly known as nitrile rubber is an oil-resistant synthetic rubber which is produced from the copolymers, unsaturated acrylonitrile and butadiene. Nitrile rubber (NBR) is considered as the workhorse of the industrial and automotive rubber products industries because of its good elongation properties, good to excellent compression set resistance, adequate resilience and tensile strength but poor flame and steam resistance. Generally NBR can be classified into 5 categories they are hot NBR, cold NBR, cross linked hot NBR, carboxylated nitrile (XNBR), bound antioxidant NBR.

### 2.2.1 Synthesis of Nitrile Rubber

Nitrile rubber can be prepared by the random polymerization of acrylonitrile with butadiene in the presence of free radical catalyst. Emulsifier/soap, monomers, catalyst, are introduced into the reaction vessel. As a result of polymerization a latex is obtained which is coagulated by using various materials (e.g. calcium chloride, aluminum sulfate). The amount of acrylonitrile present in the final copolymer varies from 15 to 50 % [1, 25].

### 2.2.2 Structure Property Relationship

Chemical structure of NBR and SBR are almost similar. Replacement of benzene ring of SBR gives NBR. Presence of this cyanogens group provides polarity to NBR rubber. As a result of this oil, heat, air permeability resistance, electrical conductivity properties of vulcanizates get enhanced. Nitrile monomer gives permeation resistance properties to a wide variety of solvents and chemicals. NBR is resistant to hydrocarbon oils, fats and solvents, because of the polar nature of acrylonitrile. Oil resistance means ability of the vulcanized rubber to resist the changes in its physical properties such as modulus, tensile strength, abrasion resistance, and dimensions when it is in contact with oil and fat. The butadiene part is responsible for the softness, flexibility of the rubber and butadiene component is involved in vulcanization process, thereby enhancing the elasticity of rubber.

ACN level, the major reason behind the polarity of the NBR rubber determines properties, of the nitrile rubber. NBR with higher acrylonitrile content has high processability cure rate with sulfur cure system, oil/fuel resistance, compatibility with polar polymers, air/gas impermeability, tensile strength, abrasion resistance, heat-aging. Lower ACN content provides improved cure rate with peroxide cure system, compression set, resilience, hysteresis, low temperature flexibility. There are some more characteristics which determines the properties of nitrile rubber. They are mooney plasticity and gel content. Normally mooney viscosity ranges from 30 to 95. Gel content ranges of nitrile rubber are 0–80 %. Permanent gel content leads to lowering in tensile strength, percent of elongation and abrasion resistance. Non permanent gels have improved green strength, reduced surface tack and poor mould characteristics [2]. Modification of nitrile rubber by carboxyl group improves abrasion resistance. Blends of NBR and PVC have much improved ozone resistance and improved flame retardant properties [6, 26].

### 2.2.3 Compounding and Processing

Pure NBR is not useful for applications. In order to get a wide variety of properties nitrile rubber should be compounded. Higher acrylonitrile NBR rubber can be chosen when high resistance applications but they have poorer low temperature properties. Lower acrylonitrile rubbers are used when low temperature and dynamic properties are important but they are poorer in oil and fuel resistance. Commonly used activators are zinc oxide at 3–5 phr level and stearic acid at 1–2 phr level. These activators are used along with sulphur curing system. With peroxide curing system cynurate (Tac) is the activator. Incorporation of different kinds of accelerators such as TMTD, TMTM, DOTG and NPV/C improve cure rate, tensile properties, resistance to oil, aging properties [27]. Antioxidants can be used in order to improve the thermal stability of all nitrile compounds. Ozone resistance of nitrile rubber can be improved by blending it with materials such as poly(vinyl chloride), chlorosulphonated polyethylene, chlorinated polyethylene, epichlorohydrin and ethylene propylene terpolymers.

For the better properties of nitrile rubber, it is reinforced with fillers. Most commonly used filler is carbon black. Many studies have been reported on the effect of different types of carbon black on the properties of nitrile rubber elastomer. Carbon black can reduce the total quality cost of nitrile compounds while improving or maintaining key properties [4, 28]. Recently for non black applications fillers like nanoclay, organosilicates, calcium carbonates are widely used [3, 5, 29]. The tribological properties of GO/NBR under dry sliding and water-lubricated condition have been reported [8, 30].

In order to get better low temperature properties and to improve processability plasticizers are used. Commonly used plasticizers are ester types, aromatic oils and polar derivatives. In summary nitrile compounds can be compounded with vulcanization system, fillers, activators, antioxidants, plasticizers etc. The processing of nitrile rubber includes mixing, extrusion, pre-forming to required shape and vulcanization. Mode of mixing of nitrile rubber is either by two roll mill or by internal mixing equipment. Mooney viscosity values of nitrile rubber determine how the material will process. Compounds can be processed by extrusion, calendaring or moulding by injection or compression.

### 2.2.4 Applications of NBR

Because of its oil and fuel resistant characteristics nitrile rubber finds its greatest market in applications where these characteristics are necessary. Applications of nitrile rubber include in the field of molded and extruded parts for the automotive, chemical, mining, home construction, printing rolls and paper industry, conveyor belts for construction and mining industry, industrial and military footwear, gasoline and oil hoses, bulk fuel transfer hose, chemical transfer hose, industrial air hose, food handling, dairy and creamery hose, seals and gaskets, brakes and friction pads, foam lines for thermal isolation. NBR is also employed in textiles, where its application to woven and nonwoven fabrics improves the finish and water proofing properties. It can be used to prepare non-latex gloves for the healthcare industry. It has special application in agriculture, in rice dehusking rollers, where they give much longer life and greater dehusking capacity than any other polymers. NBR finds enormous uses in roll covers, hydraulic hoses, conveyor belts, oil field packers, plumbing appliances etc.

## *2.3 Silicone Rubber*

Silicone rubber is an elastomer composed of silicone-itself a polymer-containing silicon atoms along with carbon, hydrogen, and oxygen. Major use of silicone rubbers is in industry, and there are multiple formulations. Silicone rubbers contains one or two part polymers, and contains fillers to improve properties or reduce cost. Silicone rubber is non-reactive and stable. It is resistant to extreme environments and temperatures from −55 to +300 °C, at the same time it is able to maintain its useful properties. Major applications of silicone rubber include automotive applications, food storage products, electronics, medical devices etc. The high permeability of silicone rubber has been attributed to the flexibility of the siloxane linkages in the polymer [31]. Silicone rubber offers good resistance to extreme temperatures, being able to operate normally from −55 to +300 °C. At the extreme temperatures, the tensile strength and tear strength and compression set can be far superior to conventional rubbers although still low relative to other materials. Silicone rubber has a very low tensile strength compared to organic rubbers. So care is needed in designing products to withstand even low imposed loads. It is also very sensitive to fatigue from cyclic loading. Silicone rubber coated composite membranes are commonly utilized in industrial membrane based separation processes such as gas separation [32], removal of organic vapors from air [31], pervaporation of various organic species [33] including aroma compounds [34], olefin separation [35] etc.

### 2.3.1 Structure and Properties of Silicone Rubber

Silicone rubber is a thermoset elastomer having a backbone made up of alternating silicon and oxygen atoms (Fig. 5). The key properties of silicone rubber are its good temperature stability, excellent electrical conductivity, available in medical quality grades and it is easy to give color. Silicone (i.e., polysiloxane)-containing

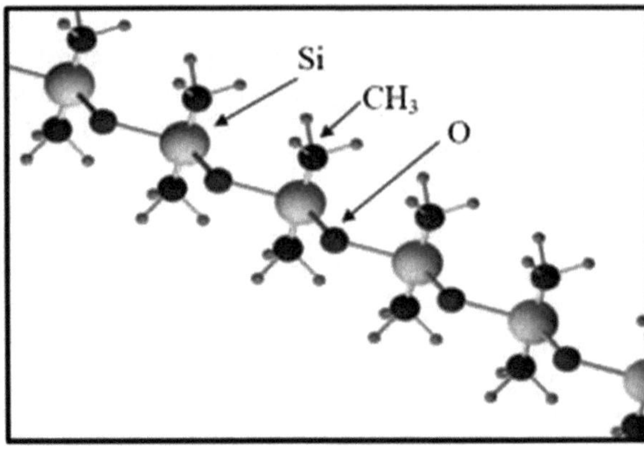


**Fig. 5** Molecular structure of silicone rubber

materials are one of the most widely used polymers in adhesives, coatings and rubber materials in the electronic circuits of microelectronic devices. The important properties of polyorganosiloxanes, which are used in preparing silicone rubber compositions, include excellent weather and thermal stability, low-temperature flexibility, and curability [36].

Silicone rubber possesses an extraordinary property that sets it apart from other elastomers its elastic behavior changes only slightly with change in temperature. Silicone rubber has excellent flame resistance. It has a flash point of 750 °C, an ignition temperature of 450 °C. Only very minor amounts of smoke are evolved during combustion. The principal combustion products are carbon dioxide, water, and ash consisting of silicon dioxide. Silicon dioxide is a good dielectric, it finds use in critical sectors to sheathe cables. It also process high oxidation resistance [37].

Silicone rubber is resistant to chemicals and it can be used in applications including contact with dilute acids and alkalies. But its chemical resistance can decrease with increase in temperature and concentration. Its resistance to various oils compares very favorably with that of organic elastomers. Silicone rubber is a highly inert material and will not react with chemicals. Because of its inertness, it is used in many medical applications and in medical implants. During manufacture, heat may be required to cure the silicone into its rubber form. It is carried out in a two stage process. It can also be injection molded. Mechanical properties of silicone rubber are given in Table 2.

**Table 2** Mechanical Properties of silicone rubber

| Mechanical properties | |
|---|---|
| Hardness, Shore A | 10–90 |
| Tensile strength | 11 $N/mm^2$ |
| Elongation at break | 100–1,100 % |
| Maximum temperature | +300 °C |
| Minimum temperature | −120 °C |

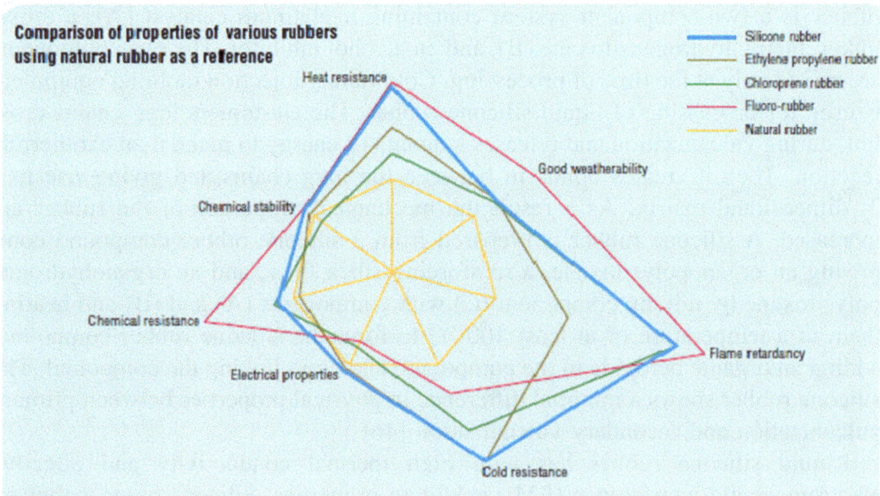


**Fig. 6** Comparison of properties of various rubbers using natural rubber as a reference

The siloxane bonds that form the backbone of silicone are highly stable. Compared to organic polymers, these silicone rubbers have higher heat resistance and chemical stability, and better electrical insulation. Silicone molecules are helical and intermolecular force is low. It results in high elasticity, high compressibility, and good resistance to cold temperatures. The methyl groups located on the outside the coil structure can rotate freely. So silicone has distinctive interfacial properties, including water repellency and good repeatability. The tear strength of silicone rubber is generally around 10 kN/m. These products are ideal for moulding large items. Comparisons of various rubbers based on properties are given below (Fig. 6).

Silicone rubber is widely used in many fields due to its high and low temperature resistance, long service life and inertness. But because of the poor mechanical properties, it cannot be used without reinforcement and this deficiency is usually overcome by adding certain reinforcing fillers [38–40], such as carbon black [41–43], fumed nano silica [44], and precipitated silica [45].

### 2.3.2 Compounding and Processing of Silicone Rubber

Both solid and liquid silicone rubber have a similar basic structure, but different cure and processing methods. High consistency rubber or solid silicone rubber is prepared in large batches and the components are mixed up at high temperatures followed by the addition of a peroxide catalyst. When cross-linking starts the reaction is interrupted before the completion of vulcanization. Then the mildly cross-linked silicone rubber is then rolled out in the form of sheets. Liquid silicone rubber is a two-component system containing a platinum catalyst (A), a cross-linker, methylhydrogensiloxane (B), and an alcohol inhibitor. The two components are mixed only at the time of processing. Cold runner injection molding equipment is used for processing of liquid silicone rubber. The elastomers long chains cross link during vulcanization and releases a quanta of energy to make it an exothermic reaction. Then it creates bonds in between the long chains and giving rise to a 3- dimensional matrix. As a result the mechanical properties of the rubber are increased. A silicone rubber is prepared from a silicone rubber compound comprising an organopolysiloxane, a reinforcing silica filler, and an organohydrogen polysiloxane by mixing component (C) with components (A) and (B) and heating them at a temperature of at least 100 °C to form the silicone rubber compound, adding an organic peroxide to the compound, and cross linking the compound. The silicone rubber shows a minimal difference in physical properties between primary vulcanization and secondary vulcanization [46].

Liquid silicone rubber has with high thermal conductivity and effective electromagnetic interference (EMI) shielding properties. Silicone-based polymers were prepared by the equilibrium polymerization of cyclic siloxane and end-blockers [47]. Magnesium ferrite nanoparticles with spinel magnetic properties were synthesized by the sol–gel method. A liquid silicone rubber (LSR) nanocomposite was prepared by compounding *a,w*-vinyl poly(dimethyl/methylpenylsiloxane) prepolymer (VPMPS), *a,w*-hydrogen poly(dimethyl/hydrogenmethylsiloxane) prepolymer (HPDHS), catalyst and magnesium ferrite .

Hydrophobic nano silica sol was added into polyvinylmethylsiloxane to prepare reinforced high temperature vulcanized (HTV) silicone rubber [48]. HTV silicone rubber filled with 40 phr hydrophobic nano silica sol showed excellent mechanical and optical properties [48].

There are three different types of crosslinking reactions in silicone rubber (Fig. 7). They are (a) addition curing (where polymer contains vinyl groups and

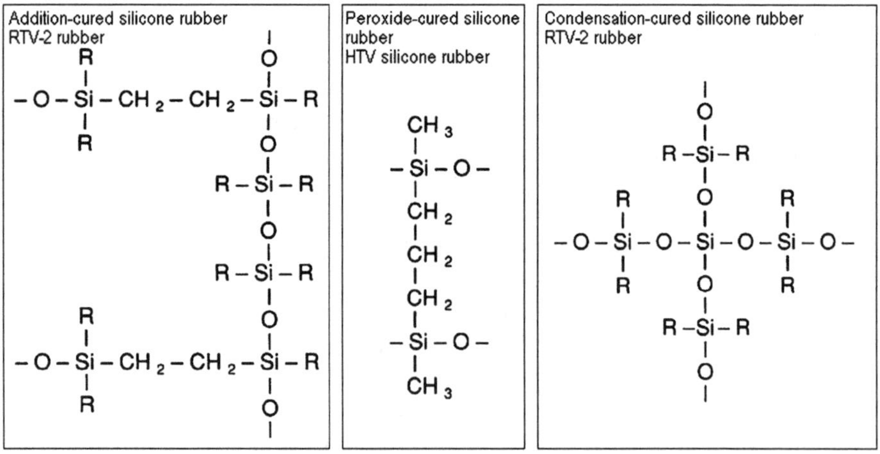


**Fig. 7** Three different types of crosslinking reactions in silicone rubber

crosslinking agent contains Si–H groups) (b) peroxide (-initiated) curing (where polymer contains vinyl groups) and (c) condensation curing (between $\alpha$, $\omega$-dihydroxypolydimethylsiloxanes and silicic acid esters). A platinum catalyst is needed for addition reaction. A tin catalyst is used for condensation curing systems. But, peroxide-initiated curing does not require a catalyst. Addition curing functions by bonding Si–H groups to double bonds. Complexes of platinum or palladium may serve as catalyst and if platinum-olefin complexes are used, curing will take place at normal room temperature. Platinum complexes with nitrogen are used for effecting addition curing at elevated temperatures (e.g. Pt-complexes with pyridine, benzonitrile or benzotriazole). To carry out peroxide curing, it is first necessary to create free radicals. It is possible either with heat or with radiation and different organic peroxides may serve as free-radical generators. Selective crosslinking and superior vulcanizates can be obtained by incorporating vinyl groups into the polymer. Typical catalysts for condensation curing are dibutyltin dilaurate and dibutyltin dioctoate. They function as the catalyst for the reaction between dihydroxypolydimethylsiloxanes and silicic acid esters. The rate of reaction depends on the crosslinking agent, and on the type of catalyst. There is no sulfur used for curing silicone rubbers.

Silicone as an organic polymer is widely used for spacecraft materials which suffered from atomic oxygen (AO) attack [49, 50]. Although the existence of atomic oxygen was known in the early days of space exploration, an awareness of the damaging effects on spacecraft materials was not well-known until the space shuttle began flying missions at much lower altitudes in low earth orbit (LEO). The surface shrinkage of the oxidized silicone leads to cracking of the partially oxidized brittle surface. Such cracks resulted in further exposure to atomic oxygen and ultimately decreased the transmittance of silicone [51–53]. For example, the refractive solar concentrator lenses are fabricated from flexible silicones. Under the radiation of charged particles in space environment, the silicone would be aged, which directly influences the reliability and lifetime of a spacecraft [54, 55].

So a reasonable transmittance of the silicone materials is very important for the solar concentrator lenses in LEO and other space environments. In the traditional silicone rubber industry, the silicone coupling agents are often used to treat inorganic additives, which were used in organic polymers, and change their hydrophilic surfaces into hydrophobic surfaces. This method can enhance the wettability of organic polymers to inorganic additive, and then form chemical bonds between them. Therefore, silicone coupling agents have the capability to develop "molecular bridge" in the interface of organic and inorganic materials. Figure 8 shows the 'molecular bridge' mechanism [56].

Silica films were deposited on flexible silicone rubber using solgel method. The surface morphology indicates that the silica sol can easily form a uniform thin film on the surface of silicone rubber pretreated by high concentration coupling agent. After depositing silica films, both AO erosion resistance and optical transmittance of silicone rubber are improved. The procedure for preparing sol–gel coating films is schematically shown in Fig. 9.

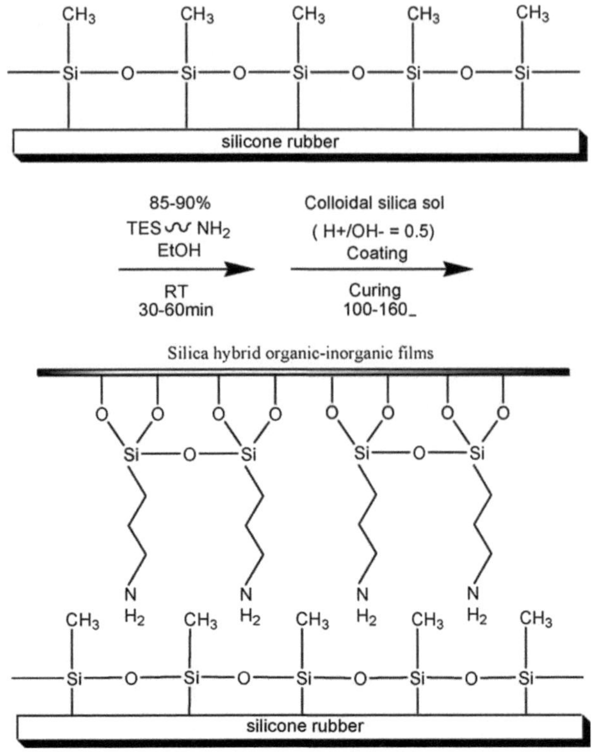


**Fig. 8** The 'molecular bridge' in the interface of silicone rubber with silica films by silicon coupling agents [56]

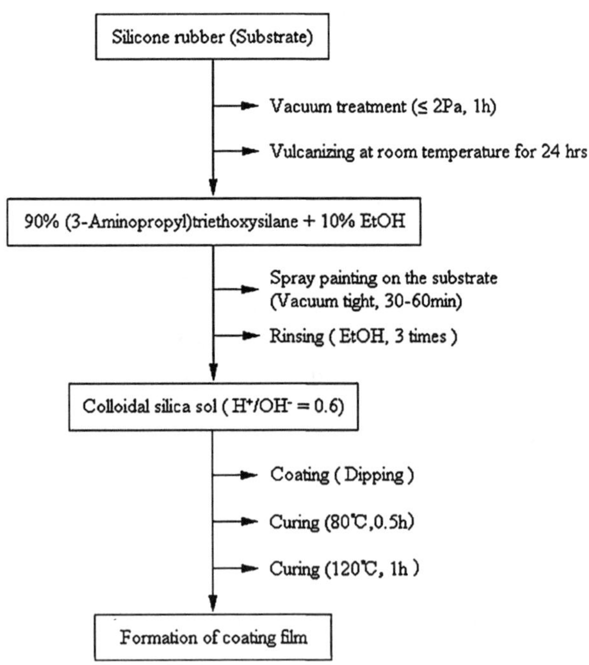


**Fig. 9** Preparation process of sol–gel coating films on the silicone rubber substrate [56]

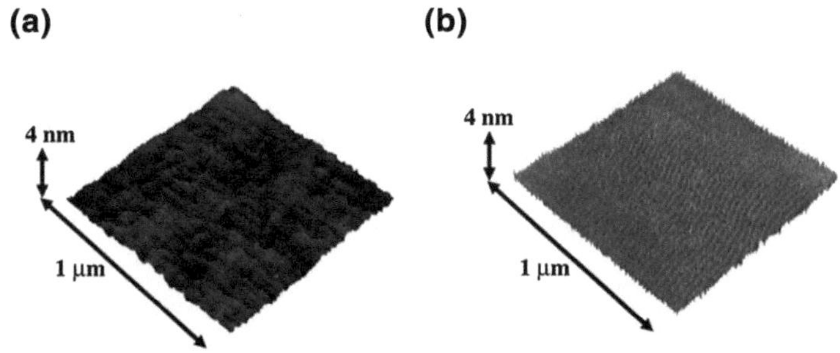


**Fig. 10** Surface morphologies of silicone rubber before and after the plasma irradiation [57]. **a**Before plasma exposure. **b** After plasma exposure by Ar/$O_2$ glow discharge, 5 min

Hino et al. [57] reported that the surface wettability of silicone rubber could be enhanced by Ar, Ar/$H_2$ and Ar/$O_2$ plasma irradiations. Silicone rubber is an electrical insulator, so that the charge-exchanged argon atom might have contributed to the enhancement of surface wettability. The mixing of oxygen in an argon plasma significantly increased the wettability, compared to Ar and Ar/$H_2$ plasmas (Fig. 10). This result suggests that the chemical reactivity to water is enhanced by the reactive oxygen species.

### 2.3.3 Applications

Silicone rubber is a synthetic compound known for its weatherability and ability to maintain useful properties over a wide range of temperatures. Silicone rubber is a thermal and electrical insulator and is resistant to oxidation, and degradation from ultraviolet radiation. These properties make silicone rubber excellent for electrical insulators and the properties of silicone rubber come from the structure of the polymer and the heat stability and resistance to oxidation and weathering are derived from the strength of the silicone-oxygen bonds. A flexible polymer chain gives low surface energy (hydrophobicity) and low temperature flexibility.

Dimethylsiloxanes can be modified by changing the carbon groups bonded to the silicone. Silicone rubber products include automotive applications to a large variety applications such as cooking, baking, food storage products, sportswear, footwear, electronics, home repair, hardware, and host of unseen applications. Organic (containing carbon) side groups attached to the silicone atoms allow for crosslinking and tailored applications. The self healing property of the silicone rubber broadens its application profile to many areas [58].

## *2.4 Fluoro Elastomers*

Fluorinated elastomers have special applications particularly due to their semi crystalline or totally amorphous nature and the unique combination of relevant properties. The first flouro elastomers discovered was polychlorotrifluoroethylene (PCTFE) with low molecular weight and polytetrafluoroethylene (PTFE) with high-molecular weight [59] respectively in 1937 and 1941 and several fluorinated homopolymers came thereafter. These synthetic elastomers have high resistance to oxidation and have excellent hydrolytic stability [60], which is attributed to the low polarizability and the strong electronegativity of the fluorine atom.

### 2.4.1 Processing and Vulcanization

Similar to other elastomers fluoro elastomers have also must be compounded and vulcanized. The curing (crosslinking) changes both the physical and chemical structure of the material and creates three-dimensional structures capable of exhibiting good mechanical performance. The fluoro elastomer additives allow the easy processing of narrow molecular weight metallocene catalyzed resins with simultaneously improving their optical and mechanical properties. The vulcanization mechanisms of these elastomers are quite similar with that of halobutyl rubber, since both involves halogen elimination step. The elimination of halogen atoms creates crosslinking sites in these elastomers. The mechanism involving in this reaction can be of two types- ionic and radical curing [61–71].

#### Ionic Curing

In this curing process, a nucleophile initiates a dehydrofluorination reaction which is characterized by the simultaneous elimination of the fluorine atom and an adjacent hydrogen atom. This causes the formation of a double bond in the elastomer backbone. Even though a large number of curatives are useful in fluoro elastomer systems, amine-based ones are most significant. The curing mechanism involved in a vinylidene fluoride ($VF_2$) site followed by an amine addition is explored by using $^{19}F$ and 1H NMR, FT-IR techniques coupled with solid-state curing experiments (where double bonds are formed by dehydrofluorination and nucleophiles are added to the double bond) [65, 72]. The amine curing agents are also used along with MgO [65, 73, 74]. The main advantage of amine based curatives is their capability to enhance rubber-to-metal bonding especially the hexamethylene crosslinks can impart better dynamic properties. Acid acceptors are used in the process of diamine curing, which play a multirole, since these compounds can affect dehydrohalogenation, assist in hydrolysis of the Schiff base and regenerate the free amine from its hydrofluoride. Similar to this system, dihydroxy cure systems were also developed which offers dramatically improved compression-set resistance, excellent heat stability and hydrolytic stability with greatly improved processing safety. Even though amine based curative system is very convenient, it has a disadvantage in finding more reactive base unlike carbamate salts of amines and other Schiff bases [75].

In addition to the amino system discussed so far, bisphenol curative is the other frequently used curing agent via ionic mechanism. The most commonly used compound is bisphenol–bis(hydroxyphenyl) hexafluoropropane) [76] due to the processing advantage. Others, like hydroquinone, substituted hydroquinones and 4, 4′-disubstituted bisphenols are also used commercially to lesser extents [77]. The bisphenol does not react with polymer in the absence of accelerators [72]. So an onium (phosphonium, ammonium, etc.) salt in combination with a metal compound is often used as an activator for this curing reaction. The onium salt acts as phase transfer catalyst and determines the phenolate-anion generation during the curing reaction. This is very important as the phenolate-ions are substituted for fluorine atoms at unsaturated carbon atoms leading to crosslinking. The formation of double bonds in the polymer backbone is the rate determining step in this reaction [64]. The quaternary salt catalyst regenerates repeatedly and so its amount in the compound is several times lower than that of the stoichiometric composition. Crosslinking continues up to the use up of bisphenol. Fluoroelastomers containing PAVE are not usually cured with bisphenols [78]. The main drawback of bisphenol cures is the presence of some unsaturation in the vulcanized parts.

Metal compounds are also useful as curatives, where they serve a dual purpose. It absorbs certain gaseous and acidic materials which are involved during curing and also attacks the fluoroelastomers chemically and weakens it. It also provides a long-term aging stability.

### Peroxide Curing

Peroxide crosslinking is basically a free radical process. It involves the thermal decomposition of peroxide to provide free radicals which abstract hydrogen from the methylene groups along the polymer chain, and the resultant polymeric radicals interact, directly or through the intermediary of radical traps to form crosslink's. Peroxide crosslinking occurs at specific sites available on the cure site monomer (CSM). The cure-site monomer and curing mechanism was originally developed for an improved low-temperature fluoroelastomer based on vinylidene fluoride ($VF_2$) and perfluoromethyl vinyl ether (PMVE).

Two different pathways are developed to increase the efficiency of peroxide cure vulcanization of fluoroelastomers. The first one concerns the increase of the content of hydrogen-containing groups in the copolymer, such as in TFE/P copolymers (tetrafluoro ethylene and propylene). The hydrogen of tertiary carbon atom in propylene units is a little screened (only from one side from fluorocarbon groups the energy of the C–H bond is lowered by the induction effect of $CH_3$ groups). So, these types of H atom are easily accepted by free radicals with a high yield of polymeric radicals and formation of crosslinks. These polymers are cured by a peroxide co reagent (i.e. radical trap) system. The reaction mechanism of peroxide curing with coagent is well known and described in the literature. These more stable radical intermediates, in turn, abstract halogen (X) atoms from the polymer, generating polymeric radicals. The driving force for such a chain reaction during propagation is the transfer of an X atom from the electron-poor fluoropolymer to the electron-rich hydrocarbon radical on the radical trap. Among the various peroxides and co reagents, a,a-bis(t-butyl-peroxy)diisopropyl benzene and triallyl isocyanurate (TAIC) and triallyl cyanurate (TAC) are most suitable [65, 78–81]. Peroxides, in the presence of acid acceptors, act as good vulcanizates. Adding tributyl stananne or triethyl silane also improves curing. It is desirable that the bromomonomers copolymerize to high conversion with minimum chain transfer and inhibition. For good curing performance, aliphatic peroxide, a suitable radical trap, and an inorganic acid acceptor are required. The introduction of iodine-containing fluoroelastomers facilitates the molding process considerably and makes possible injection molding. I/Br-fluoroelastomers combine the properties from iodine as chain end and from Br in the polymer chain.

#### Other Systems of Radical Curing

Curing also occurs via certain other mechanisms, other than the exact ionic or radical processes. For instance, although thiol-ene systems have easily led to vulcanize hydrogenated elastomers, this process has been used scarcely to cure fluoroelastomers because of the difficulty in the availability of mercapto side groups. However, an original trifluorovinyl v-thioacetate monomer (FSAc) was recently used to copolymerize with VDF [82], and the resulting copolymer was hydrolyzed to generate elastomers bearing mercapto lateral groups.

Radiation can also induce curing reactions in fluoroelastomers. Fluorinated polymers containing hydrogen (including VDF-base copolymers) and copolymers of TFE and HFP can be crosslinked with different degrees of effectiveness by high-energy radiation. Several reports have come out explaining the radiation cross-linking of fluoropolymers [83–85] and many articles detail different aspects of this process, recently reported by Logothetis [86] and Forsythe [87, 88]. The mechanism of radiation crosslinking of fluoroelastomers is similar to that proposed for other elastomers: it is a typical radical mechanism closed to peroxide curing. In this way, the total dose for optimum crosslinking can be lowered and the properties of gum can be considerably improved. Doses of 10–15 M rad are sufficient to give excellent properties and the cured parts remain clear and colorless [84]. The advantage of radiation cured materials is that they do not contain any filler or chemical, i.e. any impurity that can be extracted from the vulcanizates which is so important in the semiconductor industry.

#### Post-curing of Fluoroelastomers

For getting good physical properties researchers follow two step curing process in other words post curing. The first step of this is heating the material at high pressure (compression molding) and in the second step it is heated at atmospheric pressure for long time and it is more applied for chemical curing than for radiation

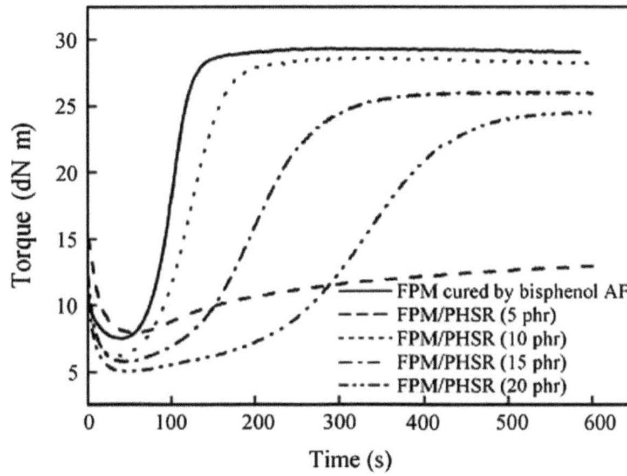


**Fig. 11** Vulcanization curves of the blends at 170 °C [95]

crosslinking [73, 77]. Smith and Perkins [89] observed that the post-curing enhances crosslinking as compared to normal curing process and it was explained by the formation of aromatic rings via Diels–Alder addition. Similarly other works also reported on post- curing [90–92]. They could explain a thermally induced bond-breaking and making process occurring during post-cure. Even though a lot of work has been performed in this area, the true mechanism behind it is not yet understood clearly. Usually the cured parts of this elastomer can be made by curing perfluoroelastomers with a high-energy electron beam followed by a post-cure. The chemical resistance and compression set properties of these materials are excellent, but the tensile properties are lower because of the absence of any reinforcing fillers in it [93].

The curing process of perfluoroolefins containing Br and I atoms [93] and those containing nitrile groups are well established. Vulcanization of perfluoroelastomers with nitrile groups is brought about by the catalytic interaction of tetraphenyltin or silver oxide on the pendant cyano group, thereby creating a triazine crosslinked structure [80, 94].

### Self Curing of Fluoroelastomers

Self-vulcanizing blends of phenol hydroxy silicone rubber (PHSR) and a fluoroelastomer (FPM) were fabricated by Wang et al. [95] and they observed smaller $S_{min}$ values and longer scorch time for FPM/PHSR blends than FPM with the same level of bisphenol curing agent. Figure 11 represents the vulcanization curves of FPM/PHSR obtained from the rheometer at 170 °C. It is clear that the increase of PHSR loading in FPM blends decreased both minimum and maximum torques and increased the scorch time and the optimum cure time except for FPM with 5 phr PHSR.

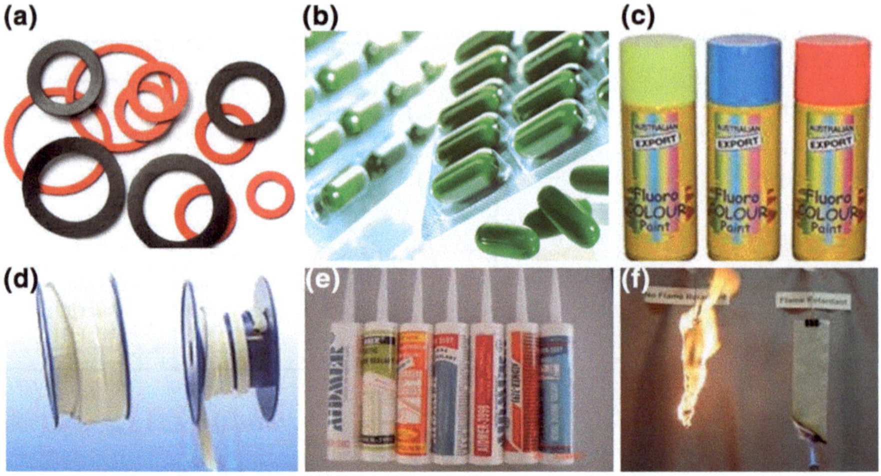


**Fig. 12** General applications of poly fluoro-elastomers (**a**) O-Rings and Gasket (**b**) Packing (**c**) Paints (**d**) Grommets (**e**) sealants and (**f**) flame retardants

### 2.4.2 Properties and Applications

In spite of their high cost (due to the unusual processing of polymerization, cost of purifying the gaseous monomers and small scale production), these polymers have found major applications in modern technologies as given in Fig. 12. These elastomers undergo a wide variety of chemical reactions and operating conditions [62, 65, 93, 96, 97] in these industries. Fluoro elastomers are thermally stable and in general can replace the three commonly used environments, aliphatic hydrocarbons (mineral oil), deionized water, and dry heat. The main characteristics of fluoro elastomers that make them applicable in industries can be summarized by the following points.

1. Outstanding high temperature resistance.
2. Improvements in broad temperature damping characteristics and low-temperature resilience [80].
3. Excellent resistance to fuel, oil, acid and chemicals.
4. Good resistance to gas penetration and radiation and flame retardancy.
5. Stable with thermal, chemical, and UV environments, so used in building industries (paints and coatings resistant to UV).
6. Poor swelling in oils and hydrocarbons, but resistant to heat, ageing, concentrated acids and alkalies.
7. Excellent resistance to heat, fluids, and oxidizing media combined with good physical properties [98–101].
8. Petrochemical and automotive industries, aerospace and aeronautics (use of elastomers as seals, gaskets, making tanks of liquid hydrogen for space

shuttles), chemical engineering (high-performance membranes) [105], static and dynamic seals, as diaphragms, valve seals, hoses, coated cloth, shaft seals, expansion joints, etc.

9. In optics as core and cladding of optical fibers, for treatment of textiles, stones (especially for old monuments), microelectronics.
10. Semiconductor processing equipments [106, 107] and to prevent equipment leakage, equipment failure and costly downtime [80].

## *2.5 Polyurethane Rubber*

Polyurethane was invented by Otto Bayer and his coworkers in 1937. Polyurethane is a synthetic elastomer. It is a combination of carbamate links. The carbamate links can be formed by a chemical reaction between isocyanate and alcohol. The polyurethane polymer comes under reactive polymers branch because it is formed by reactive groups such as iso cyanate and alcohol. Most of polyurethane elastomers are based upon segmented block copolymers of the general molecular formulae (AB)n. It contains alternating soft and hard segments [102, 103]. Generally the soft segment consists of polyether or polyester diols, and the hard segment is formed by the reaction products of diisocyanates with a chain extender.

### 2.5.1 Synthesis of Polyurethane

The polyurethane elastomer can be formed by different combinations of diols and diisocyanates in the presence of various catalysts. So, the exact property of polyurethane depends up on nature of diol and isocyanate and the catalyst. Mehdi Barikani et al. [104] prepared polyester based thermoplastic polyurethanes, and polyether based thermoplastic polyurethane by using polyol and toluene diisocyanate through prepolymer method. In this method they have used ethylene glycol (EG), 13-propane diol (PD), 1,4-butane diol (BD), 1,6- hexane diol (HD) and 1,10-decant diol (DID) as chain extenders for getting good thermal properties. They have concluded that polyester based thermo plastic polyurethanes have high tensile strength, and thermal stability than poly ether based thermo plastic polyurethanes. The physical and mechanical properties can be varied which depends up on length of the chain extender. This is due to the fact that if chain length is large then the phase separation will be more.

As we know the conventional polyurethane is insoluble in aquous media. If we want to increase the solubility of poly urethane, we have to incorporate ionic or nonionic hydrophilic segments in the back bone of polyurethane chain. These kinds of aqueous dispersible polyurethanes are playing crucial role in coatings industry.

### 2.5.2 Structure–Property Relationship

Generally, the polyurethane contains carbamate (urethane) linkages. This could be formed by the addition reaction between the isocyanate and alcohol. So much work had been carried out to understand the structure- property relationship of polyurethane elastomers. Cooper and his group [105] were the first to report on the structure property relationship in polyurethane elastomer. They have noticed that the hard and soft segments are the cause for the phase separation process. The phase separated hard and soft segments are the main reasons for enhanced properties of thermoplastic polyurethane. Pukánszky et al. reported on the two sets of polyurethanes from polyether and polyester polyols, and 4,4-methylene is(phenyl isocyanate), 1,4-butanediol is a chain extender [106]. The nature of polyol has strong influence on the structure as well as properties of polyurethane. The polyurethane prepared by polyester polyol shows high $T_g$. This is due to strong interaction between hard and soft segments in polyester polyurethane. Also the transparency and crosslink distribution in polyester polyurethane is higher than polyether polyurethane. The high transparency and crosslink distribution of polyester polyurethane is due to the formation of smaller dispersed particles of the hard phase. The property of polyurethane can be changed by addition of nano particles such as nano clay, carbon nanotube and graphene etc. Xia Cao et al. reported on the changes in structure- property relationship by the addition of organically modified nanoclay to the polyurethane [107]. The clay nano particles enhance the cell density and reduce size of particle compared to pure polyurethane. Nano clay enhances the glass transition temperature and crosslink density of polyurethane. This is due to the formation of hydrogen bonding between clay and polyurethane [108].

### 2.5.3 Processing of Polyurethane Rubber

Like natural rubber, polyurethane rubbers can be processed on mills or in internal mixers. A suitable polyurethane rubber is formed from a mixed ester of adipic acid with ethylene glycol and propylene glycol terminated by OH groups and having a molecular weight of approximately 2,000 by reaction with an excess of 4,4-diphenyl methane diisocyanate to form the NCO prepolymer and subsequent reaction with butanediol for chain extension. The polyurethane rubbers are converted into polyurethane rubber mixes by the same methods which are used for natural rubber. For polyurethane rubbers, the cross linking agent may be a diisocyanate, a peroxide or sulphur.

### 2.5.4 Applications

Polyurethane products have tremendous applications in various fields including building construction, furniture, automobile seats, house decorations, filling of spaces and cavities, water vessels, electronic components, adhesives etc.

## *2.6 Polysulfide Rubber*

Among the various synthetic and natural rubbers known to man polysulfide rubbers also come under the class of special purpose rubbers. It was discovered in 1926 by Joseph Cecil Patrik, an American chemist. Like many other prominent inventions in science, polysulfide rubber also is a child of serendipity. Patrick and his companion Mnookin were attempting to synthesise inexpensive antifreeze using ethylene dichloride and sodium polysulfide when they came across a new structure. The substance was like gum with a terrible odour which clogged a sink in the laboratory and none of the solvents they tried could remove it. Their frustration changed to delight when they realized that it was a new substance highly resistant to any kind of solvent. The elastomer became famous under the trade name Thiokol (1929) (after the Greek *theion*, "brimstone" [sulphur] and *kommi*, "gum"). The commercial polysulfide rubbers from Thiokol corporation are trademarked brands Thiokol FA, Thiokol ST and Thiokol LP. The disadvantage due to the bad odour of polysulfide polymers is overcame by the fact that it has become the basis of much advancement in technology. Many construction and aerospace industries are indebted to the technical performance and versatile properties of these polymers.

### 2.6.1 Synthesis

Polysulphide rubber (Thiokol) is synthesised by the condensation polymerization reaction involving dihalide derivatives of aliphatic hydrocarbons (e.g. 1, 2-dichloroethane) and alkali metal polysulfides (e.g. sodium polysulfide). The products obtained are thiokol and sodium chloride as shown below.

$$\underset{\text{1,2-Dichloroethane}}{Cl\text{-}CH_2\text{-}CH_2\text{-}Cl} + \underset{\text{Sodium polysulfide}}{n\ Na\text{-}S\text{-}S\text{-}Na} + \underset{\text{1,2-Dichloroethane}}{n\ Cl\text{-}CH_2\text{-}CH_2\text{-}Cl} \xrightarrow{\text{Polymerisation}}$$

$$\underset{\text{Thiokol}}{Cl\text{-}[\text{-}CH_2\text{-}CH_2\text{-}S\text{-}S\text{-}CH_2\text{-}CH_2\text{-}]_n\text{-}Cl} + \underset{\text{Sodium Chloride}}{2n\ NaCl}$$

Brief procedure is described here. 40 g NaOH is dissolved in 200 ml distilled water in a round bottomed flask and is placed in a water bath. 15 g sulphur is added to this and refluxed for 30 min. A reddish brown solution is obtained and 500 ml water is added to this. 2 ml dichloromethane is added to this and refluxed for another 45 min till the yellow colored polysulfide rubber is obtained. The contents are cooled at room temperature, washed with water and coagulated with acids.

Thiokol FA has a linear structure with hydroxyl groups at the chain ends, formed by hydrolysis of dichloromethane during polymerisation. Thiokol ST when formed first has a branched structure, after a second treatment some of the disulfide bonds are cleaved and thiol groups are left at the chain ends. This imparts good processability to the polymer.

### 2.6.2 Structure and Properties

Polysulfide rubbers typically consist of sulphur–sulphur linkages which connect short sequences of ethylene and the chains are terminated by reactive mercaptan groups that are also used for interlinking. These rubbers can be made into two main categories polytetrasulfides and polydisulfides, which have the general formulas $[-R-S-S-S-S-]_n$ and $[-R-S-S-]_n$ where R is an organic radical. Polysulfide rubbers are produced in both solid and liquid form corresponding respectively to high and low molecular-weight and also as aqueous dispersions, or lattices. The liquids are 100 % polymer itself unadultered with solvents or diluents whose viscosity ranges from 5 poise to 700 poise. It is formed by the reduction degradation of high- molecular mass polysulfides. It can be converted to elastic rubbers with properties somewhat equal to that of cured solid polysulfides [109].

The high sulphur content and absence of unsaturated bonds in the polymer are the main reasons for the special properties like high resistance to swelling in solvents, fuels and oils; resistance to sunlight, moisture and impermeability to gases. Also the saturated structure is responsible for its excellent resistance to oxidation, weathering and ozone. The thermal properties depend mainly on the polymer structure and in most cases the serviceability range is from −51 to 121 °C. The molecular weight of the solid polymer is of the order of (200–500)* $10^3$; a density of 1.27–1.60 g/cm$^3$ and glass transition point from –23 to –57 °C. The high density and high resistance to hydrocarbon oils of the polymer comes from the high sulphur content, about 80 percent by weight. Conversely the polymer has a tendency to relax and flows under pressure due to the low stability of sulphur–sulphur bonds. Because of its saturation, the polymer does not exhibit high tensile values, the upper maximum limit sometimes being 10.3–12.4 MPa. It does not possess good abrasion resistance also. Like other synthetic elastomers, polysulfide rubber also requires reinforcing fillers to achieve optimum physical properties. Chemical modification imparts improved flexibility, impact resistance, low shrinkage, less internal strain, better wetting properties and lower moisture absorption to these rubbers The polysulfide rubbers can be easily broken down to less molecular weight products due to the flexible backbone.

### 2.6.3 Compounding and Processing

FA and ST grade Thiokols are commonly subjected to compounding. Unlike natural rubber, FA rubber doesn't break down easily by milling process. It has to be softened by using plasticizers like mercaptobenzothiazoledisulfide (MBTS) and accelerated by adding diphenylguanidine (DPG). The ratio of MBTS is the key factor determining the extent of plasticization. Plasticization can be started with very small ratios of MBTS (0.3) and DPG (0.1). By varying the MBTS by 0.05 parts, the adequate amount can be found out. Even if with a small excess, the rubber would end up as a sticky, nonreclaimable mass on the mill. ST grade Thiokol posses no such difficulties in compounding and is used as such.

Carbon blacks (thermal blacks, semireinforcing and fast-extruding furnace blacks are preferred) are commonly used as fillers in polysulfide rubbers. This provides reinforcement, increases tensile strength and compression performance and also reduces the cost of the compound. Other fillers like zinc oxide, $TiO_2$ and lithopone are also used nowadays. Recently polysulfide sealants are being modified using additives of fullerenes and carbon nanotubes. The chemical, mechanical and adhesive property increased significantly by the strong effect of nanoparticles [110]. Clay is not preferred as it retards vulcanization. Addition of coumarone indene resin can improve the tackiness of the compound. FA grade requires about 10 parts ZnO while ST grade requires 5 parts of zinc peroxide and 1 part of $Ca(OH)_2$ for curing. Use of stearic acid (1 part) could prevent the sticking of stock on the mill and helps in good dispersion.

Vulcanization of polysulfide rubbers usually proceed through the oxidative polycondensation reaction between oligomer SH groups with transition metal dioxides ($MnO_2$, $PbO_2$) [111]. For vulcanizing polysulfide rubbers, different inorganic oxidants like zinc oxide, p-quinone dioxime, and a mixture of altax and diphenylguanidine (activators) are also used. They oxidise the terminal SH-groups and results in lengthening of macromolecules and the mercapto groups are crosslinked through oxidation polycondensation.

$$\sim RSH + \text{oxidant} + HSR\sim \; \rightarrow \; \sim R - S - S - R\sim \; + \; H_2O \\ + \text{ product of oxidant reaction}$$

Curing can also be effected with rf current, photocuring or using peroxides. The simplest vulcanizing agents are oxygen in air and molecular sulphur. Vulcanized polysulfide rubbers exhibit inferior mechanical properties when compared to other vulcanized synthetic rubbers.

Conventional mixing and fabricating equipments like two-roll mill or Banbury type internal mixers are used for incorporating fillers, curing agents and additives into the solid rubber. Further processing operations like extruding, calendaring, molding or steam vulcanization could be done in the normal manner but with good factory control than that of general purpose rubbers. Liquid polysulfide rubber compounds use internal mixer, three-roll paint mill, colloid mill or ball mill for mixing. Depending upon the specific characteristics of the end product, the resulting products may be applied by brushing, spraying or casting. Owing to their versatile nature in formation of end products, liquid rubbers are dominating over solid-rubber in application field. Curing of these rubbers can be done over wide temperature ranges. The high popularity of this polymer in a variety of industrial applications lies in the fact that they are able to cure even at room temperature.

### 2.6.4 Applications

Polysulfide rubbers have high resistance to petroleum solvents, organic solvents (esters, ketones etc.), aromatic fuels, oils, greases, sunlight, ozone and UV rays. Hence they are used as static seals for aircraft, building and marine industries where no other material serves the purpose. When these sealants are chemically modified an enhancement in the inherent properties occurs. Some of them includes modifications with polythio-urethane-urea [112] or with a silyl group [113]. They are also used as binder for rocket fuel, flexible moulds, dental moulding compounds and electrical encapsulations. Polysulfide polymer is absorbed in leather and the cured product is highly resistant to oil and chemicals without loss in pliability and is used for making packing seals, boots and gloves. When polysulfide and rubber are combined, durable roofing materials are produced.

Polysulfide rubbers are also used for applications like gaskets, washers, diaphragms, rubberized fuel storage tanks and various types of oil and gasoline hose high solvent resistance is required. Another use of these rubbers is in the manufacture of inks, paints and coatings. Care should be taken when they are used on plastics as they will degrade PVC and ABS. *Mostly liquid rubbers are employed in applications that include cold-setting casting* and molding such compounds exhibiting flexibility, low shrinkage and excellent dimensional stability. Liquid polysulfides may also have other industrial applications like coatings, adhesives, modifiers of epoxy resins etc. Polyaniline-thiokol rubber composite coatings when applied on steel surface is very effective in preventing corrosion [114].

## *2.7 Chlorosulfonated Polyethylene*

Chlorosulphonated polyethylene is a synthetic rubber, produced by the controlled treatment of polyethylene in solution by chlorine and sulfur dioxide. Hypalon is a trademark for chlorosulfonated polyethylene (CSPE), noted for its resistance to chemicals, temperature extremes, and ultraviolet light. It was a product of DuPont Performance Elastomers, a subsidiary of DuPont [115].

### 2.7.1 Structure and Properties

The synthetic chlorosulfonated rubber is obtained by the simultaneous chlorination and chlorosulfonation of polyethylene [116]. This chlorosulfonated polyethylene (CSPE) contains a modified polyethylene chain in its backbone with chloro and sulfonylchloride side groups as shown in Fig. 13. This rubber can also be crosslinked as in the case of all other elastomers for technological applications. The –$SO_2Cl$ groups and labile chlorine atoms present in CSPE also participate in the vulcanization process. Such materials are known for their high toughness, weatherability, and resistance to oxidation and oil/solvent and they also exhibit

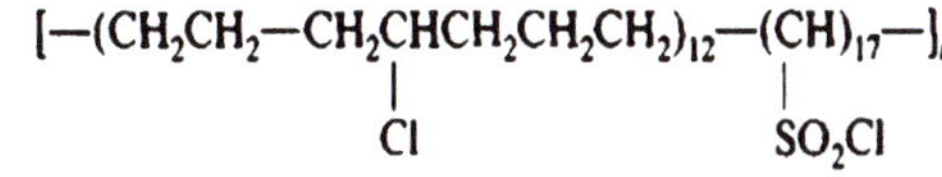

**Fig. 13** Structure of Chlorosulphonated Polyethylene

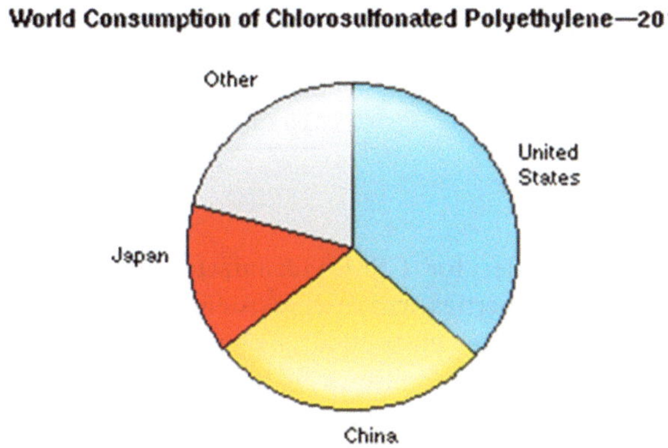


**Fig. 14** World consumption of chlorosulfonated polyethylene

polyethylene properties such as hardness, stiffness and partial crystallinity especially at low-chlorine levels [117].

The density, chlorine content and sulfur content of CSPE are 1.11–1.26 g/cm$^3$, 27–45 % and 0.8–2.2 % respectively. The resistance of CSPE to fire, oil, and microorganisms as well as the surface adhesion behaviour is attributed to the chlorine content in it. Its resistance to ozone, inorganic acids and concentrated alkalies make chlorosulfonated polyethylene superior to other rubbers. On the other hand the increase in chlorine content decreases the low temperature properties. The rubber is impermeable to gas, and has good dielectric properties. Low flammability, excellent green strength, UV resistance and stability are other notable features of CSPE. It exhibits the tensile strength of about 32 MN/m$^2$ and a relative elongation of 350–600 % in addition to the good abrasion resistance [118].

The given pie chart (Fig. 14) shows world consumption of chlorosulfonated polyethylene: If we look into the world consumption of chlorosulfonted poly ethylene rubber in the year 2011, there is a huge contribution of about 80 % from United States, China and Japan.

### 2.7.2 Compounding

The common ingredients used for compounding of chlorosulphonated polyethylene are given in Table 3. Magnesia is used as the acid acceptor and the main function of acid acceptor is to act as heat stabilizer to absorb acid byproducts of the curing reaction and to maintain sufficient alkalinity to allow effective curing reactions to proceed. The activator (pentaerythritol) increases the effectiveness of the acid acceptor apparently by solubilizing it in the polymer. Carbon black is the

**Table 3** The common ingredients used for compounding of Hypalon

| Ingredient | Function | phr |
|---|---|---|
| Hypalon-40(35 % Cl) | Polymer | 100 |
| Magnesia | Acid acceptor | 4 |
| Pentaerythritol | Activator for 2 | 3 |
| Hard clay | Filler | 80 |
| Aromatic oil | Plasticizer | 25 |
| Paraffin wax | Process aid | 3 |
| Sulphur | Vulcanizing agent | 1 |
| Tetra methyl thiuram disulphide | Accelerator for 7 | 2 |

preferred filler for CSM vulcanizates because it gives best reinforcement of physical properties and best resistance to chemical degradation, to compression set, and to water absorption. Aromatic petroleum oils are widely used as plasticizers primarily because of their low cost. Paraffin wax and polyethylene glycol are all effective process aids that do not affect scorch safety.

Two types of curing processers are available with CSM elastomers and both are effective regardless of the level of chlorine in the elastomers. Ionic cures of CSM are possible when the acid acceptor is a divalent metal oxide. CSM is also capable of undergoing covalent cross linking, and most applications for CSM make use of this type of curing. Three different systems that are commonly used in covalent cure are sulphur cure, peroxide cure and malemide cure.

Internal mixing is the most cost efficient volume system for producing well mixed CSM compounds. CSM compounds are successfully processed by compression, transfer or injection molding. Due to the marked thermoplsticity of CSM compounds, multicavity transfer and injection moulds require carefully balanced runner systems and similar flow restrictions for the gates so that uniform filling of all cavities is achieved [119, 120].

### 2.7.3 Applications

CSPE is highly applicable both in the production of industrial and household goods. It is useful for anticorrosion coatings, insulating cables, automotive hoses, machine parts, floor tiles, magnetic rubber, rubber coated cloth, gaskets, flexible tubes, rolls and linings. It is also used as a film-forming agent in varnishes and paints and as a good adhesive in reinforced concrete. Special applications include escalator handrails, diaphragms, lining for chemical processing equipment, fabricating the inflatable boats and folding kayaks etc. Conductive CSPE composites are also useful in power distribution, audio, and telephone application, as packaging and semi conducting polymeric materials. Image showing applications of chlorosulphonated polyethylene is given in Fig. 15.

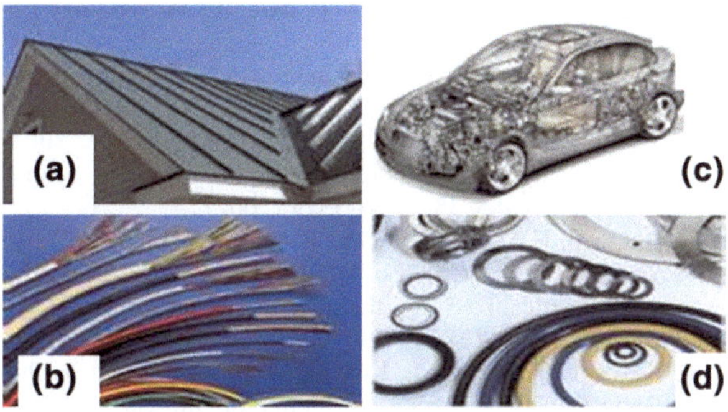


**Fig. 15** Image showing applications of Chlorosulphonated polyethylene in (**a**) Construction sector (**b**) Wire and cable sector (**c**) Automotive (**d**) Industrial products sector

## 3 Conclusion

In the coming years polymer chemists will be able to design new kind of functional elastomers fulfilling specific functions for a definite application. Such new targets are real challenges and should attract the interest of many academic and industrial researchers in this fascinating area. Indeed cure chemistry plays a major role in developing the properties of synthetic elastomers. Each synthetic rubber as discussed has its own applications and by compounding them we can get intended properties for a particular application. Incorporation of nanofillers provides fascinating functional properties to elastomers. The synthesis of new elastomers by direct radical co- or terpolymerization of monomers has been further proved to be a versatile way that offers many opportunities for obtaining wide range of synthetic rubbers. The production and demand of synthetic elastomers are having an increasing market for novel applications. This offers additional chances to a technology that already forms the basis of the commercial production of classes of chemicals which are among the most advanced and successful products. Still there are many challenges in developing new type of elastomers which include the choice of appropriate precursors that would degrade biocompatible compounds. In the rapidly advancing field of tissue engineering, polyurethanes offer numerous opportunities to develop suitable scaffolds. New synthetic protocols such as controlled radical polymerization, atom radical polymerization etc. offer new methodologies for the design of new functional elastomers. Better understanding of the correlation between structure and properties of elastomers is extremely important for the design of new functional materials for various applications.

## References

1. Puskas, J.E., Wilson, G., Duffy, J.: Synthesis of Butyl Rubber by cationic polymerization. Ullman's Encycl. Ind. Chem. **A23**, 51–57 (1998)
2. Kaszas, G., Puskas., J.E.; Baade, W.: Butyl and Halobutyl rubbers. Polym. Mater. Encycl. (1996)

3. Robinson, K.J., Feniak, G., Walker, J.: Proceedings of International Rubber Conference, Prague, 1973
4. Vukov, R.: Zinc Oxide crosslinking chemistry of Halobutyl Elastomers–a model compound approach. Rubber Chem. Technol. **57**, 284–290 (1984)
5. Scott, J., Greg. D.F., Whitney R.A.: Amine substitution reactions of brominated poly(isobutylene-co-isoprene): new chemical modification and cure chemistry. Macromolecules **35**, 3374–3379 (2002)
6. Baldwin, F.P., Buckley, D.J., Kuntz, I., Robison, S.B.: Rubber Plastics Age **42**, 500 (1961)
7. Scagliusi, S. R., Cardoso, E. L.C., Lugao, A. B.: Effect of gamma radiation on chlorobutyl rubber vulcanized by three different crosslinking systems. Rad. Phys. Chem. **81**(9), 1370–1373 (2012)
8. Razzaghi-Kashani, M., Kokabi, M.: Improvement in physical and mechanical properties of Butyl Rubber with montmorillonite organo-clay. Iranian Polym. J. **16**(10), 671–679 (2007)
9. Morrison, N.J.: The formation of crosslink precursors in the sulfur vulcanization of natural rubber. Rubber Chem. Technol. **57**, 97–103 (1984)
10. Ghosh, P., Katare, S., Patkar, P., Caruthers, J.M., Subramaniah, V.V.: Sulphur vulcanization of natural rubber for benzothiazole accelerated formulations: from reaction mechanisms to a reactional kinetic model. Rubber Technol. Handb., 2nd edn. *New York: Hanser Publishers* **76**, 592–693 (2003)
11. Morgan, B., Mcgill, W.J.: Benzothiazole-accelerated sulfur vulcanization. IV. Effect of ZnO and bis(2-mercaptobenzothiazole)zinc(II) on 2-bisbenzothiazole-2,2′-polysulfide formation in 2-bisbenzothiazole-2,2′-disulfide and 2-bisbenzothiazole-2,2-disulfide/sulfur. J. Appl. Polym. Sci. **76**, 1405 (2000)
12. Chu, C.C., Vukov, R.: Determination of the structure of butyl rubber by NMR spectroscopy. Macromolecules **18**, 1423–1430 (1985)
13. Chu, C.C., Watson, K.N., Vukov, R.: Determination of the structure of chlorobutyl and bromobutyl rubber by NMR spectroscopy. Rubber Chem. Technol. **60**, 636–646 (1987)
14. Lautenschlaeger, F.K.: Model compound vulcanization-part I. Basic studies. Rubber Chem. Technol. **52**, 213–231 (1979)
15. McSweeney, G.P., Morrison, N.J.: The thermal stability of monosulfidic crosslinks in natural rubber. Rubber Chem. Technol. **56**, 337–343 (1983)
16. Skinner, T.D.: The CBS-accelerated sulfuration of natural rubber and cis-1, 4-polybutadiene. Rubber Chem. Technol. **45**, 182–192 (1972)
17. Gregg, E.C., Lattimer, R.P.: Polybutadiene vulcanization. Chemical structures from sulfur-donor vulcanization of an accurate model. Rubber Chem. Technol. **57**, 1056–1097 (1984)
18. Kuntz, I., Zapp, R.L., Pancirov, R.J.: The chemistry of the zinc oxide cure of halobutyl. Rubber Chem. Technol. **57**, 813–825 (1984)
19. Hendrikse, K.G., Mcgill, W.J., Reedijk, J., Nieuwenhuizen, P.J.: Vulcanization of chlorobutyl rubber. I. The identification of crosslink precursors in compounds containing ZnO/ZnCl2. J. Appl. Polym. Sci. **78**, 2290–2301 (2000)
20. Jurkowska, B, Olkhov, Y.A., Jurkowski, B.: Thermomechanical study of Butyl Rubber mastication during compounding. J. Appl. Polym. Sci. **68**, 2159–2167 (1998)
21. Hofmann, W.: Rubber Technology Handbook, 2nd edn, p. 611. Hanser Publishers, New York (1989)
22. Qu, L., Huang. G., Wu, J., Tang, Z.: Damping mechanism of chlorobutyl rubber and phenolic resin vulcanized blends. J. Mater. Sci. **42**, 7256–7262 (2007)
23. Sanjay, M, Gautam, S.: Effect of maleic anhydride grafted ethylene propylene diene monomer (MAH-g-EPDM) on the properties of kaolin reinforced EPDM rubber. J. Appl. Polym. Sci. **119**, 2268–2274 (2011)
24. Enizceylan, S., Burakkaracik, Sevild. Yakan, Oyas. Okay, Anndoguzokay, Evaluation of Butyl Rubber as sorbent material for the removal of oil and polycyclic aromatic hydrocarbons from seawater. Environ. Sci. Technol. **43**, 3846–3852 (2009)
25. Galgotiya, S.: Handbook of Rubber technology, 278–292 (1998)

26. Gheno, S., Pessan, L.A.: Effect of the size of dispersed NBR phase in PVC/NBR blends on the stability of PVC to electron irradiation. In: Polymer, pp. 6833–6839 (2001)
27. Elhamouly, S.H.: Influences of accelerators on the structures and properties of Nitrile Butadiene Rubber. Mod. Appl. Sci. **4**, P47 (2010)
28. Thompson, B.: Thermax® N990 thermal carbon black in Nitrile Rubber compounds. Asia-Pacific Marketing Cancarb Limited, 2010
29. Hwang, W.-G., Wei, K.-H., Wu, C.-M.: Mechanical, thermal, and barrier properties of NBR/Organosilicate Nanocomposites. Polym. Eng. Sci. **44**, 11 (2004)
30. Li, Y., Wang, Q.; Wang, T., Pan, G.: Preparation and tribological properties of graphene oxide/nitrile rubber nanocomposites. J. Mater. Sci. **47**, 730–738 (2012)
31. Stern, S.A.: Polymers for gas separations: the next decade. J. Membr. Sci. **94**, 1–65 (1994)
32. Reddy, B.S.R., Senthilkumar, U.: Prospects of siloxane membrane technology for gas separation—a review. J. Sci. Ind. Res. **62**, 666–677 (2003)
33. Panek, D., Konieczny, K.: Pervaporation of toluene and toluene/acetone/ethyl acetate aqueous mixtures through dense composite polydimethylsiloxane membranes. Desalination **200**, 367–373 (2006)
34. Raut, A.M.: Pervaporation of aroma compounds using virgin and silicate-filled organophilic membranes: effect of aroma compound structure and comparison with distillation selectivity. Sci. Technol. **39**, 1791–1814 (2004)
35. Shi, Y., Burns, B.M., Feng, X.: Poly(dimethyl siloxane) thin film composite membranes for propylene separation from nitrogen. J. Membr. Sci. **282**, 115–123 (2006)
36. Joo, J., Lee, C.Y.: High frequency electromagnetic interference shielding response of mixtures and multilayer films based on conducting polymers. J. Appl.Phys **88**, 513–519 (2000)
37. Clarson, S. J., Semlyen, J.A.: Siloxane Polymers. Prentice–Hall, Englewood Cliffs (1993)
38. Donnet, J. B.: Nano and microcomposites of polymers elastomers and their reinforcement. Compos. Sci. Technol. **63**, 1085–1088 (2003)
39. Hepburn, C.: Filler reinforcement of rubber. Plast. Rubber Int. **9**, 11 (1984)
40. Ogunniyi, D.: Filler reinforcement in rubber. Elastomerics **120**, 24 (1988)
41. Goritz, D., Rabb, H., Frohlich, J., Marier, P.G.: Surface structure of carbon black and reinforcement. Rubber Chem. Technol. **72**, 929 (1999)
42. Frogley, M.D., Ravich, D., Wagner, H.D.: Mechanical proper- ties of carbon nanoparticle-reinforced elastomers.Compos. Sci. Technol. **63**, 1647 (2003)
43. Leopoldes, J., Barres, C., Leblanc, J.L.: Influence of filler–rubber interactions on the viscoelastic properties of carbon-black-filled rubber compounds. J. Appl. Polym. Sci. 2004, 91, 577
44. Aranguren, M. I., Mora, E., Macosko, C.W.: Compounding fumed silicas into polydimethylsiloxane: Bound rubber and final aggregate size. J. Colloid Interf. Sci. **195**, 329 (1997)
45. Schaefer, D.W., Chen, C.Y., Yang, A.J.: U.S. Pat., 20050228106 (2005)
46. Takahashi et.al.: European Report,EP 97 30 6663. 9 Jan 1998
47. Kang, D.W., Hak. G.Y., Lee, K.S.: Preparation and characteristics of liquid silicone rubber nanocomposite containing ultrafine magnesium ferrite powder. J. Inorg Organomet Polym **14**, 73–84 (2004)
48. Qian Wang, Q.Z., Huang Y.-h., Fu, Q.: Preparation of high-temperature vulcanized silicone rubber of excellent mechanical and optical properties using hydrophobic nano silica sol as reinforcement. Chinese J. Polym. Sci. **26**, 495–500 (2008)
49. Laikhtman, A., Gouzman, I., Verker, R., Grossman, E.: Atomic oxygen and UV irradiation effects on fluorosilicone rubber: comparison of RF plasma and in-flight exposure. High Perform. Polym. **20**, 447–460 (2008)
50. O'Neill. M.J., McDanal, A.J., Piszczor, M.F.: 31st IEEE Photovoltaic Specialists Conference, Orlando, FL, January, 2005
51. Soucek, M.D., Dworak. D.P., Chakraborty, R.: A new class of silicone resins for coatings. J. Coat. Technol. Res. **4**, 263–274 (2007)

52. Banks, B.A., Dever. J.A., Gebauer, L., Hill, C.M.: Atomic oxygen interactions with FEP Teflon and Silicones on LDEF. In: 1st LDEF Post-Retrieval Symposium, Kissimmee, FL, June, 1991
53. Banks, B.A., Rutledge, S. K., de Groh, K.K., Mirtich, M.J., Gebauer, L., Olle, R., Hill, C.M.: The implications of the LDEF results on space station freedom power system materials. In: 5th international symposium on materials in a space environment, Cannes–Mandelieu, France, September, 1991
54. Banks, B., Rutledge, S., Sechkar, E.,Stueber, T., Snyder, A., Hatas, C., Brinker, D.: The 8th International Symposium on Materials in a Space Environment Arcachon, France, June, 2000
55. Zhang, L.X., He, S.Y.: Damage effects and mechanisms of proton irradiation on methyl silicone rubber. Mater. Chem. Phys **83**, 255–259 (2004)
56. Xinga, A., Gao, Y.: Preparation and atomic oxygen erosion resistance of silica film formed on silicon rubber by sol–gel method. Appl. Surf. Sci. **256**, 6133–6138 (2010)
57. Hino, T., Igarashi, Y., Ymauchi, Y., Nishikawa, M.: Surface wettability of silicon rubber after irradiation with a glow discharge plasma. Vacuum **83**, 506–509 (2009)
58. Keller, M.W.: A self-healing poly(dimethyl siloxane) elastomer. Adv. Funct. Mater. **17**, 2399–2404 (2007)
59. Plunkett, R.J. P.: US Patent 2, assigned to Kinetic Chemicals. Chem. Abstr. 35, 3365 (1941)
60. Smart, B.E.: In: Patai, S., Rappoport, Z. (eds.) The chemistry of functional groups (suppl. D). Wiley, New York, 1983, chap 14
61. Ferro, R., Fiorello, G., Restelli., Extruding fluorelastomers to meet higher performance needs,. Parts 1 and 2. Elastomerics 1989
62. Améduri, B., Boutevin, B., Kostov, G.: Fluoroelastomers: synthesis, properties and applications. Prog. Polym. Sci. **26**, 105–187
63. Smith, S.: Preparation, properties and industrial applications of organofluorine compounds. In: Banks, R.E., (ed.) Fluoroelastomers, pp. 235–295 (1982)
64. Logothetis, A.L.: Chemistry of fluorocarbon elastomers. Prog. Polym. Sci. **14**, 251–296 (1989)
65. Arcella, V., Ferro, R.: Fluorocarbon elastomers. In: Scheirs, J. (ed.) Modern fluoropolymers. Wiley, New York, pp. 71–90 (1997)
66. Marshall, J., Kalrezw-type perfluoroelastomers Ð synthesis, properties and applications. In: Scheirs, J. (ed.) Modern fluoropolymers. Wiley, London, pp. 349–358 (1997)
67. Dixon, S., Rexford, D., Rugg, J.S.: Vinylidene Fluoride – Hexafluoropropylene Copolymer. Ind. Eng. Chem. **49**, 1687–1690 (1957)
68. Pattison, D.B.: US Patent 3,876,654, assigned to DuPont, 1975
69. Paciorek, K.J.L.: Chemical crosslinking of fluoroelastomers. Wiley, New York, 291–374 (1972)
70. Florin, R.E.: Radiation chemistry of fluorocarbon polymers, pp. 317–380. Wiley, New York (1972)
71. Fetters, L.J.: Nitroso fluoropolymers. In: Wall, L.A. (ed.) Fluoro polymers. Wiley, New York, pp. 175–193 (1972)
72. Schmiegel, W. W.: Crosslinking of elastomeric vinylidene fluoride copolymers with nucleophiles. Angew. Makromol. Chem. **76**, 39–65 (1979)
73. Smith, J. F.: The Chemistr. of "TViton" , Fluorocarbon Elastomer. Rubber World **142**, 102–107 (1960)
74. Paciorek, K.L., Mitchell, L., Lenk, C.T.: Mechanism of amine crosslinking of fluoroelastomers. I. Solution studies. J. Polym. Sci. **45**, 405–411 (1960)
75. Garvey, B. S.: Elastomers. Materials of Construction Review. Ind. Eng. Chem. **52**, 889–891 (1960)
76. Dick, J.S., Worm, A.: Storage stability of FKM compound based on a bisphenol af/onium cure system and its potential as a standard reference compound. Rubber World **219**, 22–28 (1999)

77. Schmiegel, W.W., Logothetis, A.: Polymers for ®bers and elastomers. In: ACS symposium series, vol. 260, pp. 159–182 (1984)
78. Van Cleeff, A.: Fluoroelastomers. In: Scheirs, J. Modern Fluo-ropolymers, Wiley, New York, pp. 597–614 (1997)
79. Banks, R.E., Smart, B.E., Tatlow, J.C.: Organofluorine chemistry: principles and commercial applications. Plenum Press, New York, pp. 373–395 (1994)
80. Cook, D., Lynn, M.: Rapra Review Reports, vol. 3, Report 32, Shrewsbury, 1995, pp. 1–27
81. Hatada, K., Kitayama, T., Vogl, O.: Perfluoroelastomers and their functionalization. In: Macromolecular design of polymeric materials. Marcel Dekker, New York, pp. 447–455 (1997)
82. Ameduri, B., Boutevin, B., Petrova, P., Kostov, G.: Synthesis and polymerization of fluorinated monomers bearing a reactive lateral group. Part 10. Copolymerization of vinylidene fluoride (VDF) with 5-thioacetoxy-1,1,2-trifluoropentene for the obtaining of a novel PVDF containing mercaptan side-groups. Des. Monomers Polym. **2**, 267–285 (1999)
83. Lyons, B.J.: The crosslinking of fluoropolymer with ionising radiation: a review. In: Second International Conference on Radiation Processing for Plastics and Rubbers, Canterbury, UK, March 1984, pp. 1–8
84. Lyons, B.J.: Radiation crosslinking of fluoropolymers—a review. Radiat. Phys. Chem. **45**, 158–174 (1994)
85. Lyons, B.J.: The radiation crosslinking of fluoropolymers, Wiley, New York, pp. 335–347 (1997)
86. Logothetis, A.L.: Crosslinking of tetrafluoro-ethylene–perfluoro-(methyl vinyl ether) perfluoro-elastomers with electron beam irradiation. Polym. Int. **48**, 993–995 (1999)
87. Forsythe, J.S.: Effect of temperature and a crosslinking promoter on the $\gamma$-radiolysis of a perfluoro-elastomer. Polym. Int. **48**, 1004–1009 (1999)
88. Forsythe, J.S., Hill, D.J.: Radiation chemistry of fluoropolymers. Prog. Polym. Sci. **25**, 101–136 (2000)
89. Smith, J.F., Perkins, G.: The mechanism of post cure of viton A Fluorocarbon elastomer. J. Appl. Polym. Sci. **5**, 460–466 (1961)
90. Apotheker, D., Finlay, J., Krusic, P.J., Logothetis, A.L.: Curing of fluoroelastomers by peroxides. Rubber Chem. Technol. **55**, 1004–1011 (1982)
91. Taguet, A., Ameduri, B., Boutevin, B.:Crosslinking of Vinylidene Fluoride-Containing Fluoropolymers. Advances in Polymer Science **184**, 127–211 (2005)
92. Kalfayan, S.H., Silver, R.H., Maezzo, A.A.: Accelerated heat-aging studies on fluorosilicone rubber. Rubber Chem. Technol. **48**, 944–952 (1975)
93. Logothetis, A.L.: Perfluoroelastomers and their functionalization. In: Macromolecular design of polymeric materials. Marcel Dekker, New York, pp. 447–455 (1997)
94. Logothetis, A.L.: Organofluorine chemistry: principles and commercial applications, pp. 373–395. Plenum Press, New York (1994)
95. Wang, Y.-m., Liu, L., Luo, Y.-f., Jia, D.: Studies on self-vulcanizing fluoroelastomer/phenol hydroxy silicone rubber blends. Chinese J. Polym. Sci. **27**, 381–386 (2009)
96. Kraskshin, M.A., K.G., Kazmina NI., Production of tyres. RTI ATI (Russ) 1 **4**, 5–8 (1975)
97. Van cleeff, A.: Modern Fluoroeeastomers. Wiley, New York, 1997, pp. 597–614
98. Sinha, N. K., Baldev R.: Basis of property limits for inflatable seal fluoroelastomers. Nuclear Eng. Des. **244**, 100–108 (2012)
99. Albano, M., Arcella, V., Chiodini, G., Saronno, V., Minutillo, A.: European Patent 0,618,241 B, assigned to Ausimont, 1994
100. Legare, J., Thomas, E., Fulford, K., Cargo, J., Santa Monica, C.A.: Characterization of elemental extractables in perfluoroelastomer and fluoroelastomer sealing parts. In: Microcontamination'93 Conference Proceedings, 1993, pp. 36–46
101. Kim, D.H., Hwang, S.H., Park, T.S., Kim, B.S.: Effects of waste ground fluororubber vulcanizate powders on the properties of silicone rubber/fluororubber blends. J. Appl. Polym. Sci. **127**, 561–569 (2013)
102. Hepburn, C.: Polyurethane Elastomers. Applied Science Publishers, New York (1982)

103. Hardianto, H.,Mayorga, V.I.: Prosiding Seminar Sehari 70 Tahun Noer Mandsjoeriah Surdia, pp. 4–19, 2003
104. Mehdi, B.: Thermoplastic polyurethane elastomers, synthesis and study of effective structural parameters. Iran. Polym. J. **5**, 231–235 (1996)
105. Cooper, S.L., Tobolsky, A.V.: Properties of linear elastomeric polyurethanes. J. Appl. Polym. Sci. **10**, 1837–1844 (1966)
106. Bagdi, K.M., Sajo, I., Pukanszky, B.: Specific interactions, structure and properties in segmented polyurethane elastomers. eXPRESS Polymer Letter, **5**, 2011
107. Cao, X.: Polyurethane/clay nanocomposites foams: processing, structure and properties. Polymer **46**, 775–783 (2005)
108. Kim, C.J., Youn, J.R.: Environmentally friendly processing of polyurethane foam for thermal insulation. Polym. Plast. Technol. Eng. **39**, 163–185 (2000)
109. Lu, Y., Zhang, J., Chang, P., Quan, Y., Chen, Q.: Effect of Filler on the compression set, compression stress-strain behaviour and mechanical properties of polysulfide sealants. JAPS, **120,** 2001–2007 (2011)
110. Zaitseva, E.I., Donskoi, A.A.: Novel polysulfide sealing compositions for aircraft industry. Polym. Sci. Series D **2**, 250–256 (2009)
111. Donskoi. A. A, Zaitsev. E. I., Sealants based on polysulfide elastomers. Polym. Sci. Series D **1**, 280–297 (2008)
112. Quan, Y., He, P., Zhou, B., Chen, Q.: Modification of polysulfide sealant with polysulfide polythio-urethane-urea. J. Appl. Polym. Sci. **106**, 2599–2604 (2007)
113. Matsui, T., Nakajima, M., Nonaka, T., Dokoshi, N.: New liquid polysulfide polymer terminated with silyl group. J. Appl. Polym. Sci. **93**, 2642–2649 (2004)
114. Dinga, K., Jiaa, Z., Mab, W., Jianga, D., Zhaoa, Q., Caoa, J., Tonga, R.: Corrosion protection of mild steel with polyaniline–thiokol rubber composite coatings. Prot. Metals **39**, 71–76 (2003)
115. Warner, R.R.: Hypalon S-2—New Elastomer. Rubber Age. **71**, 205 (1952)
116. Nersasian, A., Johnson, P.R.: Infrared spectra of alkanesulfonic acids, chlorosulfonated polyethylene, and their derivatives. J. Appl. Polym. Sci. **9**, 1653–1668 (1965)
117. Smook, M. A., Pieski, E. T., Hammer, C. F.: Derivatives of Chlorosulfonated Polyethylene and Their Infrared Spectra. Ind. Eng. Chem. **45**, 2731–2737 (1953)
118. Maynard, J. T.: Crosslinking Chlorosulfonated Polyethylenes. Rubber Chem. Technol. **36**, 963–974 (1963)
119. Johnson, P.R.: Crosslinking chlorosulphonated polyethylene with ammonia. Du Pont HYPALON Report Oct. 1964
120. Dupuis, I.C.: Selecting a curing system. Du Pont HYPALON report HP-30.1

# Compounding and Vulcanization

**R. Rajesh Babu, G. S. Shibulal, Arup K. Chandra and Kinsuk Naskar**

**Abstract** Compounding is a unique requirement of the rubber, generally not found in other material. The performance properties can be controlled by properly selecting and adjusting various compounding ingredients. The stages of rubber product manufacturing are broken down into three primary classes: selection of compounding ingredients, mixing or compounding, and vulcanization techniques or final product manufacturing process. The present chapter gives a brief introduction of the all classes with their importance. By proper selection of the variables in each class, the properties can be manipulated from virtually incompressible with a bulk modulus some thousand times greater than shear modulus, to large extent impermeable to gases and liquids and with excellent recovering and abrasion resistance etc.

## 1 Introduction

Rubbers became the essential material in our daily part of life, starting from tyres, seals to erasers, to mention just a few applications. According to ASTM 1566-03a, Rubber can be defined as "a material that is capable of recovering from large deformations quickly and forcibly, and can be, or already is, modified to a state in which it is essentially insoluble (but can swell) in boiling solvent, such as benzene, methyl ethyl ketone, or ethanol-toluene azeotrope". But the definition seems to be incomplete. The term rubber and elastomer are used interchangeable in convention. The essential requirements for a substance to be rubbery are [1, 2]:

R. Rajesh Babu · A. K. Chandra
Apollo Tyres Ltd., Limda Plant, Waghodia, Vadodara, Gujarat, India

G. S. Shibulal · K. Naskar (✉)
Rubber Technology Centre, Indian Institute of Technology, Kharagpur, India
e-mail: knaskar@rtc.iitkgp.ernet.in; kinsuk@mailcity.com

P. M. Visakh et al. (eds.), *Advances in Elastomers I*,
Advanced Structured Materials 11, DOI: 10.1007/978-3-642-20925-3_4,

1. The material must be of high molecular weight (macromolecular).
2. Must be amorphous in nature, at ambient condition.
3. Glass transition temperature (Tg) must be well below the room temperature.
4. Must have low secondary forces between molecules so as to obtain the requisite flexibility.
5. Crosslinking sites are essential to make a moderate degree of crosslinking to establish an elastomeric network.
6. The main chain backbone should be free of weak links which could be the site of unwanted chain breakage (chain scission).

Although, they are long chain molecules their final properties are inferior for major end user applications (most of the engineering applications). So it is essential to convert unusable pristine rubber into more useful final products. The processing of adding or incorporating ingredients into the rubber is called compounding. In other words, rubbers (natural or synthetic) are blends with other materials (ingredients) to balance the chemical and physical properties of the final product and also to improve and modify the flow characteristics during processing. Hence the process of compounding is a complex multidisciplinary science necessitating knowledge of organic, inorganic and physical chemistry, polymer physics and chemical reaction kinetics. Compounding is a unique requirement of the rubber, generally not found in other material. The final properties can be dictated by selecting and adjusting the ingredients [3]. The stages of rubber product manufacturing are broken down into three major classes:

1. Selection of compounding ingredients
2. Mixing or compounding
3. Vulcanization techniques or final product manufacturing process.

The objective of this chapter is to give brief introduction of the all classes with their importance. By proper selection of the variables in each class, the properties can be manipulated from virtually incompressible with a bulk modulus some thousand times greater than shear modulus, to large extent impermeable to gases and liquids and with excellent recovering and abrasion resistance. Hence rubbers are found in applications such as tyres, tubes, conveyor and V-belts, large dock fenders, building foundations, rubber bearings for bridges, automotive engine components, and a wide range of domestic and engineering appliances. In most applications for rubber products there are no alternative materials except other rubbers.

## 2 Compounding Ingredients

The general compounding ingredients are broadly classified into six categories: it is unfortunate that, the selection of an additive may be beneficial for one property but may have adverse effect on another. So the selection and optimization of quantity of ingredients plays a vital role in the final performance properties.

1. Polymer—Natural or synthetic rubber
2. Vulcanization system—Sulphur, peroxide, metal oxide, resin etc.
3. Fillers—Reinforcing and non-reinforcing fillers
4. Age resistors—Antioxidants and antiozonants
5. Processing aids—Plasticizers, oils, tackifiers
6. Miscellaneous—Blowing agent, colorants, flame retardants etc.

## *2.1 Polymers*

As Barlow mentioned, "Rubber is to rubber compound as flour is to bread: although the choice and amount of additives can give considerable variation to the end product, the main characteristic determinant is the kind of rubber (or flour) used" [4]. Hence the ultimate end use properties largely depend on the selection of the rubber or rubber blends. The general uses of rubber in engineering applications were broadly classified as natural and synthetic based on their origin.

### 2.1.1 Natural Rubber

Natural rubber (NR) is *cis*-1,4 polyisoprene and is present as latex in a large variety of plants in many regions of the world particularly in Thailand, Malaysia, Indonesia and India. The most important source is the tree *Hevea brasiliensis.* About 10 million hectares (100,000 km$^2$) are currently planted, producing around 6.7 million tonnes of NR annually. Global natural rubber consumption is split among tires (75 %), automotive mechanical products (5 %), nonautomotive mechanical products (10 %), and miscellaneous applications such as medical and health-related products (10 %). In general, the availability of NR is broadly categorized into three different groups:

1. Visually inspected—Ribbed smoke sheet, pale latex crepe, sole crepe and field coagulum crepe
2. Technically specified—Standard Malaysian Rubber (SMR), Standard Thai Rubber (STR), Indian Standard Natural Rubber (ISNR)
3. Speciality or modified NR
    a. Physically modified—Oil extended, thermoplastic NR, deproteinized NR.
    b. Chemically modified—Methylmethacrylate grafted NR (MMA-g-NR), styrene grafted NR, Epoxidized NR, superior processing NR.

Natural rubber is a general purpose elastomer. Its high resilience, low heat build-up and excellent dynamic properties coupled with outstanding processability, make it an ideal rubber for automotive tyres. The most engineering applications of NR involve its use as a spring. In comparison with metal springs, NR springs require no maintenance, have high energy storage capacity and non-linear load deflection characteristics, can accommodate a certain amount of misalignment and are easier to install. The other typical engineering applications of NR include antivibration mountings, flexible couplings, and bearings for buildings for protection from earthquakes, dock fenders and rail pads [5, 6].

### 2.1.2 Synthetic Rubber

The synthetic rubber was mainly divided into two groups

1. General purpose
2. Special purpose

General purpose synthetic rubbers are basically hydrocarbon polymers. These includes styrene butadiene rubber (SBR), Polybutadiene (BR), Polyisoprene (IR) etc. These rubbers contain substantial chemical unsaturation in their backbones "diene", causing them to be more susceptible to attack by oxygen, and especially by ozone. On the other hand, speciality rubbers have innate characteristics like oil resistance, ozone resistance, and high temperature resistance determined by the repeating units [7].

## *2.2 Vulcanizing System*

Vulcanization or crosslinking is the formation of chemical crosslinking along the polymer backbone or macromolecules to decrease its plasticity, tackiness, cold and hot flow and also to improve its end use properties like strength, elasticity and stability. The process of vulcanization leads to the formation of three dimensional elastic networks [9]. In other words, plastic (visco-elastic) material is transformed into elastic (elasto-viscous) material. The process of vulcanization is depicted graphically in Fig. 1. In this diagram, the polymer chains are represented by the lines and the cross-links by the black circles. Materials that are able to form crosslinks between polymer chains may generally be defined as vulcanizing agents. The properties that result from a vulcanizate depend on the number and type of the crosslinks formed. The following paragraphs deal with the various vulcanizing agents used in the field of rubber technology.

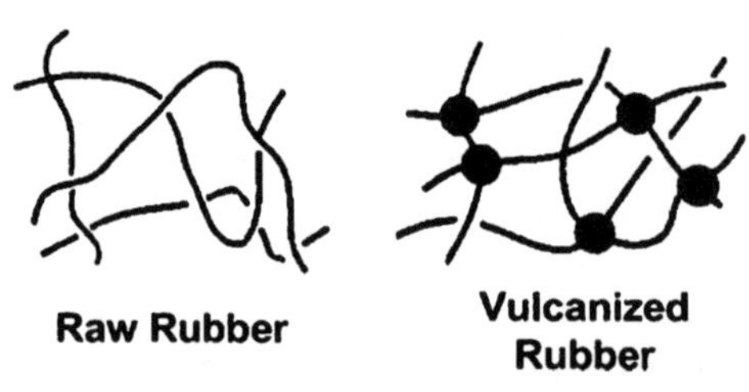


**Fig. 1** Vulcanization of raw rubber into crosslinked rubber (*lines*: polymer chains, *solid circles*: crosslink points) [9]

### 2.2.1 Sulphur

Charles Goodyear was the first to invent the process of vulcanization in 1839. The process involves heating rubber with sulphur, which was first successfully used in Springfield, Massachusetts, in 1841. Thomas Hancock used essentially the same process after one year in England. However, the vulcanization time was too long (>5 h) and the vulcanizates suffered from disadvantages such as ageing, elastic properties [10]. Since then, a large number of research groups and works were focused in the field of vulcanization chemistry to improve the process by reducing the time of cure and also to improve the performance properties.

Ground sulphur is the oldest and most widely used crosslinking agent in the rubber industry, often referred to as rhombic sulphur or rubber makers' sulphur. The molecular structure of rhombic sulphur comprises an eight membered ring and is crystalline in nature. It has a melting point of 115 °C and is soluble to a limited degree in elastomers; for example, around 1 % w/w in natural rubber at room temperature, increasing to a level of the order of 7 % at 100 °C. Relatively low solubility of sulphur in the rubber compound leads to the formation of sulphur bloom on the surface on storage and in service, if used in excess. Sulphur vulcanizates provide an outstanding balance of cost and performance, exhibiting excellent strength, flexibility and durability at very low cost. One major limitation imposed upon the use of sulphur as a vulcanizing agent is that the elastomer must contain some chemical unsaturation [11].

#### Accelerated Sulphur Vulcanization

Accelerators can be defined as organic chemical ingredients, which speed-up the vulcanization reaction in terms of shorter time and/or at a lower temperature. Oenslager was the first to study the effect of aniline as crosslinking accelerator in the sulphur curing system. However, aniline is too toxic for use in rubber products. Its less toxic reaction products with carbon disulfide i.e., thiocarbanilide, were introduced as an accelerator in 1907. The use of accelerators in concentrations as low as 0.5 phr reduce the time to as short as 1–3 min. As a result, elastomer vulcanization by sulphur without accelerator is no longer of much commercial significance. It had been reported that the physical properties of the vulcanisate and its resistance to aging were also improved by the use of accelerator or combination of accelerators. There are wide varieties of accelerators available to the compounder. Accelerators may be classified in several ways: (a) inorganic or organic, (b) acidic or basic, (c) by chemical type, or (d) by speed of the cure, giving rise to the terms slow, medium, semi-ultra and ultra accelerator. Functionally they are classified as primary and secondary accelerators [12].

*Primary accelerators*: Primary accelerators are mercapto based accelerators; generally provide considerable scorch delay, medium fast cure, and good modulus.

Examples are sulfenamides and thiazoles.

*Secondary accelerators* (*booster*): Secondary accelerators are usually scorchy, and cause very fast cure. They are normally used along with the primary accelerators. A primary at about 1 phr and a secondary (or booster) at 0.1–0.5 phr, exhibit to give faster vulcanization than each product separately.

Examples are guanidines, thiurams, dithiocarbamates, dithiophosphates, etc.

Some of the widely used accelerators in the rubber industry are given in the Table 1. A typical cure characteristic of sulphenamide accelerators is shown in Fig. 2. In order to rationalise this extensive range of materials and its advantages the readers are advised to refer the review articles by Datta and others [8–11]. However, it is worth noting here about the two most common accelerators as well as sulphur donors i.e., the tetramethylthiuram disulfides (TMTD) and dithiodimorpholine (DTDM). Although they increase the rate of cure, but also increase the level of mono- and di-sulfidic cross-links, which are reversion-resistant and more stable toward oxidative degradation. But there is a continuous urgency to reduce the use of accelerators based on secondary amines, which can react with nitrogen oxides to form suspected carcinogenic nitrosamines. This is especially a problem with dithiocarbamate-type accelerators. Dibenzylamine-derived dithiocarbamates and those based on sterically hindered amines are reported to exhibit very less carcinogenic effect [8, 12].

The general reaction path of accelerated sulphur vulcanization is depicted in Fig. 3 and thought to be as follows [9]:

- Accelerator reacts with sulphur to give monomeric polysulphides of the structure Ac-Sx-Ac, where Ac is an organic radical derived from the accelerator
- The monomeric polysulfides interact with rubber to form polymeric polysulphides; e.g. rubber-Sx-Ac. During this reaction, MBT is formed if the accelerator is a benzothiazole derivative and if the elastomer is natural rubber.
- Finally the rubber polysulphides react, either directly or through an intermediate, to give cross-links, rubber-Sx-rubber.

## Activated-Accelerated Sulphur Vulcanization

The effect of activators is to increase the crosslinking efficiency of the accelerated sulphur vulcanization system. Often, the rate of vulcanization increases to more than threefold with the addition of small amount of activators. Zinc oxide is the most important and common inorganic activator of sulphur crosslinking system, but magnesium and lead oxide also find use. In addition to that, zinc oxide has been used as a pigmentation ingredient where it is particularly effective in absorbing ultraviolet rays. With high thermal conductivity and heat capacity in comparison to other compounding materials, zinc oxide can be used in considerable quantity for thick article vulcanization process. It can, at reasonable addition levels, also assist in reduction of mould shrinkage values. Basic zinc carbonate which is more soluble than zinc oxide can be used in the production of transparent goods. As an activator of vulcanization, zinc oxide requires the presence of fatty

**Table 1** Commonly available accelerators for sulphur vulcanization [10]

| Compound | Abbreviation | Structure |
|---|---|---|
| *Benzothaizoles* | | |
| 2-Mercaptobenzothaizole | MBT | 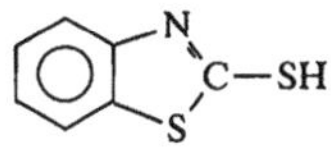 |
| 2-2′-Dithiobisbenzothaizole | MBTS | 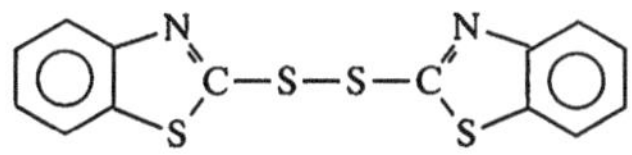 |
| *Benzothaizolesulfenamides* | | |
| *N*-Cyclohexylbenzothiazole-2-sulfenamides | CBS | 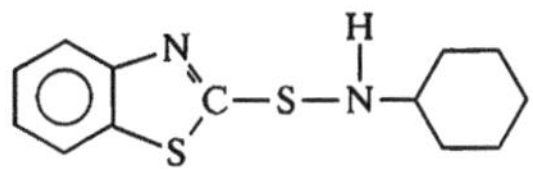 |
| *N*-t-Butylbenzothiazole-2-sulfenamides | TBBS | 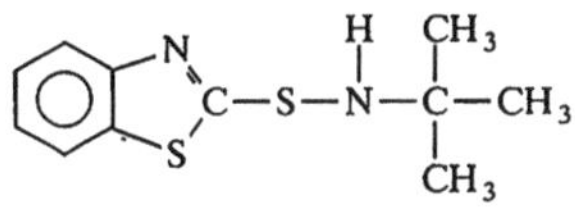 |
| 2-Morpholinotiobenzothiazole | MBS | 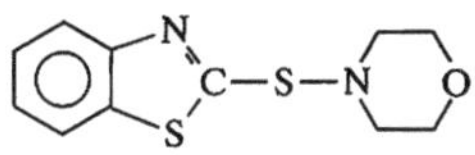 |
| *N*-Dicyclohexlbenzothiazole-2-sulfenamides | DCBS | 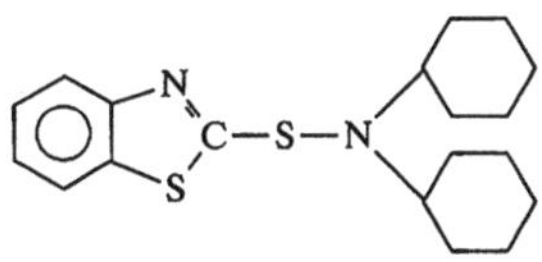 |
| *Dithiocarbamates* | TMTD | |
| Tetramethylthiuram monosulfide | | 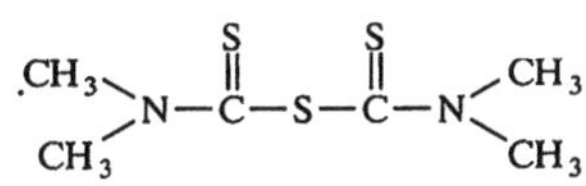 |
| Tetramethylthiuram disulfide | TMTM | 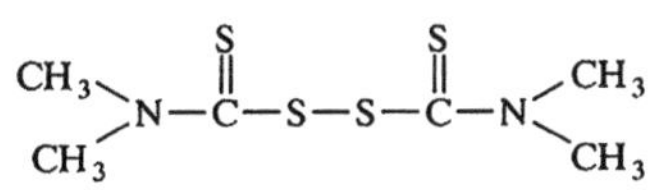 |
| Zinc diethyldithiocarbamate | ZDEC | 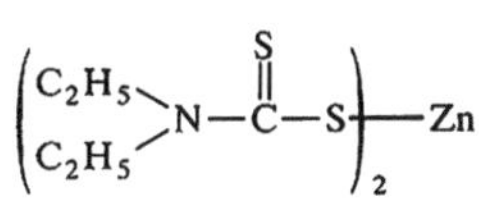 |
| *Amines* | | |
| Diphenylguanidine | DPG | 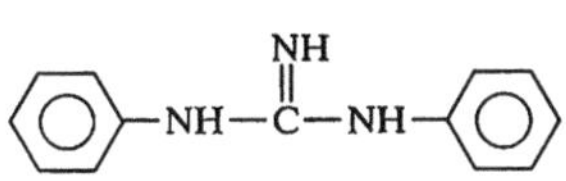 |
| Di-*o*-tolyguanidine | DOTG | 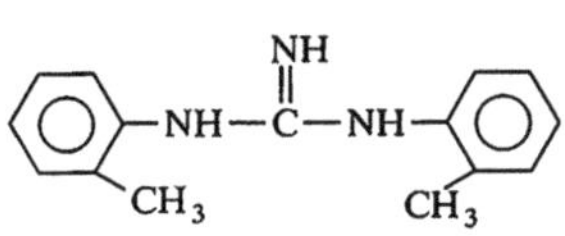 |

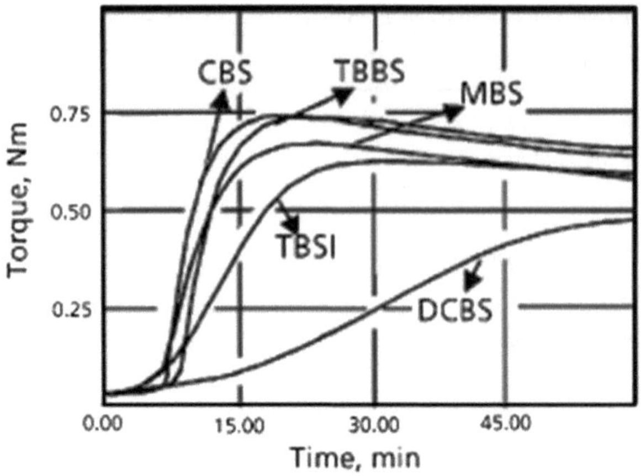


**Fig. 2** Cure characteristics of various sulphenamide accelerators for NR compounds, (Natural rubber-100; Zinc oxide-5.0; Stearic acid-2.0; Sulphur-2.5; Sulphenamide-0.6; composition expressed in phr) [10]

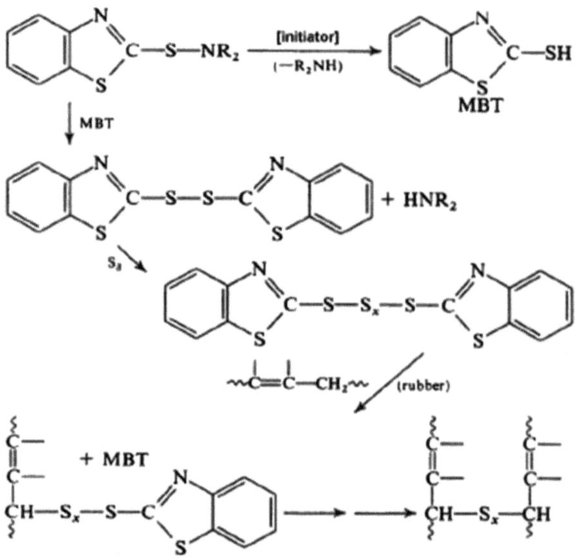


**Fig. 3** Reaction scheme of MBT accelerated natural rubber vulcanization [9]

acids, thereby converting the zinc into a rubber soluble form. Stearic acid is the most commonly used fatty acid. Addition of fatty acids also provides enhanced compound processing characteristics together with improved dispersion of fillers and chemicals.

Recently, zinc content of rubber especially in the tyre sector, has come under increased scrutiny because of environmental concerns. It is reasoned out that, zinc is a heavy metal and has a potential detrimental influence on aquatic organism. The concern for tyres is that a notable amount of zinc dissipates into the environment from the rubber dust due to abrasion of tyres during service condition. To till date, there has been no complete replacement of zinc oxide in terms of price

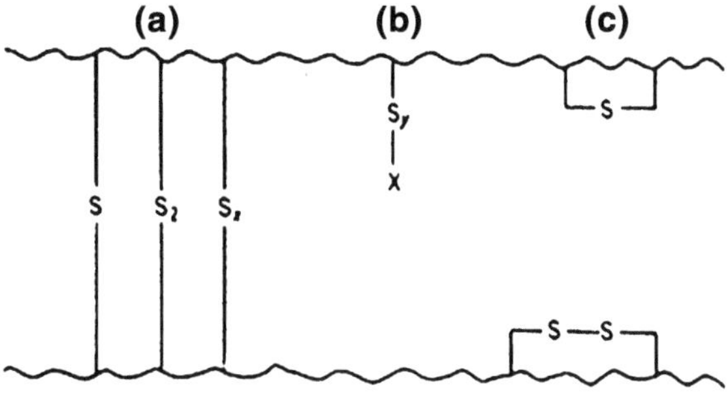


**Fig. 4** Types of sulphur crosslink: (*a*) monosulfidic, disulfidic, and polysulfidic; (*b*) pendant sulphur; (*c*) intramolecular linkages. Types (*b*) and (*c*) are wasteful of sulphur [13]

**Table 2** Various sulphur vulcanizing system (in phr) [12]

| Ingredient | CV | Semi-EV | EV |
|---|---|---|---|
| Rubber | 100 | 100 | 100 |
| Zinc oxide | 5.0 | 5.0 | 5.0 |
| Stearic acid | 1.0 | 1.0 | 1.0 |
| Sulphur | 2.5 | 1.25 | – |
| MBT | 1.0 | 1.25 | – |
| TMTD | – | – | 3.0 |

and performance. However, attempts have been made to reduce the amount of zinc by using nano-zinc oxide, activated zinc oxide and zinc soaps.

Figure 4 illustrates the various sulphur crosslinks that can be formed during vulcanization by sulphur. However, the nature and amount of crosslinks produced depend on the accelerator type and accelerator/sulphur ratio. Polysulfidic crosslinks give better fatigue properties, but poor compression set and aging properties. Monosulfidic crosslinks give better aging properties and compression set, but poor fatigue properties [13].

Over the years three types of sulphur cure system are commonly employed:

1. Efficient vulcanization systems (EV)—low sulphur/high accelerator (S/Acc $< 1$)
2. Semi-efficient vulcanization systems (Semi-EV)—intermediate sulphur and accelerator (S/Acc $\approx 1$) and
3. Conventional vulcanization systems (CV)—high sulphur/low accelerator (S/Acc $> 1$)

EV systems are those where a low level of sulphur and a correspondingly high level of accelerator are employed in vulcanizates for which an extremely high heat and reversion resistance is required. In the conventional curing systems, the sulphur dosage is high and correspondingly the accelerator level is low. The CV systems provide better flex and dynamic properties but worse thermal and reversion resistance. For optimum levels of mechanical and dynamic properties of vulcanisates with intermediate heat, reversion, flex and dynamic properties, the so-called Semi-EV systems with an intermediate level of accelerator and sulphur are employed. The levels of accelerator and sulphur in CV, Semi-EV and EV systems are shown in Table 2.

Apart from the sulphur itself, sulphur bearing compounds that liberate sulphur at the vulcanization temperature can be used as vulcanizing agents. These are termed as “sulphur donors”, which can directly substitute sulphur. Some of the commonly used chemicals are dithiodimorpholine (DTDM), 2-morpholino-dithio-benzothiazole sulfinamide (MBS), Dipentamethylenethiuram tetrasulfide (DPTT). Others, like tetramethylthiuram disulfide (TMTD) can act simultaneously as vulcanization accelerator and sulphur donor. Sulphur donors may be used when a high amount of sulphur is not tolerable in the compounding recipe, for example, high temperature vulcanization of rubber. They are used in EV and Semi-EV systems. Sulphur donors are used to generate a network capable of resistance to degradation on exposure to heat. Generally sulphur donors convert initially formed polysulfides to monosulfides which is characteristic for EV or Semi-EV systems [10, 11].

### 2.2.2 Peroxide Vulcanization

In 1914, Ostromyslenski, a Russian researcher was the first to discover that natural rubber could be transferred into a crosslinked state after treatment with dibenzoyl peroxide. However, serious interest in vulcanization with various peroxides commenced only with the commercial introduction of dicumyl peroxide (DCP) by Hercules in the late 1950s. DCP was the first commercially available peroxide to combine high efficiency, good vulcanizate properties with low cost and broad spectrum utilities [13].

Peroxides interact with polymers in a variety of ways, initiate polymerization, modify their rheological properties (vis-breaking), alter polarity or attach functional groups (grafting) and enhance high temperature performance (vulcanization). The effect that peroxide has on a polymer depends on the nature of the polymer type and concentration of the peroxide and reactivity of other component present. The driving force of peroxide vulcanization is free radicals. Radicals are atoms or molecular fragments that contain an unpaired electron. These unstable radicals energetically react in ways that allow the electron to pair with another [14]. Although all radicals are unstable, some are more reactive than others. The major advantages of using peroxide crosslinking system are:

1. Improved high temperature resistance
2. Reduced compression set
3. Ability to crosslink saturated as well as unsaturated rubber
4. Rapid vulcanization without reversion
5. Formation of C–C bond, which is similar to the bond strength to every other bond in polymer backbone.

The basic peroxide crosslinking reaction consists of three main steps as shown in Fig. 3. In this process, thermal energy causes the homolytic cleavage of the peroxide molecule to create radicals. These radicals then abstract hydrogen from the polymer to form polymer radical. Eventually two polymer radicals combine to form a covalent crosslink. In addition to the above productive reactions, numerous

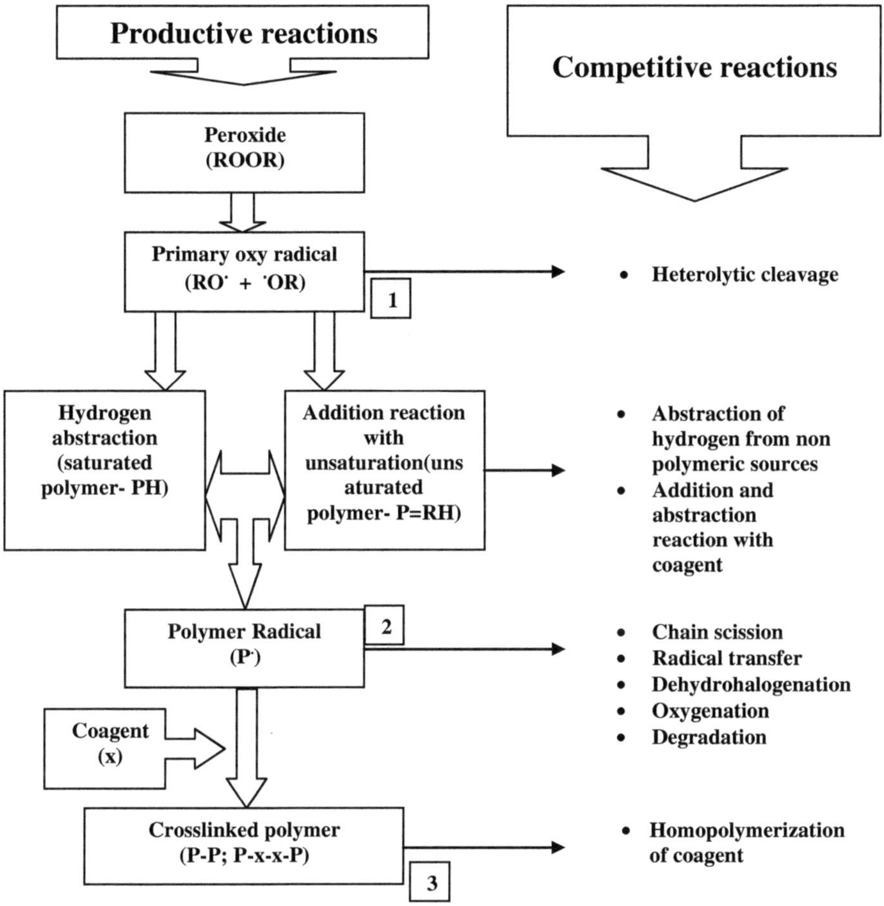


**Fig. 5** Various productive and competitive reactions involved in coagent assisted peroxide crosslinking system [15]

side reactions may take place during the course of the reaction, which reduces the efficiency as shown in Fig. 5. Peroxide crosslinking needs special attention system for the selection of compounding ingredients, such as plasticizers, oils and acidic fillers (such as silica) otherwise it might affect the crosslinking efficiency [13, 14].

*Criteria for the selection of peroxide* [13]:

The following are the some of the important criteria in selecting peroxide (Geo Specialty Chemicals, Technical bulletin)

1. It should be efficient in vulcanizing the elastomer
2. The decomposition temperature should be moderate to have scorch safety and fast cure rate
3. It should not be vulnerable to side reaction, which reduces the efficiency
4. It should be non-volatile, non-toxic and non-irritating

5. Neither the peroxide nor its decomposition products should cause degradation of the polymer.

Depending on the chemical structure, organic peroxides are classified into different groups such as Hydroperoxides, Ketone peroxides, Peroxyacids, Dialkyl peroxides, Peroxyesters, Peroxycarbonates, Diacylperoxides, Peroxydicarbonates, Peroxyketals, Cyclo keton peroxides etc.

Compounds of the dialkyl peroxide class belong to the most thermally stable category, whereas the peroxy-esters possess only a limited thermal stability: for a rubber compound this greatly restricts the practical suitability of these materials for crosslinking. In order to get desired property and process safety during vulcanization, it is necessary to select proper peroxide. To choose the most suitable peroxide for vulcanization purpose, it is necessary to know the characteristics of the peroxide. The following are the most important characteristics of peroxide.

*Active oxygen content*: Active oxygen content represents not only the amount of free radicals produced from the peroxide, but also the concentration and purity of peroxide. Peroxide with low active oxygen content is less efficient in vulcanization, simply because the peroxide group is more dilute on a molecular level.

*Half life period*: Half life period is defined as the time required for the decomposition of one half of a given quantity of a peroxide in dilute solution at a given temperature. Decomposition is a first order reaction and characterized by the given temperature. For convenience to compare the stability of peroxides in dilute solutions, they are commonly listed accordingly to the temperature at which they have half lives of 10 or 1 h. Commonly, the half-life time is determined by the differential scanning calorimetric-thermal activity monitoring (DSC-TAM) of a dilute solution of the peroxide in monochlorobenzene. The dependence of the half-life time on temperature can theoretically be described by the Arrhenius equation.

$$k_d = A.e^{-\frac{E_a}{RT}} \tag{1}$$

$$t_{1/2} = \frac{\ln 2}{k_d} \tag{2}$$

where

$k_d$ rate constant for the peroxide decomposition ($s^{-1}$)
A Arrhenius frequency factor ($s^{-1}$)
Ea activation energy for the peroxide decomposition (J/mol)
R 8.3142, gas constant, (J/mol·K)
T temperature (K)
$t_{1/2}$ half-life time (min).

*Activation energy*: Activation energy is defined as the energy required to activate the decomposition mechanism. Although all peroxides are thermally viable to decompose to yield free radicals, but their relative stabilities vary considerably. Volatility is also an important determinant in shelf-life and transportation. Most stable peroxide includes dialkyl, diaryl and diaralkyl types, whereas the peroxy esters possess only a limited thermal stability.

## Coagent Assisted Peroxide Crosslinking

Coagents are multifunctional vinyl monomers, which are highly reactive towards free radicals. Since all the co-agents contain terminal unsaturation, it is expected that addition/polymerization is the principal mechanism by which they react in a rubber compound. Addition of co-agent to these polymer formulations suppresses the side reactions and allows crosslinking to occur via tighter network formation. These additives are used to improve the physical properties and the processability of peroxide cured elastomers. Properties which are generally enhanced by the use of co-agents are: improved heat resistance, better mechanical properties, improved resilience, increased abrasion resistance, improved rubber to metal adhesion. Co-agents are normally classified into two broad categories, based on their state and rate of cure [16].

*Type I coagents*: Type I coagents increase both the rate and state of the cure. Most of these contain readily accessible unsaturation without allylic hydrogen. So it mainly undergoes addition reaction rather than hydrogen abstraction. They include multifunctional acrylates and methacrylate esters, bismalamides, etc. The zinc salt of acrylic and methacrylic acids also belong to this group.

*Type II coagent*: Type II coagents increase only the state of cure. Coagent of this type contains both readily accessible vinyl unsaturation and abundant number of easily abstractable allylic hydrogens. They include polybutadiene, allylated species like triallyl cyanurate (TAC), Triallyl isocyanurate (TAIC) and diallyl phthalate (DAP) etc.

Generally in a coagent assisted peroxide cure system, it is well established that the most favorable reaction is crosslinking via coagent molecule (CM). However, several competing reactions like formation of coagent bridges or domains (CD) (homopolymerized coagent compound) may also occur between two effective crosslink points and pendent coagent bridge grafted on the polymer chains [17]. When a coagent domain is grafted to the polymer chain, the effect can be similar to the addition of reinforcing filler with high modulus. The resulting measurable outcome is high crosslink density. A conceptual model of crosslink formation and coagent bridge formation of coagent assisted peroxide system is shown in Fig. 6.

The actual state is a combination of the above mentioned various reactions. However, the balance of formation of coagent domains (homopolymerization) over polymer grafting of coagent domains would presumably depend on concentration of coagent, polarity difference between coagent and polymer (solubility parameter) and adequacy of mixing [18].

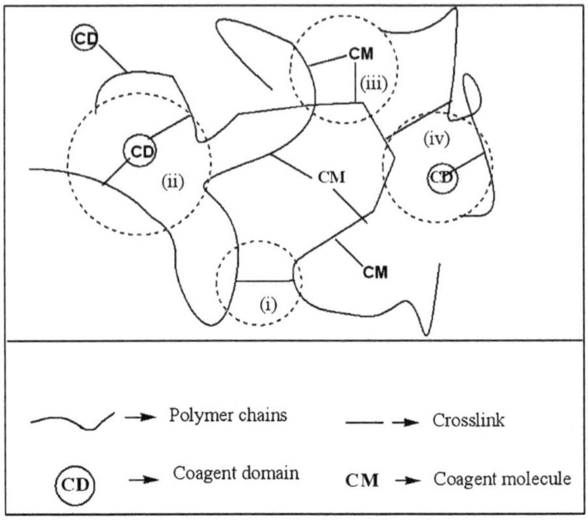


**Fig. 6** A conceptual model of coagent assisted peroxide cured polymer network [14]. (*i*) Crosslink derived from peroxide only. (*ii*) Crosslink derived through coagent domain formation (inefficient use of coagent). (*iii*) Crosslink derived through coagent molecule (efficient use). (*iv*) Coagent domain grafted to polymer network (behaves as reinforcing filler grafted to polymer chains)

Sulphur as Coagent in Peroxide Vulcanization

Despite the improvement in peroxide cure properties achieved by the use of conventional coagents, dynamic and some mechanical properties of peroxide vulcanisates are still inferior to those of typical sulphur vulcanizates. For this reason, there is a limitation in the use of peroxide cured rubber products, particularly in dynamic applications. However, if the above mentioned properties of peroxide cured rubber can be improved, the range of applications can be significantly widened. To achieve this sulphur was employed as coagent in peroxide vulcanization system. The idea is to hybrid the vulcanizate with sulphidic and C–C linkage. The effect of sulphur as an additive in peroxide vulcanization in NR rubber was studied and found that sulphur reduces the crosslinking efficiency of the peroxide. Accordingly, it belongs to the Type II co-agents. They suggested that sulphur may interfere with the free radical crosslinking reaction in a way to decrease the network density, increase the work at break, tensile strength and worsen the aging properties.

### 2.2.3 Metal Oxide

Combination of zinc and magnesium oxide is often used to vulcanize the chloroprene rubbers (CR). CR can be vulcanized in the presence of zinc oxide alone; however, magnesium oxide is necessary to give scorch resistance [19]. A mechanism that has been proposed for the vulcanization of CR by the action of zinc oxide and magnesium oxide is shown in Fig. 7. The accelerator that has been

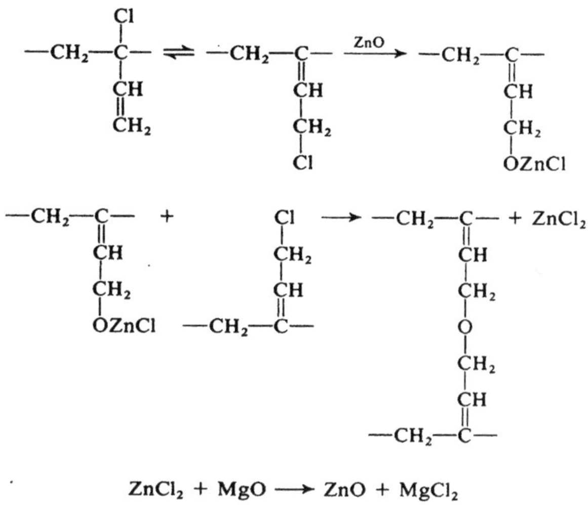

**Fig. 7** Metal oxide (Zinc oxide and Magnesium oxide) vulcanization of CR [8]

widely used with metal oxide cures is ethylene thiourea (ETU) or 2-mercapto-imidazoline. However, extensive use of ETU in vulcanization of CR is restricted because of its carcinogenic nature.

### 2.2.4 Resin Cure

Difunctional compounds are generally used in resin cure system [8, 20].

1. Butyl, chlorobutyl, bromobutyl and EPDM are elastomers that can be cross-linked with reactive phenol–formaldehyde resins (PF). PF resin cure requires activation by a halogen-containing material, preferably tinchloride ($SnCl_2$). The practical importance of using PF resins is to cure butyl rubber, which is widely used for the bladders and the curing bags used in the retread industry. Resin curing of SBR and BR imparts excellent cut growth and abrasion resistance. Resin cured nitrile rubber shows high fatigue life and high relaxation, while resin-cured butyl rubber shows outstanding ozone and age resistance [21].
2. Diamine or polyamine is the widely used crosslinking agent for acrylate and fluoroelastomers. Amine-cured acrylates tend to stick and to have poor storage stability; skin irritations are another drawback. In the fluorocarbon rubbers, diamines or polyamines give poor processing safety. Thus blocked amines, in which the inert molecule splits into active components at high temperatures, were developed and used.
3. Blocked diphenyl methane diisocyanate to produce urethane crosslinks is applied in curing variety of rubbers [11, 12]. The chemistry lies in the

dissociation of the compound into two quinonedioxime molecules and one diphenyl methane diisocyanate. The quinone reacts with the rubber via a nitroso group and forms crosslinks via a diisocyanate group. The performance of this system in NR is characterised by excellent age resistance, fatigue resistance and outstanding reversion resistance. But problems may be with lower scorch resistance and rate of cure.

4. *p*-Benzoquinone dioxime can crosslink various elastomers such as natural rubber, SBR, and EPDM. However, it is of more technical importance only in butyl rubber. The cure rate increases with the degree of unsaturation in the butyl, and the addition of sulphur raises the modulus to limited extent.

### 2.2.5 Radiation Crosslinking

Polymeric substances, which are predominantly high molecular weight organic compounds such as plastics and elastomers (rubber), respond to radiation in several ways. Both ultraviolet (UV) and electron beam (EB) radiations are classified as electromagnetic radiations along with infrared (IR) and microwave (MW). Different polymeric materials respond to irradiation in different ways which depends on the nature and type of chemical structure. Many will be modified by the formation of a cross-linked network, and by changing their surface properties or structure, some will be degraded. Another field applicable to polymeric systems is polymerization and grafting. The process involves the electronic excitation or ionization of organic molecules after absorption of energy. For the transformation of organic molecules from the ground state to the excited state, energies typically in the range from 2 to 8 eV are required. The excited molecules are able to enter into chemical reactions leading to chemically reactive products that initiate the polymerization, cross-linking and grafting reactions [11, 12, 22].

Both UV and EB are widely used in the rubber technology. Without any doubt, UV irradiation process is the cheaper because the equipment is simpler, smaller and considerably less expensive to purchase and operate. But the process requires the use of suitable photoinitiators for curing the polymers. Because of limited penetration of UV radiation, this technique is widely applicable in coating or paint industry. EB curing is greatly employed in the cable and wire industry. EB irradiation involves the use of high energy electrons. It usually reacts with materials by knocking off electrons and creating ions or free radicals. Therefore, EB is classified as ionizing radiation. On the other hand, microwave and infrared, normally heat the materials by enhancing their molecular motion and as such are non-ionizing radiations.

In view of the nature of irradiation, most of the events initiated by UV irradiation occur near the surface, because the absorption is governed by the Beer-Lambert Law, whereas the reactive species produced by EB are dispersed randomly throughout the entire thickness of the material as shown in Fig. 8. The reactions

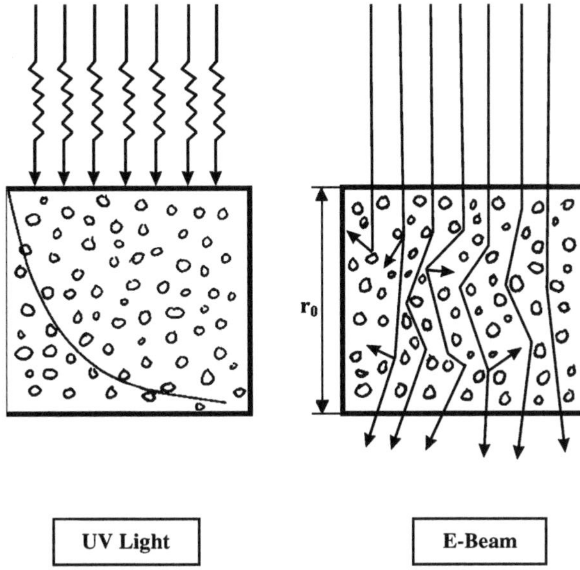


**Fig. 8** Nature of UV and EB irradiation with reference to the depth cure gradient [22]

initiated by the formation of radicals are the most important ones in the electron beam curing process. Electron energy plays a major role in determining the ultimate physical properties. Two competing reactions occur simultaneously when polymers are exposed to the electron energy: chain coupling and chain scission. It is a matter of fact, at high electron dosage chain scission dominates the chain coupling effect and finally leads to give poor physical properties [22].

In order to increase the rate and state of radiation crosslinking, radiation promoters are used. They are classified as direct and indirect promotors. The former compound enters directly into the cross-linking reaction and become the part of actual connecting molecular links. On the other hand, indirect promotors do not involve in the crosslinking process, but enhance the formation of free radicals and reduce the secondary reactions.

1. Indirect promotors—chlorinated aliphatic compound, nitrous oxide, sulphur monochloride
2. Direct promotors—Maleimides and dimaleimides, thiols, polyacrylic and polyallylic compounds.

On the other hand, the protection of polymers against radiation damage leads to the introduction of large number of effective protective agents, often referred to as antirads. A considerable reduction of crosslinking was found by the use of aromatic amines, quinones, aromatic hydroxyl sulphur and aromatic nitrogen compounds. The degree of protection offered by these compounds increases with their concentration, but reaches a limiting value at a concentration of a few weight percent.

### 2.2.6 Mixed Vulcanization

As mentioned previously in this chapter, the cure behavior of peroxide vulcanization and sulphur vulcanization is different and leads to different properties of the cured compounds. If it was possible to obtain a cured rubber with both sulphur bridges and C–C bonds, it might be possible to achieve the good property characteristics from sulphur vulcanizates and also the good ones of peroxide vulcanizates, by means of combination of both vulcanization systems. The obtained rubber product can be called as mixed vulcanizate. If mixed vulcanization is just defined as vulcanization which leads to both sulphidic and C–C crosslinks, then the use of sulphur as co-agent in peroxide vulcanization, as above, can be considered as mixed vulcanization [23, 24].

However, some authors have argued that mixed vulcanization as a combination of two different vulcanization systems in separate vulcanization steps. In that case peroxide cure system which uses sulphur as co-agent would not fulfill the concept and definition of mixed vulcanization. The mixed vulcanization of EPDM with sulphur and peroxide in two vulcanization steps was studied by van der Burg [23]. The vulcanizates were found to have lower crosslink density and intermediate tear strength and compression set. Van Bevervoorde [24] studied the combination of electron beam irradiation curing, which leads to C–C bonds between polymer chains, and sulphur curing of EPDM, as a mixed vulcanization system, viz. in two curing steps. Vulcanizates obtained without filler in the formulation show similar, or slightly higher strength properties and lower compression set when compared to sulphur vulcanizates. When carbon black was added to the formulation, tensile and tear strength increased substantially while the compression set was similar to the mixed vulcanizates without filler.

It can be summarized that by mixed vulcanization it is possible to improve properties of rubber compounds to a certain extent, with the penalty of other properties like scorch safety and low cure rate. Although, the process seems to be interesting and attractive but the concept not widely accepted from an industrial point of view.

### 2.2.7 Dynamic Vulcanization

The concept of dynamic vulcanization was first introduced by Gessler in 1962. Earlier work of Fisher in PP/EPDM TPVs with peroxide as crosslinking agent resulted in the commercialization of "Uniroyal TPR" thermoplastic rubber in 1973. Greater industrial attention was generated only after extensive study of TPVs based on various blend components by Coran and Patel in 1980s [9, 25]. The process involves the crosslinking of a rubber phase while it is being mixed with a thermoplastic to form a homogenous melt, the so called product is termed as thermoplastic vulcanizates (TPVs). The process needs to be carried out under high shear and above the melting temperature of the thermoplastic component and at sufficient high temperature to activate and to pursue the process of vulcanization.

As a result, products are obtained which consist of cosslinked rubber particles dispersed in a continuous thermoplastic matrix, that explains both elasticity and melt processability. Now TPVs are widely used in automobiles, wires/cables, biomedical and soft-touch applications etc. [26].

The concept of dynamic vulcanization is extended for rubber–rubber blends. Tubeless tire has the inner liner for air impermeable layer for which butyl rubbers are often used. Butyl rubbers hardly adhere to the other tire components consisting of diene rubber compounds, e.g. blends of natural rubber, styrene-butadiene rubber and butadiene rubber. To solve this problem, the concept of dynamic vulcanization for halogenated butyl rubber/natural rubber blends is applied by using suitable crosslinking agent for butyl rubber. Higher adhesive strength (three times) without sacrificing the low air permeability results by the generation of unique morphology of butyl rubber-rich particles in the NR-rich matrix [27]. Blends of NR and EPM could enjoy the economic advantage over current EPM formulation as EPM are relatively expensive material, provided without loss of aging resistance. For a blend of NR/EPM, to exhibit excellent ozone resistance, EPM should form the matrix phase with NR phase as dispersed. Due to viscosity ratio, conventional blend results in the NR as matrix and EPM as dispersed for 50/50 blend ratio. The concept of DV is applied to inverse the morphology by crosslinking the NR phase without affecting the EPM phase. The insertion of crosslinks in the NR phase cause an increase in viscosity and thereby prefer to remain as the dispersed phase. The resulting morphology of EPM as matrix as crosslinked NR as dispersed is found to give promising aging and ozone resistance [28].

## *2.3 Fillers*

In general, filler is something that is used to fill up the gaps. The term filler in rubber technology is often misleading, implying a material that is primarily intended to fill the compound in the direction to reduce the cost of the more costly rubber. But in fact, fillers change one or more of these properties: optical properties and color, improve surface characteristics and dimensional stability, change thermal, magnetic and electrical properties, improve mechanical properties, durability, and rheology, chemical reactivity, biodegradability etc. It is necessary to define precisely what filler in the polymer technology is: filler is a solid material capable of changing the physical and chemical properties of elastomeric materials by surface interaction and by its own physical characteristics. The performance of filler is mainly determined by its innate characteristics. The following list summarizes some of the major characteristics of the filler [8, 9]:

1. Physical state
2. Chemical composition
3. Particle shape, size and internal structure
4. Aspect ratio and density

5. pH (acidic and basic nature)
6. Refractive index and color
7. Moisture and oil absorption
8. Thermal, electrical and magnetic properties.

Though many types of fillers such as fibers, whiskers, particulates etc. are widely used in rubber industry, but still today particulate fillers forms a major share. The particulate fillers are broadly classified as reinforcing and non-reinforcing depending on whether they enhance the performance characteristics of the final product.

### 2.3.1 Non-Reinforcing Fillers

Fillers such as calcium carbonate, clay, chalk powder, etc., which do not exhibit any reinforcing effect but only increase viscosity of the compound. These are often called as extenders, which are used to reduce the cost of rubber goods. The most widely used fillers are china clay and calcium carbonate.

Clay is probably the most commonly used non-reinforcing fillers and is classified as hard clay and soft clay depending upon their particle size (soft clay, >2 μm and hard clay, <2 μm). Clay is hydrated aluminum silicate produced by decomposition of alumina-silica minerals such as kaolin. They are low cost fillers which can be used at high volume loadings to provide economical compounds. However, clays impart improvement in properties like hardness, abrasion resistance and also reduce mould shrinkage.

Calcium carbonate also known as whiting that is extensively used in rubber formulations primarily to impart color and cheapen the product. The two main types are ground limestone and precipitated calcium carbonate. The first one is made by grinding mineral limestone and the second is obtained by chemical precipitation from salt solution. Ground whiting gives low tear resistance, but the finer particle size precipitated materials exhibit smooth extrusion characteristics and have been proved to impart good hot tear strength.

Other non-reinforcing fillers include barytes (barium sulphate), mica, titanium dioxide ($TiO_2$) and silicates of calcium and zinc. Barytes are generally used in medical applications because of their unique chemical resistance and inertness. Mica is used in composites where high thermal expansion and high resistivity are needed. Titanium dioxide is incorporated in rubber to mask the inherent color of the matrix.

### 2.3.2 Reinforcing Fillers

The basic feature that distinguishes non-reinforcing and reinforcing fillers is enhancement of performance characteristics of the filled compound. Reinforcement in vulcanized rubber can be defined as simultaneous increase in stiffness and resistance to fracture by the addition of filler. Particulates like carbon blacks and silica are the most widely used reinforcing fillers in rubber industry.

a. **Carbon black**

Carbon black (CB) reinforcement becomes a subject of scientific interest only in the 1940s due to the growing use of synthetic rubbers in demanding applications, namely automotive and truck tires. Carbon black is essentially an elemental carbon obtained by incomplete combustion or thermal decomposition of hydrocarbons usually obtained in the form of near-spherical particles coalesced into aggregates of colloidal size. Carbon black is manufactured by a variety of processes namely: Furnace, thermal, acetylene and channel. The characteristics of the carbon blacks and its reinforcing nature in a polymer matrix are widely dependent on the type of the manufacturing process. As mentioned previously, the size of the filler is probably one of the most important characteristics for reinforcement and hence the available grades of carbon black are classified according to the particle size and listed in Table 3 [29].

Physical Characteristics of Carbon Black

1. *Carbon black particle*: The CB particles are hollow semi-spherical particles composed of two to four turbostratic graphitic layers (Fig. 9a). Aromatic groups of molecules adhere to form a spherical colloidal particle. In other words, they are small spheroidal, non-discrete component of a carbon black

**Table 3** Types of carbon blacks classified on the basis of particle size [29]

| Type | Symbol | Mean particle size (nm) | Surface area ($m^2/g$) | pH | Nature |
|---|---|---|---|---|---|
| Super abrasion furnace | SAF | 14–20 | 120–140 | 9–10 | Reinforcing |
| Intermediate super abrasion furnace | ISAF | 18–24 | 110–120 | 8.5–9 | |
| High abrasion furnace | HAF | 24–28 | 75–95 | 8–9 | |
| Hard processing channel | HPC | 22–25 | 100–110 | 3.7–4 | |
| Medium processing channel | MPC | 25–29 | 90–105 | 3.8–4 | |
| Easy processing channel | EPC | 29–33 | 80–90 | 3.8–5 | |
| Fine furnace | FF | 40–45 | 55–70 | 9–10 | Medium reinforcing |
| Fast extrusion furnace | FEF | 30–50 | 45–70 | 9–10 | |
| General purpose furnace | GPF | 50–55 | 45–50 | 9–10 | |
| High modulus furnace | HMF | 45–65 | 30–60 | 9.5–10 | |
| Semi-reinforcing furnace | SRF | 60–85 | 25–45 | 9.5–10 | |
| Lamp black | LB | 100–150 | 13–25 | 4–4.2 | |
| Superconductive furnace | SCF | 16–20 | 120 | 9.5–9.8 | Conductive |
| Conductive furnace | CF | 24 | 110 | 9–9.3 | |
| Conductive channel | CC | 17–23 | 100–150 | 3.5–4 | |
| Acetylene | – | 40–45 | 40–70 | 7–9 | |
| Fine thermal | FT | 120–200 | 15–35 | 8.5–9 | Others |
| Medium thermal | MT | 250–500 | 5–10 | 7–9 | |

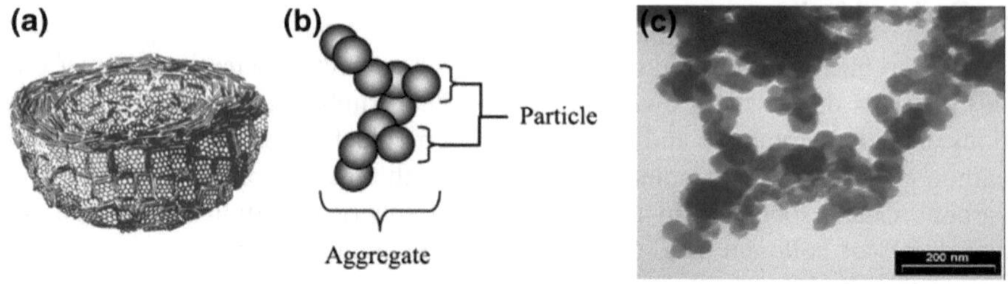


**Fig. 9** Physical characteristics of carbon black [30, 31]. **a** Structure of a primary particle in carbon black: Aromatic groups of molecules adhere to form a spherical colloidal particle. **b** Schematic representation of carbon black particle and aggregates. **c** TEM images of the agglomerates formed by the physical adsorption of aggregates

aggregate. Particle diameters can range from less than 20 nm in some furnace grades to a few hundred nanometers in thermal blacks.

2. *Carbon black aggregates*: Carbon black particles are fused together to form a discrete, rigid, colloidal entity of coalesced particles as shown in Fig. 9b. It is the smallest dispersible unit of carbon black that cannot be separated by any means. Aggregate dimensions measured by the Feret diameter method can range from as small as 100 nm to a few micrometers. Primary particle size distribution has been estimated by Transmission Electron Microscope (TEM) image analysis, but carbon black surface area is usually more efficiently obtained by adsorption methods such as nitrogen, Brunauer—Emmett—Teller method (BET), N-cetyl-N,N,N'-trimethylammonium-bromide (CTAB) and Iodine absorption techniques.
3. *Carbon black agglomerate*: Agglomerates are clusters of physically bound and entangled aggregates. Agglomerates can vary widely in size from less than a micrometer to a few millimeters in the pellet. TEM, Scanning Electron Microscope (SEM) and mercury porosimetry can be used to qualitative the state and nature of agglomeration (Fig. 9c).

### Chemical Characteristics of Carbon Black

Manufacturing of carbon black in the conventional techniques, often result to the incorporation of some organic and inorganic impurities. Organic impurities are mainly due to the incomplete combustion of fuels that has been reabsorbed on to the carbon black particle. They are mainly polyaromatic hydrocarbons. However, it has been demonstrated that organic impurities have no significant effect on carbon black reinforcement. Mineral impurities mainly absorbed or adsorbed during quench and pelletization steps in the carbon black production process. Although the water is purified, the traces of mineral salts precipitated in the carbon black surface. Mineral impurities don't seem to alter carbon black reinforcement properties but they have a significant effect on rate of vulcanization, which increases with the pH value of carbon black. Notable chemical functions in the carbon black are illustrated in Fig. 10. Oxygenated functions such as hydroxy, acidic groups on carbon black surface were found to exist in the carbon black surface [8–10]. These groups were believed to be produced by surface oxidation in the production process, probably during pelletization followed by drying. In addition to the oxygenated functions, double bonds may also found to exist in the surface. Such double bonds could react with sulphur, and radicals to provide chemical bonding between carbon black surface and polymer, but their direct quantitative determination has never been obtained. However, the content and reactivity of hydrogen present on graphitic edges have been determined by isotopic exchange and correlated to carbon black reinforcement ability.

b. **Silica**

Silicon as a carrier of ''inorganic life'', occurs in nature almost exclusively in the form of crystalline solids in approximately 800 different siliceous minerals. Precipitated silica is the highest reinforcing non-black filler and is closest to carbon black in terms of reinforcing tendency. The formation of precipitated silica is a chemical reaction of sodium silicate (water glass), and sulphuric acid. Precipitated silica progresses from a simple filler used for producing coloured rubber compound to an important additive in terms of improving the tyre performance. Silica is essential for maintaining a good balance of heat build up, wear resistance and cut-chip resistance in tyre. Since the advent of silica compounds in passenger tyres in the early 1990s, consumption of silica has grown rapidly. Silica, as principle filler in passenger tread compounds, gives higher wet traction and better rolling resistance, at a reasonable wear resistance level compared to carbon black. However, silica has significant limitation in terms of raw material cost and process cost [32].

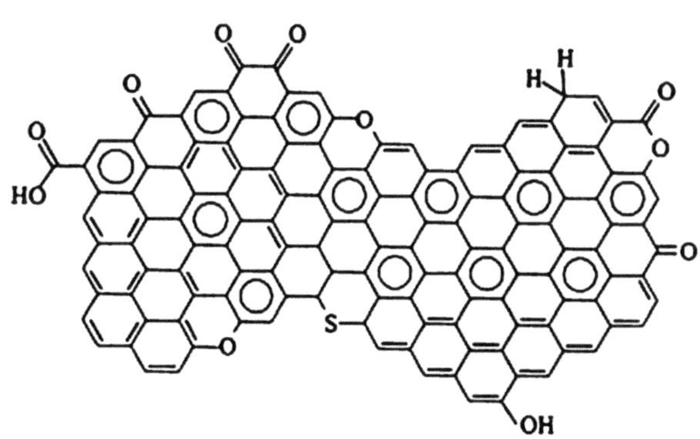


**Fig. 10** Surface functional groups of carbon black [9]

#### Physical Characteristics of Silica

1. *Silica particle size*: Particle size of silica ranges from 0.02 to 0.1 μm. It depends mainly on the route in which it is prepared, also on the process variables. Precipitated silica particles are much more porous than carbon black. Silica fillers are characterized by a high specific surface energy with low dispersive component of surface energy which results in a strong filler–filler interaction and a difficult processing behaviour and consequently poor reinforcing tendency. The specific surface area of silica is generally determined by using two methods: the $N_2$-adsorption, so-called BET, and the adsorption of CTAB, however, with certain limitations [8–10].
2. *Silica aggregates and agglomerates*: Primary silica particles remain condensed in the form of aggregates and with typical dimensions of about 100–200 nm, which are the real reinforcing species in rubber compounds. When the particles are close together, an interaction between the primary particles can take place. The degree of condensation in the direction to form aggregates (commonly designated by structure) determines the inter-particle void volume and pore diameter within the aggregates. The measurement of this "structure" can be performed by dibutylphthalate (DBP) adsorption. Conventional silica has a DBP value of typically 175/100 g; whereas for highly dispersible silicas, the DBP value is typically 200/100 g or above.

#### Chemical Characteristics of Silica

The surface of silica is covered by a layer of acidic silanol groups and different siloxane groups: geminal, vicinal, clustered and isolated groups. The functional groups in silica are randomly distributed over the whole surface whereas in carbon black the functional groups are preferably located on the edges of the crystallites. The silanol groups on the surface of different silica particles interact with each other, resulting in strong agglomerates due to the hydrogen bonds between the silanol groups. The moieties on the silica surface also interact with basic accelerators, resulting in reduced curing rates and lower crosslink densities. They can react with other chemical compounds such as stearic acid, polyalcohols and amines. The silica surface also has a tendency to absorb moisture due to the hydrophilic character of the filler. These adversely influence the curing reaction and hence the properties of the final product.

As the compatibility between silica and rubber is low, a reduction of the polarity differences is required. This can be done by silane coupling agents such as bis(triethoxysilylpropyl)tetrasufide (TESPT), which is capable of reacting with the silica surface and the polymer, and is commonly used in silica-filled rubber compounds. TESPT is composed of a poly-sulphide part which can react with the rubber and ethoxysilyl-groups on the silicon atom which can react with the hydroxyl groups present on the silica. The polysulfide part of TESPT releases reactive sulphur moieties in silica-filled compounds during rubber processing due to splitting of the TESPT. Therefore, TESPT is unstable at high shear or high temperature conditions, resulting in a "sulphur donor" effect of TESPT. An alternative silane, bis(triethoxysilylpropyl)disulphide (TESPD) has also been introduced. TESPD is actually not a pure disulfide but rather a mixture of polysulphides. The advantage of TESPD is higher stability at high shear conditions or high temperatures compared to that of TESPT and therefore less scorch sensitivity. However, due to its sulphur content, additional elemental sulphur is required to achieve comparable reinforcement to TESPT [32].

The reaction between silane and silica, the so-called silanization, is a two step process as shown in Fig. 11. The primary step is the reaction of the first alkoxy group of the silane coupling agent with silanol groups on the silica surface. After this primary reaction, an intermolecular condensation between silanes on the silica surface, the so-called secondary reaction takes place caused by unreacted ethoxy groups of the silanes.

Newly developed NXT® silanes can help improve rolling resistance while retaining or improving wet traction. The NXT family of silanes has a unique molecular structure that can dramatically improve silica dispersion, improve processing, reduce hysteresis by shortening bond distance, and increase bond strength. These performance attributes typically can be achieved simultaneously with dynamic properties representative of equivalent or improved wet traction.

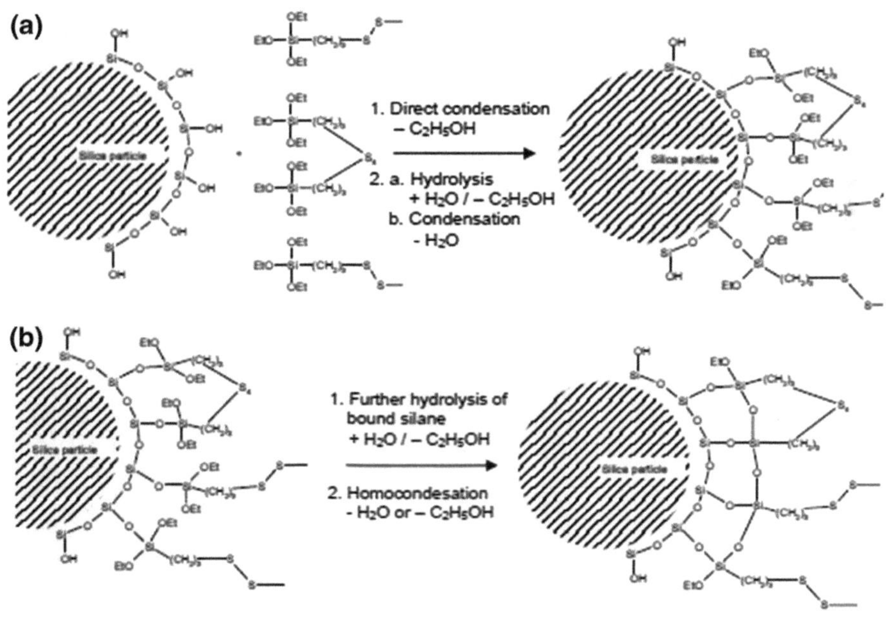


**Fig. 11** Silanization reaction (**a**) primary reaction (**b**) secondary reaction [32]

c. **Carbon–Silica Dual Phase Filler**

Carbon–Silica dual phase filler (CSPDF) is new generation reinforcing filler which consists of two phases, a carbon phase with a finely divided silica phase (domains) dispersed therein. This unique filler was mainly produced by co-fuming process of carbon and silicon-containing feed stocks in the carbon black reactor [33, 34]. Microstructure of dual phase filler reveals that carbon phase is turbostratic in nature similar to conventional carbon black and the silica particles occupy the interstitial positions of the carbon black. Doping of this foreign substance in carbon black surface creates more surface defects in the graphitic crystal lattice and/or smaller crystallite dimensions and this act as active centres.

Recently, Cabot Corporation commercialized two groups of CSDPF: CSDPF 2000 series and other is CSDPF 4000 series. While both new materials contain both silica and carbon black, some of the differences are found in the distribution of silica, silica surface coverage and silicon content [35]. Figure 12 shows the variation of distribution of silica in CSDPF 2000 and 4000 series. The following paragraph gives overview of the process and potential advantage of CSDPF 2000 and 4000 series.

CSDPF 2000: It is produced by a co-fuming process in which the silicon-containing feedstock is injected into the hydrocarbon feedstock. The silica domains are distributed finely throughout the aggregates.

CSDPF 4000: This method involves introducing a carbon black yielding feedstock into a first stage of a multi-stage reactor, and introducing a second, silicon-containing feedstock into the reactor at a location downstream of the first stage. It is apparent that the silica domains are intimately attached on the CB surface with higher surface coverage. In other words, CSDPF 4000 has much higher silica surface coverage relative to CSDPF 2000 which is due to the silica domain distribution and high silicon content.

In terms of practical advantage, they exhibit reduced hysteresis while maintaining or increasing abrasion resistance and unlike silica, offer less abrasive to the processing equipments. Although from the chemical composition point of view the dual phase filler is between CB and silica, but interestingly CSDPFs give the lowest

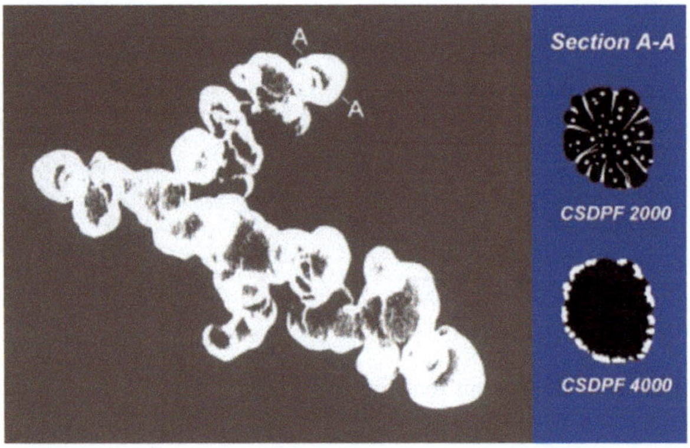


**Fig. 12** General view CSDPF: Variation in silica distribution in CSDPF 2000 and 4000 series [35]

Payne effect. A still lower Payne effect is observed when the coupling agent TESPT (bis(3-triethoxysilyl propyl)tetrasulfide) is employed. The better micro-dispersion of dual phase filler in relation to the other two conventional fillers (carbon black and silica) has been mainly attributed to the lower filler-filler interaction. In the tyre sector, these dual phase fillers with suitable coupling agent shows low rolling resistance with better traction and skid resistance in tire tread compound. However, the compound cost is the major hindrance for the complete replacement of conventional carbon black usage.

### Physical Characteristics of Dual Phase Filler

1. Particle size: Dual phase filler consists of carbon and silica phases with primary particles ranging from 1 to 20 nm (particle size range of silica) and 20 to 50 nm (particle size range of N220 carbon black). Since dual phase filler consists of carbon and silica phase, the surface areas derived from both nitrogen ($N_2$) and iodine ($I_2$) adsorption methods can give the information about the surface area of individual phases. Nitrogen adsorbs onto both carbon and silica surfaces, whereas iodine does not adsorb onto the silica phase. Hence, the difference in $N_2$ and $I_2$ adsorption values can be roughly correlated to the surface area of individual phases. However, there are some limitations in this method as iodine number is influenced by surface impurities such as oxygen and sulphur groups and also pore size distribution.
2. Structure: The structure of dual phase fillers can be measured from the difference in the values of dibutyl phthalate adsorption. The higher value for dual phase fillers shows that these fillers are of high structure than conventional carbon black having similar surface area (N234). The structure of dual phase filler increases with increase in silica content. The pore size distributions calculated from argon adsorption isotherm by using density function theory (DFT) reveals that the pore size increases with an increase in silica content of dual phase fillers. The micropore volume for dual phase fillers is higher than conventional carbon black (N234).

### Chemical Characteristics of Dual Phase Filler

As mentioned previously, this filler consists of a silica phase finely distributed in the carbon phase along with 90–99 % elemental carbon, oxygen and hydrogen as in traditional carbon black. Depending on the silicon content and surface area, dual phase fillers contain substantially more functional groups available to react with organosilane coupling agents than carbon black. The active hydrogen content of the dual phase fillers as measured by Lithium aluminium hydride ($LiAlH_4$) are higher than that for carbon black, and increase with increasing silicon content. The pH of the dual phase fillers are more acidic than carbon black, and decrease with increasing silanol content. ESCA and FT-IR experiments demonstrate that the carbon–silica dual phase fillers are comprised of composite aggregates, each containing carbon and silica phases. In these dual phase aggregates, the silica phase is intimately distributed within the carbon phase. Quantitative ESCA reveals that these dual phase fillers have excess oxygen over the expected stoichiometric amount for the silica present. This excess oxygen is likely due to oxygen groups on the carbon phase.

### Nano Fillers

Nanofiller is a class of new-generation fillers, which have at least one characteristic length scale in the order of nanometer with varying shapes ranging from isotropic to highly anisotropic needle-like or sheet-like elements. A nanocomposite is defined as the composite of two materials, one having the dimension of nanometric level at least in one dimension. In polymer nanocomposites (PNC), the fillers are dispersed on a nanolevel. Uniform dispersion of these nanosized particles can lead to better interfacial contact (interfacial area) between a polymer and the fillers. This large interfacial area between the filler and a polymer and the nanoscopic dimension differentiate nanocomposites from traditional composites. The major characteristics, which control the performance of nanocomposites, are nanoscale-confined matrix polymer, nanoscale inorganic and organic fillers, and nanoscale arrangements of these constituents. In 1980, the Toyota Research Group of Japan invented nylon 6-based nanocomposite with smectite-group clay particles dispersed as fillers in the matrix by in situ polymerization method. This invention explored an entirely new area of research in polymer composite field. These clays having a characteristic structure that can form nanocomposites by the breakdown of the clay layers in nanometer range. Afterwards, various nanocomposites have been prepared based on various polymers and nanofillers having different structures. Nano-dispersion of these fillers enhances the degree of improvement in properties [36, 37].

The principle advantages of nanocomposites over conventional composites are:

1. Lighter weight due to low filler loading,
2. Affordable cost due to lesser amount of filler used,
3. Improved properties compared to these conventional filler-based composites at very low loading of filler,
4. Combination of specific properties (barrier, adhesion, flammability, etc.).

### Classifications of Nano-Composites

Depending on the nature and type of nanofiller used, the nanocomposites can be divided into subclasses.

1. Clay-based nanocomposites
2. Silica-based nanocomposites
3. Polyhedral oligomeric silsesquioxane (POSS)-based nanocomposites
4. Carbon nanotube-filled nanocomposites and
5. Nanocomposites based on other nanofillers like metal oxides, hydroxides, and carbonates.

Different clays having different structures and compositions give different types of nanocomposites.

Types of Nanofillers

Depending on the innate structure, nanofillers are classified as:

1. Layered carbon and graphite are nanofillers having platelet structure.
2. Carbon nanotubes and nanofibers owing to their very high aspect ratio show a directional property and they are also conductive in nature.
3. Silica is globular. When their loading is higher, they form clusters. Nanoclays and graphite have platelet-like structure. The layers of these nanofillers break down in a polymer matrix as the polymer gets into the layers more and more. The typical thickness of a single layer is $\sim$1 nm.
4. Aluminosilicates have two-dimensional giant structures, the aspect ratio in this case is very high.
5. Graphite having layered structure can also be intercalated by polymers and can form nanocomposites by exfoliation of the layers.

Comprehensive review of the various types of nanocomposites was given by several authors [36, 37]. Detail description of the micro and nano rubber composites is dealt with in later chapters.

## 2.4 Age Resistors

Most unsaturated rubbers like NR, SBR, NBR and BR etc. are more prone to oxidation and ozonation during service owing to presence of double bonds in the main polymeric backbone. Hence to extend the service life of the rubber compound, it is essential to added age resistors. Age resistors are often further classified as antioxidants and antiozonants, designated accordingly in terms of the protective nature on the vulcanized rubber compound to which they are added. Antioxidants are used to protect rubbers from the effects of thermal oxidation whereas antiozonants reduce the influence of ozone attack on rubber compounds. The vast majority of compounds will contain one or more of the age resistors. Retardation of oxidative degeneration and the effects of ozone attack can be mitigated but not totally overcome by the use of chemicals. It is unfortunate that most effective age resistors will carry the penalty of causing staining of the rubber compound on the surfaces [38].

### 2.4.1 Antioxidants

Oxidative ageing of rubbers is limited by the rate of diffusion of oxygen into the rubber product and is usually confined to the outer or exposed region. Oxidative degradation of the polymers is a free radical process. This oxidation process, known as auto-oxidation, consists of three steps: initiation, propagation, and termination, as depicted in Fig. 13. Antioxidants inhibit auto-oxidation by reducing the rate of auto-oxidation during processing, storage, and service. There are two major groups of antioxidants, commonly known as primary and secondary antioxidants. Primary antioxidants act as chain terminators, and secondary antioxidants act as hydroperoxide decomposers. The primary antioxidants remove the chain-carrying species (R* and ROO*), while the secondary antioxidants convert the hydroperoxides to non-radical species. Commercially available antioxidants can be divided into three categories: phenolic antioxidants, aromatic amine antioxidants, and hydroperoxide-decomposing antioxidants. Each antioxidant performs in a specific way to protect polymers and rubber compounds from oxidation. The performance of each antioxidant is related to its chemical reactivity, rate of staining or migration to the surface of vulcanized rubber and volatility.

Phenolic: The phenolic antioxidants are less staining but are also less effective for heat aging and do not confer protection against ozone attack. These

Initiation

$$RH \xrightarrow{\text{energy}} R^{\bullet}$$

$$2RH + O_2 \longrightarrow [RH\cdots O_2] + RH \longrightarrow 2\,R^{\bullet} + H_2O_2$$

$$ROOH \longrightarrow RO^{\bullet} + {}^{\bullet}OH$$

$$2\,ROOH \longrightarrow RO^{\bullet} + ROO^{\bullet} + H_2O$$

Propagation

$$R^{\bullet} + O_2 \longrightarrow ROO^{\bullet}$$

$$ROO^{\bullet} + RH \longrightarrow ROOH + R^{\bullet}$$

Termination

$$2R^{\bullet} \longrightarrow \text{Nonradical products}$$

$$R^{\bullet} + ROO^{\bullet} \longrightarrow \text{Nonradical products}$$

$$2ROO^{\bullet} \longrightarrow \text{Nonradical products}$$

**Fig. 13** Bolland Mechanism of polymer degradation by oxidation mechanism [8]

antioxidantas are not as powerful as some staining antioxidants; nevertheless they are important class of antidegradants due to their non-staining and non-discolouring nature, mainly used in white and coloured vulcanizates. This group includes monophenols, bisphenols, thiobisphenols, polyphenols and hydroquinone derivatives. The most widely used phenolic antioxidant is 2,6-di-*t*-butyl-*p*-cresol, frequently referred to as BHT [38].

Amine: Amine antioxidants in general are better antioxidants than phenolic antioxidants. However, most amine antioxidants are discoloring and staining and have limited approval for food contact use. The materials can be categorised as: naphthylamines, diphenyl amine derivatives, dihydroquinolines and paraphenylenediamines. Most commonly used secondary amine is polymerized 1,2-dihydro-2,2,4-trimethylquinoline (TMQ). An antioxidant mechanistic of TMQ with the resonance structures and the transformation products is shown in Fig. 14.

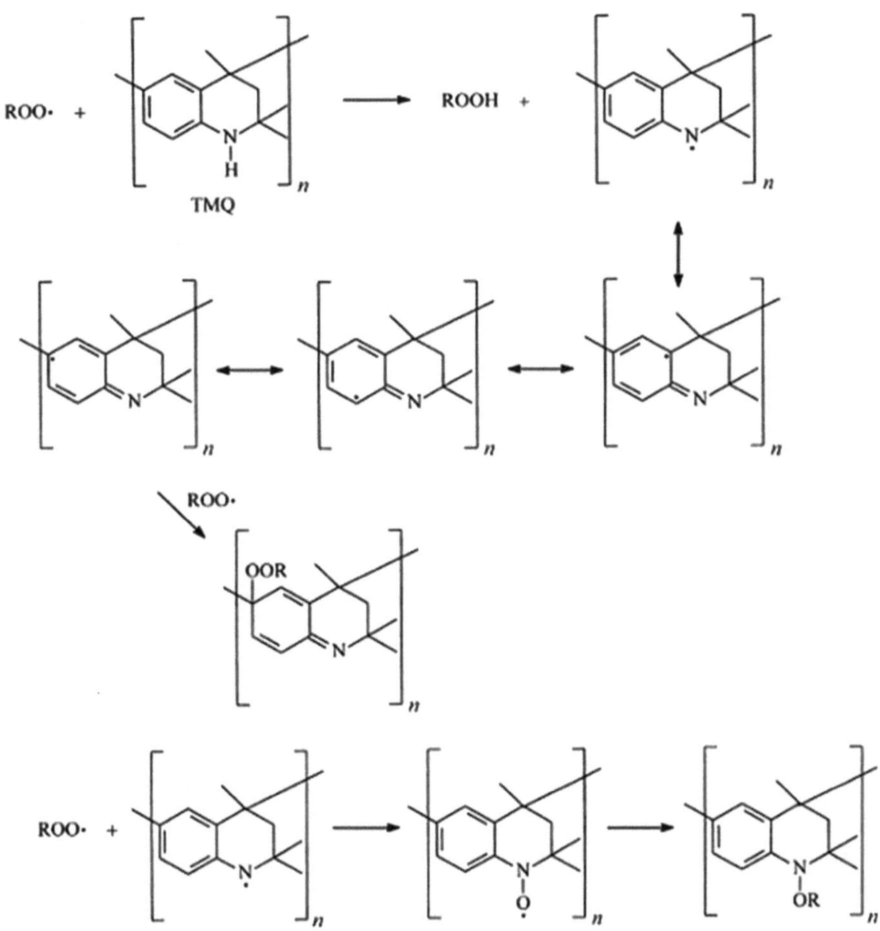


**Fig. 14** Antioxidation mechanism of 1,2-dihydro-2,2,4-trimethylquinoline (TMQ) [8]

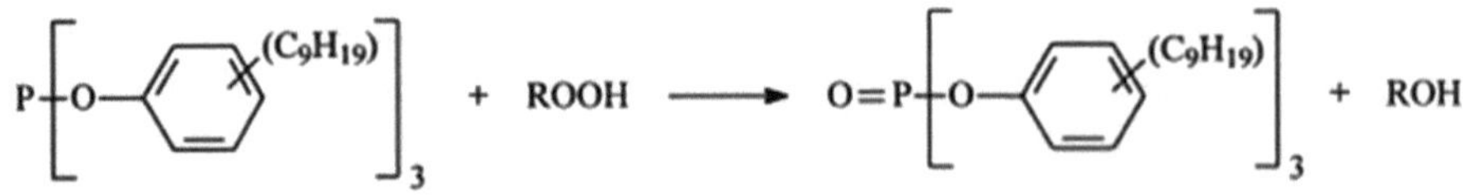

**Fig. 15** Antioxidation mechanism of tris (nonylphenyl) phosphite, (TNP) [8]

Hydroperoxide decomposer: Hydroperoxide-decomposing antioxidants reduce the rate of chain initiation by converting hydroperoxide, ROOH, into non-radical products. Two major classes of the hydroperoxide-decomposing antioxidants are organic phosphite esters and sulphides. The most commonly used material of phosphate class is tris(nonylphenyl)phosphite (TNP). However, it suffers from a tendency to hydrolyse during prolonged contact with moisture and it is destroyed during vulcanization. Its hydroperoxide-decomposing mechanism is depicted in Fig. 15.

### 2.4.2 Antiozonants

Ozone attack occurs mainly at the olefinic double bond of a unsaturated (diene) rubber and, if not protected against, will result in loss of physical integrity for thin sectioned articles and surface cracking on thick sectioned products [4, 5].

There are two principal methods by which ozone attack can be prevented:

1. Having an impermeable layer on the rubber surface
2. Incorporating a chemical antiozonant to hinder the auto ozonation reaction.

The first method involves the addition of waxes (microcrystalline wax, paraffin wax, etc.,) as a compounding ingredient, which in service will bloom on the rubber surface. This layer will act as physical barrier for the ozone to attack the unsaturation site. Two important properties are intended to be important in case of selection of waxes: rate of migration and melting point. Above the melting point of the wax, its protective action will nullify and the rate of migration determines the how fast the layer is forming. It is a common practice to use the blends of waxes to provide a protection against wide range of temperature. In addition to that, most widely used antiozonants are the substituted *para*-phenylenediamines. They not only offer protection against ozone but also function as antioxidants and as outstanding antiflex agents. One major drawback is that they are discolouring and staining. They can be conveniently grouped into three categories based on the chemical nature of substituents: diaryl-*p*-phenylenediamines (DAPD), dialkyl-*p*-pheylenediamines (77PD) and alkyl-aryl-*p*-phenylenediamines (IPPD, 6PPD), whose structures are shown in Fig. 16.

The second method is the addition of chemical ingredients which will hinder the reaction path of ozonation. Benzofuran derivatives and enolethers are offered as chemical antiozonants for light coloured rubber compounds. Benzofuran derivatives are used in CR and its blends with other rubbers and give ageing

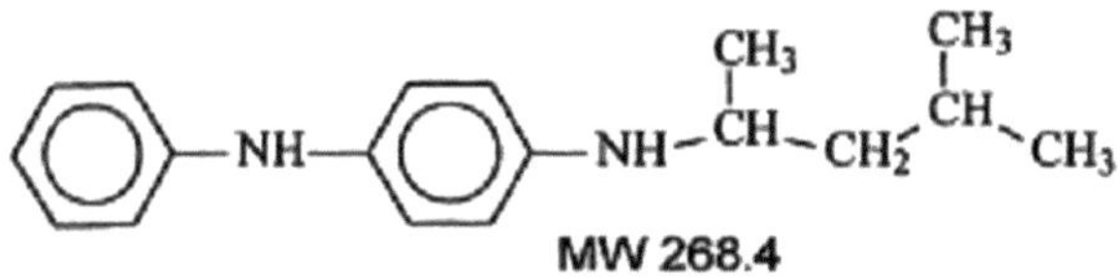


*N*-(1,3-Dimethylbutyl)-*N*'-phenyl-*p*-phenylenediamene (6PPD)

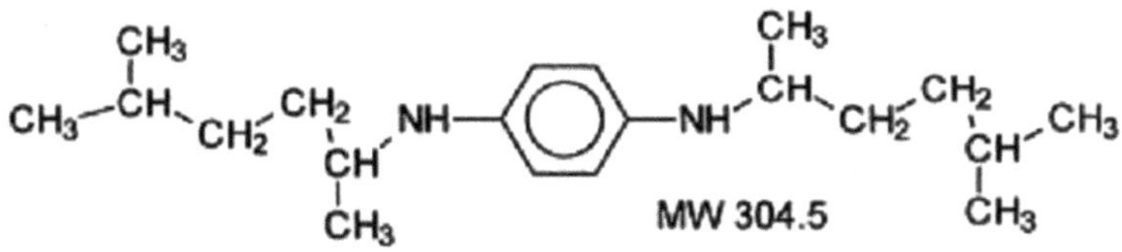


*N*,*N*'-Bis-(1,4-dimethylpentyl)-*p*-phenylenediamine (77PD)

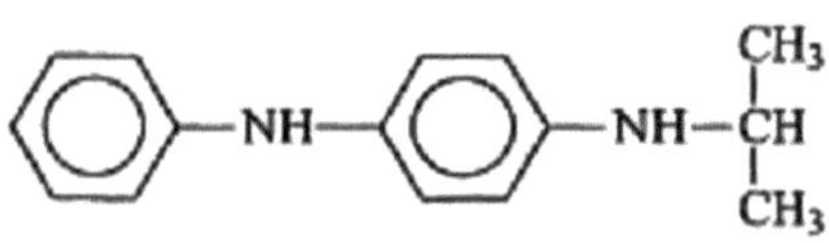

*N*-Isopropyl-*N*'-phenyl-*p*-phenylenediamine (IPPD)

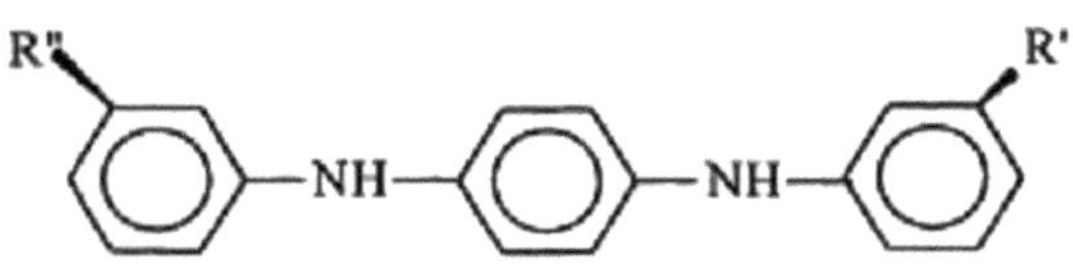

*N*, *N*'-Phenyl-*p*-phenylenediamine (DAPD)

**Fig. 16** Various commercially available antiozonants [4]

protection in addition to ozone protection. The enolethers give ozone protection but not ageing protection with NR, IR, SBR, BR, but are less effective in NBR. Enolethers act synergistically with microcrystalline waxes. Triazine derivative such as (*N*-1,4-dimethylpentyl-*p*-phenylenediamino)-1,3,5-triazine, although an active antiozonant is not used extensively in the rubber industry due to its limited solubility in synthetic rubbers and its high cost. It functions also as an antioxidant and is non-staining.

## 2.5 Processing Aids

There are numerous additives on the market that can have an appreciable influence on processibility [39]. Table 4 gives the overview of different process additives

Table 4 Commonly used processing aids [8]

| Processing aids | Action | Examples |
|---|---|---|
| Peptizer | Reduces the polymer viscosity by chain scission | 2,2′-Dibenzamidodiphenyl-disulfide, Zinc salt of pentachlorothiophenol |
| Dispersing agent and Lubricating agnet | Improve filler dispersion<br>Reduces mixing time and energy<br>Improves compound flow and release | Mineral oils, fatty acid esters, fatty alcohols, metal soaps, fatty acids |
| Homogenizing agent | Improve the compatibility between polymer blend | Resin and its blends |
| Stiffening agent | Increase hardness | High styrene rubber, pheonolic resin |
| Softening agent | Lowers hardness | Mineral oils |
| Mould release agent | Easy product release from mould | Metal soaps, fatty acid amides |

used in the field of rubber technology. Although various processing additives are used, oil, tackifier and plasticizers are the most common and crucial ingredients to improve either processing or product properties.

### 2.5.1 Mineral Oils

Mineral oils are high boiling fractions obtained in refining crude oil. Mineral oils serve three major functions:

1. Improve processing during mixing, milling, and extruding
2. Modify the physical properties of the rubber
3. Reduce the cost of the rubber compound.

They are classified according to the predominant chemical structure as: aromatic, naphthenic and paraffinic and their basic properties are enumerated in Table 5. It is important that the process oil selected for a rubber compound should be compatible with the rubber. Poor compatibility can cause bleeding of the oil, poor distribution of pigment and poor physical properties [8–10]. Good compatibility results in more efficient mixing, better cure characteristics and improved physical properties. In order to have good compatibility, the oil and the rubber need to have similar molecular units as well as optimized viscosity and molecular weight levels. Aromatic oils are very compatible with SBR because the aromatic molecules and the phenyl group on the SBR backbone are both polar. Paraffinic oils, which are the least polar, are more compatible with rubber such as EPDM, which has a non-polar backbone. As a general guide, naphthenic oils are preferred in EPDM, CR, SBR, BR, and IIR compounds. In addition to the compatibility, stability also plays a significant role in the final product performance. The most stable oils are the paraffinics because of their ability to resist oxidative attack. The

**Table 5** Properties of different oils used in rubber industry [8]

| Property | Paraffinic | Naphthenic | Aromatic |
|---|---|---|---|
| Specific gravity (ASTM D1250) | 0.85–0.89 | 0.91–0.94 | 0.95–1.00 |
| Pour point (°C) (ASTM D97) | −18 to −9 | −40 to −18 | +0 to +32 |
| Refractive index (ASTM D1747) | 1.48 | 1.51 | 1.55 |
| Aniline point (°C) (ASTM D611) | 95–127 | 65-105 | 35–65 |
| Molecular weight (ASTM D2502) | 320–650 | 300–460 | 300–700 |
| Aromatic content (%) (ASTM D2007) | 19–30 | 20–40 | 65–85 |
| Chemical groups | High level of isoparaffinic molecules | High level of saturated rings | High level of unsaturated single- and multiple-ring compounds |
| Oxidative stability | Good | Moderate | Lower |
| Odour | Lower | Moderate | Higher |

oxidative instability of oil is related to the presence of polar structures, such as nitrogen and sulphur heterocyclic structures. The higher the aromaticity of oil, the worse will be its resistance to oxidation.

Addition of oil is an unavoidable part for the tire manufacturers. Basically, mixture of aromatic, naphthanic, paraffinic, polycyclic aromatic (PCA) or poly-aromatic hydrocarbon (PAH) materials are used in practice. Mostly, 75 % of extender oils are used in the tread, subtread, and shoulder; 10–15 % in the side-wall; approximately 5 % in the inner liner; and less than 10 % in the remaining parts for a typical PCR tire. In total, one passenger tire can contain up to 700 g of oil. But, due to environmental issues tyre industries have major concerns in the usage of nitrosamine (TRGS-522) and poly aromatic hydrocarbons (PAH) free containing materials in rubber compounds. Oils contain high levels of PAH are identified to cause tumours in various mice skin painting tests. A lot of studies since 1970 have been done that proved the carcinogenic potential of PCAs. Therefore European Union (EU) with directive 2005/69/EC prohibits the usage of PAH-rich tyre extender oils or tyres containing these oils on the EU market from January 1, 2010. The directive defines that the extender oils allowed after the directive steps in force will have to meet two criteria. The content of total ECHA defined poly aromatic hydrocarbons (PAH) cannot exceed 10 ppm, which is assumed to be met if the DMSO extractable amount in the oil is below 3 % by method IP346. Tyres will be checked with method ISO 21461, which utilizes NMR and measures bay area protons in polycyclic aromatic compounds (PCA). It analyzes an extract from the tyre compound. The result correlates to the amount of PCA in the extender oil. A maximum limit of 0.35 % has to be met for the tyre to be approved for marketing in the EU [40].

Keeping the toxicity index as concern, petroleum oil companies involved in the development of different extender oils which have very low PCA (poly cyclic aromatic) content particularly for tire industry. These nontoxic oils are commonly Distillate Aromatic Extract (DAE), Treated Distillate Aromatic Extract (TDAE), and Mild Extraction Solvates (MESs). Existing extender oils can be replaced by Residual Aromatic Extract (RAE), Hydrogenated Naphthenic Oils (HNAP), and blends of highly immobile asphaltenic hydrocarbons. Low-molecular weight polymer, which acts as a process aid during mixing, processing, and gets cured along with the rubber and offers better performance, is a potential candidate that can replace the conventional process oil. Use of vegetable oils from different sources is an area where lot of developmental work is going on owing to toxicity and scarcity of mineral oil.

### 2.5.2 Tackifiers

Resins from natural as well as petroleum origin have been used as tackifier for many years to increase the tack of the uncured compound. It facilitates the building operation of the tyre compounds in terms of maintaining structural integrity of different components like sidewall, tread, carcass during manufacturing. Tack is defined as the ability of two uncured rubber compound surfaces to adhere together or resist separation after being in contact under moderate pressure for a short period of time. Two types of tack may be considered: autohesive tack, in which both materials are of the same chemical composition, and heterohesive tack, where the materials have different compositions [8]. Tack properties must be optimized; too high a tack value will cause difficulties in positioning the components during the building operation and may lead to trapped air between tire parts, giving after-cure defects. A factor inherent in tack is compound green strength, the resistance to deformation and fracture of a rubber stock in the uncured state. Tackifiers are products that may occasionally act as homogenizing agents. They comprise rosin, coumarone-indene resins, alkylphenol-acetylene, and alkylphenol-aldehyde resins. Other hydrocarbon resins such as petroleum resins, terpene resins, asphalt, and bitumen can also be included, although their effectiveness is generally not very high. The solubility of any resin in the rubber system is an important factor in determining the effect upon the properties of the rubber compound.

### 2.5.3 Plasticiziers

As a processing and property modifier, the term plasticizer is most commonly referred for the synthetic liquids or ingredients added for the purpose of depressing the second order transition (glass transition temperature; $T_g$). The static modulus and tensile strength are lowered in most cases, and correspondingly a higher elongation at break and better low temperature flexibility result. A variety of plasticizers with molecular structures containing polar groups are used, most of

**Table 6** Various elastomers and plasticizers according to polarity difference [8]

| Elastomer types | Plasticizers |
|---|---|
| NBR (high ACN) | Phosphates |
| AU, EU | Dialkylether aromatic esters, dialkylether diesters |
| NBR (medium ACN), ACM, AEM | Tricarboxylic esters, Polymeric plasticizers |
| CO,ECO | Polyglycol diesters |
| CSM,CR | Aromatic diesters, aromatic triesters |
| NBR (low ACN), HNBR, SBR,NR,BR | Aliphatic diesters |
| EPDM,EPR,IIR, | Alkyl monoesters |

these being ester based [41]. Compatibility plays an important role in the selection process of a plasticizer for a particular polymer for which it is used, otherwise bleeding will occur. Exudation of plasticizer in this way can be a problem in subsequent manufacture and leads to failure if it occurs at a rubber/metal interface during service [38]. A homogeneous and stable mixture of plasticizer and elastomer is achieved if their polarities are nearly the same. In any case, sufficient compatibility is required to achieve the processability and physical properties advantage. Liquid elastomers are also plasticizers that can undergo co-crosslink during vulcanization and cannot be extracted. The vulcanizate properties are insignificantly changed, but hysteresis tends to be slightly higher. Among the synthetic plasticizers, ester based are widely used in rubbers such as NBR, CR and CSM. For compatibility reasons they are mainly added in polar rubbers. Their main function is to modify properties rather than to improve processing. In many cases, they enhance low temperature flexibility and the elasticity of the vulcanizates. Table 6 summarizes the various plasticizers according to the polarity difference and the compatibility with the different elastomers. Flame retardant ester plasticizers (phosphate esters) are of relevant importance, because halogen-containing products, such as the chlorinated paraffins, are not generally permitted in use.

## *2.6 Miscellaneous Ingredients*

A large number of additional compounding ingredients are available, although not as widely used as those mentioned previously but have specific importance and usage for a particular property [1–4]. Some of them are discussed below:

### 2.6.1 Pre-Vulcanization Inhibitor

Vulcanization inhibitors have been used in rubber compounds for many years as a means of increasing processing safety. An efficient vulcanization inhibitors, increase scorch safety whilst having no adverse effect on the rate of vulcanization.

The major and most widely used *N*-(cyclohexylthio)phthalimide (CTP), often termed pre-vulcanization inhibitor (PVI). It is effective with a wide range of polymers, accelerators and other compounding ingredients. It neither affects vulcanisate properties nor causes staining or porosity. Although most effective in sulphenamide accelerated stocks, it is also used with both MBT and MBTS. In most applications 0.1–0.3 phr is sufficient to give remarkable scorch safety. There is a linear relationship between scorch resistance with the dosage of PVI. However, above 0.5 phr is not preferable in most applications. Furthermore the addition of PVI permits processing at elevated temperatures, thereby increasing productivity. In general PVI is not effective in thiuram based compound, resin and metal oxide vulcanization system.

More recently, prevulcanization inhibitors based on thioketals have been introduced:

- Bis-isopropylthio)acetoacetanilide: An effective retarder for mercaptobenzothiazole (MBT) but less effective for *n*-cyclohexyl-2-benzothiazole sulphenamide (CBS) and dibenzothiazole disulphide (MBTS).
- *N*-isopropylthio-*N*-cyclohexylbenzothiazyl-2-sulphenamide (iso-PCBS): Alternative to CTP for the control of sulphur/sulphenamide vulcanization reactions. Its ability to react with the amines liberated by the accelerator dissociation during the cure reaction and the autocatalyst MBT (also liberated) gives iso-PCBS a strong retardation capability. Byproducts are thioamines and isopropylthiobenzthiazole (iso-PBDT), the latter being a powerful retarder in its own right [38].

### 2.6.2 Blowing Agent

Blowing agents used in the rubber industry are capable of evolving relatively high volumes of gas when heated to a prescribed moulding temperature. They are of two main types: inorganic and organic. The inorganic blowing agents, usually ammonium carbonate or sodium bicarbonate, evolve carbon dioxide on decomposition. The organic agents mainly evolve nitrogen, but may also give off carbon dioxide, carbon monoxide, ammonia and water according to their chemical composition. The choice of particle size of the inorganic blowing agent dictates to a large extent the size of the pores of the resulting sponge, the larger particle types giving larger, more irregularly sized cell geometry. Sodium bicarbonate is the primary blowing agent, it is common practice to use with a weak acid, such as stearic or oleic acid, whose function is to trigger the reaction and assist in the uniform decomposition of the bicarbonate. The organic blowing agents, usually Azodicarbonamide (ADC), p,p′-Oxybis(benzenesulphonyl hydrazide) (OBSH), Dinitroso pentamethylene tetramine (DNPT) usually decomposes to give gaseous products. ADC is activated by zinc oxide, zinc stearate (strongly) and urea (slowly). Barium stearate, calcium stearate and triethanolamine, when added at 10 phr, moderately activate gas evolution of ADC. DNPT is flammable, so it must be stored carefully to avoid contact with heat sources. It must also be kept dry as moisture reduces its activity.

### 2.6.3 Antireversion Agent

Reversion characteristics of rubber compounds are of great concern. Lot of novel chemicals has been introduced to increase the reversion resistance of the rubber compounds. Figure 17 shows some of the commonly used reversion resistance chemicals. These additives are mainly intended to reduce the reversion characteristics of the sulphur cured vulcanizates. In general reversion resistance is a result of the formation of a crosslink network containing a higher proportion of monosulphidic crosslinks. Blends of aliphatic and aromatic zinc carboxylic an acid salt is commercially available that provides reversion resistance by promoting the formation of sulphidic crosslinks of lower sulphur rank. The most common and well known is bis-(3-triethoxysilylpropyl)-tetrasulphide (TESPT), interestingly it acts as coupling agent for silica filled vulcanizates and sulphur donors during vulcanization. 1,3-bis(citraconimidomethyl)benzene (Perkalink 900®), functions by a crosslink compensation mechanism. It compensates for the loss of polysulphidic crosslinks during the process of reversion with crosslinks based essentially on a carbon–carbon structure [42]. The material reacts at the onset of reversion and can be added to existing compounds with no further change in formulation. Duralink HTS and Vulcuren KA 91 88 are commonly used as post vulcanization stabilizers. These materials will enhance the life of the product, enable the users for more retreading for tire tread compound, and thereby reduce the material demand.

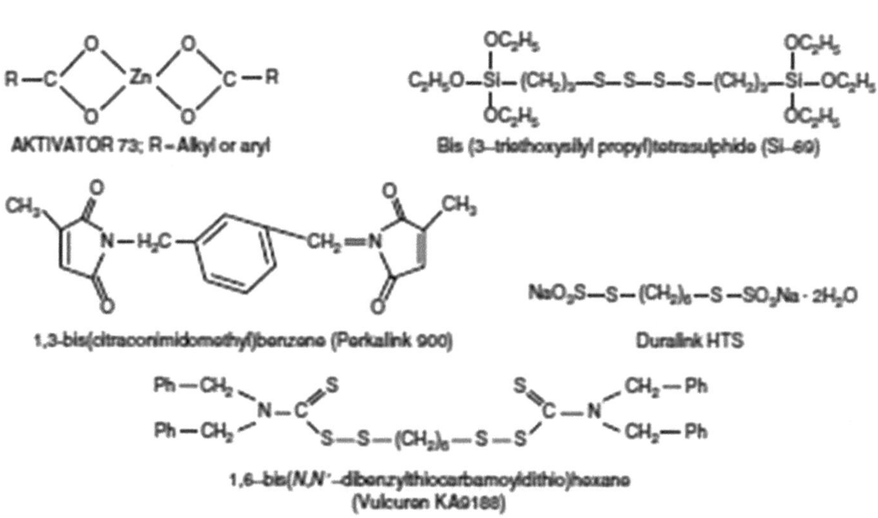


**Fig. 17** Chemical structures of commonly used anti-reversion agents [36]

### 2.6.4 Flame Retardant

Flame retardants (FRs) are the chemicals that are incorporated in organic polymers (both natural and synthetic) by blending in compounding stage or insitu preparation stage. About 350 different FR substances (both organic and inorganic) are described in literature and are available in the market, with no specific reference to fire regulations. But the FRs most abundantly used at the present time is based largely on six elements: chlorine, bromine, phosphorus, aluminum, boron, antimony. The hydrated alumina/aluminum hydroxide or aluminum trihydrate (ATH) claims as one of the superior FR additives due to its inherent nature of char formation with less toxic gases. Halogen containing FR materials are very important class of FR chemicals. The order of the thermal stability for the different halogen compounds is F > Cl > Br > I. However, now-a-days usage of halogen containing FR is restricted because of toxic gas emission and smoke generation. Phosphorus FRs include elemental red phosphorus, water-soluble inorganic phosphates, insoluble ammonium polyphosphate, organophosphates and phosphonates, phosphine oxides, and chloroaliphatic and bromoaromatic phosphates are also widely used in elastomer compounding. Zinc borate can be employed alone or synergistically with antimony trioxide, aluminium hydroxide and halogen-containing additives. Zinc borate enhances the char formation and can significantly reduce the smoke volume produced. It can also be used as the sole flame retardant or as a partial replacement for antimony trioxide. Zinc borate has a low toxicity rating. Due to their relatively low cost, ease of handling, and low toxicity, inorganic FR such as magnesium hydroxide ($Mg(OH)_2$) and Aluminum trihydrate (ATH) are gaining interest in recent years.

### 2.6.5 Biological Control

These ingredients are used to enhance the resistance to microorganisms such as fungi, bacteria, and other microbial agents for certain applications. Additives are available to help reduce or stop such attack. Such additives, however, cannot be used in applications that entail contact with potable water and food. Ethylene oxide (ETO) is the most commonly used for chemical sterilization. ETO chemically reacts with amino acids, proteins, and DNA to prevent microbial reproduction. Ethylene oxide can be used with a wide range of rubbers and plastics (e.g. petri dishes, pipettes, syringes, medical devices, etc.) and other materials without affecting their integrity.

### 2.6.6 Pigment

In order to establish a permanent colour for the compound it is necessary to use the correct ‘white’ filler, and to establish a good base white pigmentation as a foundation for the colour of the product. Both organic- and inorganic-based pigments can be used in the compounding of rubber. Organic pigments confer greater brightness but can change their hue (colour) upon heating; they are also more expensive. Inorganic pigments are less expensive and more heat stable, and more resistant to chemicals. Correct colour matching is a skill, and needs consideration of the effects of the ingredients of the compound on the final colour of the vulcanized compound. Some of the commonly used chemicals are recited with reference to the required colour in rubber compound:

White pigments—titanium dioxide, lithopone, zinc oxide etc.
Coloured pigments—Iron oxide (black), copper phthalocyanine (blue), chrome oxide/cobalt green/polychloro copper phthalocyanine (green), cadmium sulphide/selenide (orange), chrome rutile yellow/cadmium sulphide/zinc sulphide (yellow) etc.

## *2.7 Double Networks in Rubber Vulcanizates*

The concept of a double network was introduced more than 50 years ago by Tobolsky and co-workers. Double network rubber refers to an elastomer that has been crosslinked twice, the second time in a deformed state. These result in materials with unusual and enhanced properties which have been termed "double network elastomers". The deformation employed may be uniaxial tension, biaxial tension, torsional, bending etc. A general schematic representation of process involved in the preparation of the double networking of rubber is shown in Fig. 18. Double networks exhibit anisotropic properties, and are usually characterized in terms of their residual stretch ratio, equal to the ratio of their length along the stretch direction of the second cure to the initial length (prior to the second curing, or equivalently, in the uncured state). However, stretch ratio does not uniquely define a double network, as different cure strains and crosslink apportionment between the two networks can yield the same residual strain, but different mechanical properties.

From the kinetic theory, the elasticity of rubber has been attributed to the changes in the conformations and configurational entropy of a system of long-chain molecules. When the chains undergo deformation, the internal energy is considered to remain constant. Hence, when loading and unloading the network, the heat exchange with the surroundings is mainly by entropic contributions. Heat is given out from the elastomer to the surroundings while loading, and is absorbed while unloading. This heat exchange governs the mechanical and thermal properties of these networks. If a partially cured elastomer is first deformed and then subjected to additional crosslinking in the deformed state, a second crosslinked network can be formed within the initial network, as already depicted in Fig. 18. The resulting properties arise from a competition between these two networks during small deformations, where heat released by one

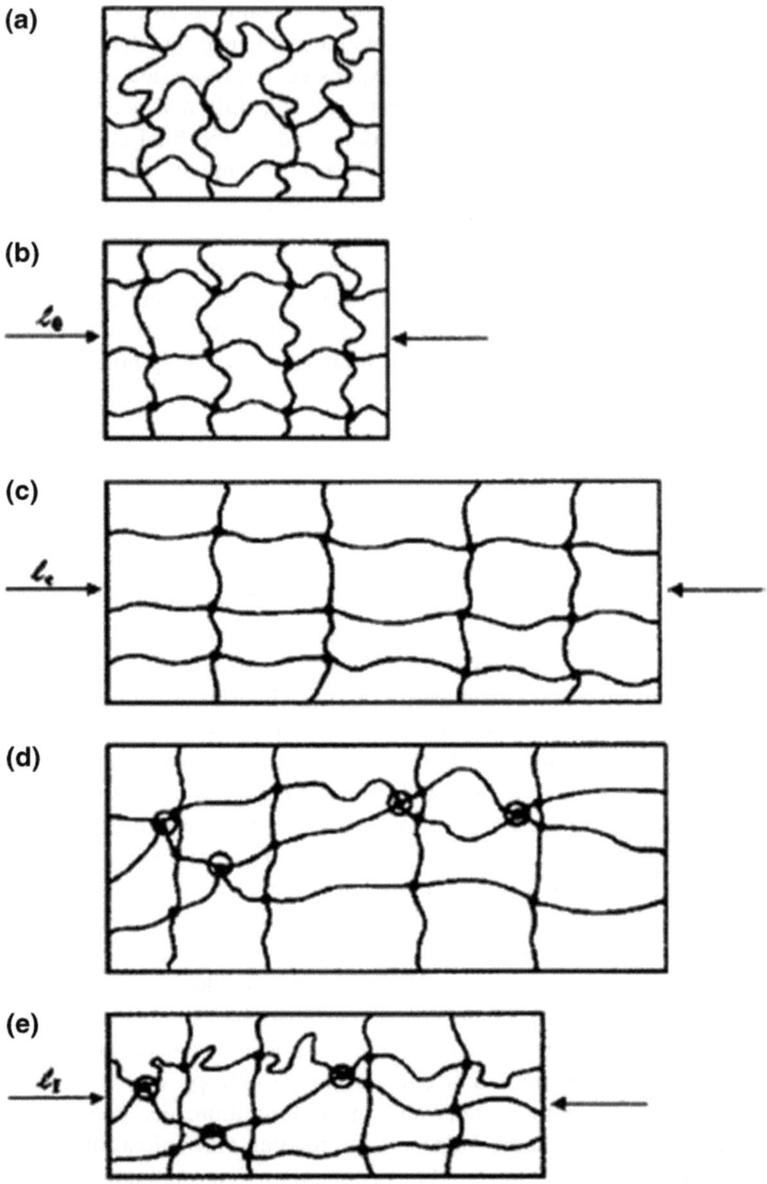


**Fig. 18** Formation of double network rubber vulcanizate [45]. **a** Uncrosslinked network; **b** initial (first) crosslinked network; **c** extended state (uniaxial extension); **d** second crosslinked network; **e** relaxed state after double network formation [45]

network is being absorbed by the other network. Thus, a lower modulus is expected in this low strain, competitive regime relative to a single network system. At higher elongations these networks work in parallel to each other to provide a collaborative behavior, and consequently a higher modulus than in a single network system is expected.

Roland and co-workers [43, 44] showed that gum natural rubber double network crystallizes at lower strain and can have enhanced tensile strength and fatigue life compared to that of conventional single networks. Rubber vulcanizates with permanent chain orientation can be produced by preparing double networks or composite networks. They can be viewed as interpenetrating networks in which the same chain segments belong to both networks and, more important, the component networks are oriented. It is this orientation that gives rise to enhancement and anisotropy in properties. The properties of an elastomeric network depend not only on the density of junctions but also on the distribution and orientation of the chains when the junctions are formed. Double networks can also arise spontaneously by chain scission, by strain-induced crystallization or in the presence of reinforcing fillers [45].

Orientation has a significant effect on the mechanical response of elastomers, especially rubbers which undergo strain induced crystallization. Although this orientation is usually not stable in a flexible chain polymer above its glass transition temperature, permanent orientation can be achieved via a "double network". After an initial, isotropic crosslinking, the rubber is stretched and crosslinked a second time, which leads to anisotropic nature. Control of the orientation and crosslink apportionment yields higher modulus. For strain-crystallizing rubbers, the benefits are even greater, substantial improvements in crack-growth and fatigue performance can be realized.

## *2.8 Industrial Compounding and Vulcanization Techniques*

### 2.8.1 Compounding

Mixing or compounding as a general operation may be considered as four basic processes occurring simultaneously [46].

1. Incorporation—Subdivision of filler and coherent mass formation
2. Dispersion—Fracture and particle size
3. Distribution—Homogenization
4. Plasticization—To achieve the correct viscosity.

Rubbers are high molecular weight species; they are innate with high viscosity. In order to achieve a good dispersion of ingredients, viscosity of the rubber should be reduced. Thomas Hancock in 1819 was the first to introduce the mastication of gum rubber by the 'Pickle' method. It was purely a mechanical operation by breaking its nerviness. Advantages of mastication are manifolds:

1. Decreases viscosity of rubber
2. Promotes good incorporation and dispersion of fillers and chemicals

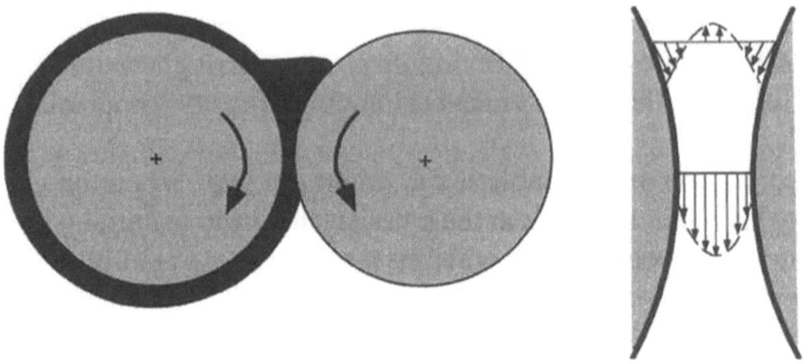

**Fig. 19** Two roll mill

3. Increases compound plasticity and building tack
4. Improves die swell and calender shrinkage.

Mixing or compounding is accomplished by two roll mill or internal mixer or continuous compounder to prepare it for final vulcanization step. The following paragraphs brief the conventional equipments for compounding.

### Batch Compounding

#### *Two Roll Mill*

The two roll mill represents the earliest form of rubber processing machine, used by the rubber manufacturer being developed from the masticator of Hancock. The roll mill (Fig. 19) consists of two rolls placed horizontally next to one another, which are peripherally drilled for heating or cooling. In general, a so-called stock blender is placed above these rolls, and this not only serves the purpose of cooling the sheets, but has an additional homogenizing effect due to the steady reversing of rubber sheet on the mill. The sheet, in its full width, is continuously turned between two rotating coils. The raw rubber is masticated and the ingredients are added step-by-step during the band formation by adjusting the nip gap. In the modern factory conventional two roll mills are used for compound blending and for 'sheeting off' of compound mixed by other means [38]. Mill mixing is an efficient method and widely used for breaking down agglomerates and to give good homogeneity. Two important variables are found to be more important: nip gap and friction ratio.

### Internal Mixer (Banbury and Kneader)

Fernley H. Banbury was the first to introduce the internal mixing of rubbers using intermeshing rotors. Since then internal mixer gained considerable importance in terms of economy, security, service, and ecology. It has thus displaced the open two roll mixing for most compounding operation, but to a limited level. The various parts of the Banbury mixer are shown in the Fig. 20. Raw rubber is dropped through the hopper to the mixing chamber, where it is fitted with the rotors. The ram or press is forced down either by pneumatic or hydraulic controlled cylinder, whose pressure is adjusted to control the state and mode of mixing process. The mixed or compounded rubber is discharged from the machine through a discharge door at the bottom of the mixer [10, 12]. Effective mixing requires both dispersive and distributive processes to take place during the mixing cycle. To achieve that, control of temperature in the mixing chamber, rotor design and its speed and ram pressure are to be kept control, to improve the quality of mixing.

## Continuous Compounding

In the era of development of compounding rubbers, machines were developed which would carry out the function of continuously mixing rubbers and filler powders. Extruders with modified barrels and screws have been used to carry out the function of a continuous mixer of rubber compounds. The so called "Multi-Cut Transfermix" as shown in Fig. 21, gives a more concentrated mixing and plasticising of the compound in only one transfer zone, avoiding the lengthening of extruders in response to increasing demands of processing [46].

**Fig. 20** Batch internal mixer

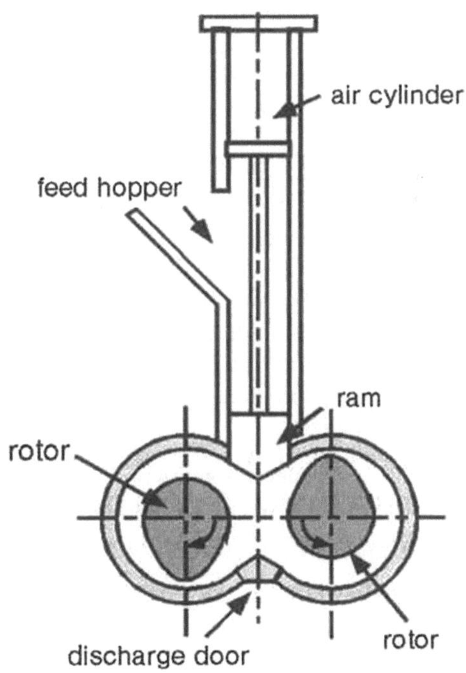

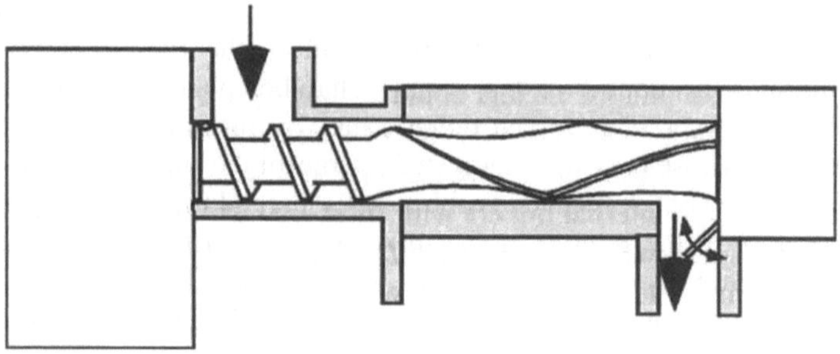

**Fig. 21** Continuous internal mixer

### 2.8.2 Vulcanization Techniques

Following the process of compounding, the compound, made of uncured rubber and of curing agents as well as of additives, is heated in a mould up to a temperature at which the reaction starts, the shape of the final material being given by the mould. During the cure, an irreversible reaction takes place, leading to a three-dimensional molecular network, and the plastic material is converted into an elastic one. Various vulcanization techniques are discussed below:

#### Batch Vulcanization Techniques

In the batch vulcanization process, vulcanization step is done in batch wise in single cavity or multicavities mould. The following paragraphs will brief about the various techniques of batch vulcanization process.

#### Compression Moulding

Compression moulding is the most fundamental method for moulding rubber. It is done by placing a cut-to-weight quantity of uncured rubber into the cavities of an open mould. The mould is then closed, as far as possible, and placed between the hot platen of the hydraulic press. Pressure from the press platen causes the uncured rubber to deform into the shape of cavity or cavities, as heat from the press platens vulcanize the product.

Advantages:

Moulds are generally simple—Inexpensive moulding
Less compound waste—there is no transfer pot
Excellent for low production runs and bulky parts.

Disadvantages:

Longer curing cycle
Labour intensive process.

## Transfer Moulding

Transfer moulding is an extension of compression moulding optimized for higher-volume production runs. Unlike compression moulding, transfer moulding uses a transfer pot with top platen, in addition to the product cavity plates, to push a quantity of uncured rubber into the product cavities. In its most basic form, a transfer mould consists of a top plate, which acts as a plunger that pushes uncured rubber into the cavities. In operation, uncured rubber is placed into the transfer pot and the mould is closed. When pressure is applied by a hydraulic press, the plunger forces the uncured rubber in the transfer pot, through the sprues, into the product cavities where heat from the press vulcanizes the product [10].

Advantages:

Shorter production cycle
Suitable for complicated geometry products.

Disadvantages:

Expensive tooling
More compound waste.

## Injection Moulding

Injection moulding incorporates the automatic feeding, heating and plasticization of a rubber mix and its subsequent injection into a mould at a temperature as close as possible to the vulcanization temperature. A measured quantity of the mix is injected through a narrow orifice, by a high pressure ram into a tightly closed mould, in which vulcanization takes place. The clamping force of an injection press is what keeps the mold tightly closed during the injection and moulding cycle. The whole operation being carried out by a single machine specially designed for the purpose. Most of the injection moulding machines are designed according to the First-In, First-Out (FIFO) principle. The compound first plasticised will be the first material to be injected in order to maintain the thermal history of the entire shot volume.

Advantages:

Shorter production cycle due to plasticization followed by injection process
High clamping pressure on mould results to give very little or no flash products
Labour extensive process
Uniform curing of complicated parts.

Disadvantages:

Requires large production runs
Tooling cost considerably higher.

### Types of Injection Moulding Machines

Injection moulding machines can be in a horizontal or vertical mode, with or without tie bars, and with a C frame structure for easy mould fitting and access. They also differ in their mode of heating method of plasticisation and delivery of the mix to the mould cavity. Focusing on the method of delivery, they are further classified into ram and screw type. Simple ram machines are cheaper than screw machines having equivalent shot volumes, but not widely used [38]. A higher and more uniform compound temperature and thus a more uniform viscosity is achieved in screw type.

Most important screw machines can be subdivided into:

(a) Reciprocating screw (screw-ram or screw-plunger) machine in which the screw itself acts as a ram
(b) Screw-machine-with-separate-ram' or 'pre-plasticising screw' or 'extruder-plunger' type in which the screw extrudes into a separate chamber and a separate ram injects the compounded rubber into the mould cavity.

Furthermore, in the way to increase the production rate, various modifications were done on the processing techniques. Some of them are discussed below:

Sliding table: The standard machine is equipped with two-sided sliding tables, which use a common top plate and two identical bottom plates.

Shuttle table: The shuttle machine has two complete moulds attached to a forming station. The moulds into which rubber is injected are heated up to about two-thirds of the cure time in the central unit. The station is then opened (but not the mold), and the mold is shifted to the final forming station for the rest of the cure time. By using the shuttle table, the cycle time can be reduced to about 50 %.

Multistation rotary table: The reciprocating screw type of injection unit is used for injection the same rubber compound into moulds in each station of the rotary table. This setup features a station each for unloading product, for cleaning molds and loading inserts.

Multiple injection units: They are mainly used to make products that require two different colors or compounds of different hardness

### Reaction Injection Moulding

Reaction Injection Moulding (RIM) involves the rapid mixing of two or more highly reactive low molecular weight compounds before the injection of the mixture in a closed mould. Injection moulding of silicone and polyurethane rubbers differs, because the material mixed and injected is usually liquid. RIM is important for urethane rubbers because the process involves the mixing of diisocyanae and diol (usually in the form of liquids) and is very energy efficient.

### 2.8.3 Continuous Vulcanization Techniques

In the continuous vulcanization process shaping, curing and finishing are done in a single or continuous operation [12, 38]. The basic components are extruder, vulcanizer and down streamer. The following are the some of the widely used techniques in the field of continuous vulcanization techniques:

#### Steam Tube

Steam tube vulcanization is widely employed in the cable industry. The cable after covering is passed through a steam tube connected to the extruder crosshead. The end of the tube connected to the extruder is fitted with a seal system. The cable is allowed to pass through the steam tube, where it gets vulcanized, and the vulcanized sheath emerges through a sealing system is cooled and then coiled. The length of the steam tube depends on the size of cable to be vulcanized, as it is necessary for full cure to be achieved before the cable emerges from the end distant from the extruder or the sealing system. Gas tubes are also available for dry curing.

#### Hot Air Tunnel

Vulcanization using hot air systems remains the most important production system for small cross section profile manufacturing. Three most important parameters play significant role: length of tunnel, temperature and speed of the hot air and time of cure which can be controlled by extrudate throughput transfer speed. The systems usually consist of modular units which can be built up into the required length. Air speed can usually be controlled and the compound throughput can be adjusted from speeds of 2.5 up to 20 m/s. These systems are sufficient for thin section profiles, but thicker articles will require slower transport speeds or additional curing device. Major disadvantage of the hot air tunnel is poor heat transfer and consequently requires a longer production line. Addition of infrared and microwave systems to these units can be employed to overcome the heat transfer problem.

#### Fluidized Bed

This technique uses the principle of fluidization of particles by the passage of air through the mass. In its simplest form the powder, usually small spherical glass beads (0.15–0.25 mm diameter) are kept in constant motion by a preheated gas introduced through porous tiles in the base of the bed. Accountable amount of the heat transfer problem can overcome by this process. The limitation of this process is that the glass beads often tend to stick to the rubber profiles. After removal of the vulcanized product it is necessary to remove the glass beads from the product surface. Ultrasonic cleaning units can be used for this purpose. It is also necessary to clean the glass beads from time to time to remove vulcanization products from their surface.

### Liquid Curing Method

An Liquid Curing Method (LCM) unit is an insulated tank filled with a heated liquid medium: eutectic salt, glycol, oil, or a low melt-metal alloy (bismuth/tin). The tanks were usually heated by electric heater and care should be taken for proper insulation. The heat transfer coefficient for LCM is higher than hot air and fluidized bed (2,770 kJ/m$^2$/h/°C vs. 70 kJ/m$^2$/h/°C for hot air and 1,480 kJ/m$^2$/h/°C for fluidized bed), which indicates rapid heat transfer to rubber. As a result, high line speed and short cure cycle are possible. Eutectic salt is by far the best, since it does not involve contamination and high cost. The operating temperature is generally 120–280 °C. Commonly, salt baths comprise of a stainless steel trough or tank and a hold-down device which also acts as a conveyor. The pressure exerted by the difference in specific gravity of the rubber compound and the salt medium create an upward pressure forcing the extrudate against the steel conveyor. This pressure can cause severe product distortion in soft and sponged products.

Generally four important factors are considered during compounding: (1) selection of a fast vulcanization system with adequate processing safety and reversion resistance, (2) ingredients should not liberate any toxic product and should be safe enough with the salt at higher temperature, (3) elimination of porosity, and (4) choice of fillers and process aids for good extrusion characteristics. During processing, accurate synchronization and optimization of the speed and temperature in the line is essential for control of product dimensions.

### Microwave Curing

Microwave was defined as electromagnetic spectrum having frequencies greater than 1,000 Mz ($10^9$ cycles/s). Current use includes frequencies from 100 to 3,000 MHz and wavelengths 0.1–10 cm in the microwave region. The frequency bands 915 and 2,450 MHz are used for industrial and domestic heating applications. Magnetrons of up to 6 kW can be used and their life expectancy is in the region of several thousand hours. The actual lifetime depends upon the dielectric characteristics of the compounds being processed, i.e., the more polar the compound, the longer the life of the magnetron. Non-polar rubbers do not absorb as much energy as polar rubbers and consequently unused energy gets reflected back. Some of the reflected energy is captured by special absorbers but some still reaches

**Table 7** Comparison of performance of conventional continuous vulcanization techniques [12]

| Property | Hot air | Fluid bed | Salt bath | Microwave |
|---|---|---|---|---|
| Heat transfer | Poor | Fair | Good | Excellent |
| Distortion | Fair | Good | Good | Excellent |
| Surface oxidation | Yes | Yes | No | Yes |
| Energy efficiency | Poor | Fair | Good | Excellent |
| Potential hazard | No | No | Fire | Radiation |

the magnetrons. This reflected energy reduces the life time of the magnetron. Hence nonpolar rubber materials must be compounded with dipolar and electrically semiconducting materials such as metal oxides and carbon black to compensate for the lack of dipoles in the rubber compound. In fact, the elastomeric compounds to be cured or preheated by microwave should be properly designed to provide adequate heating rate and uniform cure. Additionally, because of the reduced energy absorption the magnetrons are required to operate continually at their maximum output, which also affects their life-span.

The advantages of microwave heating are reduced tooling cost, reduced labour and floor space, cleanliness of operation, improved quality control, greater throughput, and chances for automation. Microwaving also eliminates such undesirable effects as cure-in-bend, water marks, and talc contamination in steam autoclave curing, and salt drag-out in molten salt bath curing. The disadvantages are difficulty in formulating a suitable compound (which must balance such factors as cost and machine characteristics), the delicate adjustment of UHF power required for a uniform cure, localized undercure and warping, and higher capital investment and maintenance cost. Microwave curing was mainly used in the vulcanization of automotive profiles, cables, tubes, small rubber parts through conveyer belt, and preheating of thick rubber profiles immediately before vulcanization, followed by curing tire treads. Table 7 summarizes the comparison of performance of different conventional vulcanization techniques [12].

## References

1. Cheremisinoff, P., Cheremisnoff, P.N.: Elastomer Technology Handbook. CRC Press, NJ (1993)
2. Hofmann, W.: Rubber Technology Handbook. Hanser Publications, Munich (1989)
3. Brydson, J.A.: Rubbery Materials and Their Compounds. Elsevier Applied Science, London (1988)
4. Barlow, F.W.: Rubber Compounding: Principles, Materials and Techniques, 2nd edn. Marcel Dekker, New York (1993)
5. Brydson, J.A.: Rubber Chemistry. Applied Science Publishers, London (1978)
6. Gent, A.N.: Engineering with Rubber-How to Design Rubber Components. Hanser, Munich (2001)
7. Blow, C.M.: Rubber Technology and Manufacture. Butterworths, London (1971)

8. Rodgers, B.: Rubber Compounding-Chemistry and Applications. Marcel Dekker, NewYork (2004)
9. Mark, J.E., Erman, B., Eirich, F.R.: Science and Technology of Rubber. Academic Press, NewYork (1978)
10. White, J.R., De, S.K.: Rubber Technoloist's Handbook, Rapra Technology, Shawbury (2001)
11. Datta, R.N.: Rubber Curing System, Rapra review reports, vol. 12(12), Shawbury (2002)
12. Bhowmick, A.K., Hall, M.M., Benary, H.A.: Rubber Products Manufacturing Technology. Marcel Dekker, CRC Press, New York (1994)
13. Loan, L.D.: Mechanism of peroxide vulcanization of elastomers. Rubber Chem. Technol. **40**(1), 149–176 (1967)
14. Akzo Nobel company brochure: Crosslinking Peroxide and Coagent: 1.XL.0.3.A/06-91
15. Dluzneski, P.R.: Peroxide vulcanization of elastomers. Rubber Chem. Technol. **74**(3), 451–492 (2001)
16. Costin, R.: Selection of coagents for use in peroxide cured elastomers. Application Bulletin 5519, Sartomer Company, USA (2004)
17. Dikland, H.G.: Coagents in peroxide vulcanization of EP(D)M rubber. PhD Thesis, University of Twente, The Netherlands (1992)
18. Dikland, H.G., Ruardy, T., Van der Does, L., Bantjes, A.: New coagents in peroxide vulcanization of EPM. Rubber Chem. Technol. **66**(5), 693–711 (1993)
19. Morton, M.: Rubber Technology, 3rd edn. Van Nostrand Reinhold, New York (1987)
20. Coran, A.Y.: Vulcanization. In: Mark, H., Bikales, N.M., Overberger, C.G., Menges, G. (eds.) Encyclopedia of polymer science and engineering, vol. 17, 2nd ed. Wiley, New York (1989)
21. Van Duin, M., Souphanthong, A.: The chemistry of phenol-formaldehyde resin vulcanization of EPDM: part I. Evidence for methylene crosslinks. Rubber Chem. Technol. **68**, 717 (1995)
22. Drobny, J.G.: Radiation Technology for Polymers. CRC Press, Boca Raton (2003)
23. van der Burg, T.H.: Kunststof en Rubber **46**(3), 26 (1993)
24. Van Bevervoorde, M.: Improving mechanical properties of EPDM rubber by mixed vulcanization. PhD, University of Twente, Netherlands (1998)
25. Coran, A.Y., Patel, R.: Rubber-thermoplastic compositions. Part II. NBR-nylon thermoplastic elastomeric compositions. Rubber Chem. Technol. **53**, 781 (1980)
26. Naskar, K.: Thermoplastic elastomers based on PP/EPDM blends by dynamic vulcanization. Rubber Chem. Technol. **80**(3), 504–519 (2007)
27. Yasuhisa, M., Aklhiro, N., Inoue, T.: J. Soc. Rubber ind. **73**, 62–69 (2000)
28. Tinker, A.J., Jones, K.P.: Blends of Natural Rubber. Chapman and Hall, London (1998)
29. Donnet, J.-B., Voet, A.: Carbon Black, Physics, Chemistry, and Elastomer Reinforcement. Marcel Dekker, Inc., New York (1976)
30. Donnet, J.B., Bansal, R.C., Wang, M.J.: Carbon Black, 2nd edn. Marcel Dekker, Inc., New York (1993)
31. Wang, M.J., Wolff, S., Tan, E.-H.: Filler-elastomer interactions. Part VIII. the role of the distance between filler aggregates in the dynamic properties of filled vulcanizates. Rubber Chem. Technol. **66**(2), 178 (1993)
32. White, J., De, S.K., Naskar, K.: Rubber Technologist's Handbook. vol 2. Rapra, Shawbury (2009)
33. Wang, M.-J., Lu, S.X., Mahmud, K.: J. Polym. Sci., Part B: Polym. Phys. **38**(9), 1240 (2000)
34. Wang, M.-J., Patterson, W.J., Brown, T. A., Moneypenny, H.G.: Rubber and Plastics News, **27**(14), 12 (1998)
35. Wang, M.J.: Billerica, Massachusets, Kautschuk, Gummi, Kunststoffe, **58**(12), 626–637 (2005)
36. Bhowmick, A.K.: Current Topics in Elastomers Research. CRC press, Baco Raton (2008)
37. Thomas, S., Stephen, R.: Rubber Nanocomposites: Preparation, Properties and Applications. Wiley, Singapore (2010)
38. Simpson, R.B.: Rubber Basics. Rapra, Shawbury (2002)

39. Franta, P., Vondracek, B.M., Duchacek, V.: Elastomer and Rubber Compounding Materials. Elsevier, Amsterdam (1988)
40. Chandra, A.K.: Paper presented at International workshop on, "Climate Change & Rubber Cultivation : R&D—Priorities", RRII—Kottayam, India, 28–30 July 2010
41. Doolittle, A.K.: Mechanism of plasticization. In: Bruins, P.F. (ed.) Plasticizer Technology. Reinhold, New York (1965)
42. Kumar, N.R., Chandra, A.K., Mukhopadhyay, R.: Effect of 1,3-bis(citraconimidomethyl) benzene on the aerobic and anaerobic ageing of diene rubber vulcanizates. J. Mat. Sci **32**, 3717 (1997)
43. Santangelo, P.G., Roland, C.M.: Rubber Chem. Technol. **67**, 359 (1994)
44. Roland, C.M., Warzel, M.L.: Rubber Chem. Technol. **63**, 285 (1990)
45. Aprem, A.S., Joseph, K., Thomas, S.: Studies on double networks in natural rubber vulcanizates. J. Appl. Polym. Sci. **91**, 1068–1076 (2004)
46. Funt, J.F.: Mixing of Rubber. Rapra, Shawbury (2009)

# Elastomer Processing

**M. A. Fancy, Reethamma Joseph and Siby Varghese**

**Abstract** Generally elastomer processing involves two major steps. First one is the designing of a mixing formulation for a specific end-use and the second one is the production process by which rubber compound is transformed into final product. When designing a mixing formulation the compounder must take account not only of those vulcanisate properties essential to satisfy service requirements but also cost of the raw materials and the production process. There should always be a compromise between cost of production and quality of the product. This chapter is an attempt to deal with different processing techniques normally used in the rubber industry.

## 1 Introduction

The processing of a rubber formulation is a very important aspect of rubber compounding [1]. The raw polymer can be softened either by mechanical work termed mastication or by chemicals known as peptisers. Under processing conditions various rubber chemicals, fillers and other additives can be added and mixed into the rubber to form an uncured rubber compound. These compounding ingredients are generally added to the rubber through one of the two basic type of mixers; two roll mill or internal mixers.

## 2 Two Roll Mill

The first use of the two roll mill was in the 1830s in USA. Hancock's Pickle was patented in 1837, although models had actually been in use from the early 1820s.

M. A. Fancy · R. Joseph · S. Varghese (✉)
Technical Consultancy Division, Rubber Research Institute of India,
Kottayam 686 009 Kerala, India
e-mail: sibyvarghese100@yahoo.com

P. M. Visakh et al. (eds.), *Advances in Elastomers I*,
Advanced Structured Materials 11, DOI: 10.1007/978-3-642-20925-3_5,

**Fig. 1** Two-roll rubber mixing mill (Website for this image: http:// gdrubbermachine.com)

The first machine that appears suitable for rubber was a twin rotor design patented by Paul Pfleiderer in 1878/1879 [2].

Two roll mill consists of two horizontal, parallel heavy metal rolls which can be jacketed with steam and water to control the temperature [1–17]. These rolls are connected to the motor through gears to adjust the speed. Rolls turn towards each other with a pre set adjustable gap or nip to allow the rubber to pass through to achieve high shear mixing (Fig. 1).

## 2.1 Friction Ratio

The speed of the two rolls is often different [41–46]. The back roll usually turns at a faster speed than the front roll, this difference increases the shear force. The difference in roll speeds is called friction ratio, which is dependent upon the mill's use. For natural rubber mixing a ratio of 1:1.25 for the front to back roll is common [3].

## 2.2 Cooling

Cooling is employed either through cored rolls or through peripherally drilled rolls. The principal one employs cored rolls i.e., water is sprayed onto the outside of an axially drilled central core.

## 2.3 Other Attachments

Mills are fitted with a metal tray under the rolls to collect droppings from the mill. Guides are plates which are fitted to the ends of the rolls to prevent the rubber from contamination with grease etc. Safety measures are also attached to the mill for protecting the operator as well as the mill.

## 2.4 Mixing Process

There are five stages in the mixing process [4, 5]. They are:

1. Banding the rubber on the first roll.
2. Viscosity reduction by mastication or peptisation [5].
3. Incorporation of ingredients.
4. Distribution.
5. Dispersion.

When a highly elastic rubber of high molecular weight is fed into the mixer, it must be converted to a state in which it will accept particulate additives. This stage is called viscosity reduction. It is achieved either by a physical mechanism called mastication or by chemical means called peptisation. Now the rubber is ready to flow around the additives, incorporating and enclosing them in a matrix of rubber. Incorporated additives are then available for distribution. For better incorporation and distribution, with the help of a cutting knife give suitable cuts from either sides of the front roll.

During distributive mixing the rubber flows around the filler particle agglomerates and penetrate the interstices between particles in the agglomerate and the rubber mix becomes less compressible and its density increases [19, 20, 23]. The rubber which has penetrated the interstices becomes immobilised and is no longer available for flow. This immobilisation reduces the effective rubber content of the mixture. The incompressibility of the mixture allows high forces to be applied to the particle agglomerates, causing them to fracture. This action is called dispersive mixing, which serves the purpose of separating the fragments of agglomerates once they have been fractured. The addition of plasticizers facilitates easy incorporation of the fillers. Curatives are added at the end of the mixing cycle. After thorough incorporation of all the ingredients the mix is homogenised and the batch is then sheeted out. For best mixing procedure the temperature is kept at 75–80 °C by careful adjustment of flow of cooling water through the rolls. The sequence of mill mixing is as follows [4]:

1. Band the rubber
2. Mastication/peptisation
3. Addition of cure activators
4. Half of the filler and oil
5. Rest of the fillers
6. Curatives
7. Homogenisation
8. Sheeting out the compound.

It is better to keep the rubber compound at ambient temperature for one day, for better consistency in properties.

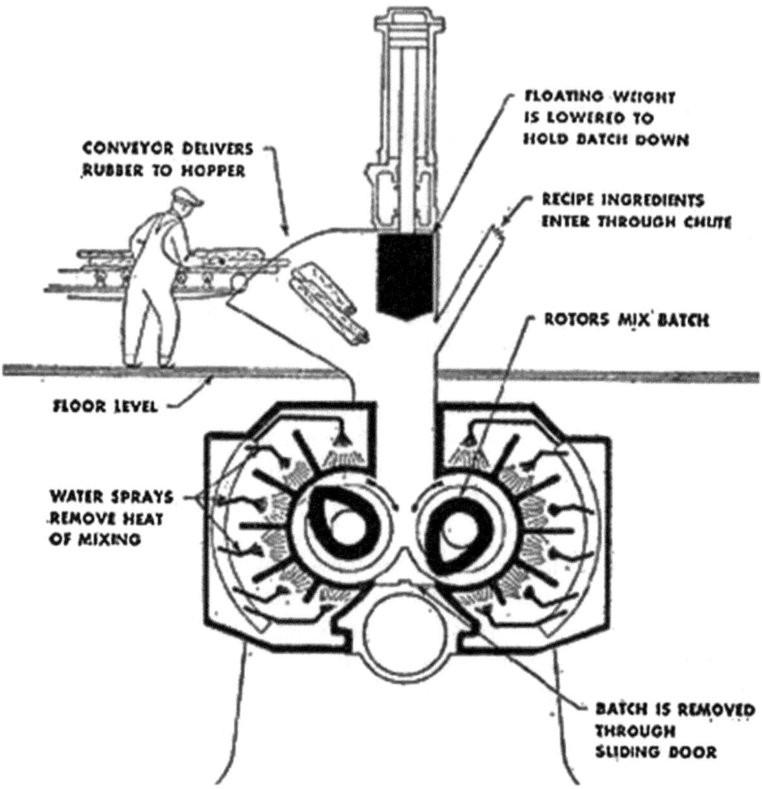


**Fig. 2** Diagrammatic section of Banbury mixer (Website: http://bouncing-balls.com)

## 3 Internal Mixers

The internal mixers were initially developed by Fernley H [1, 2, 29, 30] Banbury from 1916 onwards. Both two roll mills and internal mixers are batch mixers, mill mixing is relatively a slow process, and the batch size is limited. Internal mixers overcome these problems by ensuring rapid mixing and large output [6, 18, 21–23]. An internal mixer consists of two horizontal rotors with wings or protrusions, encased by a jacket (Fig. 2).

### *3.1 The Intermix [6]*

The concept of Intermix was developed in the UK during the early 1930s by an unknown engineer of the ITS Rubber Company. Construction and detailed design

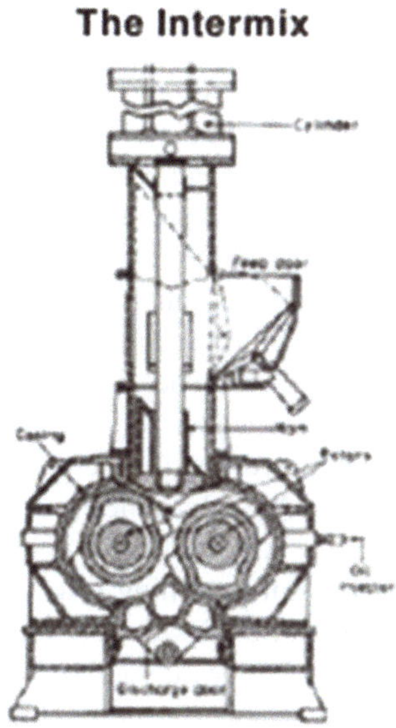


**Fig. 3** Diagrammatic section of Shaw Intermix (Courtesy Francis Shaw & Co, Ltd)

**Fig. 4** Rubber kneader (www.rubber-machinary.in)

of the Intermix was contracted to Francis Shaw a company of Manchester, who eventually acquired and patented the design (Fig. 3).

## *3.2 Rubber Kneaders [7]*

There are two different types of internal mixers used in the industry at large. The first type is more commonly known as a "Banbury®" type intensive mixer and the second type is known as a "Kneader" [7]. The primary difference between the two types of mixers is rotor, throat, chamber and floating weight design. The former also discharges the batch through a bottom door where as the kneader tilts to discharge the batch (Fig. 4).

Conventional Kneaders have two tangential non-intermeshing rotors as well as pneumatic operated floating weights. With the conventional kneader design the temperature in a batch can not be sufficiently controlled to achieve 1 pass mixing. With conventional kneaders the batch temperature after the primary kneading stage is high because of poor temperature transfer from the mixing contact surfaces to the batch. Therefore, the batch has to be either cooled down or transferred to another kneader for the final kneading stage. This additional step is cost prohibitive as well as time consuming.

The MXI-Intermeshing Kneader imparts superior dispersion by reducing filler particle size during the kneading process. The reduction of particle size is achieved by the intermeshing rotor design.

Traditional kneaders have two counter rotating rotors with each mixing rotor having two wings affixed on it. The two wing rotors typically rotate at two different speeds through connecting gears. The wings move material from one portion of the chamber to the other while also providing material movement along the rotor axis. These kneader do not have intermeshing rotors and therefore can have differential rotor speeds.

Conventional Kneaders have a one piece rotor design which includes a rotor shaft with two wings welded on the shaft. Water cooling is provided through a passage in the rotor shaft and small jackets in each wing. This cooling method is not sufficient for single pass mixing. The MXI-Kneader consists of a two piece rotor design. An over-sized rotor shaft and a cast blade shell portion. The cast blade shell is provided with a spiral water passage which is close to the material contact surface. The assembled rotor has a much larger outside diameter than conventional kneaders. This allows for more cooling surface as well as larger mixing surfaces.

Conventional kneader rotors have a shaft and two wings one wing is typically shorter than the other, to have adequate material movement inside the mixing chamber. The MXI-Kneader has a rotor shaft with one long wing (blade) and two nogs (small blades) for mixing. The conventional kneader's blades are typically long high and narrow. The new MXI-Kneader has much wider land width and stubby in shape. The much wider rotor tip (land width) greatly enhances the dispersion effect. The materials are subjected to a larger smearing action of the batch against the rotor tip to chamber wall as well as the rotor tip to rotor shaft. In the non-intermeshing type kneader no mixing occurs between the rotor tip and rotor shaft due to the non intermeshing design.

Conventional kneaders use pneumatic pressure to push the batch down into the rotors and mixing chamber with a floating weight (ram). This pneumatic system is unreliable and inconsistent. The pneumatic ram moves completely uncontrolled and the ram position is controlled by the rotor dragging force as well as the size of rubber pieces it is trying to force into the rotors. The hydraulic ram exerts positive pressure on the batch and can be accurately controlled in the desirable position which leads to better batch to batch consistency.

## *3.3 Intermix and Banbury*

In Tire factories and large rubber factories the internal mixers has practically replaced the two roll mill for the preparation of compounds. Both these machines are used in rubber industry [1]. Compared to tangential system in Banbury the Intermeshing system in Intermix have more effective temperature control, drive power is around 10–20 % higher. But optimum fill level is 5 % lower because of the narrow intermeshing zone.

The basic difference between the two machines lies in the rotor design. The intermix is an example for the interlocking type rotors and the Banbury is of non-interlocking type. In both cases the rotors run at even speed and the nogs or wings are designed to produce a friction ratio between the rotors. In the Banbury, the mixing process is carried out between the rotors and the jacket. In the Intermix the work is done between the rotors. Both the machines are fitted with a ram to ensure that the rubbers and powders are in contact [8, 24, 34, 35].

### 3.3.1 Machine Sizes

A range of sizes of machines are available. The most popular size of machine is one with a batch load of about 200 kg of compound.

### 3.3.2 Rotor Speeds [3]

Internal mixers of 200 kg size can be obtained with rotor speeds in the range of about 20–66 rev/min. To carry out a mixing, certain number of rotor revolutions are needed, then the mixing time is directly proportional to the rotor speed.

### 3.3.3 Ram Thrust and Fill Factor

The thrust applied to the ram affects the output of the internal mixer. Increase in the ram thrust reduces the voids in the machine [3, 32]. For efficient mixing the fill factor is also important.

For a given rotor speed and ram pressure, there is a correct volume of the compound to give efficient mixing. If this is divided by the volume of the chamber the fill factor is obtained. For rubber compounds normally it will be 70–80 %. The remaining corresponds to the voids in the mixture. If the fill factor is accurate, large output of better quality will result. Increase in the ram thrust increases the rate of increase of temperature, and reduces the mixing cycle, and gives more rapid ingredient absorption, and gives greater reproducibility in mixing.

### 3.3.4 Cooling Arrangement

Drilled sides are now common for cooling arrangements. The drilled sides comprise cooling passages drilled under the surface of the body. If cooling is higher, slippage can occur between the rubber and the rotors. As this gives inefficient mixing, warm water is circulated through the machine. The temperature of the water can be regulated by cooling water when heat is being generated and by electrical heaters when the machine is too cold.

When a mix has been completed in the internal mixer, cool it as quickly as possible. The batch of compound is either dropped into an extruder, or to a two roll mill. The compound is then treated with anti-tack prior to cooling and storing. Various degrees of automation are possible for these systems.

### 3.3.5 Mixing Procedure

The following facts should be noted for better mixing:

1. Generally the efficiency of mixing depends on the sequence of material input to the mixer. The steps for ingredient addition should be minimum. For each addition the ram should be raised, with the ram up, there is no pressure on the mix, which leads to little effective mixing.
2. Particulate fillers should be added at the earliest stage of mixing. This helps to achieve good dispersion as a result of the high viscosity at the initial low temperature. A higher viscosity will lead to an increased shear stress at a given rotor speed. For the same reason plasticizers should be added at the later stage. Oils may coat the rotors and chamber wall and cause slippage and reduce mixing efficiency. They are therefore added together with fillers to reduce this action.
3. The curative package should not be added at he elevated temperature stage. Batches are usually dumped from an internal mixer on to a mill where they may be further worked while being cooled. Curatives are added at this point.

### 3.3.6 Upside Down Mixing

This method involves adding all the dry ingredients other than the elastomer to the mixer first, then all the liquids, and finally the elastomer [9].

### 3.3.7 Advantages

1. It is faster and simplest.
2. It is employed when the polymer content is less than 25 %, and also for polymer having poor self-adhesion.
3. It is effective for compound having large volume of liquid plasticisers and large particle size fillers.

### 3.3.8 Disadvantages

1. Small particle size carbon blacks cannot be effectively mixed by this technique, as it does not provide a high level of dispersion.
2. If the polymer is of high viscosity upside down mixing will result in the development of temperature and lead to poor dispersion.
3. Clays which are difficult to wet due to low surface energy do not incorporate well in this method.

### 3.3.9 Take-Off Systems

After the mixing the batch has to be cooled and converted into strips or sheets suitable for feeding to the next process. This is done by dumping the batch through the drop door at the bottom of the mixer on to a cool mill capable of handling the entire batch. A three or four roll calender is used when uncured rubber is applied to a textile fabric or steel cord as a coating. Extruders are used when the uncured stock is to be shaped into a tyre tread, a belt cover, or hose tube for example. Extruders with lower screw length to screw diameter (L/D) ratios are considered hot feed extruders while one with high ratio are called cold feed extruders.

## 4 Continuous Mixers

The continuous mixing of rubber compounds is very much in its infancy. The earliest machines used for continuous processing of true curable materials were the Extruding, Venting and Kneading (EVK) machine made by Wernar and Pfliederer and Mixing, Venting, Extruding (MVX) machine made by Farrell Bridge. EVK primarily used in EPDM extrusion compound area using powdered polymer, MVX for cable compounding and in the production of tire compounds using granulated Polymer [2, 10]. Recent work on continuous mixing of rubbers has centred on modified twin screw compounders that have been used for some considerable number of years for in the plastic compounding industry. Because of the numerous compounding ingredients used in the tire industry in their varying physical form, an economic and sufficiently accurate proportioning of the compounds in a continuous mixer is barely possible. Now-a-days continuous mixers are used in rubber compounding only for partial operations such as making batches either consisting of elastomer and filler or another one containing chemicals.

Both mills and internal mixers are batch mixers. In order to replace the batch mixing process, attempts were made since world war II to develop a continuous mixing technique. Examples are Double R mixers from Francis Shaw in the late 1940s, the continuous mixer from Farrel's Corporation in the 1960s, and the Transfer mix from U.S Rubber Company in the late 1960s. Continuous mixing normally demands that solids are fed in a particulate form [11]. In order to convert rubber bales into pellets; disintegrators are needed. All continuous mixing operations face the problem of how to weigh continuously the multiplicity of very variable weights of rubbers and their compounding ingredients with the accuracy required. Disintegrating rubber will consume power and powdered rubbers cost more than baled rubber.

## 5 Trouble Shooting the Mixing Process [9]

Problems in the mixing process are usually due to inadequate dispersion, contamination, poor processability on the dump mill, scorchy compound and batch to batch variation. Once the problem has been identified, corrective action is often simple. The list below suggests possible causes of such problems.

1. Inadequate dispersion or distribution
   - Insufficient work input, mixing time
   - Order of ingredient addition not proper
   - Batch size too large or too small
   - Insufficient ram pressure, wrong rotor speed, wear of rotors and chamber wall
   - Cold polymer (this applies especially to natural rubber, EPDM and butyl)
   - Excessive moisture in fillers
   - Oils added at temperature below pour point.
2. Scorchy compound [33]
   - Too high a heat history after addition of curatives
   - Accelerator added too soon
   - Inadequate distribution of curatives
   - Too high a rotor speed
   - Materials added at too high a temperature
   - Inadequate cooling of compound after take-off.
3. Contamination
   - Physical contamination of one or more ingredients
   - Insufficient clean-out between batches of different base polymers

- Oil-seal leak.

4. Poor handling on dump mill

- Incorrect roll temperatures, speed and friction ratio
- Too high a loading of clay fillers, viscous plasticizer
- Poor distribution or dispersion
- Scorchy compound
- Compound left on mill too long.

5. Batch to batch variation

- Variation in initial loading temperatures, ram pressure, cooling water flow, or temperature
- Variation in compounding ingredients
- Variation in dump time, temperature, or energy input
- Variation in milling time or mill settings
- Variation in amount of cross-blending on mill.

# 6 Extrusion Process

## *6.1 Introduction*

In rubber processing, extruder is mainly used for shaping the rubber compound into the desired profile before it is finally processed. There are two type of extruder—ram extruder and screw extruder. Ram extruder has high operating cost and lower output. Now, the screw extruder is used mainly for the production of tubing, channel, tire treads and for the insulation of the wire and cables (Fig. 5) [16].

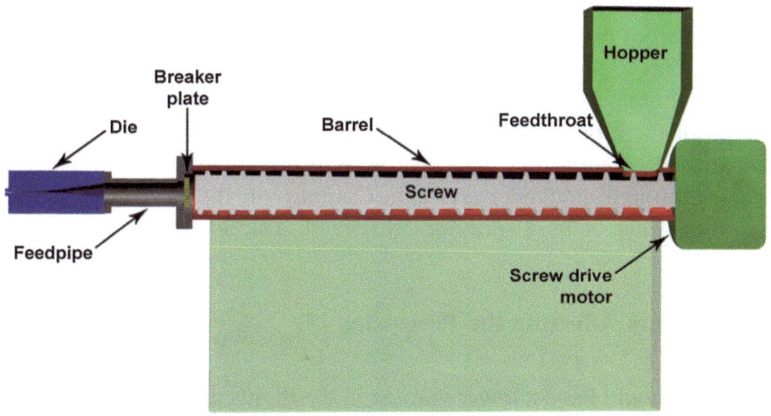


**Fig. 5** Schematic representation of a rubber extruder (www.http.wikipedia.org/wiki/Extrusion)

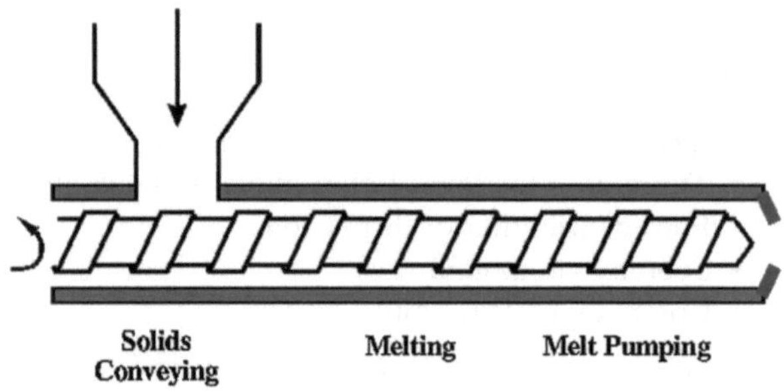


**Fig. 6** Screw extruder (http.www.polymerprocessing.com/operations/screw/big.html)

## *6.2 Screw Extruder*

The extruder consists of a feed hopper, cylindrical barrel, rotating screw, head attachment and a die [3, 14]. The screw is driven by the an eletric motor through appropriate reduction gear system. The compound to be shaped is fed into the machine through the feed hopper. The width of the compound strip fed is slightly less than the width of the feed hopper and the thickness of the strip should be slightly less than or equal to the depth of the flight of the screw. The barrel is usually made of hardened steel and is jacketed for the circulation of steam or cold water. Heating of the barrel is necessary in the early stages, when it is started temperature is developed inside in this stage, steam supply is cut off and cold water is circulated to maintain the temperature (Fig. 6).

The head attachment of the extruder varies in shape according to the purpose for which it is used [25, 27, 29]. The design of the head is very important to get free movement of the compound at equal pressures and speeds from all side of the head into the die. Any point within the head where the compound doesn't move is known as dead spot. Provision for heating and cooling should be provided at the head attachment for better control of temperature.

The die of the extruder shape the compound into the desired profile. For better shape and finish of the extrudate, the design of the dies is very important. The die is the hottest part of the extruder and is usually heated initially by a gas flame. The cross sectional area of the die should never be lower than 5 % less or greater than 30 % more of the cross sectional area of the extruder.

The extrudate coming out from the die is usually carried to the next stage of processing, through conveyor system. Cooling of the extrudate is done by immersion in water or by a spray of cold water. Talc is applied to the extrudate.

### 6.2.1 Parameters Affecting the Processing [3]

1. The screw should have a lower volume in the flights at the out going than at the in going end. It is most important that an extruder screw is full at the discharge end, otherwise dimensional changes in the extrudate may occur.

2. The design of the head is very important, the head equalizes the pressure from the screw and barrel and the compound moves smoothly to the die at equal pressure and speed.
3. The last stage is die, which forms the compound into the desired shape. Die should be designed to operate under conditions of minimum stress and at pre determined running speed and temperature. The extrudate should be produced under these conditions. The lower the viscosity of the compound, the greater is the through put to be expected in unit time. Dimensional variations is at minimum when the compound has the minimum entrapped stress.

### 6.2.2 Hot Feed, Cold Feed and Vacuum Extruder

Depending on the design of the screw and barrel, extrudate can be three types namely, hot feed, cold feed and vacuum extruders. The screw of the vacuum zone there is provision at the barrel for connecting it to a vacuum device. In the vacuum zones, the screw is either deeper or more widely cut or the cylinder in that zone is slightly bigger than in the other zones. Vacuum extruder helps to remove any traces of moisture or entrapped air from the compound. Hence it is used in shaping articles for open steam cure, hot air cure, molten salt cure and fluidized bed cure. Since moisture and air trapped in the compound is removed during vacuum extrusion, the product will be free from porosity.

Depending on the design of the screw, the extruder may be used for hot or cold feeding of the compound. The hot feed extruder has got a short barrel, the length to diameter ratio of the screw is low, in the range of 5:1 and the compression ratio is nearly equal to unity. In the case of cold feed extruder the length to diameter ratio of the screw is high, in the range of 20:1 and the compression ratio is greater than unity [28].

In hot feed extruders, pre-milled rubber is fed into the extruder, output will be uniform and the production equilibrium can be attained within a short time. As its name refers cold rubber compound can be fed into cold rubber feed extruders. Out put depends on the nature of the compound and it takes longer time to attain production equilibrium. Hot feed extruder has high operating cost and higher output than cold feed extruder. Product of consistent quality can be obtained only if the compound fed into the machine is uniform viscosity, temperature and volume. Maintain the temperature of the barrel, screw, head and die constant.

### 6.2.3 Defects in Extrusion

1. Die swell

As the compound comes out from the die, it shrinks along its length resulting in slight increase in overall dimension of the extrudate. Nature of the compound, uniformity of the speed stock and speed of the screw and take off conveyer systems can affect die swell.

2. Rough surface

Poor finish of the extrudate may be due to poor dispersion of the compounding ingredients, very high money viscosity of the polymer, very low temperature and pressure of extrusion. By proper adjustment of this good finish of the product can be obtained.

3. Porosity

This is due to the presence of excess moisture in the compounding ingredients, use of high volatile compounding ingredients and presence of entrapped air. Proper drying of fillers before using, addition of material like calcium oxide in the compound and use of vacuum extruder can reduce porosity in the extrudate.

4. Collapse of the material

Collapse of the extrudate occurs when the quality of the polymer used is poor, viscosity of the compound is very low and when the processed material is recycled several times.

#### 6.2.4 Troubleshooting the Extrusion Process [9]

Despite the many feedback microprocessor control system available on the market, it is still often the skill, experience, and understanding of the extruder operator that determines the success or failure of an extrusion operation. Success or failure has to be measured in economic terms, that is the hourly production rate of the process. This depends on minimizing scrap and downtime both in start-up.

## 7 calendering

### *7.1 Introduction*

Calenders are used in the rubber industry primarily to produce rubber compounds and sheets of various thicknesses, coating textiles or other supporting materials with this rubber sheet or frictioning fabrics with rubber compound.

### *7.2 Machinery*

A variety of products like sheeting for lining, hospital bed sheets, films, frictioning of tire fabrics for Bicycle/motor cycle/Auto tires, hoses beltings, profiling, cushion gum, single coating, laminating doubling etc. Special Calenders for profiles and Inner liner can be done. The Rubber Calender Machines are made in a wide range of sizes, from small laboratory unit up to the largest production calender machines (Fig. 7).

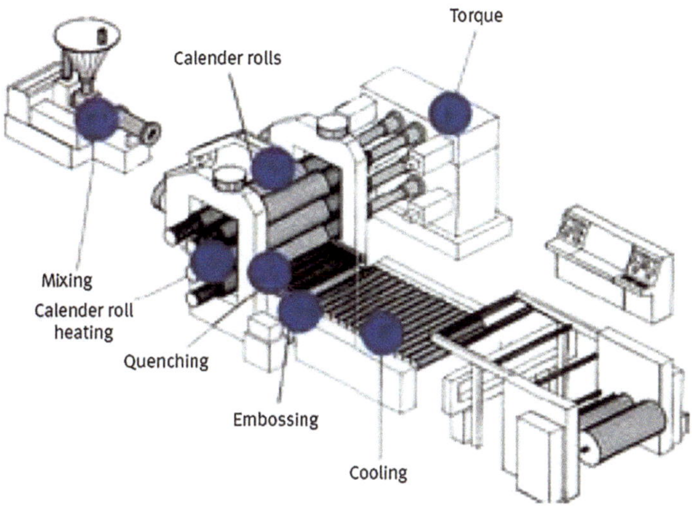


**Fig. 7** Calendering unit (www.http:/wikipedia.org/wiki/calendering)

## *7.3 Calendering Unit*

Rubber calenders are differentiated by the number of rolls, their arrangements, and their size (diameter and width). They can have two, three, or four rolls in a variety of configuration. For the production of tire stock, belting, and sheeting the 3 roll vertical calender with 24″ diameter, 68″ width rolls, and four-roll Z and L calenders with 28 × 78″ rolls are standard. Four-roll calenders are used for applying compound to both sides of tire cord fabrics in one operation.

### 7.3.1 Types [11]

There are three main types of calender—the I type, L type and Z type (Fig. 8).

**Fig. 8** Roller setup in a typical 'I' type calender (www.appropedia.org/polymer-calendering)

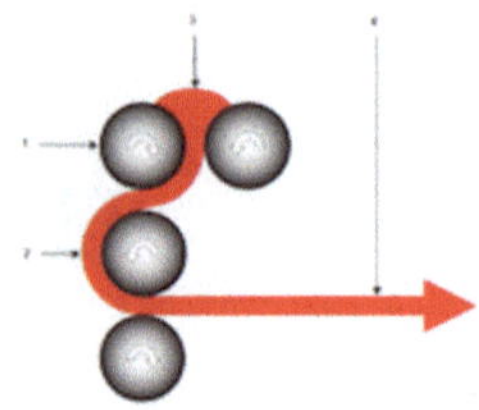

**Fig. 9** Roller setup in a typical inverted 'L' type calender (www.appropedia.org/polymer-calendering)

**Fig. 10** Roller setup in a typical 'Z' type calender (www.appropedia.org/polymer-calendering)

The "I" type, as seen in Fig. 1, was for many years the standard calender used. It can also be built with one more roller in the stack [8, 12]. This design was not ideal though because at each nip there is an outward force that pushes the rollers away from the nip (Fig. 9).

The L type is the same as seen in Fig. 2 but mirrored vertically. Both these setups have become popular and because some rollers are at 90° to others their roll separating forces have less effect on subsequent rollers. L type calenders are often used for processing rigid vinyls and inverted L type calenders are normally used for flexible vinyls (Fig. 10).

The z type calender places each pair of rollers at right angles to the next pair in the chain. This means that the roll separating forces that are on each roller individually will not effect any other rollers. Another feature of the Z type calender is that they lose less heat in the sheet because as can be seen in Fig. 10 the sheet travels only a quarter of the roller circumference to get between rollers. In most other types this is about half the circumference of the roller.

### 7.3.2 Feeding [9]

To ensure steady operations of the calender, and to control shrinkage, the compound has to be preheated to around 93 °C, and thoroughly fluxed and plasticized before being fed to the calender.

Calenders use rolls with axial drilling about 50 mm under the surface, through which water at a pre-set temperature is continuously circulated. The rolls with the axial hole through the centre are known as cored rolls and the latter are periphery drilled rolls. Periphery drilled rolls without controlled—temperature water going through them are unsatisfactory in operation, since hot and cold water used alternately increase and decrease the roll temperature too rapidly. Periphery drilled rolls are normally heated and cooled at a speed of 1 °C/min i.e., it give much better control.

### 7.3.3 Sheeting [25]

This process is carried out on a three—roll calender with thickness control by a feed back system from the product. The rolls are crowned to compensate for deflection under load, and so to maintain a constant roll gap across the width of the sheet.

### 7.3.4 Frictioning [25, 26]

This is impregnating a textile or metallic fabric between two rolls running at different speeds so that the rubber compound is forced into the interstices of the substrate.

### 7.3.5 Spreading [3]

The main working part of a typical spreading machine as shown in (Fig. 11).

A roll of dried or pre-treated fabric is fitted on to location A and the leader cloth is fed through the rest of the machine until finally taken up on roller J. from A, the cloth passes over a spreader bar B to ensure that all creases are removed from the fabric and to keep it under the correct lateral tension. The smooth tensioned fabric is then fed over the bearer roller C and under the doctor blade D, which is pre-set to give the correct build-up of dough on the fabric surface. The angle between the blade and the fabric and the distance between them control the coat thickness and the degree of 'strike through' (degree of penetration) of the dough. The greater the angle at which the blade meets the moving fabric, the greater the degree of penetration.

The fabric then enters the steam chest area, where the solvent is driven off and removed by means of the extraction unit F; the speed of travel of the fabric is dependent on the rate of solvent removal. On emerging from the end of the steam

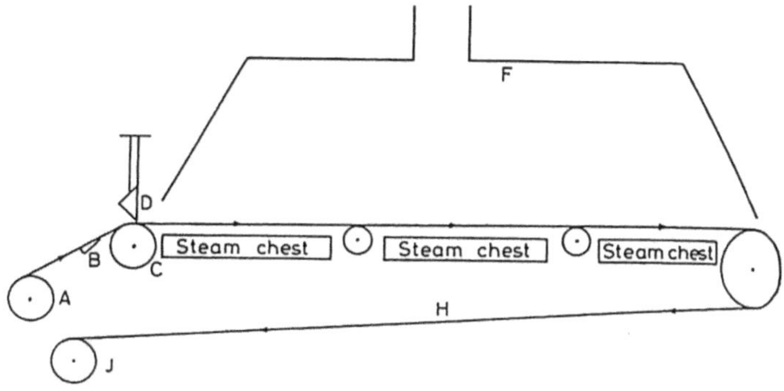


**Fig. 11** Diagrammatic sketch of spreading machine. Courtesy ICI dyestuffs division

chest, spread fabric requires cooling before it is rolled up on roller J. This is achieved by means of the festooning device, placed at H, which may consist of a single or several rollers; if the dough is of a sticky nature, it may also necessary to use a liner cloth or dust the surface with talc to prevent blocking together of the rubber-fabric laminate during storage.

Once a machine has been set up to run, it is necessary to carry out the spreading of the first few yards at the low speed to check the coating thickness against specification, either by means of a vernier gauge or electronically. Normal running speed of the order of 10 m/min.

### 7.3.6 Skim Coating or Topping by Means of the Calendering

The operation of applying a substantial thickness of rubber to fabric on a calender is termed skim coating or topping. In this method, compound is fed around a calender roll from a calender nip, and the sheeted compound is applied to the fabric at a second nip. The rubber sheeting must be travelling at the same speed as the fabric at the point where it is laid on to the fabric; however, sheeting can be produced from roll which run at the same speed or with a friction ratio in the nip (Fig. 12).

### 7.3.7 Temperature Effects

The temperature of the fluid melt has been found to be highest at the rollers. This happens for two reasons:

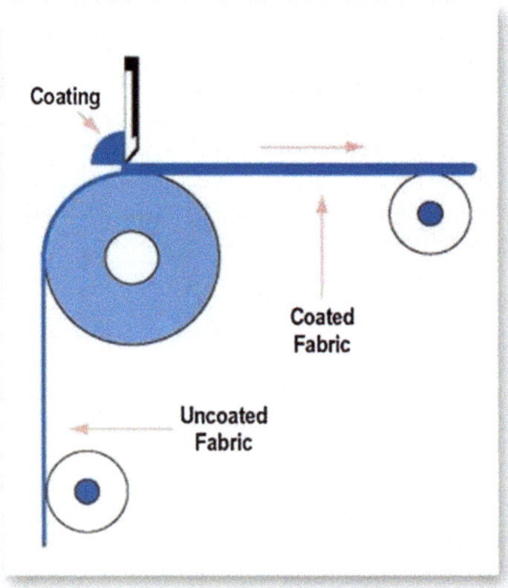


**Fig. 12** Topping by means of a calender (www.appropedia.org/polymer-calendering)

1. The shear is highest at the sides in laminar flow and therefore friction and heat is also highest there.
2. The heat is added to the system through the rollers, and the fluid doesn't conduct it very well.

The effects of this tend to grow in magnitude with more and more viscus the fluid is. If one were to raise the rolling temperature there would be changes in the above fluid mechanics. It would decrease the viscosity; consequently decreasing the power input, pressure and roll separating forces in the fluid. It would also lower the chances of a fracture in the fluid and make the surface finish better, but this all comes at the price and increases the chances of thermal degradation.

### 7.3.8 Velocity Effects on Final Product

The calender is able to produce the polymer sheeting at a fast rate. It can produce sheeting at a rate between 0.1 and 2.0 m $s^{-1}$. By increasing the speed the heat has even less time to spread throughout the fluid from the rollers causing an even greater temperature variation. It also causes an increase in shear forces in the fluid at the rollers, which increases the chances of surface defects like fractures. The speed clearly needs to be chosen very carefully in order to produce a quality product.

### 7.3.9 Roll Bending

In calendering the rollers are under great pressures, which can reach up to 41 MPa in the final nip. The pressures are highest in the middle of the width of the roller and due to this the rollers get deflected. This deflection causes the sheet being made to be thicker in its center than it is at its sides. There are three methods that have been developed to compensate for this bending:

1. Roll crowning
2. Roll bending
3. Roll crossing.

Roll crowning uses a roller that has a bigger diameter in its center to compensate for the deflection of the roller. Roll bending involves applying moments to both ends of the rollers to counteract the forces in the melt on the roller. With roll crossing the rollers are put at a slight angle to each other and because of this the force of the rollers on the melt is higher in the middle where the rollers are on top of each other more, and less force is applied on the edges where the rollers are not directly over top of each other.

### 7.3.10 Roll Cambering

The calender rolls are usually cambered and are not parallel to compensate for variation in thickness across the sheet. This is an ideal solution for a calender which produces one gauge of sheet from one compound. Calenders can be re-cambered in a few hours to accommodate a permanent compound change or to take up the wear of the roll; if more than one compound is processed then some other device for resetting the crown will be needed.

### 7.3.11 Calendering Technology

Rubber compound behave as viscous non-Newtonian liquids. If a uniform gauge of sheeting is to be produced, then the viscosity of the compound must be constant.

In order to achieve uniform viscosity, the temperature of both the compound and the calender must be controlled, as the viscosity of rubber compound is affected very considerably by temperature (Fig. 13).

When an unsupported sheet is taken from a calender nip, it shrinks along its length and increases in the thickness and the width. This results in rubber sheets having a crown, i.e., they are thicker in the centre than at the edges.

### 7.3.12 Unsupported Sheeting

Two-, three-, or four—roll calenders are used for the production of sheeting containing no textile fabric reinforcement.

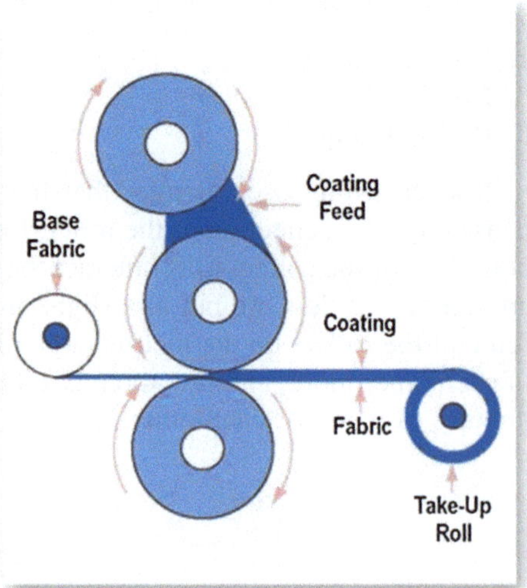


**Fig. 13** Rubber–fabric coating by calendering (www.appropedia.org/polymer-calendering)

For precision gauge control, a sheet is produced in a first nip, and this is then fed round a roll to a second nip. The second nip gives either less blistering or a thicker sheet for the same amount of blistering. The quality of sheeting is more dependent upon the quality of feed than on any factor other than the temperature of the calender rolls. Occasionally, three nips are used for calendering, i.e., a four-roll.

### 7.3.13 Application of Rubber from Solvent Dispersion or Dough

When it is necessary to apply a coating of rubber to a fabric which is too delicate for the calendering process, or when the compound is not suitable, the technique of spreading is used. This process consists of the application of the compound dispersed in solvent at high concentration in the form of dough, as it is termed. A stationary blade (commonly called a doctor) regulates the thickness applied to the fabric as it is passed underneath the blade. The fabric then drawn over a heated chest where the solvent is evaporated and usually recovered for re-use by adsorption on an active carbon.

### 7.3.14 Troubleshooting Problems in Calendering [9]

1. Scorch
   - Poor temperature control
   - Running speed too fast, leading to excessive shear heating
   - Stock warmed upon mill too long.
2. Blistering
   - Roll temperature too high
   - Feed bank too large, resulting in entrapped air
   - Sheet thickness too high.
3. Rough or Holed sheet
   - Inadequate stock warm-up
   - Amount of material in bank too small, or too large, to form rolling bank
   - Varying stock temperature.
4. Tack
   - Temperature of rolls too high
   - Incorrect stock feed temperature.
5. Bloom
   - Low solubility of some ingredients in formulation.

# 8 Continuous Vulcanization System [9, 31]

As a rubber compound containing a curative system is held at the curing temperature the production of cross-links causes it to change from a viscoelastic fluid to an elastic solid. As it leaves the die the compound is still a fluid, and as the stresses built up in the passage through the head and the die relax, the dimension of the profile, originally those of the die, change.

## *8.1 Pressurized Steam Systems*

These are commonly used for products having a core or other reinforcement and profile that are easy to seal, such as wire, cable, and hose. The time required for heat to penetrate to the centre of the cross-section depends on the diameter. The weight of rubber per unit length is proportional to the square of the diameter.

## *8.2 Hot Air Curing Systems*

These consists, basically, of an insulated tunnel, a metal mesh conveyor to support and move the profile through, and a counter current of air, heated to up to 300 °C. Heat transfer is poor (coefficient 70 $kJ/m^2/h/°C$) and so lines of 100 ft are required to complete the curing process. With compounds (e.g., EPDM), which are not readily susceptible to oxidation ultrahigh temperature, shorter ovens can be used.

## *8.3 Microwave System*

Microwave system provide quick and uniform heating throughout the profile, which is especially useful for thick profiles, profiles of varying thickness, and for sponge. The compound has to be microwave receptive (i.e., polar), which most polymer are not. However, many carbon black are, and if necessary, it is also possible to add other chemicals specifically to increase microwave receptiveness. Usually, a short microwave section is used immediately after the die to boost the extrudate temperature to curing temperature, followed by a hot air tunnel to maintain temperature until the profile is cured. There is much less heat loss with a microwave system than with the system described previously because the heat is generated in the rubber itself. This high energy-efficiency makes the use of electricity, a more expensive source, economically feasible.

# 9 Moulding [3]

Moulding is the operation of shaping and vulcanising the plastic rubber compound by means of heat and pressure in a mould of appropriate form. There are three general moulding techniques:

1. Compression
2. Transfer
3. Injection moulding.

## *9.1 Compression Moulding [9]*

In compression moulding a pre-weighed, pre-formed piece is placed in the mould. The mould is closed, with the sample under pressure as it vulcanises. For the satisfactory large scale production of components, it is necessary to use carefully designed and well constructed steel moulds, suitable hardened and finished depending upon the surface quality required for the product. Cavity pressure is maintained by slightly overfilling the mould and holding it closed in a hydraulic press. Heat is provided by electricity, hot fluid or steam.

Compression moulding is the oldest and most universally used technique and for many products the cheapest process because of its suitability for short runs and because of the low mould cost. The press used for conventional compressional moulding has two or more platens, which are heated either electrically or by saturated steam under pressure. The platens are brought together by pressure applied hydraulically to give a loading from 75 to 150 kgf/cm$^2$ of projected mould cavity area (Fig. 14).

## *9.2 Advantages [4]*

1. Moulds have low investments cost.
2. Due to the simplicity in the mould design, it is suitable for curing thick rubber articles. Hence large rubber articles (tyres, belts etc.,) and small articles (gaskets, washers etc.,) maybe cured by this process.

## *9.3 Transfer Moulding*

Transfer moulding involves the transfer of a rubber compound from a heated reservoir or transfer pot, through a narrow gate called sprue or runner into the closed cavity of a mould by a piston where the compound is cured at pre-determined temperature and pressure (Fig. 15).

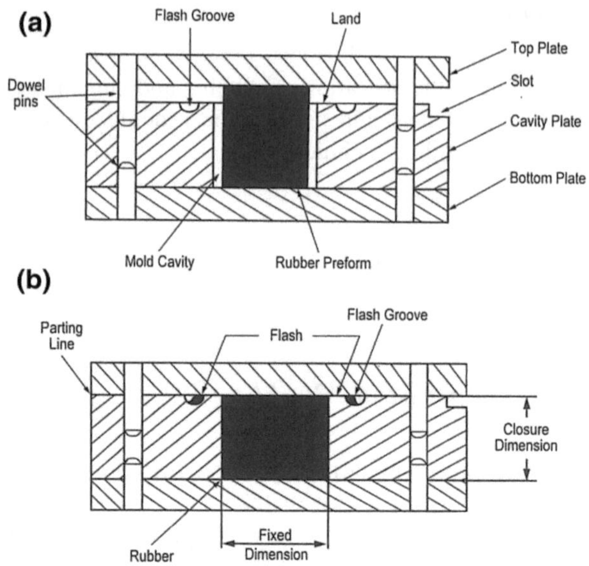


**Fig. 14** Compression mould containing a rubber prefoam. **a** Before closing and **b** after closing (Courtesey of J.Sommer and Rubber Division ACS)

Transfer mould consists of three parts. The upper part called piston, the lower part the mould itself. Both upper and lower parts are attached to hydraulic press. The centre part which contains the cylinder, the injection nozzle is removable. Though the moulds are more expensive, this process permits better heat transfer. Important points to be remembered during transfer moulding.

1. Clearance between the transfer pot and plunger (piston) should be optimised [4].
2. The plunger should not tilt in the pot.
3. The plunger face area should be larger than the projected cavity area.
4. Compound should be optimised to reduce the vulcanised scrap in the mould cavity and the sprue.
5. Changes in pressure should be monitored by pressure transducers, since pressure variations during the transfer process can open the mould cavity to flash.

## 9.4 Advantages

1. As the mould is closed before the rubber charge is forced into it, closer dimensional control is achievable.
2. In the transfer process fresh rubber surfaces are produced. This allows the development of a strong rubber to metal bonding with any insert in the mould.

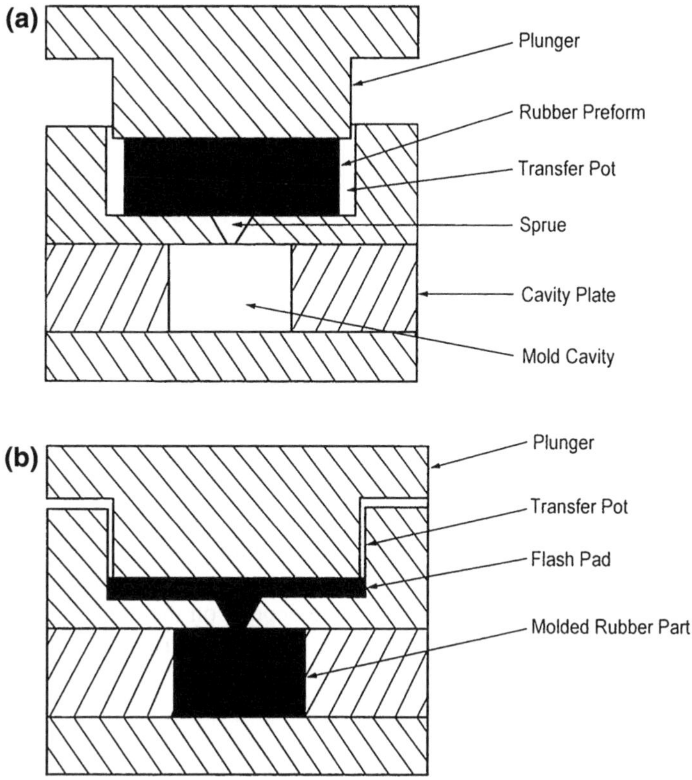


**Fig. 15** Transfer mould showing. **a** preform transfer pot before closing **b** after closing (Courtesey of J.Sommer and Rbber Division ACS)

3. Production cost is lower due to shorter cure times as a result of heating the rubber due to flow through sprue, runner and gate and shorter downtime between runs as only one charge blank is necessary even if a multi-cavity mould is used.

## *9.5 Injection Moulding [9, 13, 14]*

Injection moulding is now a well established process in rubber industry. The operation of an injection moulding machine requires feeding, fluxing and injection of a measured volume of a compound, at a temperature close to the vulcanisation temperature into a closed and heated mould. The process also requires a curing period, demoulding and if necessary mould cleaning or metal insertion before the cycle starts again. For maximum efficiency almost all of the above operations should be automatic. The difference between transfer and injection process lies in

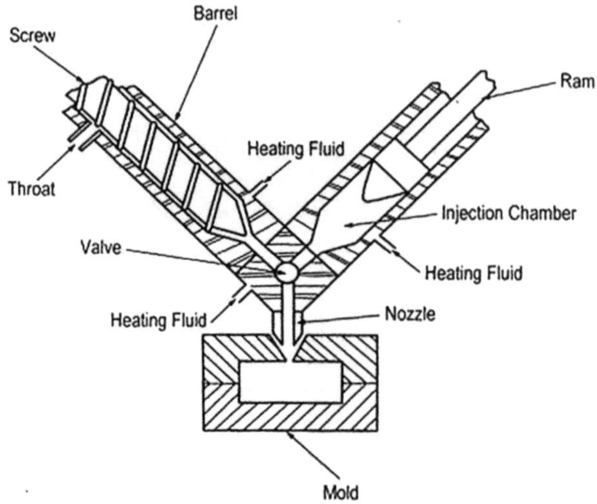


**Fig. 16** Injection moulding machine with separate ram and screw (Courtsey o J. Sommer and rubber Division ACS0)

the degree of automation. In the injection moulding process there is always a reserve of material being heated and plasticized during the vulcanisation step.

The types of injection machines available depend on the method of heating and peptisation of the compounded rubber. The main types used are:

1. The ram type
2. The reciprocating screw type
3. Screw-ram type.

Among the three simple ram type machines cost less than screw machines [15]. The mix receives heat only by thermal conduction from the barrel, high injection temperature and thermal homogeneity are difficult to achieve and they are not widely used.

In the reciprocating screw type the screw acts both as an extruder and a ram. In this type of machine, the mix is heated and plasticised as it progresses along a retractable screw. When the necessary shot volume has accumulated in front of the screw. it is injected by forward ramming action of the screw. With this system more uniformly controlled feeding of the material can be achieved together with more rapid heating of the stalk, due to mechanical shearing, and a greater degree of thermal homogeneity. However, during the induction stage as the screw act as ram there is some leakage back past the flights of the screw and this limits the injection pressure. This type of machine is only possible for low shot volumes, or for very soft compounds.

The preferred basic design for the rubber injection moulding is the screw-ram type because they combine the advantages of both screw and ram type (Fig. 16). In the standard 'V' configuration the plasticated compound is fed through a check valve into an accumulation chamber. One disadvantage is that the first rubber fed through the check valve is the last one to be injected. This can lead to adhesion and build up of rubber on the face of injection piston, which can cure, break off and cause rejects and moulding problems. So modifications have developed to overcome this. In the first in-first out system the screw and ram though separate are inline. Initially the injection ram is in the forward position and the injection chamber is empty. As compound enters through the ram it is forced back by incoming material until a limit switch controlling short volume is activated. Injection then takes place through a special ball type torpedo which completes the plastication and thermal homogenisation. As the material does not reach the final injection temperature until it reaches the nozzle temperature earlier in the system can be relatively low (approximately 70 °C. Such inline systems have gained popularity in recent years.

Multi stage rotary press units can be fitted to many of the injection moulding system, thus enabling continuous moulding to take place. In this injection unit that feeds a number of moulds carried on a rotating carrousel. This system is economic and practical. Such machine may have automatic ejection of parts and runner system, cleaning and spraying of the moulds and automatic loading of metal inserts.

## 9.6 Compounds for Injection Moulding [9]

Compounds for injection moulding must have sufficient scorch safety to flow through the nozzle, runners and gates without scorching but still cures rapidly in the mould. For maximum productivity, the compound has to injected rapidly into the mould at near vulcanisation temperature. This area has to be optimised within which a particular material will flow well and cure effectively without the danger of scorching. The three major areas in which data on a compound are required are rheological behavior, rate of vulcanisation and heat flow into and through the compound.

## 9.7 Advantages [4]

1. A high output of production can be obtained.
2. Automation is the process can lead to cost saving and high quality.
3. Cure time may be reduced due to pre-heating of rubber.
4. Uniform curing of variable thickness component is possible.
5. Flash trimming is eliminated.
6. There is no bumping.
7. The finished products can be removed more rapidly.
8. Complete filling of mould cavity is ensured.
9. Feeding is much easier in the strip form and is more economical.

RIM involves the rapid mixing of two or more highly reactive low molecular weight compounds before the injection of the mixture in a closed mould. The reaction may be polymerisation or molecular network formation in a very short time, approximately 30 s The total cycle time is 1–2 min. RIM is important for urethene rubbers because the process is very energy efficient.

## *9.8 Trouble Shooting the Moulding Process [9]*

### 9.8.1 Scorchiness of the Compound

Compounds must have sufficient processing safety to flow through the nozzle, runners and gates without scorching but still cure rapidly in the mould. For this the balance of viscoelastic and the curing characteristics of the compounds are extremely important. Most rubber compounds that will compression mould can be satisfactorily injection moulded, provided that they flow well enough and are not too scorch sensitive. For maximum productivity the compound has to be injected rapidly into the mould at near vulcanisation temperature.

### 9.8.2 Shrinkage

On cooling both the mould cavity and the moulded part contract usually by a differential amount because the metal and rubber have different coefficients of thermal contraction. Shrinkage is defined as the difference between the dimensions of the mould cavity and those of the moulded part, when both are measured at room temperature. The amount of shrinkage has to be allowed for in mould design. It will vary depending on the polymer, cure temperature, time and pressure.

### 9.8.3 Adhesion

Adhesion has two aspects, adhesion to the mould surface, which is not wanted, and adhesion to a metal insert in the part, which is wanted. Mould release agents are used to prevent the one, and adhesion promoters to ensure the other.

### 9.8.4 Backrinding [13]

This is the term applied to the torn look that occurs at the mould parting line of compression moulded parts and at the gates of transfer and injection moulds. It is caused by thermal expansion of the rubber after cross-linking, which can force the cross-linked rubber into the space at the parting line or gate, causing it to rupture. The best way to minimise this is to minimise the shot weight commensurate with filling the cavity. Increasing scorch time can also help because it ensures that the mould is filled before curing begins as the injection temperature can be raised.

### 9.8.5 Mould Fouling

The build-up of material in a mould especially in the corners is a major problem. The cause is usually deposition of chemicals and their subsequent oxidation or degradation. These agents may originate in the rubber, in fillers or from release agents. Thus, there are a wide variety of deposits whose severity varies from compound to compound and also depends on injection rate and mould temperature. Mould cleaning is often done by blasting with some abrasive particulate material such as glass beads, plastic or metal beads.

### 9.8.6 Orange Peeling

This is usually caused by the initial layer of rubber in contact with the heated mould surface having cross-linked before succeeding layers have filled the mould. Usually occurs in injection moulding and the remedy is to increase the scorch time of the compound.

### 9.8.7 Porosity

This is due to under cure and the presence of volatiles, especially moisture in the compound. Higher injection and mould temperature or longer mould closed time should resolve this.

### 9.8.8 Blisters

Air entrapped in the rubber compound is the usual cause. This can be eliminated by a higher back pressure, slower injection rate, or effective venting of the mould.

## References

1. Dick, J.S.: Rubber Technology-Compounding and Testing for Performance, p. 17–18. Carl Hanser publishers, Munich (2001)
2. http:/tirenews4u.wordpress.com
3. Blow, C. M., Hepburn, C.: Rubber Technology and Manufacture, p. 263–290. Published for the Plastics and Rubber Institute, London (1982)
4. Bhowmick, A. K.: Rubber Products Manufacturing Technology, p. 347. Marcel Dekker Inc, New York (1994)

5. Freekley. P.K.: Rubber Processing and Production Organisation, p. 46. Plenum Press, New York (1985)
6. Wood, P.R.: Rubber Mixing. Rapra Technology Ltd., Shawbery (1996)
7. Kim, P.S., White, J.L.: Rubber Chem. Technol. **67**, 880–891 (1994)
8. www.kneadermachinery.com
9. Johnson, P.S.: Rubber processing, Hanser publishers, Munich (2001)
10. Salma, S.R.: Operation and maintenance of mixing equipment. In: Grossman, R.F. (ed.) The Mixing of Rubber, Chapman and Hall, London (1997)
11. White, J.L.: Rubber Processing, Technology, Materials, Principles, p. 484–485. Carl Hanser, Munich (1995)
12. www. appropedia.org/polymer- calendering
13. Nakajima, N.: In: The science and practice of rubber mixing, p. 356. Rapra technology Ltd, UK (2000)
14. Sommer, J.G.: Rubber Chem. Techno. **58**, 672 (1985)
15. Wheelans, M.A.: Injection moulding of rubber. Hastead press, London (1974)
16. Hofmann, W.: Rubber Technology Handbook. Carl Hanser Publishers, Munich (1989)
17. Peterson, A.: Rubber Mixing Technology Course. Center for Continuing Engineering Education, Univ. Of Wisconsin, Milwaukee (1999)
18. Valsamis, L.N., et al.: Evaluating the performances of internal mixers. In: Grossman, R.F. (ed.) The Mixing of Rubber, Chapman and Hall, London (1997)
19. Tokita, N., White, J.L.: Appl Polym. Sci. **10**, 1011 (1966)
20. Tokita, N.: Rubber Chem. Technol. **52**, 387 (1979)
21. Banbury Mixer. Bulletin No. 224-C Farrel Corporation, Ansonia, CT
22. Shaw Intermix Mark 5 Series Farrell Corporation, Ansonia, CT
23. Asai, T. et al.: Presented at Proceedings of International Rubber Conference, Paris (1983)
24. Nortey, N.O.: Rubber World. **49** (1999)
25. Sheehan, E., Pomini, L.: Rubber World. **50** (1997)
26. Rubber machinery sourcebook Rapra Technology Ltd. Shawbury, UK (2000)
27. Smith, T.: Tire Technol. Int. **53** (1996)
28. Lovegrove, J.G.A.: Extrusion of Rubber Rapra Technologies Technical Report (1989)
29. Christy, R.L.: Rubber World **180**, 100 (1979)
30. Lambright, J.: In: Long, H. (ed.) Basic Compounding and Processing of Rubber, ACS Rubber Division, Akron (1985)
31. Cappelle, G.: Rubber Products Manufacturing Technology Marcel Dekker Inc. New York (1994)
32. Bhowmick, A.K., Mangaraj, D.: Electron beam processing of rubber. In: Bhowmic et al. (ed.) Rubber Products Manufacturing Technology, Marcel Dekker Inc, New York (1994) Division, Akron (1985)
33. Wood, P.R.: Tire Technol. Int. **134** (2000)
34. Dolezal, P.T., Johnson, P.S.: Rubber Chem. Technol. **53**, 253 (1980)
35. Freekley, P.K., Patel, S.R.: Rubber Chem. Technol. **58**, 751 (1985)

# Immiscible Rubber Blends

**C. M. Roland**

**Abstract** Most polymer blends are thermodynamically immiscible, leading to a phase-segregated morphology. Control of this morphology, including the domain sizes and interfacial regions, along with partitioning of compounding ingredients such as filler and curatives between the phases, provides opportunities for achieving properties that are otherwise unattainable. This chapter reviews fundamental aspects of phase-separated rubber blends, with a survey of the important literature on the topic.

Due to the vanishingly small entropy gain accompanying the mixing of high polymers, most polymer blends are phase-separated; there is no mixing at the segmental level, and the morphology is heterogeneous. The few thermodynamically miscible rubber blends include those having components exhibiting specific interactions (e.g., chlorinated polymers with epoxidized rubber [1, 2]); trivial blends of copolymers (siloxanes [3], polyolefins [4–6], nitrile rubbers (NBR) [7], ethylene-propylene rubbers [8, 9], butyl and polyisobutylene [10, 11], 1,4-polybutadiene (PBD) and styrene-butadiene rubber (SBR) [12, 13]); and miscellaneous cases such as 1,2-polybutadiene/1,4-polyisoprene (NR) [14], polyisobutylene/head-to-head polypropylene [15], polyepichlorohydrin/poly(vinylmethylether) [16, 17], and acrylate rubber/fluorocarbon copolymers [18]. The focus of this chapter is immiscible blends, in which the components are segregated into spatially distinct domains. These domains can range in size from a few hundred nm to microns, and usually have a very broad size distribution (Fig. 1) [19]. Except at the interface of these phases, the dynamics of the components are essentially the same as for the pure materials. However, immiscible blends can still yield novel and useful properties, provided the components are "compatible", a term loosely defined as a blend that does not undergo macroscopic phase separation and has some advantageous properties. Unlike miscible blends, the properties of which are roughly the average of

---

C. M. Roland (✉)
Chemistry Division, Naval Research Lab, Code 6120, Washington DC 20375-5342, USA
e-mail: mike.roland@nrl.navy.mil

P. M. Visakh et al. (eds.), *Advances in Elastomers I*,
Advanced Structured Materials 11, DOI: 10.1007/978-3-642-20925-3_6,

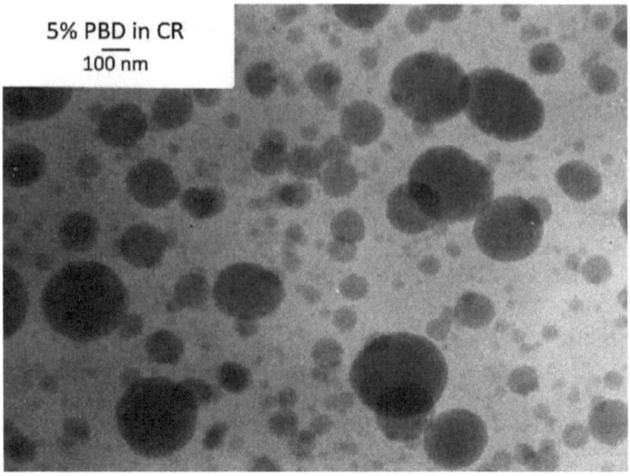


**Fig. 1** Transmission electron micrograph of a blend of 5 % 1,4-polybutadiene in polychloroprene. The mean diameter of the dispersed particles is 80 nm, with a very broad size distribution. A 100 nm scale bar is shown in the *upper left corner*. From Ref. [19]

those of the pure component, phase-separated blends can exhibit behavior not otherwise attainable. Some aspects are sensitive to the size of the domains, as well as the composition and interconnectedness of the interfacial regions. An important variable in heterogeneous blends is the spatial distribution of crosslinks, filler, stabilizers, etc. The ability to alter the phase morphology and the distribution of compounding ingredients offers the potential for performance benefits, and many commercial elastomers are phase-separated mixtures.

## 1 Morphology and Properties

The phases of an immiscible blend can be co-continuous, or one component can be dispersed within a continuous matrix of the other. The former is favored by equal concentrations and equal viscosities; that is, $\eta_1 \approx \eta_2$ and $\phi_1 \approx 0.5$ [20, 21]. This is illustrated in Fig. 2 for a blend of PBD and ethylene-propylene-diene terpolymer (EPDM) [20]. Most blends consist of discrete particles in a continuous phase, with the latter usually the lower viscosity component, provided it is present at a sufficient concentration. During mechanical mixing, domains of the lower viscosity material deform and encapsulate the higher viscosity phase, to produce a "globular" morphology. However, for an immiscible blend the morphology is never at equilibrium (which would correspond to macroscopic phase separation). The size distribution of the dispersed phase represents a steady-state balance between the breakup of the particles and their coalescence (Fig. 3) [19, 22–24], processes that continue throughout mixing and processing. Since to a first approximation the breakup of particles by the flow field is independent of particle concentration, whereas the coalescence probability increases with concentration, the expectation is that the dispersed phase size increases with $\phi$, in general accord with experimental results. The final particle size distribution depends both on the rheological

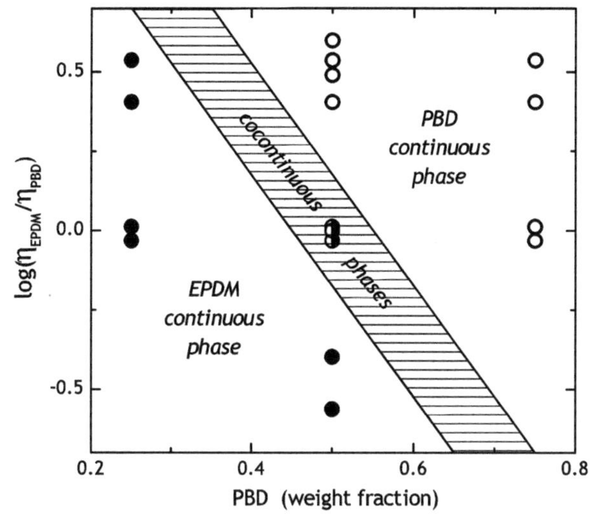


**Fig. 2** Dependence of the phase morphology on the viscosity ratio and composition for blends of PBD and EPDM. Dispersed PBD and EPDM particles are indicated respectively by *open* and *filled symbols*; *half-filled symbols* indicate a co-continuous morphology. Data from Ref. [20]

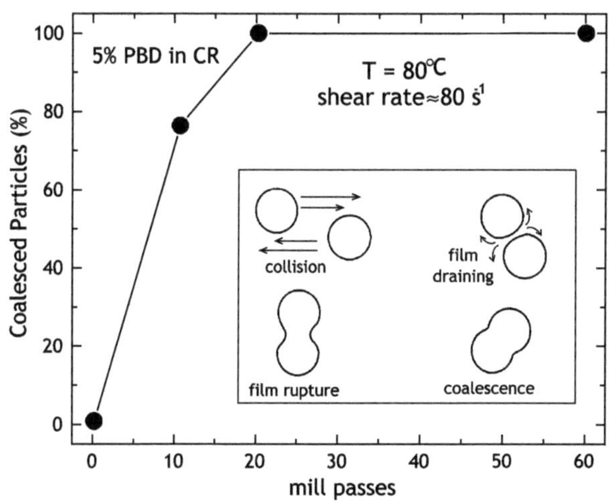


**Fig. 3** Amount of PBD ($M_w = 89$ kg/mol) dispersed in polychloroprene ($M_w = 255$ kg/mol) that has coalesced with other particles, as a function of the number of passes through a two-roll mill at the indicated conditions. The mean particle radius was 40 nm. The coalescence was monitored from the small-angle neutron scattering intensity; initially half the dispersed PBD was deuterated, so that coalescence reduced the scattering. Despite the highly viscoelastic nature of the materials and the brief contact time of the particles during flow, coalescence readily occurs and thus exerts a major effect on the blend morphology. In the *inset* is an idealized depiction of shear-induced coalescence of dispersed particles. From Ref. [19]

properties of the components and the type of mixing. Stretching flows are more effective at dispersion than shear fields (the latter a combination of stretching and rotational flow), and generally high stresses and strain rates produce a more finely dispersed phase. Nevertheless, the particle size distribution usually has only a modest effect on bulk properties, as seen in mixing rules for blend properties, which consider only the pure component properties and the relative amounts of the components [25, 26]. These include (written for the viscosity) series

$$\eta_{12} = \varphi_1\eta_1 + (1 - \varphi_1)\eta_2 \tag{1}$$

and parallel representations

$$\eta_{12}^{-1} = \varphi_1\eta_1^{-1} + (1 - \varphi_1)\eta_2^{-1} \tag{2}$$

and a log-additivity rule

$$\eta_{12} = \eta_1^{\varphi_1}\eta_2^{1-\varphi_1}. \tag{3}$$

These expressions are strictly empirical and can be generalized with a power-law expression [27]

$$\eta^n = \phi\eta_1^n + (1 - \phi)\eta_2^n. \tag{4}$$

Equations (1)–(4) ignore the effect of phase size and connectedness. The lack of predictive capability limits their utility [28], fitting to actual data often requiring additional adjustable parameters. Even for the simple case of Newtonian fluid mixtures, the viscosity can depend on the particle size, which in turn depends on the mixing or flow conditions [29–31]. Figure 4 [32] shows the variation of the

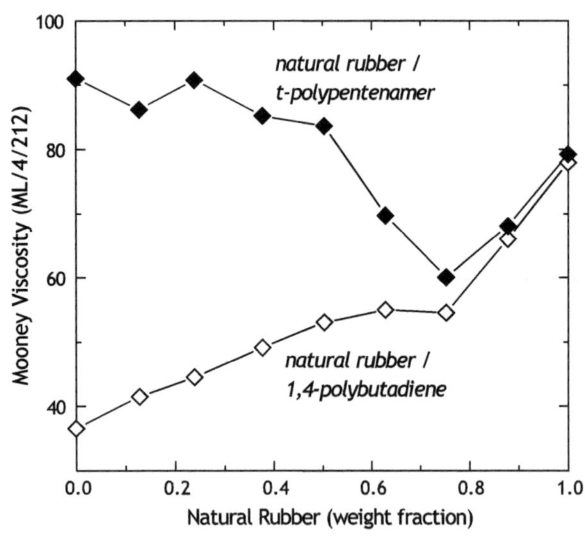


**Fig. 4** Mooney viscosity of blends of natural rubber with 1,4-polybutadiene (*filled squares*) and with *trans*-polypentenamer (*open circles*). Data from Ref. [32]

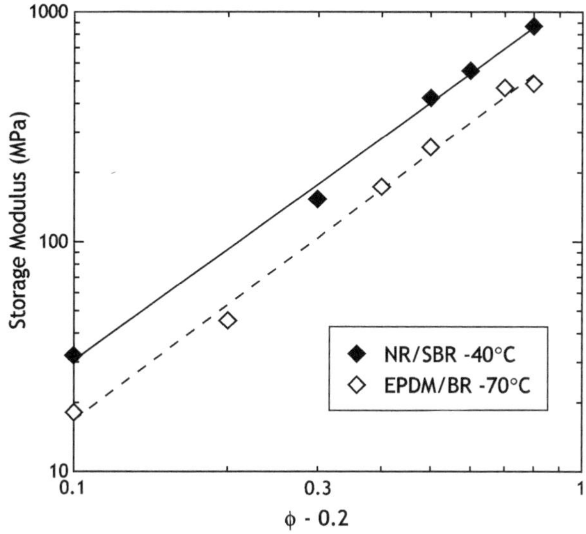


**Fig. 5** Dynamic storage modulus of blends of natural rubber and styrene-butadiene copolymer (*filled symbols*) and EPDM and 1,4-polybutadiene (*open symbols*) at the indicated temperatures plotted versus the concentration of the glassy component minus 0.2 (the latter representing the weight fraction at the percolation threshold). Vitrification of the higher $T_g$ component governs the magnitude of $G'$. Data from Ref. [35]

viscosity with composition for a blend of PBD and polypentenamer; the dependence is very complex and not obviously described by any mixing rule.

Expressions similar to Eqs. (1)–(4) are used to describe the modulus of blends. If the temperature is intermediate between the component $T_g$'s, the phase morphology can affect the blend stiffness in interesting ways, as any applied strain is manifested very differently for each phase. If the discrete particles are glassy, their influence on the modulus is similar to that of a conventional filler. Rubbery particles dispersed in a continuous glassy phase represents the morphology of rubber-toughened plastics [33, 34]. A continuous phase of higher $T_g$ gives rise to a dramatic increase in the modulus, as seen in blends of NR and PBD in Fig. 5 [35].

The failure properties of rubber blends are more sensitive to the details of the domain structure than other mechanical properties. Clarke et al. [36] obtained greater tensile and tear strengths in blends of NR and PBD when mixing was sufficient to reduce the domain size below 1 μm; no further improvement in properties was observed with further mixing (Fig. 6). Blends of a fluoropolymer with hydrogenated nitrile rubber exhibited the highest strength for intermediate compositions, associated with co-continuity of the phases (Fig. 7) [37].

The strain-crystallizability of NR can govern the cut growth behavior and other failure properties of its blends (Fig. 8) [38–40], in particular when the NR is present as a continuous phase [41]. Interestingly, one study found that NR/PBD blends exhibited a maximum in elongation to break for roughly equal concentrations of the components [42], which presumably yields a co-continuous phase morphology. Blends of polychloroprene (CR), which is also strain-crystallizable, with synthetic 1,4-polyisoprene (which has lower cis content and reduced crystallizability than natural rubber) exhibited greater tear strength with increasing CR content [43]. When dispersed as small particles, crystallization of a polymer proceeds more slowly than in the bulk, although the ultimate degree of crystallinity

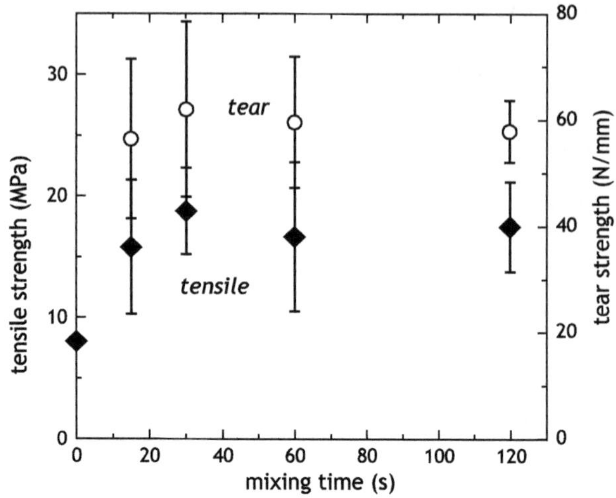


**Fig. 6** Effect of mixing time on tensile and tear strength of 50/50 blend of natural rubber and 1,4-polybutadiene [36]

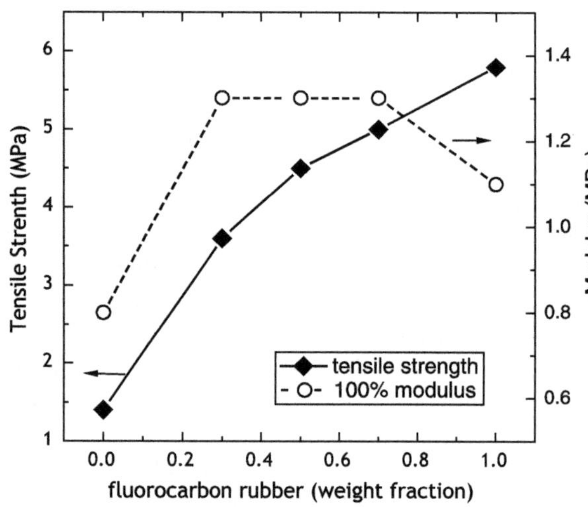


**Fig. 7** Stress at break (*diamonds*) and at 100 % extension (*circles*) for a blend of a perfluoromethyl vinyl ether polymer with hydrogenated nitrile rubber. Data from Ref. [37]

is the same [44, 45]. This effect is ascribed to a reduced nucleation rate, similar to the slower crystallization of NR latex when the rubber particles are smaller than 1 μm [46].

## 2 Compatibilization

Improving the compatibility of the blend components yields smaller and/or more interconnected phases, both of which can potentially improve the properties. This can be achieved through the use of compatibilizing agents or chemical modification of the components. Compatibilizers are surfactants that modify the

**Fig. 8** Fracture strength of NR, PBD, and a 50/50 blend as a function of size of edge cracks introduced into the test specimens. For cracks beyond about 2.2 mm, the material fails before reaching stresses sufficient to induce bulk crystqallization of the NR. Data from Ref. [39]

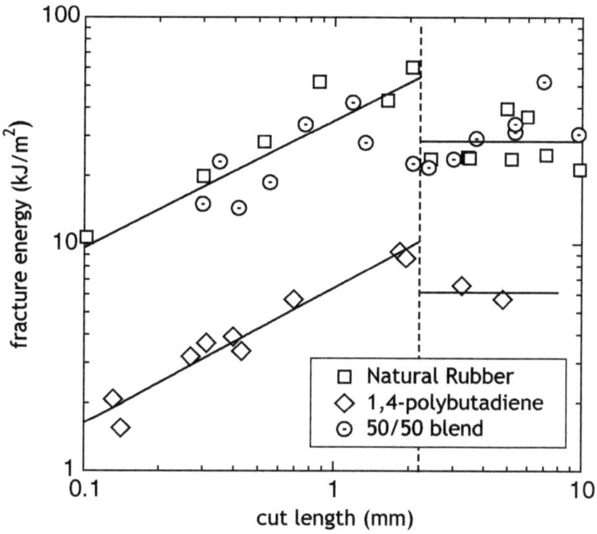


interfacial tension to decrease the dispersed particle size. Generally higher concentrations of modifiers yield smaller domains sizes up to the saturation point [47]. The compatibilizer may have other functions. For example, immiscible blends of NR and SBR intended for pressure-sensitive adhesives show a single glass transition in dynamic mechanical spectra when plasticized by large quantities of tackifying resin, the latter evidently functioning as a compatibilizer, in addition to improving adhesive performance [48]. Block copolymers can be used to compatibilize phase-segregated blends, by reducing the surface tension [49] and simultaneously enhancing the strength of the interfacial regions [50–52].

Examples of chemical modification to achieve better homogeneity include: EPDM modified with maleic anhydride [53]; acrylamide-grafted polydimethylsiloxane (PDMS) with sulfonated EPDM [54]; and mercapto-modified ethylvinyl acetate rubber in blends with NR [55]. Similar efforts have been made to compatibilize EPDM with PDMS [56, 57], NR [58] and PBD [59]. A variation on this approach is to obtain interphase crosslinking by oxidatively crosslinking the blend components. Molding at very high temperatures (200 °C) for extended times have been shown to compatibilize certain rubber blends [60].

Compatibilizers can have an indirect effect on blend morphology and properties when they function as plasticizers. Plasticizers and processing aids are used to reduce the resistance to flow of polymers by lowering the internal friction (viscosity), as well as the friction with the walls of mixers, extruders, roll mills, etc. By inducing slippage at the interface with the mixing vessel, shear flow is suppressed without affecting the extensional flow that most effectively disperses the components and any filler particles. (Note that dispersive mixing refers to breakup of these constituents into smaller sizes, and is different from distributive mixing, which gives a more spatially uniform concentration of the ingredients). However, lubricants can be depleted and

Fig. 9 Viscosity of an SBR with 50 phr N326 carbon black as a function of added PDMS measured by oscillatory (*open symbols*) and continuous (*filled symbols*) shearing at a rate equal to 4.5 $s^{-1}$. Data from Ref. [64]

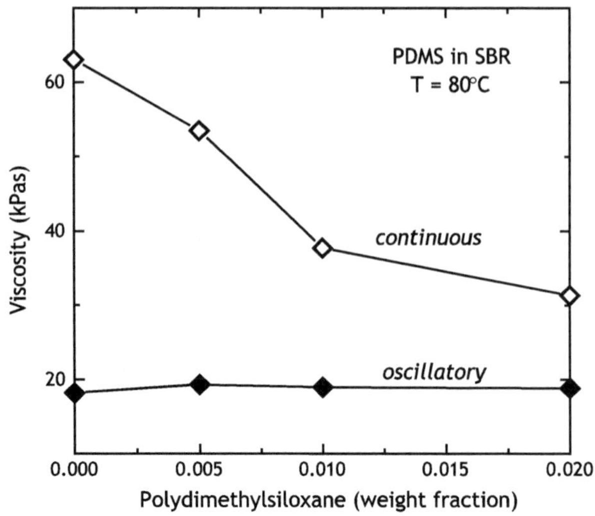


at high concentrations may affect the properties of the cured rubber. An alternative is to add small quantities of a polymeric component of lower viscosity. During processing, the lower viscosity component of a phase-separated blend tends to accumulate at the surface, where the shear rates are largest [61, 62]. The phenomenon of the morphology arranging to accommodate the applied stresses is referred to as the principle of minimum viscous dissipation [63].

Figure 9 [64] shows the viscosity measured dynamically and during continuous shear flow of an SBR containing low levels of a PDMS. The silicone has a viscosity ten times lower than the SBR (due to the lower molecular weight and higher entanglement molecular weight of the silicone polymer). Since the strain in the oscillatory experiment is spatially homogeneous, the PDMS is uniformly distributed and has negligible effect on the dynamic viscosity, given its low concentration. However, the apparent viscosity for flow through a capillary die is much lower for the blend, because the material at the surface becomes enriched with the PDMS [64]. The consequence is a non-uniform velocity profile and lower resistance to flow. Note that since the extensional flow is unaffected (being determined by the geometry of the die), the dispersion of carbon black in this compound was unaffected by the PDMS [64]. And since the total amount of the PDMS is low, bulk properties were also unchanged by its presence.

This segregation of two incompatible polymers can continue over time under quiescent conditions, governed by the diffusive mobility of the polymers. For example, Bhowmick et al. [65] observed that PDMS diffuses to the surface during aging of blends. (The opposite phenomenon—miscible components spontaneously interdiffusing—has also been reported [1].) Surface accumulation of one component can also result from interaction (chemisorption) with the walls of the processing equipment. This can lead to contamination of the walls by the adhering polymer. If there is strong incompatibility with the main component, the result can

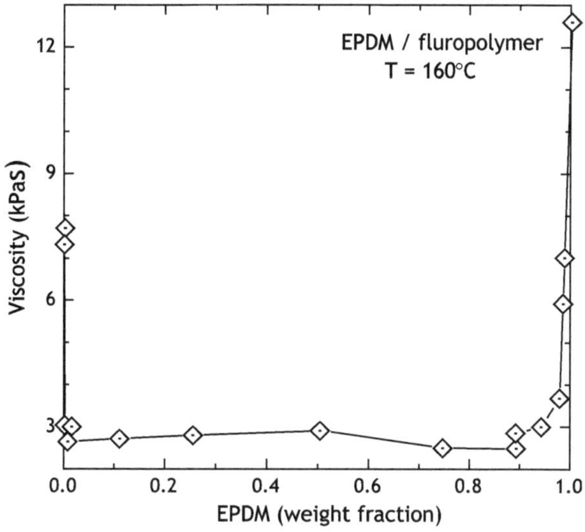


**Fig. 10** Viscosity of a blend of EPDM and a Viton® fluoroelastomer measured in a capillary rheometer at a nominal shear rate equal to 14 $s^{-1}$. Data from Ref. [66]

be slippage at the interface. This is seen in the enormous reduction in apparent viscosity in blends of EPDM and a fluorelastomer (Fig. 10) [66]. The blend exhibits a resistance to flow almost an order of magnitude smaller than for either of the neat components.

## 3 Distribution of Crosslinks

The distribution between the phases of curatives, plasticizers, antioxidants, fillers, etc. can have an effect on the cured properties of blends. Thus, non-uniformity of plasticizers can affect the phase morphology, through their influence on the component viscosities. Unequal partitioning of the antioxidants in a blend can result in inferior resistance to degradation and a shortened service life. A major issue in heterogeneous blends is the distribution of crosslinks. Uniform crosslinking generally gives the best properties, with severe imbalances resulting in over- or undercured material. Since crosslinking increases the glass transition temperature, a disparity in crosslinking of the components can alter their respective $T_g$'s in the blend (Fig. 11) [67]. Achieving uniform states of cure can be difficult, since the components may have different affinities for the curatives, common for blend components differing in polarity or degree of unsaturation. Even components of similar polarity and unsaturation can have cure imbalances if their crosslink reactivities are different, as has been observed in vulcanization and peroxide curing of NR/PBD blends [68, 69]. Curative depletion in the faster-reacting phase can induce diffusion from the other component [70]. The levels of sulfur and accelerators are usually below their solubility limits, which engenders curative migration [71–75].

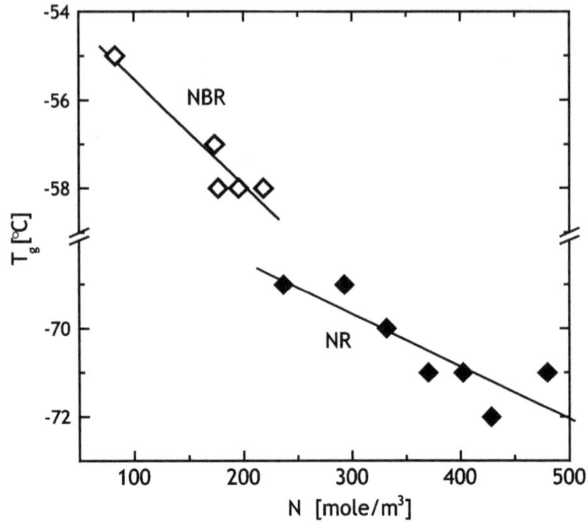


**Fig. 11** Glass transition temperature of each component in blends of natural and nitrile rubbers as a function of the number of repeat units between crosslink junctions. The cure system, tetramethylthiuram monosulfide with either *bis*-alkylphenol disulfide or free sulfur, preferentially resides in the more polar NBR phase, causing the latter to be two to five times more crosslinked. Data from Ref. [67]

An obvious solution to crosslink imbalances is to use curing chemicals with more nearly equal solubility and reactivity for the components, for example through judicious selection of vulcanization temperature [76] and accelerators [77, 78]. For example, changes in the accelerator altered the tensile strength of NR blends with acrylic rubber by more than a factor of two [77]. The curatives can also be premixed into the components at optimal concentrations prior to blending of the rubbers [79]; however, this increases the potential for prevulcanization (scorch) and does not address curative migration. One study found that when uniformly distributed initially, curatives are prone during mixing to take up residence within the continuous phase [79]. Precuring the more saturated component prior to blending can alleviate crosslink misapportionment [80, 81], although this may be impractical and will affect the phase morphology. Covalent bonding of the curatives to the polymers, of course, precludes migration [58, 82, 83].

For strength properties there is an additional requirement of achieving interfacial crosslinking in order to mutually adhere the domains. This is difficult when the blends have different reactivities, a common example being EPDM with NR or PBD. One approach to circumvent the problem is grafting accelerators to the components [82].

## 4 Distribution of Filler

Non-uniform filler distribution is an important issue with blends, since reinforcement of both phases is necessary to optimize physical properties. The affinity of carbon black, for example, varies among polymers (usually being higher in the more polar or unsaturated component [84]), so that non-uniform distributions in

**Fig. 12** Carbon black content of each component in immiscible blends of SBR with (*top*) 50 % epoxidized natural rubber and (*bottom*) natural rubber. The filler preferentially incorporates into the more polar phase. Data from Ref. [85]

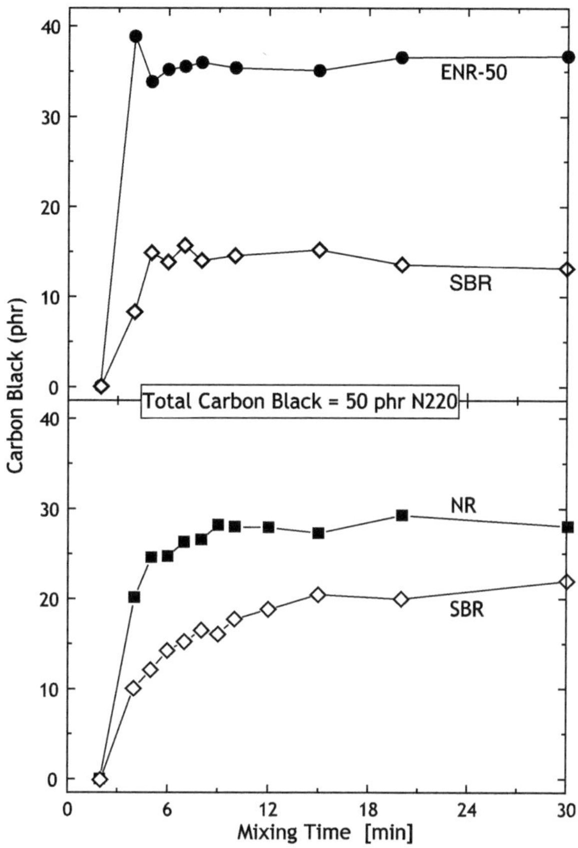


blends are common. This is illustrated in Fig. 12 [85], showing the different uptake of carbon black in blends of SBR with NR, a relatively non-polar polymer, and with a polar, epoxidized natural rubber. The carbon black distribution is more inhomogeneous for the latter.

At least for carbon black, the filler distribution obtained during mixing is irreversible, because the polymer chains adhere strongly to the particle surface. This can be taken advantage of by sequential addition of the filler or by adding the filler to the components prior to their blending; thus, the distribution between blend components is controlled by the mixing procedure [86]. Silica bonds weakly to polymers, so that the particles can transfer between phases during mixing. This is illustrated in Fig. 13 [87], in which the bound rubber was measured for each component as a function of mixing time. The higher rate of wetting of silica by NR leads to high silica content initially; however, over the course of the mixing some silica transfers to the SBR component.

Since the concentration of filler affects the melt viscosity, its inhomogeneous distribution can influence indirectly the phase morphology (see Fig. 1). Properties such as the elasticity and hysteresis depend non-linearly on filler concentration, so

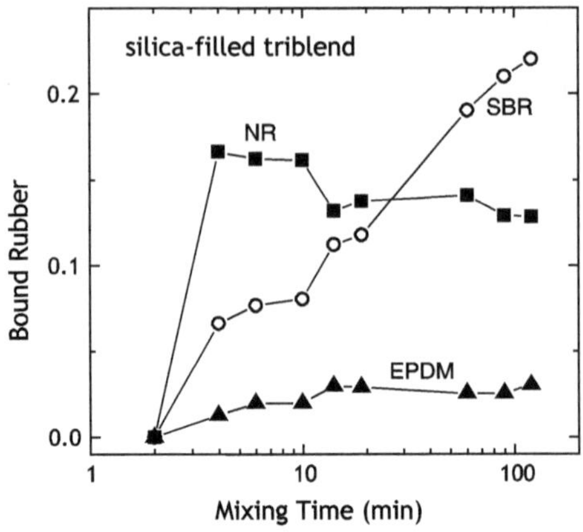


**Fig. 13** Bound rubber measured for each of the components of a blend of NR, SBR, and EPDM during mixing. *Note* transfer of the silica from the NR to the SBR. Data from Ref. [87]

that a non-uniform distribution can be exploited to achieve properties that depart from the mean of those of the components. This approach has been demonstrated to afford control of the die swell and elastic rebound of rubber blends [88–90]. Carbon black particles located at the interface between the components can give rise to electrical conductivity higher than achievable with neat elastomers [91–93], which can be useful for antistatic or shielding applications.

## 5 Summary

Among commercial elastomers, the volume of blends is probably as large as the volume of pure rubbers, and certainly the vast majority of the former are phase separated; i.e., thermodynamically immiscible but not incompatible. The problems with rubber blends outlined in this brief review are well-known, and various solutions have been developed. Given the difficulty and expense of producing new polymers, blends will continue to be an attractive source of new properties. One promising approach that is largely unexplored is nanoconfinement. It is well-established that unusually large surface to volume ratios and the intrusion of an external dimension on the length scale of polymers, including their coil size, changes the behavior from that in the bulk state [94–97]. It has been reported that in a blend, one component can experience nanoconfinement by the other, leading to large changes in the dynamic behavior [98, 99]. The effect of confinement on polymers is complex and there are many anomalies; however, with this complexity is the expectation of unique properties.

**Acknowledgment** This work was supported by the Office of Naval Research.

# References


1. Nagode, J.B., Roland, C.M.: Polymer **32**, 505 (1991)
2. Roland, C.M., Santangelo, P.G., Baram, Z., Runt, J.: Macromolecules **27**, 5382 (1994)
3. Kuo, C.M., Clarson, S.J.: Polymer **41**, 5993 (2000)
4. Lohse, D.J., Garner, R.T., Graessley, W.W., Krishnamorti, R.: Rubber Chem. Technol. **72**, 569 (1999)
5. Reichart, G.C., Graessley, W.W., Register, R.A., Lohse, D.J.: Macromolecules **31**, 7886 (1998)
6. Crist, B., Hill, M.J.: J. Polym. Sci., Polym. Phys. Ed. **35**, 2329 (1997)
7. Bartenev, G.M., Kongarov, G.S.: Rubber Chem. Technol. **36**, 668 (1963)
8. Ellul, M.D., Gent, A.N.: J. Polym. Sci. Polym. Phys. Ed. **23**, 1823 (1985)
9. Dikland, H.G., Van Duin, V.: Rubber Chem. Technol. **76**, 495 (2003)
10. Ellul, M.D., Gent, A.N.: J. Polym. Sci. Polym. Phys. Ed. **23**, 1953 (1984)
11. Hamed, G.R., Ogbimi, F.: Rubber Chem. Technol. **59**, 347 (1986)
12. Maier, T.R., Jamieson, A.M., Simha, R.: J. Appl. Polym. Sci. **51**, 1053 (1994)
13. Sakurai, S., Izumitani, T., Hasegawa, H., Hashimoto, T., Han, C.C.: Macromolecules **24**, 4844 (1991)
14. Tomlin, D.W., Roland, C.M.: Macromolecules **25**, 2994 (1992)
15. Krishnamoorti, R., Graessley, W.W., Fetters, L.J., Garner, R.T., Lohse, D.J.: Macromolecules **28**, 1252 (1995)
16. Alegria, A., Elizetxea, C., Cendoya, I., Colmenero, J.: Macromolecules **28**, 8819 (1995)
17. McGrath, K.J., Roland, C.M.: Rubber Chem. Technol. **67**, 629 (1994)
18. Kader, M.A., Bhowmick, A.K.: Rubber Chem. Technol. **73**, 889 (2000)
19. Roland, C.M., Bohm, G.G.A.: J. Polym. Sci. Pt. B-Polym. Phys. **22**, 79 (1984)
20. Avgeropoulos, G.N., Weissert, F.C., Biddison, P.H., Bohm, G.G.A.: Rubber Chem. Technol. **49**, 93 (1976)
21. Paul, D.R., Barlow, J.W.: J. Macromol. Sci. Rev. Macromol. Chem. **C18**, 108 (1980)
22. Peng, X., Huang, Y., Xia, T., Kong, M., Li, G.: Eur. Polym. J. **47**, 1956 (2011)
23. Sirisinha, C., Saeoui, P., Guaysomboon, J.: J. Appl. Polym. Sci. **90**, 4038 (2003)
24. Chen, X., Xi, J., Guo, B.H.: J. Appl. Polym. Sci. **102**, 3201 (2006)
25. Nielsen, L.E.: Predicting the Properties of Mixtures. Marcel Dekker, New York (1978)
26. Utracki, L.A., Kamal, M.R.: Polym. Eng. Sci. **22**, 96 (1982)
27. Nielsen, L.E.: Mechanical Properties of Polymers and Composites. Marcel Dekker, New York (1974)
28. Han, C.D.: Multiphase Flow in Polymer Processing. Academic, London (1981)
29. Frankel, N.A., Acrivos, A.: J. Fluid Mech. **44**, 65 (1970)
30. Choi, S.J., Schowalter, W.R.: Phys. Fluids **18**, 420 (1975)
31. Grizzuti, N., Buonocore, G., Iorio, G.: J. Rheol. **44**, 149 (2000)
32. Folt, V.L., Smith, R.W.: Rubber Chem. Technol. **46**, 1193 (1973)
33. Riew, C.K. (ed.): Rubber Toughened Plastics, Adv. Chem. Ser, Amer. Chem. Soc. Washington **222** (1989)
34. Collyer, A.A. (ed.): Rubber Toughened Engineering Plastics. Chapman and Hall, London (1994)
35. Kluppel, M., Schuster, R.H., Schaper, J.: Rubber Chem. Technol. **72**, 91 (1999)
36. Clarke, J., Clarke, B., Freakley, P.K.: Rubber Chem. Technol. **74**, 1 (2001)
37. Hirano, K., Suzuki, K., Nakano, K., Tosaka, M.: J. Appl. Polym. Sci. **95**, 149 (2005)
38. Hamed, G.R., Kim, H.J., Gent, A.N.: Rubber Chem. Technol. **69**, 807 (1996)
39. Hamed, G.R., Kim, H.J.: Rubber Chem. Technol. **72**, 895 (1999)
40. Lee, M.-P., Moet, A.: Rubber Chem. Technol. **66**, 304 (1993)
41. Kim, H.J., Hamed, G.R.: Rubber Chem. Technol. **73**, 743 (2000)
42. Taghvaei-Ganjali, S., Motiee, F., Fotoohi, F.: Rubber Chem. Technol. **81**, 297 (2008)
43. Kardan, M.: Rubber Chem. Technol. **74**, 614 (2001)

44. Gent, A.N., Kawahara, S.: Rubber Chem. Technol. **71**, 837 (1998)
45. Kawazura, T., Gent, A.N., Kawahara, S.: Rubber Chem. Technol. **76**, 1164 (2003)
46. Wathen, G.D., Gent, A.N.: J. Nat. Rubber Res. **5**, 178 (1990)
47. Favis, B.D.: Polymer **35**, 1552 (1994)
48. Copley, B.C.: Rubber Chem. Technol. **55**, 416 (1982)
49. Virgilioa, N., Desjardinsb, P., L'Espérancec, G., Favis, B.D.: Polymer **51**, 1472 (2010)
50. Teyssie, P., Fayt, R., Jerome, R.: Makromol. Chem. **187**, 837 (1986)
51. Galloway, J.A., Jeon, H.K., Bell, J.R., Macosko, C.W.: Polymer **46**, 183 (2005)
52. Puskas, J.E., Kaszas, G.: Rubber Chem. Technol. **74**, 583 (2001)
53. Kole, S., Roy, S., Bhowmick, A.K.: Polymer **37**, 625 (1996)
54. Kole, S., Roy, S., Bhowmick, A.K.: Polymer **35**, 3423 (1994)
55. Jansen, P., Gomes, A.S., Soares, B.G.: J. Appl. Polym. Sci. **61**, 591 (1996)
56. Kole, S., Bhattacharya, A., Tripathy, D.K.: J. Appl. Polym. Sci. **48**, 529 (1993)
57. Kole, S., Santra, R., Bhowmick, A.K.: Rubber Chem. Technol. **67**, 119 (1994)
58. Coran, A.Y.: Rubber Chem. Technol. **61**, 281 (1988)
59. Go, J.-H., Ha, C.-S.: J. Appl. Polym. Sci. **62**, 509 (1996)
60. Mukhopadhyay S., De S.K.: J. Appl. Polym. Sci. **45**, 2283 (1991); **45**, 181 (1992)
61. White, J.L., Ufford, R.C., Dharod, K.R., Price, R.L.: J. Appl. Polym. Sci. **16**, 1313 (1972)
62. Everage, A.E.: Trans. Soc. Rheol. **17**, 629 (1973)
63. MacLean, D.L.: Trans. Soc. Rheol. **17**, 385 (1973)
64. Roland, C.M., Nguyen, M.: J. Appl. Polym. Sci. **35**, 2141 (1988)
65. Bhowmick, A.K., Konar, J., Kole, S., Narayanan, S.: J. Appl. Polym. Sci. **57**, 631 (1995)
66. Shih, C.K.: Polym. Eng. Sci. **16**, 742 (1976)
67. Tinker, A.J.: Rubber Chem. Technol. **63**, 503 (1990)
68. Ellul, M.D., Patel, J., Tinker, A.J.: Rubber Chem. Technol. **68**, 573 (1995)
69. Mangaraj, D.: Rubber Chem. Technol. **75**, 365 (2002)
70. Bhowmick, A.K., De, S.K.: Rubber Chem. Technol. **53**, 960 (1980)
71. Gardiner, J.R.: Rubber Chem. Technol. **41**, 1312 (1968)
72. Van Amerongen, G.J.: Rubber Chem. Technol. **37**, 1065 (1964)
73. Shershnev, V.A.: Rubber Chem. Technol. **55**, 537 (1982)
74. Huson, M.G., McGill, W.J., Wiggett, R.D.: Plast. Rub. Proc. Appl. **5**, 319 (1985)
75. Bauer, R.F., Crossland, A.H.: Rubber Chem. Technol. **61**, 585 (1988)
76. Groves, S.A.: Rubber Chem. Technol. **71**, 958 (1998)
77. Wootthikanokkhan, J., Clythong, N.: Rubber Chem. Technol. **76**, 1116 (2003)
78. Mastromatteo, R.P., Mitchell, J.M., Brett, T.J.: Rubber Chem. Technol. **44**, 1065 (1971)
79. Leblanc, J.L.: Plast. Rub.: Proc. Appl. **2**, 361 (1982)
80. Suma, N., Joseph, R., George, K.E.: J. Appl. Polym. Sci. **49**, 549 (1993)
81. Sahakaro, K., Naskar, N., Datta, R.N., Noordermeer, J.W.M.: Rubber Chem. Technol. **80**, 115 (2007)
82. Baranwal, K.C., Son, P.N.: Rubber Chem. Technol. **47**, 88 (1974)
83. Hashimoto, K., Miura, M., Takagi, S., Okamoto, H.: Int. Polym. Sci. Tech. **3**, 84 (1976)
84. Callan, J.E., Hess, W.M., Scott, C.E.: Rubber Chem. Technol. **44**, 814 (1971)
85. Le, H.H., Ilisch, S., Kasaliwal, G.R., Radusch, H.-J.: Rubber Chem. Technol. **81**, 767 (2008)
86. Varughese, S., Tripathy, D.K.: J. Appl. Polym. Sci. **49**, 1115 (1993)
87. Le, H.H., Ilisch, S., Heidenreich, D., Osswald, K., Radusch, H.-J.: Rubber Chem. Technol. **84**, 41 (2011)
88. Lee, B.L.: Polym. Eng. Sci. **21**, 294 (1981)
89. Hess, W.M.: Rubber Chem. Technol. **64**, 386 (1991)
90. Sircar, A.K., Lamond, T.G., Pinter, P.E.: Rubber Chem. Technol. **47**, 48 (1974)
91. Sircar, A.K.: Rubber Chem. Technol. **54**, 820 (1981)
92. Miyasaka, K., Watanabe, K., Jojima, E., Aida, H., Sumita, S., Ishikawa, K.: J. Mater. Sci. **17**, 1610 (1982)
93. Ibarra-Gomez, R., Marquez, A., Ramos de Valle, L.F., Rodriguez-Fernandez, O.S.: Rubber Chem. Technol. **76**, 969 (2003)

94. G.B. McKenna, J. Phys. IV **10**, Pr7-53 (2000)
95. Fakhraai, Z., Sharp, J.S., Forrest, J.A.: J. Polym. Sci. Polym. Phys. Ed. **42**, 4503 (2004)
96. Koh, Y.P., Simon, S.L.: J. Polym. Sci. Polym. Phys. Ed. **46**, 2741 (2008)
97. Roland, C.M.: Macromolecules **43**, 7875 (2010)
98. Schwartz, G.A., Colmenero, J., Aglegria, A.: Macromolecules **40**, 3246 (2007)
99. Tyagi, M., Arbe, A., Colmenero, J., Frick, B., Stewart, J.R.: Macromolecules **39**, 3007 (2006)

# Rubber/Thermoplastic Blends: Micro and Nano Structured

**Cristina Cazan and Anca Duta**

**Abstract** Research on recycling, scarcely visible only a few decades ago, is now a very active, fastgrowing discipline, particularly focusing on wastes re-use as second raw materials. This chapter presents an overview on the state-of- art in recycling, the most recent technologies, and recent developments. Rubber and PET are the most frequently recycled polymers, and are particularly addressed to within this chapter. Recent results are presented on rubber/thermoplastic-based micro/ nano blends, along with their manufacturing and characterization methods. There are described methods to obtain the rubber-PET composites, based on ground discarded tires as a matrix composites, using as fillers plastic materials (PET, HDPE, and LDPE) and inorganic oxides (CaO, ZnO, and fly ash). Based on the structural and output properties and the chapter outlines the role of various components in the polymer composites. It is demonstrated that inorganic materials in the polymer composites allow obtaining performances unrecorded by pure polymer composites. However, the control of the inorganic material (type, quantity, particle size, and molecular structure) dispersed in polymeric matrix is essential in achieving the expected performance. Using different recipes, the composites can be tailored for various indoor and outdoor applications, as building materials as paving slabs, as thermal and electrical insulators, etc.

C. Cazan (✉) · A. Duta
Faculty Product Design and Environment, Transilvania University of Brasov,
500036 Brasov, Romania
e-mail: c.vladuta@unitbv.ro; crisvladuta@yahoo.com

A. Duta
e-mail: a.duta@unitbv.ro

P. M. Visakh et al. (eds.), *Advances in Elastomers I*,
Advanced Structured Materials 11, DOI: 10.1007/978-3-642-20925-3_7,

# 1 Introduction to Thermoplastic Blends

Polymers play a vital role in the field of material science nowadays. In performance characteristics, applications and diversity, they offer a degree of versatility, not found with any other kind of materials. Depending on the mechanical behavior, polymers are generally classified into three main categories: *rubbers or elastomers, thermoplastics and thermoset resins.*

*Thermoplastics* are resins that repeatedly soften when heated and harden when cooled. This is associated with the absence of chemical cross-links in their structure. Many are soluble in specific solvents and burn to some degree. Compared with thermosets, thermoplastics generally offer higher impact strength, easier processing and better adaptability to complex designs [1]. The thermoplastic can be sub-divided into two classes, commodity plastics (e.g. polyethylene PE, polypropylene PP, polystyrene PS) and engineering plastics (e.g. polyoxymethylene POM, polyamide PA, polycarbonate PC).

Most thermoplastics are relatively soft and ductile, consisting of linear and branched polymers with flexible chains. Thermoplastics may be (re)fabricated a number of times by the simultaneous application of heat and pressure without major changes in their properties. As the temperature increases, secondary bonding forces are diminished by increased molecular motion so that the relative movement of adjacent chain is facilitated when stress is applied. However, irreversible degradation occurs whenever the temperature of melting thermoplastics is raised to a point at which molecular vibrations become large enough to break the primary covalent bonds [2, 3].

*Thermoplastic polymer composites* are falling under increasing scrutiny due to their potential to be easily repaired and/or reshaped, making them easier to recycle and reuse compared with thermosetting matrix composites.

# 2 Introduction to Rubber: Thermoplastic Blends

A polymer blend such as rubber—thermoplastic can be obtained either when rubber-rich mixtures form soft thermoplastic elastomers, or when plastic-rich blends produce rubber toughened thermoplastic. Rubber toughened thermoplastics with flexible and high impact properties can be used as economical alternatives for ordinary plastic materials.

The market for these materials has grown dramatically because of their ability to be recycled and reprocessed, using conventional thermoplastic machinery. However, many studies show that the additions of rubber to a thermoplastic matrix results in a significant overall deterioration in mechanical properties [4, 5].

The properties of *rubber-thermoplastic blends* are reported to be depend on the type of rubber/thermoplastic used in the composition and on the adhesion between the phases. Further, the adhesion between the materials depends on the melt viscosity of the matrix phase, the shape and size of the dispersed phase, and the processing conditions. In addition, blend properties also vary with the moulding techniques used to make test samples [6].

*Rubber-thermoplastic blends* can be broadly classified into three types:

1. Impact resistant rubber toughened thermoplastics;
2. Blends of vulcanizable rubbers, which contain various amounts of resins that can act as reinforcing or stiffening agents;
3. Blends showing thermoplastic elastomeric behavior, commonly known as thermoplastic elastomers (TPEs) [7, 8].

Microstructured polymer blends represents one of the most interesting field of research in material science. In this context, the development of the first thermoplastic elastomers (TPEs) in the 1950s [9] provided a new dimension to the field of polymer science and technology.

## 2.1 Classification of TPEs

TPEs may be rationally divided into the following classes according to their chemistry and morphology:

1. (Soft) triblock copolymers; e.g., styrene–butadiene–styrene (S–B-S) or styrene-ethylenebutylene-styrene (S-EB-S);
2. (Hard) multiblock copolymers; e.g., thermoplastic polyurethane (TPU), thermoplastic polyamide (TPA);
3. Blends of rubbers with thermoplastics subject of dynamic vulcanization; e.g., polypropylene/ethylene propylene diene rubber (PP/EPDM), un-crosslinked or partially or fully crosslinked.

The field of TPEs based on polyolefin rubber—thermoplastic compositions, has grown along two distinctly different product-lines or classes: one class consists of simple blends and is commonly designated as thermoplastic elastomeric olefins (TEO) according to ASTM D 5593 [10]. In the other group, the rubber phase is dynamically vulcanized, giving rise to a thermoplastic vulcanizate (TPV) or dynamic vulcanizate (DV) according to ASTM D 5046 [11]. Numerous literature data are available on this subject [12–14], especially on thermoplastic vulcanizates (TPVs) or dynamic vulcanizates (DV).

Morphologically, TPVs are characterized by the presence of finely dispersed cross-linked rubber particles distributed in a continuous thermoplastic matrix [15]. If the rubber particles are sufficiently vulcanized, the physical and chemical properties of the blend are generally improved.

TPEs have both advantages and disadvantages in their practical use over conventional vulcanized rubbers [16, 17]. They are as follows:

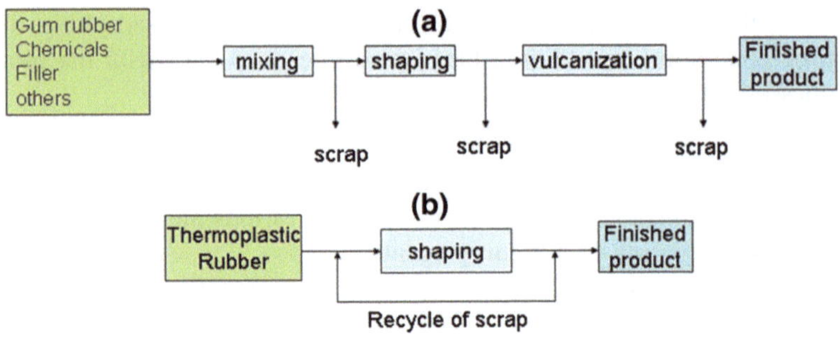


**Fig. 1** Processing scheme of vulcanized rubbers (**a**) and thermoplastic elastomers (**b**) disadvantages

*Advantages:*

- No vulcanization and very little compounding are required.
- Amenable to methods of thermoplastic processing, like injection molding, blow molding, thermoforming, heat welding etc., which are unsuitable for conventional rubbers; short mixing and processing cycles and low energy consumption. A common processing scheme for vulcanized rubbers and for TPEs is presented in Fig. 1.
- Scrap can be recycled and reused without significant performance deterioration;
- Properties can easily be manipulated by changing the ratio of the components;
- Better quality control;
- Good colorability;
- TPEs soften or melt at elevated temperature above which they lose their mechanical properties;
- Rubbery behavior;
- TPEs show creep behavior on extended use.

## 3 Rubber: Thermoplastic Based Micro Blends

In recent years, TPEs and TPVs have replaced conventional rubber in a variety of applications including appliance, automotive, medical, engineering, etc.

TPEs are obtained by copolymerization and by blending thermoplastics with rubber component. TPVs, on the other hand, are obtained by dynamically vulcanizing the rubber component in a rubber—thermoplastic blend during mixing. Both materials are processed like thermoplastics and are recyclable [18].

In most cases, TPEs are block copolymers consisting of soft and mobile 'rubbery' blocks with a low glass transition temperature ($T_g$), and rigid or hard 'glassy' blocks with a high melting temperature ($T_m$) and/or high $T_g$.

The rubber—thermoplastic blends are immiscible and their structure and morphology depend on many factors among which the flow history and the interfacial properties are the most important. At high dilutions and at low flow rates the morphology of polymer blends is controlled by three dimensionless microrheological parameters. Thus, the interfacial and rheological properties are the keys for the morphology development in polymer blends, which in turn is the controlling factor for their performance.

To improve performance of immiscible blends, usually they need to compatibilize. There are three aspects for compatibilization:

- Reduction of the interfacial tension that facilities fine dispersion;
- Stabilization of morphology against destructive modification during the subsequent high stress and strain processing;
- Enhancement of adhesion between phases in the solid state, facilitating the stress transfer, hence improving the mechanical properties of the product.

To insure compatibility as well as strong interfacial adhesion between the plastic and rubber phases still remains a major challenge in producing high performance TPE. The compatible and miscibility in a blend is not easy to achieve due to the different characteristics of each component that represents a separate phase with low attraction force along the phase boundaries. However, some miscible blends have been reported and variously interpreted as results of specific interaction, such as hydrogen bonding, dipole–dipole interaction, ion–dipole interaction or repulsive interaction. Some of the blends are also produced from chemical reaction, such as transesterification and the formation of covalent bonds within the constituents of the blends. Thus to improve the tensile properties of such blends, it is important to develop a proper control of phase morphology and better interfacial adhesion between the blend constituent via chemical or process approach by using compatibilizing agents or by dynamic vulcanization. There are three ways to do this by *physical*, *technological* or *chemical compatibilization.*

## 3.1 Physical Compatibilization

The major factors that control the compatibility in physical blend are the polymer chain entanglement, the high material viscosity at low processing temperature and by controlling the shear rate. Increased processing temperature may influence the thermodynamic miscibility. Increased shearing rate will decrease the size domain within the limits allowed by melt viscosity. However, extreme shearing force especially at low temperature and high viscosity may break polymer molecules into reactive macroradicals. Cross-combination of these radicals can then produce block or grafted polymers and prepare for the later technological compatibilization, [19].

### 3.2 Technological Compatibilization

Technological compatibilization of polymer blends can be produced or enhanced by simple physical addition of monomeric or polymeric material without expecting spontaneous chemical reactions to produce the desired properties. Addition of monomeric materials includes solvent(s), plasticizer(s), surfactant(s), and fillers which have been reported able to increase compatibility.

Another method of controlling the technological compatibilization is the addition of polymeric ingredients, based on the use of suitable block or grafted copolymer which are located at the interface between the phases of an immiscible blend and acted as emulsifying agent. However, this method cannot be applied for all types of polymer blends; it is based on the in situ formation of block or graft copolymer at the interface due to chemically reactions during melt mixing; this method is also called *reactive compatibilization* [20].

Reactive compatibilization of immiscible or incompatible blend can also be performed by the proper selection of blending ingredients, where third component addition is obviously miscible with one of the blend components and reactive with the other blend component. As a result, the emulsifier is produced in-situ and located at interface, and interacts with the phases via chain entanglement. High physicochemical affinity at both phases can strongly modify the morphology, interfacial adhesion, and final mechanical properties of the blends.

### 3.3 Chemical Compatibilization

Polymer blends are usually prepared by melt mixing process. Only a few of the available basic polymers are able to perform compatible blend in practical blending environment, without any chemical reaction. Thus, polymer modification during original polymerization reaction or modification after polymerization (post polymerization reaction) could be conducted in order to prepare such materials for compatibilization via block copolymerization, random copolymerization, attachment of terminal functional group and control of molecular weight [21].

## 4 Rubber/Thermoplastic Based Nano-Blends

The definition of nanocomposite materials has expanded significantly to encompass a variety of systems, such as the one-dimensional, two-dimensional, three-dimensional and amorphous materials, made up of separate components combined at nanoscale. Significant efforts have been concentrated for the control of nano structures through innovative synthetic approach. The properties of nanocomposites depend not only on the components properties, but also on the morphology and interface properties.

This area continuously generates the development of novel materials, with surprising properties. These properties result from combining the properties of primary components in a single material, most of the time expecting synergic effects. Experimental work showed that virtually all types and classes of nanocomposite materials lead to improved properties compared to their macrocompozite lines, and can be synthesized using rather simple and inexpensive techniques. Therefore, nanocomposites promise new applications in many sectors: low-mass components and mechanical properties improved non-linear optics, cathode ion batteries, nano-wires, sensors and other systems.

Nanotechnology is now one of the most promising opportunities for technological development in the 21st century. For research materials, development of polymer nanocomposites is rapidly becoming a multidisciplinary activity whose results may broaden the application of polymers, with benefits for many industries.

Polymer nanocomposites (PCN) have a polymer matrix (thermoplastics, thermosetting or elastomer) consolidated or reinforced with small amounts of nanoparticles (less than 5 % by weight). Nanocomposites are referred in this context to polymer systems containing inorganic particles with sizes in the nanometer range. Polymer nanocomposites have been developed in the late 80s, both in private research organizations and academic laboratories.

Most commercial applications focused on thermoplastics. Thermoplastics can be divided into two categories: frequently used resins (low cost) and high performance resins (expensive). Embedding nano-particles in a low cost matrix can provide low/average cost composite resins, with similar or better properties, comparing to the high-cost resins.

Thermoplastics combined with nanoscale materials may differ in properties from the bulk conventional composites (micro-sized or higher dimensions of the dispersed phase). Some properties of the nanocomposites, such as increased tensile strength can be obtained using a larger amount of conventional fillers, with the risk of an increased weight. Other nanocomposites properties, such as clarity or improved barrier properties can be obtained only by using the appropriate fillers, with the appropriate dimensions.

Rubber/thermoplastic based nano-blends are a new class of nanocomposites, which are particle-filled polymers for which at least one dimension of the dispersed particles is in the nanometer range. One can distinguish three types of nanocomposites, depending on how many dimensions of the dispersed particles are in the nanometer range [22].

- When the three dimensions are in the order of nanometers, we are dealing with *isodimensional nanoparticles*, such as spherical silica nanoparticles obtained by *in situ* sol gel methods.
- When two dimensions are in the nanometer scale and the third is larger, forming an elongated structure, we speak about *nanotubes* or *whiskers* as, for example, carbon nanotubes or cellulose whiskers which are extensively studied as reinforcing nanofillers yielding materials with exceptional properties.

- The third type of nano-fillers is characterized by only one dimension in the nanometer range. In this case the filler is present in the form of sheets of one to a few nanometer thick to hundreds to thousands nanometers long. This family of composites can be gathered under the name of polymer-layered nanocomposites (as example silica layers—polymer composites).

Owing to the nanometer-size particles obtained by dispersion, these nanocomposites exhibit markedly improved mechanical, thermal, optical and physical–chemical properties when compared with the pure polymer or conventional (microscale) composites.

## 5 Recent Studies on Rubber/Thermoplastic Based Micro-Blends

Polymer blending provides another path for developing new materials, leading to a significant number of thermoplastic—elastomeric products (TPE), commercialized during mid to late of 1980s [23].

The improved interfacial adhesion, morphology and mechanical properties of the blends are believed to be the result of the *in situ* formation of copolymer(s), at the interface of two immiscible polymers caused by energy treatment with or without the use of any chemicals. Recently, the blends of isotactic polypropylene (iPP) and uncured ethylene–propylene diene rubber (EPDM) treated by high power ultrasonic waves during extrusion have been investigated by Feng and Isayev [24].

The yield strength, elongation at break, tensile strength, and toughness of ultrasonically treated PP/EPDM blends were improved at certain conditions of ultrasonic treatment compared to those of untreated blends. These blends has been commercialized with trade name such as Santoprene and Geolast which posses high TPE oil resistance and performs with the versatility of rubber properties [25].

For some of the immiscible blends, with technological compatibilizations, the addition of small amount of compatibilizing agent in the blend during melts mixing could improve mechanical properties. It acts as a macromolecular surfactant and allows the formation of very small droplet of elastomer that will become small particles of vulcanized rubber when cured with dynamic vulcanization. There are limited publications concerning compatibilizing immiscible blends of TPE.

Natural rubber (NR)/HDPE are typically immiscible blends and to achieve NR/HDPE blend with practical value, various compatibilizers were used. The use of modified phenolic resin as a compatibilizer improved the mechanical properties of NR/HDPE blends due to chemical reactions between HDPE and the phenolic molecules [26]. Supri and Ismail [27] also reported, that the use of glycidyl methacrylate (GMA) on recycle polyvinyl chloride/acrylonitrile butadiene rubber blends (GMA + rPVC/NBR) has improved the mechanical properties and thermal stability, leading to a low swelling index. On the other hand, EPDM/nylon blends with maleic anhydride grafted EPR (MAH-g-EPR) also show better mechanical performance than other compatibilizing agent containing acid group due to better interfacial adhesion [24]. This again shows the beneficial effect of compatibilizer to the immiscible TPE blend.

There are many papers presenting results on the addition of functional polymers as compatibilizers. Basically, a polymer chemically identical to one of the blend components is modified to contain functional or reactive units. These functional units have some affinity for the second blend component, and have the ability to chemically react with it. However, other types of interaction such as ionic are also possible. The functional modification may be achieved in a reactor or via an extrusion modification process. Functionalised polymers (usually maleic anhydride or acrylic acid grafted polyolefins) are commercially available at acceptable cost to be used as compatibilizers. For example, the grafting of maleic anhydride (or similar compounds) to polyolefins, results in the formation of pendant carboxyl group which have the ability to form a chemical bridge with polyamides, via their terminal amino groups [28].

Papke and Karger-Kocsis [29] have studied the compatibilizing effect of ethylene/propylene rubber grafted with glycidyl methacrylate (EPR-g-GMA) in the thermoplastics elastomers/poly(ethylene terephthalate) blends. It was found that the blend compatibility with PET is strongly improved when the EPR-g-GMA is used as compatibilizer in the blends.

Papadopoulou and Kalfoglou [30] have studied the efficiency of maleic anhydride modified polyolefins on the PET/PP blend. Aravind [31] have studied the compatibilizing effect of maleic anhydride grafted ethylene propylene rubber (EPMg- MA) in ethylene propylene diene rubber/Poly(trimethylene terephthalate) (PPT) blends. It was found that the addition of EPM-g-MA reduces the domain size of the dispersed phase followed by a leveling off at higher concentrations of the compatibilizer. The addition of EPM-g-MA to the blends tends to decrease the free volume showing its compatibilizing effect.

Ethylene vinyl acetate copolymer (EVA) is obtained through chemical modification of polyethylene (PE) with vinyl acetate as co-monomer which reduces the PE crystallinity. Thus it has many characteristic of thermoplastic elastomers, which depends on the vinyl acetate content. EVA provides good mechanical properties, excellent ozone resistance, good weather resistance and relatively lower material cost [32, 33]. Epoxidized natural rubber (ENR) is a modified natural rubber having properties resembling those of synthetic rubber rather than natural rubber. ENR has unique properties such as good oil resistance, low gas permeability, improved wet grip and rolling resistance, coupled with high strength [34, 35].

There are several references on EVA blends with different types of rubbers such as nitrile rubber (NBR) [36], natural rubber (SMR 10) [37], styrene butadiene rubber (SBR) and etc. having potential use in various applications such as films, footwear, tubes and hoses.

Macaúbas and Demarquette [38] added styrene–butadiene–styrene (SBS) and styrene–ethylene/butylene–styrene (SEBS) triblock copolymers as compatibilizers in polypropylene/polystyrene system. The addition of compatibilizers to the PS phase resulted in a reduction of the interfacial tension following an emulsion curve. It was shown that for both compatibilizers the concentration at which the interfacial tension essentially levels off is smaller than the concentration at which the average radius of the dispersed phase essentially levels off.

The morphological, viscosity and interfacial tension results showed that SEBS is a better compatibilizer for the PP/PS blend than is SBS. Chen et al. [39] have investigated the compatibilization effect of triblock copolymers of poly[styrene-b- (ethylene-*co*-butylene)-b-styrene] and the diblock copolymer of poly[styrene-b- (ethylene-*co*-butylene)] in high density polyethylene/syndiotactic polystyrene (sPS) blends. Morphology observation showed that phase amount of dispersed sPS particles was significantly reduced on addition of the copolymers and the interfacial adhesion between the two phases was dramatically enhanced. Tensile strength of the blends increased at lower copolymer content but decreased with increasing the copolymer content. The elongation at break of the blends sharply increased with copolymers addition.

## 6 Recent studies on Rubber/Thermoplastic Based Nano-Blends

The nanoscale of dimensions is the transition zone between the macrolevel and the molecular level. Recent interest in polymer matrix based nanocomposites has emerged initially with interesting observations involving exfoliated clay and more recent studies with carbon nanotubes, carbon nanofibers, exfoliated graphite (graphene), nanocrystalline metals and a host of additional nanoscale inorganic filler or fiber modifications.

While the reinforcement aspects of nanocomposites are the primary area of interest, a number of other properties and potential applications are important including barrier properties, flammability resistance, electrical/electronic properties, membrane properties, polymer blend compatibilization.

The structure–property relationships of thermoplastic olefin (TPO)-based nanocomposites prepared by melt processing are reported with a main focus on the ratio of maleic anhydride-grafted polypropylene (PP-g-MA) to organoclay. The morphological observations by transmission electron microscopy, atomic force microscopy, along with X-ray diffraction are presented in conjunction with the mechanical and rheological properties of these nanocomposites, [40].

The effect of different nanofillers and compatibilizers (maleic anhydride-grafted-polypropylene and maleic anhydride-grafted-ethylene propylene diene terpolymer rubber) on the morphology, mechanical, mechanodynamical and thermal characteristics of thermoplastic olefins based on polypropylene and ethylene propylene diene terpolymer rubber blends has been analyzed. A better affinity with the matrix and a better dispersion of the nanoparticles is observed in rubber rich matrices. The maleic anhydride-grafted-ethylene propylene diene terpolymer rubber is a better compatibilizer for organo-clay nanocomposites based on rubber rich matrices [41].

Mirzadeh et al. [42] describe the effects of different polypropylene-g-maleic anhydride polymers, as compatibilizers, on the degree of exfoliation and co-continuity of thermoplastic elastomer (TPE) nanocomposites based on polypropylene/ethylene-propylene-diene monomer (PP/EPDM). X-ray diffractometry patterns and transmission electron microscopy (TEM) micrographs show nanocomposites range from an intercalated structure to a coexistence of intercalated tactoids (relatively ordered stacks) and exfoliated layers. The significant increase in crystallization temperature ($\sim$20 °C) could be beneficial for molding applications because of the faster solidification and shorter cycle time. The relaxation time obtained by stress relaxation experiments shows that the key parameter that determines the dispersion level in nanocomposites is the mobility of the compatibilizer.

Thermoplastic elastomer nanocomposites (TPE nanocomposites) based on PA6/NBR/Cloisite 30B were prepared through a direct melt mixing process in an internal mixer by Mahallati et al. [43]. The effects of NBR content (10, 30 and 50 wt. %) and nanoclay loading (3, 5 and 7 wt. %) on morphology and mechanical properties of the nanocomposites have been studied and compared with unfilled PA6/NBR blends as well. The TPE nanocomposites were characterized by X-ray diffraction (XRD), transmission electron microscopy (TEM), scanning electron microscopy (SEM), differential scanning calorimeter (DSC) and mechanical properties. Results suggested that the nanoclay is exfoliated into the TPE nanocomposite matrix. By adding the nanoclay, improved modulus of the prepared TPE nanocomposites was achieved.

Ganguly and Bhowmick [44] used atomic force microscopy for qualitative phase morphological mapping as well as quantitative investigation of surface forces measured at constituting blocks and clay regions of a thermoplastic elastomeric nanocomposite based on triblock copolymer: poly(styrene-ethylene-co-butylene- styrene) (SEBS) and organically modified nano-clay. The roughness and power spectral density analyses of surface topography provided the increment in random roughness of the nanocomposite surface compared to pristine SEBS surface. The same surfaces were examined by means of single point force-distance, and force-volume measurements. Large adhesive force of 25 nN and contact force of 260 nN were found in soft polyethylene (PEB) segments and higher cantilever deflection of 210 nm was found for clay regions of SEBS-clay nanocomposite.

There has been a great discussion on tribomaterials of polymeric nanocomposites in recent years. However, little study has been done to develop tribomaterials suitable for micro size devices with sufficient balances between tribological and mechanical properties. A study was conducted to develop new tribomaterials with high performance for micro mechanic and electric devices, made by blending vapor grown carbon fiber filled polybutylene terephthalate with various functionalized thermoplastic elastomer (TPE) based on styrene butadene elastomer. The tribological and mechanical properties of these injection-molded polymer blends were examined by Nishitani et al. [45], using three types of TPE based on styrene butadiene elastomer.

Excellent improvement in mechanical properties like tensile strength, elongation at break, and modulus was observed on incorporation of the nanofillers in the rubber/thermoplastic based nano blends. When the nanofillers were added to the plastic phase, the mechanical reinforcement is comparatively poorer due to partial destruction of the crystallinity.

## 7 Different Manufacturing Methods of Rubber/ Thermoplastic Based Micro-Blends

Rubber—thermoplastic blends have become technologically interesting for obtaining thermoplastic elastomers. Thermoplastic elastomers have many properties of the rubber, but can be processed as thermoplastics [46]. They do not need to be vulcanized during fabrication or finishing. Due to its unique fabrication ability and properties, they offer designers new flexibility in applications requiring soft-touch features, seals against fluid environment, impact protection and improved ergonomics. In addition, their general reputation of light weight, recyclability or reproducibility, chlorine free as well as environmental friendly behavior, they attracted special interest as alternative materials in several fields, such as cap and closures in house wares, sport appliances, wires and cables in automotive, electrical and electronic industries, footwear, wheels, etc. [47]. Thermoplastic vulcanisates (TPVs) are a particular family of TPEs, which are produced via dynamic vulcanization of non-miscible blends of rubber and thermoplastics, i.e. the selective crosslinking of the rubber with simultaneous mixing with the thermoplastic melt [48].

Bhowmick et al. [49] considered that the plastic acts as a continuous phase and allows the melt of the TPEs, whereas the dispersed rubber phase is responsible for the rubber elasticity. Various types of thermoplastics are used to prepare TPEs. These include polypropylene, low-density polyethylene, ultra-low-density polyethylene, linear low-density polyethylene, chlorinated polyethylene, polystyrene, polyamide, ethylene vinyl acetate (EVA) copolymer and poly(methyl methacrylate).

The first step in the design of a rubber—thermoplastic composite material is an appropriate choice of the fabrication method. There are many widely used methods to prepare rubber—thermoplastic composite materials, several of which are outlined below.

### *7.1 Injection Molding*

Injection molding is by far the most used technique in TPE processing due to its high productivity and because it is a clean process with no waste formation. Injection molding is the process by which flow state plastic is brought under adequate pressure and then is introduced into a mold. It is used in a large variety of applications, ranging from tubes or foams to finished articles; it can be applied to the co- or insert-injection. The use of hot runner methods in injection molding was reviewed by Lachmann [50] and this is an interesting diversification of the conventional technique; it maintains the flow ability of the melt during transportation to the individual cavities or to the individual gates. During injection molding, TPEs behave as the other thermoplastics in hot runner, without major problems.

## 7.2 Compression Molding

Compression molding is less used than injection molding. Compression modeling applied pressure on the exact dosage pressed powder, locked in a heated mold, when under heat influence runs a continuous polycondensation process and results in the products' stiffening, [51, 52].

## 7.3 Extrusion

Extrusion is the continuous processing whereby materials in the plastic state are forced to pass through a channel that gives them shape.

The extrusion of TPEs was reviewed by Knieps [53]. This processing technique is essential in the shaping of many different profiles; the use of single-screw extruders is predominant, but some other extruders are also used, such as those equipped with three-section or barrier screws. Extrusion is also applied to other shapes: foams, tubes, sheets, etc.

## 7.4 Blow Processings

The blow processing is a method that uses a gas (air, $N_2$) to expand a hot preform against a mold pattern to produce a hollow object.

Extrusion and injection blow molding of TPEs are particularly important whatever the shape: bottles, boots, etc. They were reviewed by Nagaoka [54], who showed that the parameters controlling the processing are similar to those which control extrusion and injection molding. Blow processes are also used to prepare TPE foams.

## *7.5 Thermoforming*

Thermoforming represents shaping of a hot plastic sheet or plate, usually to obtain a desired object shape. The number of references relative to the thermoforming of TPEs drastically increases, particularly in the last three years; most of them are patents and apparently this TPE processing technique did not yet reached its maturity [55].

## *7.6 Calendering*

Calendering is a process of rolling the material in plastic state, among warmed up cylinders that rotate in opposite direction. The distance between the cylinders regulates the sheet thickness.

# 8 Different Manufacturing Methods of Rubber/ Thermoplastic Based Nano-Blends

One of the barriers in the design and production control of nanomaterials with predefined properties is the lack of understanding of fundamental processes at nanoscale. To develop new materials by nano-level design should be able to define the relationship between the (bulk) molecular structure and how they behave in reality (surface, boundary). This requires understanding how small changes affect the macroscopic properties and the ability to reproduce these models in strictly controlled samples. The polymer nanocomposites properties are mainly depending on the polymer-nanofiller interface and on the nanofiller volume fraction.

Diverse manufacturing technologies of these materials involve machines and processes for: the procurement of the polymeric matrix, preparing the reinforcing components, fibers/fillers impregnation or treatment, making the reinforcement (network, netting, web, etc.), making the adequate composition through injection, extrusion, compression—molding, other proceedings. Basically, for every type of polymeric nanocomposite material a distinct technology, with specific operations and devices is necessary. Nanocomposite materials procurement technologies are a lot more different than the ones applied in the case of classic materials (metals, glass, ceramic, etc.). The choice of the optimum technology is made according to the shape and size of the pieces, the size of the fabricated products, and the nature of polymeric matrix, along with the product's quality and costs, [56].

# 9 Characterization Methods of Rubber/Thermoplastic Based Micro-Blends

The preparation of TPEs is closely related to the structure and morphology control, as they are structurally complex systems, requiring accurate, efficient, and rapid investigation techniques. The characterization techniques are grouped into analytical branches, but it is essential to keep in mind that most of them are relevant when associated.

## *9.1 Chromatography*

For a long time, the use of chromatography was mainly limited to molecular weight determination and to the qualitative estimation of the samples heterogeneity. At present, it is a highly efficient technique, giving very accurate values [57, 58]. Copolymers are complex macromolecular systems, characterized by two distributions in molecular weight and chemical composition; liquid chromatography used at the critical point of adsorption allows the determination of the molecular heterogeneities (chemical and molecular weight distributions) [32]. The association of chromatography with other techniques allows, for instance, the analysis of the molecular mass distribution of each block, in one run, as well as the respective chemical distribution.

## *9.2 Spectrometric Techniques*

*FT-IR spectroscopy* allows establishing the structure of an unknown compound, outlining different groups of atoms and their mobility restrictions. Chemical structure problems can be solved based on the characteristic vibration frequency of certain groups of atoms, which change in a relatively small extent under the influence of neighboring atoms and bonds. FT-IR spectroscopy is often associated with NMR [59] or X-ray diffraction [60].

*NMR spectroscopy* is essential in TPE characterization; it is often associated with other techniques. This technique allows a deep insight in the structure of the block copolymer chains. NMR is often used as a routine technique because it is efficient in the control of the TPE purity. The evaluation of accurate values of the copolymers molecular weights and their distribution, as well as of the functionality of di-functional oligomers remains a difficult problem. NMR is probably one of the most efficient tools in the evaluating the end-group concentration. In some cases, the end-units are converted into groups, which are fluorescent or adsorbing in the visible or UV light; sometimes they are converted into chemically titratable groups.

## 9.3 Scattering Techniques

*X-ray diffraction* is an important tool in assessing the crystalline composition, for instance in the investigation of the semicrystalline [61] or liquid-crystalline blocks [62] and of the stretching behavior of TPEs [63, 64].

## 9.4 Microscopies

*Atomic force microscopy* (*AFM*) has an important development in the structural analysis of TPEs. It was applied to different problems: thermo-oxidative stability and morphology, copolymers with arborescent blocks, thermoplastic vulcanizates, blends and morphology and orientation during deformation studied both by AFM [65–67].

*Transmission electron microscopy* (*TEM*) is most useful to characterize the structure and morphology of TPEs. It is almost always associated with other techniques, (Atomic Force Microscopy, AFM). TEM is used in studies of interpenetrating networks, morphology and crystallinity of hard blocks, structural evolution of segmented copolymers under strain, and blends [68–70].

## 9.5 Thermal Techniques

Differential scanning calorimetry (DSC) and other thermal analyses are frequently used in TPE characterization, often as routine methods. DSC is a powerful technique to improve the knowledge of the microphase structure. DSC and thermomechanical analysis (TMA) are often complementarly used to other techniques (particularly spectroscopic analyses) or for providing additional information.

## 9.6 Contact Angle Measurements

The study of the interfacial phenomena is important for composite materials because the interface is a frequent cause of damages that compromises the composites reliability. The evaluation of the surface energy (polar and non-polar components) can be done by contact angle measurements.

An important role in optimal adhesion is played by the chemical properties of the solid surface, influencing the surface tension, [71]. Surface tension, contact angle and surface properties are a consequence of the intermolecular interactions on the surface of the uncompensated condensed phase (S, L) with a fluid (liquid, vapor or gas). The surface tension depends on both participants.

## 10 Characterization Methods of Rubber/Thermoplastic Based Nano-Blends

In order to characterize the properties of the polymer nanocomposites and understand their responses to external stimuli, such as stress, thermal or concentration gradients, it is important to have an insight into their structure and morphology. The properties of rubber—thermoplastic nanocomposites are mostly governed by two major factors:

- Dispersion and distribution of the nanofillers within the polymer matrix;
- Interactions between the polymer chains and nanofillers.

Both factors play important roles in deciding on possible applications for the final product. Over the years, various techniques have been devised to investigate the nanocomposites structure. A list of the commonly used techniques that are widely used in nanocomposites studies are further presented [72]:

(a) Structural property characterization

- *Scattering techniques*: X-ray scattering, small angle light scattering, small angle neutron scattering;
- *Microscopic techniques*: electron microscopy, atomic force microscopy;
- *Spectroscopic techniques*: FTIR, UV–VIS, nuclear magnetic resonance;
- *Chromatography*: gel-permeation chromatography.

(b) Bulk property characterization:

- *Melt state rheometry*: shear rheology, extensional rheology
- *Solid state analysis*: mechanical testing, barrier properties, optical properties.

(c) Thermal property characterization:

- *Calorimetry and others*: differential scanning calorimetry, thermal gravimetric analysis, heat distortion temperatures.

Those characterization methods can also be meet at micro-level characterization of the polymer composites.

## 11 Applications

Disposal of used rubber into landfills has become increasingly prohibitive due to high cost, legislative pressures, public opinion, especially high environmental stress. Researchers pay much attention to recycling. Many rubber products are now recycled, such as used tires.

One important step is processing or reduction of whole tires into useful parts. Sadhan [18] noted that the processes used for rubber grinding are based on cutting, shearing, or impact, depending on the equipment and the grinding conditions. There are mentioned: cutting, milling, extrusion, ambient and wet grinding, cryogenic grinding, etc.

Shredded tires are used as fillers in highways construction because they increase the abrasion resistance and enhance the resilience. Concentrations of tire rubber used for these purposes range from 10 to 20 % [18].

The best way to recycle rubber products would be to devulcanize and reuse them in the rubber industry. Processes for devulcanization, including chemical, thermal, thermomechanical, and ultrasonic, have been worked out, but they are costly and not always suitable for commercial application, particularly in manufacturing highly engineered produces like tires.

Sadhan [18] summed up that another alternative is to blend the crumb or ground rubber with a material having the ability to flow under heat and pressure, so that it can be shaped into useful articles at a reasonable cost. This can be accomplished by mixing finely ground rubber with plastics, along with necessary additives.

A rubber—thermoplastic composite is one of the most important materials and prevails in a wide region. In rubber composites products such as tires, belts, and so on, fibers are used as reinforcements. Among all fiber reinforcing materials, PET seems to be a promising material because it has good mechanical properties, particularly in high-modulus and low shrinkage. However, PET fiber is a relatively inert polymer, and as a result, it merely possesses insufficient adhesion to rubbers. Therefore, a third component is necessary to enhance adhesion between PET fibers and rubber [50].

Good interfacial adhesion and reduced interfacial tension between the components can be obtained for compatibilized PET and rubber materials. The compatibility can be obtained by the addition of other polymers, such as high density polyethylene (HDPE), low density polyethylene (LDPE) and/or inorganic compounds ($TiO_2$). Inorganic oxides (CaO, ZnO and fly ash) were also tested for enhancing the mechanical properties.

Using different recipes, the composites can be tailored for various indoor and outdoor applications. The chapter presents composites (in terms of physical, chemical and mechanical properties) for applications as building materials as paving slabs, as thermal and electrical insulators, etc.

## *11.1 Development of Novel Nanocomposites: PET—Rubber—Metal Oxides*

Composites with inorganic structures embedded in the polymer matrix represent a new class of materials with very interesting specific properties. It is demonstrated that inorganic materials in polymer composites result in performances unrecorded by pure polymer composites. However, inorganic material type, quantity, particle size and molecular structure when dispersed in polymeric matrix are essential in achieving the expected performance [73, 74]. Thus, inorganic oxides (CaO, ZnO) of nanoscale dimensions can be used; as an oxide mixture, another waste, fly ash, was also tested.

Composites based on recycled materials, obtained by compression molding were investigated; the sample composition was rubber:polyethylene:HDPE:metal oxide = 59.75:35:5:0.25.

Differences occurring between the samples with different fillers are related to their influence on the interface and on the thermo-oxidative stability. Specific characteristics recorded during mechanical testing proved the superiority of the composites containing CaO. Tensile tests, compression and impact resistance show that samples containing CaO develop significant organic–inorganic physical interactions, without forming any new chemical species.

The fly ash collected from the electro-filters of a CPH has no significant influence on the thermal behavior of the samples, without modifying the transition temperatures, but the presence of a porous, heterogeneous structure can be noticed as an advantage in further developing good organic–inorganic interfaces.

## *11.2 Obtaining Rubber: PET Composites*

"Blending of recycled wastes for developing materials with controlled properties, embedded in novel products represents a process which has both, economic and environmental advantages. The main focus is to find solutions for recycling those polymeric wastes that are traditionally difficult to use in re-processing, as PET. Therefore, rubber—PET composites represent a tempting alternative. Tire rubber is a low cost second raw material that can easily act as matrix, but the amount of PET able to be mixed with rubber is limited by the low compatibility between these two components, as result of the low intermolecular forces between the non-polar rubber and polar PET macromolecular chains. These structural differences have also consequences on the different thermal behavior (significantly different Tg and Tt), therefore process optimization represents another issue to be solved. Many studies were done aiming at improving the PET-rubber compatibility. Obvious compatibility agents are low molecular compounds with hydrophilic-hydrophobic dual behavior; although efficient, these are expensive and toxic, increasing the costs and the environmental burden. Therefore we propose to test various other polymeric wastes as compatibility agents, up to a high PET content, [75]".

To obtain composites based on recycled materials there were used:

- Poly-ethylene terephthalate (PET)—obtained from soft drink bottles;
- Rubber—from tires and bicycle tubes;
- High density polyethylene (HDPE)—from yogurt containers, etc.;
- Low density polyethylene (LDPE)—from packaging.

The mechanical properties and the environmental behavior can be tailored by dispersing different oxide powders. Efficient dispersing occurs when the powders

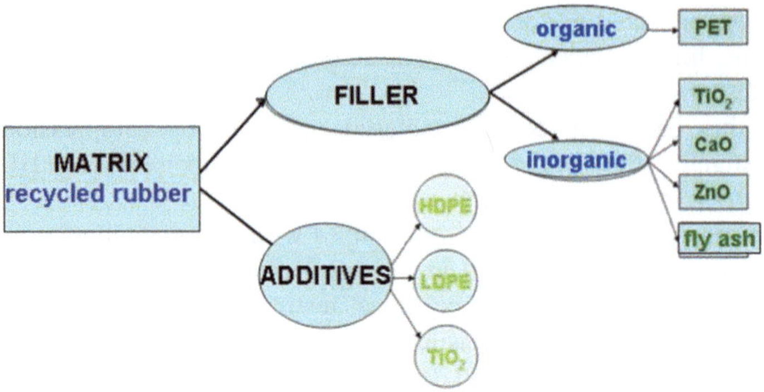


**Fig. 2** The composition of composites based on recycled rubber

are in the range of nano- or mezzo- scale, resulting hybrid inorganic–organic composites. Therefore, blends were developed using:

- $TiO_2$ powder (Degussa P25), with an average size of 21 nm primary and specific surface 50 $m^2/g$ $TiO_2$ nanoparticles consisting of 75 % anatase and 25 % rutile.
- CaO powder (Reagent Bucharest, Romania, purity >90 %)
- ZnO powder (Scharlau, purity >99 %)
- fly ash powder with the following composition: $SiO_2$ (53.32 %), $Al_2O_3$ (22.05 %), $Fe_2O_3$ (8.97 %), CaO (5.24 %), MgO (2.44 %), $K_2O$ (2.66 %), $Na_2O$ (0.63 %), $TiO_2$ (1.07 %), MnO (0.08 %), the rest being unburned carbon, (CPH Brasov, Romania)

It is important to notice that these effects are obvious if the nano-particles do not agglomerate, forming micro- sized clusters.

## *11.3 Composites Structure*

These materials have different roles in the composites, as shown in Fig. 2

## *11.4 The production Method of Composites*

To obtain rubber—PET composites based on recycled materials, there have been followed the steps outlined in Fig. 3:

- Cleaning by washing the recycled rubber and plastics, to remove impurities and labels;

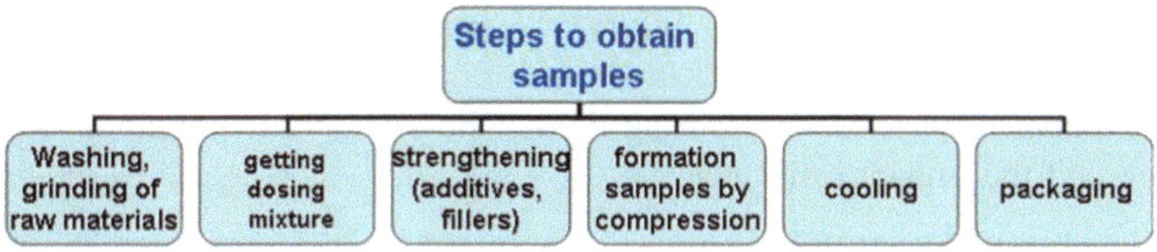


**Fig. 3** Steps to obtain samples

- Grinding materials: grinding was performed using cutting tools (guillotines, shears) to obtain pellets about 1–2 mm size and further using a centrifugal mill ZM 200 to obtain particulate sizes of 0.5–1 mm.
- Components mixing, dosed by various recipes and strengthening using additives;
- Forming composite—produced by compression molding, resulting in samples of rectangular shape (10 × 10 × 120 $mm^3$);
- Cooling—for completion the samples shape;
- Removal of the mold from the samples;
- Product packaging.

## *11.5 Techniques for Characterization of Rubber—PET Composites Laboratory*

The samples were tested and analyzed in the Department for Renewable Energy Systems and Recycling within the Transilvania University of Brasov, following the characterization methods included in Fig. 4.

The samples were analyzed in terms of:

- *Physico-chemical properties,* by FTIR and contact angle measurements, DSC, XRD and AFM analysis, following the structural changes and possible physical

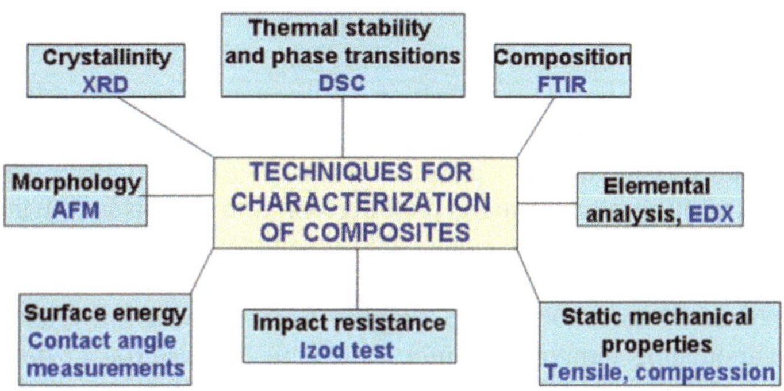


**Fig. 4** Techniques for characterization of composites

and chemical interactions at interface matrix—filling material, with proper amendments;
- *Mechanical properties*—measuring the tensile strength, the compression resistance and the impact resistance.

For each experiment three samples were tested, considering the value obtained by mediating three test results.

## 11.6 Steps to Optimize Experimental Parameters Composition and Technological Parameters

The optimal conditions for obtaining composites based on recycled rubber and plastic products were identified by studying the influence of the following parameters:

- composition parameters—the type of recycled rubber, PET percentage of composite, additives used, percentage of oxide in the inorganic—organic composites;
- process parameters—molding duration, temperature, distribution patterns of the composite components.

The composition and process parameters, were investigated in the following sequence:

- Optimization the composites composition, using rubber as matrix;
- Establishing the optimum molding duration;
- Choice of the recycled rubber type;
- Distribution of the components in composite;
- Molding temperature;
- Maximal PET percentage in the rubber—PET—HDPE composites;
- Choice of the additive;
- Optimizing $TiO_2$ percentage in inorganic—organic composites;
- Development of rubber—PET—HDPE—metal oxides composites ($TiO_2$, CaO, ZnO, fly ash).

Optimizing the composites composition with recycled rubber matrix was done by testing several sets of samples:rubber—PET, rubber- HDPE and rubber—PET—HDPE. The initial composition was considered:rubber:PET:HDPE = 85:10:5.

Starting from this composition, molding duration for obtaining the samples was optimized. The study was conducted on samples obtained in a temperature domain ranging from 180 to 260 °C, at processing durations of 45 min and, respectively of 60 min. Obtaining samples from recycled material requires a longer compounding duration (60 min.) allowing the material to turn from the solid into a high-elastic state; the optimal temperature depends on the recycled materials and mainly on their transition temperatures ($T_g$, $T_m$). For the three recycled materials (rubber, PET and HDPE) the optimal temperatures range from 220 to 260 °C. Samples with 85 % of rubber have as optimal the temperature of 220 °C.

Recycled rubber can result from different products, with different compositions (mainly with a different content of sulfur, oxides, etc.). Due to these micro-elements, the rubber matrix behavior can be very different. Therefore, the type of rubber was selected by preserving the previously optimized parameters, using two different waste sources: from the tires and from bicycle tubes. Following the EDX analysis, the sulfur content was found higher in the tires. The samples were mechanically tested, proving that samples with *high content of sulfur* and *zinc* (rubber from tires) obtained at temperatures above 220 °C, show high tensile strength and compression resistance, as result of crosslinking processes. Study on the influence of rubber type on mechanical properties of PET-rubber composites indicates that certain fillers present in the recycled rubber (S, Si, Zn, carbon black) must be controlled, typically because of their crosslinking potential, actually sometimes unwanted. According to the mechanical properties obtained, composites based on rubber tires, can be used as construction materials, such as paving slabs, while the rubber composites, with lower sulfur content (more elastic) can be used as sound/noise absorbing materials, for road impact damping places or for children playgrounds.

To test the influence of the components distribution in the composite on the mechanical properties, two sets of samples were developed—randomly mixed and layered. For the layered samples there were recorded high levels of compression forces however, the risk of delamination is high, so these composites can be recommended as pavement materials for indoor dry spaces, while randomly mixed composites can be used for outdoor applications.

The effect of additives on the rubber—PET interface was investigated based on the interface properties, preserving the processing conditions. Samples were obtained with the rubber composition:PET:additive = 70:25:5, testing as additives polymers (HDPE, LDPE) and inorganic fillers ($TiO_2$). Based on the experiments, it was found that HDPE's develops stronger interfaces comparing to other additives, and the composites can be used as construction materials [76].

While rubber recycling has numerous technological variants, PET recycling for product development raises plenty of questions because re-processing is quite limited. Therefore, embedding a higher amount of PET in the composite represents a target set for the next studies. Various samples with a different PET percentage in the composites were prepared with a PET content ranging from 10 to 45 %. Stable composites were obtained, which show no delamination, by integrating up to 35 % PET in the rubber matrix, using HDPE as a compatibility agent. These composites can be used as construction and insulating materials.

It was aimed at optimizing the percentage of $TiO_2$ in inorganic—organic composites with the generic composition:rubber:PET:HDPE:$TiO_2$ = (60–x):35:5:(x < 2), considering the well known experience from rubber processing, where inorganic fillers strongly improve the mechanical properties [77]. Samples with low $TiO_2$ content (0.25–0.5 %) were obtained with improved mechanical properties, being thus recommended for building materials like paving slabs for outdoor landscaping, courtyards, protection pillars, etc.

Based on these results, different samples were obtained with the composition rubber:PET:HDPE:metal oxide = 59.75:35:5:0.25. Differences which arise between samples obtained with different fillers can be attributed to their nature, primarily their influence on the interface. The thermal analysis run on these samples shows that nano-fillers have a favorable effect on the thermo-oxidative stability of the polymeric composites. The specific characteristics recorded during the mechanical testing show that the highest tensile strength corresponds to composites containing CaO. If metal oxides addition has a beneficial effect, adding a complex of oxides could preserve this effect. Therefore, another waste was tested as additive—the fly ash resulted from coal burning.

Fly ash is a mixture of crystalline, polycrystalline and amorphous oxides with carbon black. Adding fly ash in the composite does not significantly influence the thermal behavior of the samples, but promotes the development of porous, heterogenous structures, with voids having an irregular distribution. Considering the outstanding records for these samples containing only recyclable materials, we recommend their use as paving material in indoor applications.

## *11.7 The Influence of Environmental Factors on Polymer Composites with Inorganic Fillers*

Environmental factors may influence, alone or associated the functional properties of the composite materials both on microscopic and macroscopic level; they cause the composite "aging" and shorten their lifetime at a minimal level of performance. The aging process involves chemical bond cleavage as a result of oxidation reactions, cross-linking, polymerization/depolymerization, decomposition [78].

Accelerated aging tests are designed to estimate the relative resistance of the composites over time. Assessment of the aging rate is done by maintaining the composites under a controlled amount/flow of aging factor(s) for limited durations, after which the main properties are measured and compared with those corresponding to the samples kept in non-aggressive media.

The most influential factor in changing the thermal transitions of the polymer nanocomposites is the interaction between the filler/additive and the polymer matrix that influences the spatial constraints or the confinement. By reducing the vacant volume of the interface matrix rubber—additive, a decrease in the chain flexibility will result, having as effect a reduction in the glass transition temperature. As result, the accelerated aging tests will target these interactions which can be more easily destroyed by an energy input (thermal, UV, etc.).

The polymer composites optimized with inorganic filler material (rubber:PET:HDPE:inorganic additive = 59.75:35:5:0.25) were maintained in harsh environments: temperature variations, UV irradiation and salty mist. Their characterization has been done by FTIR analysis (bond formation/constraints), differential scanning calorimetry (phase transitions), X-ray diffraction (crystalline phase amount and structure), atomic force microscopy (morphology), contact angle measurements (surface charge); the results were confirmed by mechanical tests.

The materials used were composites, having:

- The matrix—recycled rubber powder (rubber from tires, 0.5 mm);
- Organic filler material—PET powder (0.5–1 mm);
- Additive—HDPE powder (0.5–1 mm);
- Inorganic filler material—metal oxide nanoparticles—$TiO_2$, ZnO, CaO and fly ash respectively.

### 11.7.1 Influence of Low Temperatures

The fillers in the polymer nanocomposites structure have as effect an increase in the thermal stability of the composite material. Interactions between polymer and inorganic filler materials are powerful and increase the molecular cohesion energy and hence the heat needed to activate the mechanism of thermal degradation [78].

Four series of composite material were developed, based on rubber, HDPE and PET and one oxide was added as inorganic filler: CaO, ZnO, $TiO_2$ or fly ash. Samples marked with K−1 ($TiO_2$), K−2 (CaO), K−3 (ZnO), K−4 (fly ash) were analyzed. For each test three samples were analyzed and the representative results were considered. The samples were kept in the laboratory at low temperatures between −40 and −10 °C for 500 h, then tested and analyzed following physico-chemical and mechanical properties. The results were compared with samples kept at room temperature, considered standard.

#### FTIR Analysis

The series of samples were analyzed, the bands' values obtained from the analysis were compared with those of the samples kept at room temperature and those of the individual components, including the filling agents.

The IR spectra of the composite material containing $TiO_2$ showed displacements of several bands: from 1,720 to 1,714 $cm^{-1}$—corresponds to C=O stretching vibration and indicates the presence of oxidation components on rubber surface; 1,524–1,537 $cm^{-1}$—attributed to stretching vibrations of C=C of aromatic nucleus; 1,411–1,410 $cm^{-1}$—C=O group vibration resulting from the polymers oxidation; 1,376–1,375 $cm^{-1}$; 1,341–1,342 $cm^{-1}$—are due to bond-stretching vibration of –C–C– from $-CH_3$ and to the deformation and vibration of the styrene nuclei; 907 $cm^{-1}$ la 906 $cm^{-1}$ existing bands due to deformation vibration of carbon-hydrogen bond to the aromatic nuclei; 730 $cm^{-1}$ la 729 $cm^{-1}$—predominant PET bands in this period are moved to lower wave numbers, corresponding to weaker bonds.

For samples containing CaO there were recorded bands' shifting from 1,261 to 1,260 $cm^{-1}$ and from 869 to 865 $cm^{-1}$ (C–C valence vibration). No bands were noted at the wave numbers: 1,714, 1,219, 1,494, and 1,862 $cm^{-1}$; they formed a band with low intensity at 1,409 $cm^{-1}$ corresponding to the strain in the plane of the group –OH alcohol resulting from the oxidative degradations of composite material components.

For samples containing ZnO, due to possible chemical interactions between the components, new bands were registered: 1,587 $cm^{-1}$ (assigned to the stretching vibrations of C=C of aromatic nucleus) 1,492 $cm^{-1}$, and band shift were recorded, from 2,117 to 2,113 $cm^{-1}$; from 1,263 to 1,257 $cm^{-1}$; from 723 to 719 $cm^{-1}$ as result of the physical interactions between the components in the composite material.

For samples containing fly ash some characteristic bands were no longer recorded: at wave numbers of 1,302, 1,141 $cm^{-1}$ (CH deformation in plane C–C valence vibration) and a new band was formed at 1,720 $cm^{-1}$ corresponding to aldehyde C=O group vibration. Also bands shift from 932 to 958 $cm^{-1}$ can be obtained (corresponding to the deformation vibration of carbon-hydrogen bond to the aromatic nucleus).

All these aspects indicate interactions developing bonds, ranging from chemical (strong) to physical (weak) that contribute to the interfaces formation; the strongest interfaces correspond to the samples with CaO and ZnO, while the samples with $TiO_2$ have developed pure physical interfaces.

## Differential Scanning Calorimetry

The comparative Differential Scanning Calorimetry (DSC) study on the thermal behavior of the samples without and with $TiO_2$ is presented in Table 1, for the samples kept at low temperatures. The results show that the melting temperatures ($T_t$) are slightly changed by incorporating the inorganic filler. The heats of fusion, however, decrease when adding the fillers. This change is explained by the lower degree of crystallinity and by the increase in the crystallites size, indicating a possible supra-organization around the oxide powder.

## X-ray Diffraction

The X-ray diffraction spectra of the samples containing $TiO_2$, kept at low temperatures are presented in Fig. 5 and present a typical halo, characteristic to the

**Table 1** Thermophysical parameters for the samples kept at low temperatures

| Thermophysical properties | $T_m$ (°C) | $\Delta H_m$ (J/g) | $\chi_c$ | $T_g$ (°C) | $\Delta C_p$ (J/g*°C) |
|---|---|---|---|---|---|
| Composites with $TiO_2$ | 122.84 | 2.50 | 0.08 | −35.61 | 1.158 |
| | 247.73 | 5.14 | | | |

**Fig. 5** Diffractometry for samples containing $TiO_2$ kept $T_{low}$

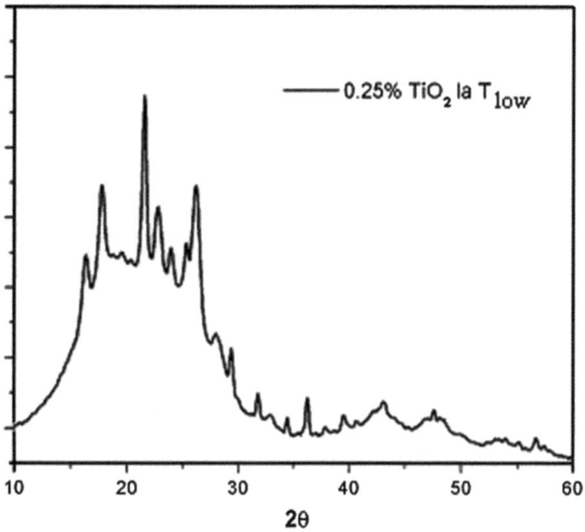


amorphous phase, given mainly by rubber; the crystalline phase is mostly related to the other components in the composite.

No new peaks were observed (supplementary to those corresponding to the single components forming the composite assembly), proving that the interface does not contain new crystalline compounds. The crystallites sizes were calculated using the Scherrer equation (241 nm) and the crystalline degree was evaluated at 0.1, which is consistent with DSC analysis result. There is an increase of the crystallites sizes from 154.13 nm (samples kept at room temperature) to 276 nm (samples kept at low temperatures).

## Atomic Force Microscopy

The surface characteristics of the composites were evaluated based on Atomic Force Microscopy (AFM) analysis Fig. 6. Rather inhomogeneous surfaces are resulting, which is expected considering the initial dimensions of the component pellets, also reflected by the high average roughness. It is important to notice that in performing AFM on composites with components having initial dimensions much larger than the scanning area, special attention must be devoted in choosing the analysis spot. Otherwise, results without any significance are actually obtained.

## Contact Angle Measurements

Contact angle measurements were made using water and 3.5 % NaCl solution, for 10 min, at a liquid flow into the droplet of 1μL/s; the recorded data are presented in the Figs. 7 and 8.

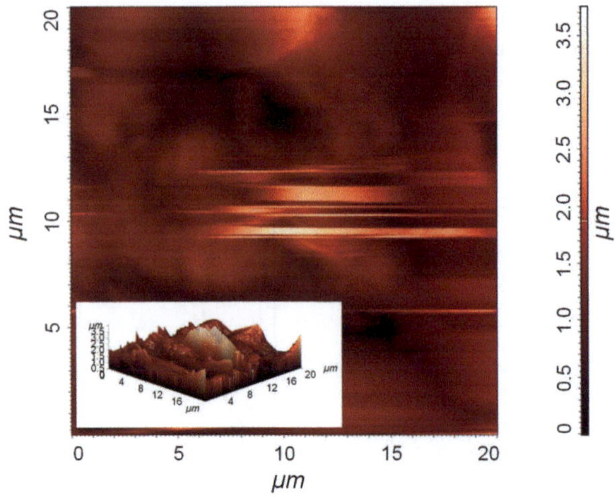


**Fig. 6** AFM images for samples kept $T_{low}$

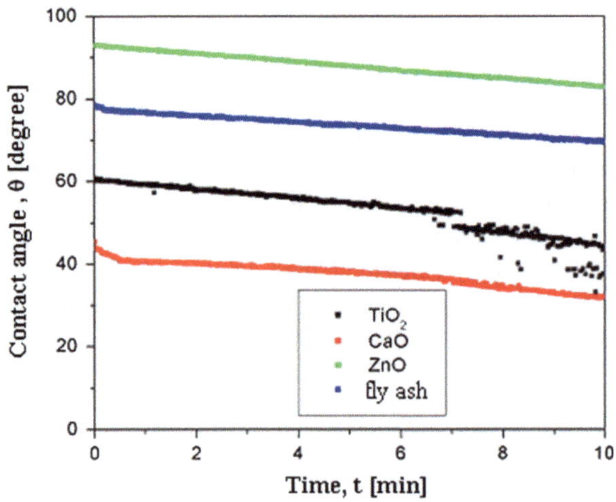


**Fig. 7** $\Theta_{water}$ angle variation with time for samples kept at low temperatures

A linear dependence of the water contact angle could be observed, with the specific fitting parameters presented in Table 2. Two types of kinetics were checked in the water/composite heterogeneous systems and, for all samples, the pseudo-second order kinetics is predominant, resulting in high values of the rate constants, k.

A comparative analysis on the contact angle values of the samples kept at low temperatures and at room temperature shows higher values for the latest. This means that at low temperatures changes have occurred, as confirmed by the AFM analysis, increasing the roughness (from 699.54 nm for samples at room temperature to 1364 nm for samples kept at low temperatures). With decreasing temperature, the materials tend to develop a glassy phase, the main component responsible being rubber (which crystallizes at about −25 °C).

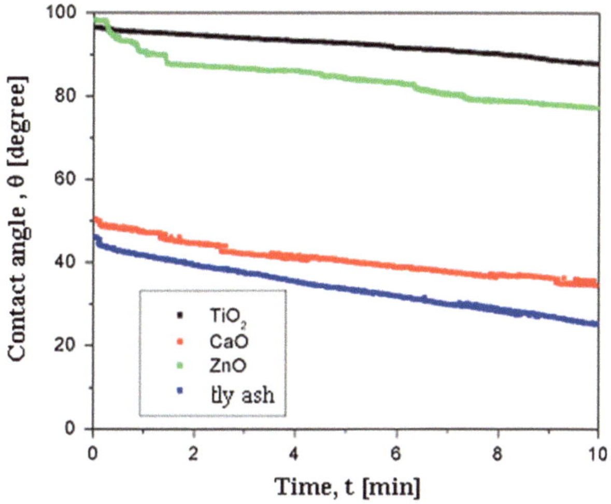


**Fig. 8** $\Theta_{salt}$ angle variation with time for samples kept at low temperatures

**Table 2** Linear dependence of specific fitting parameters $\theta$apa for samples from $T_{low}$

| Type samples | First-order pseudo-kinetics | | Second-order pseudo-kinetics | | |
|---|---|---|---|---|---|
| | k | $R^2$ | $k_2$ | $\theta_e$ | $R^2$ |
| K−1 | 0.0003 | 0.9705 | 0.1030 | 52 | 0.9983 |
| K−2 | 0.0266 | 0.9768 | 0.0678 | 32 | 0.9958 |
| K−3 | 0.0114 | 0.9996 | 0.0654 | 83 | 0.9995 |
| K−4 | 0.0106 | 0.9979 | 0.0831 | 70 | 0.9996 |

**Table 3** Surface energy values for nanocomposites

| Composites with | Surface tension (mN/m) | Dispersive component (mN/m) | Polar component (mN/m) |
|---|---|---|---|
| $TiO_2$ | 68.81 | 0.65 | 63.16 |
| CaO | 53.26 | 21.79 | 31.47 |
| ZnO | 16.21 | 4.29 | 11.92 |
| Fly ash | 66.28 | 65.32 | 0.95 |

In the contact angle measurement using NaCl solution, the values are lower, proving slightly polar surfaces (low charged). It is not only the surface charge but also the surface pores/voids responsible for these values. The samples with ZnO and CaO have an uneven surface, favoring a fast absorption/adsorption.

Surface energies were estimated using the contact angles values and the specific surface tensions of the two liquids, Table 3. Surface energy data confirms the predominant influence of polarity on the contact angle value; ash containing composites have a significant amount of surface energy, almost exclusively as result of the dispersive interactions, once again confirming the low surface polarity as result of a good mixing of the fly ash in the composite bulk. On the other hand,

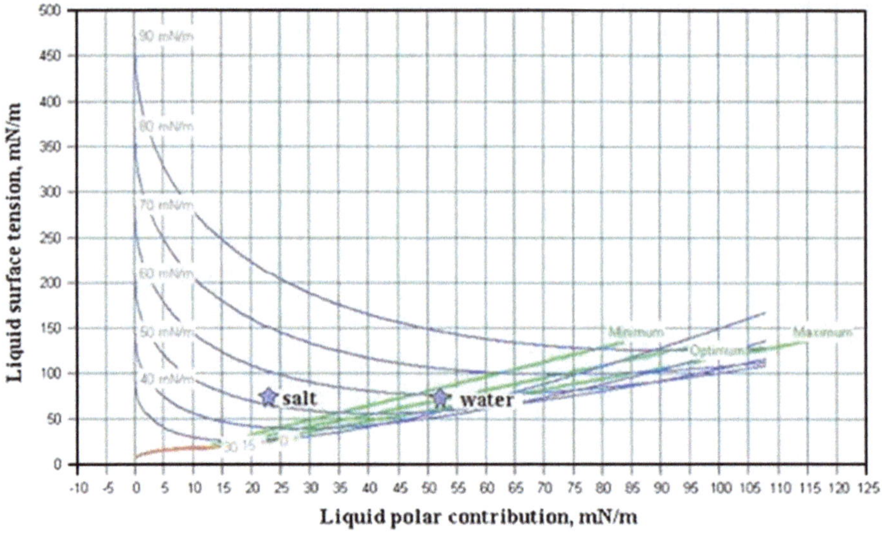


**Fig. 9** Adhesion work diagram for the sample containing ZnO kept at $T_{low}$

the $TiO_2$ composite has a high value of the polar component indicating a surface segregation. These data confirm that even in the absence of active interfaces, the presence of inorganic components produces reorganizations of polymeric components and a different behavior at low temperatures.

The work of adhesion of the liquid on the solid phases was estimated based on the polar and dispersive contributions of the surface tensions of the two testing liquids, Fig. 9. For the samples containing ZnO, with a low surface energy, the adhesion work diagram was plotted. The chart shows that the two liquids are not within tolerance lines (red lines) so no liquid wet the surface, confirming the high values of the contact angles ($\theta > 90$).

According to the mechanical properties, it can be concluded that the best resistance at low temperatures was exhibited by the samples containing fly ash, with compression forces up to 8,000 N and impact resistance of 9.92 kJ/m$^2$. This composite material can be used as building material as plates/slabs/paving mats outdoors in winter.

### 11.7.2 The High Temperatures Influence

Aging can also be the result of high temperatures action. The samples were kept at temperatures between 10 and 50 °C, in laboratory conditions for 500 h, in order to monitor the physico-chemical and mechanical changes and for choosing the optimal applications. In the samples analyzed were contained following fillers: $TiO_2$, CaO, ZnO, and fly ash.

## FT-IR Analysis

The FTIR spectra show no new bands and no missing peaks, only low energy bands shift proving that, in this range, temperature does not strongly affects the bulk and interface bonds.

## DSC Thermal Analysis

The DSC thermogram was developed for the sample with $TiO_2$. Two melting temperatures can be identified, similarly with the low temperatures behaviour, corresponding to HDPE and PET in the composite, Table 4. Compared to the standard sample (kept at room temperature), there is a variation of the transitions energies and a correspondent increase in the crystallinity degree, as result of possible structural reorganizations, without developing new types of structures and associations.

## X-ray Diffraction

These data are confirmed by the XRD pattern of the samples containing $TiO_2$, Fig. 10, where no changes are observed in the peaks position, allowing the conclusion that this average temperature domain (usually corresponding to summer temperatures) has no strong impact on the composite structure.

## AFM Analysis

If not the temperature as such, the thermal treatment duration (500 h) could be an aging cause. A change in the microphase morphology resulted in a decrease in the average roughness (RMS) at values down to 430 nm, Fig. 11. The possible mechanisms may include macromolecular chains rearrangements, forming denser structures. The sample has a porous structure, mainly as result of the $TiO_2$ nanoparticles agglomeration.

## Contact Angle Measurements

Interaction between the liquid and the surface was estimated by contact angle measurements. Values of the $\theta$ angle recorded for 10 min are presented in Figs. 12

**Table 4** Thermophysical properties of samples containing $TiO_2$ kept $T_{high}$

| Thermophysical properties | $T_m$ (°C) | $\Delta H_m$ (J/g) | $\chi_c$ | $T_g$ (°C) | $\Delta C_p$ (J/g*°C) |
|---|---|---|---|---|---|
| Composites $TiO_2$ | 123.03 | 12.08 | 0.67 | −45.01 | 0.764 |
| | 243.78 | 9.41 | | | |

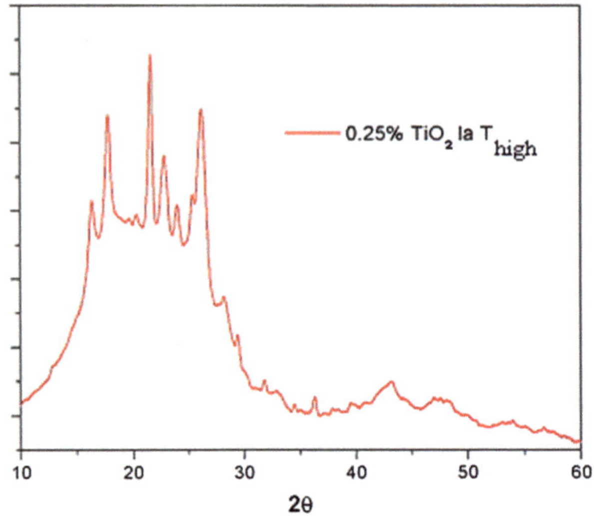


Fig. 10 Diffractometry for samples containing $TiO_2$ kept at $T_{high}$

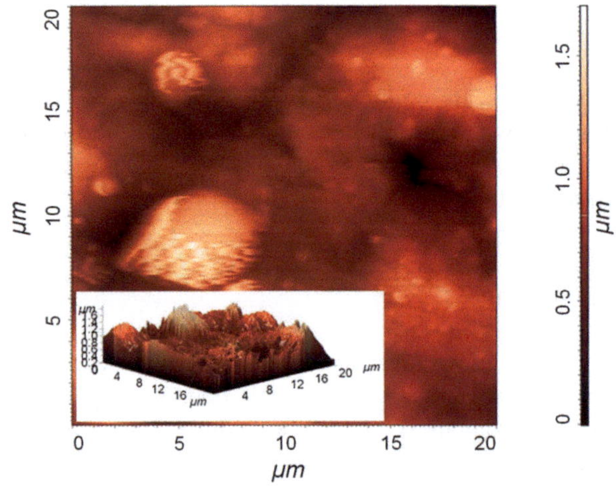


Fig. 11 Images of atomic force microscopy for samples kept at $T_{high}$

and 13. It is observed the linear dependence of the water angle and the high values, over 90° indicating that wetting is not a likely mechanisms because of the surface energy and/or uniformity and/or compact structure of the samples. Exceptions are the samples containing with ZnO and 77.5° angle, having a porous surface, making wetting possible. This sample presents however a high value of the contact angle when using NaCl solution, (96.7°), confirming the bulk infiltration of the ZnO nanoparticles.

Compared with the samples preserved at room temperature, the samples tested at higher temperatures show: higher $\theta_{water}$ values for samples with CaO, $TiO_2$ and lower values for those containing ash and ZnO and higher $\theta_{salt}$ values for samples with CaO, ash, ZnO and lower values for those containing $TiO_2$. As in all

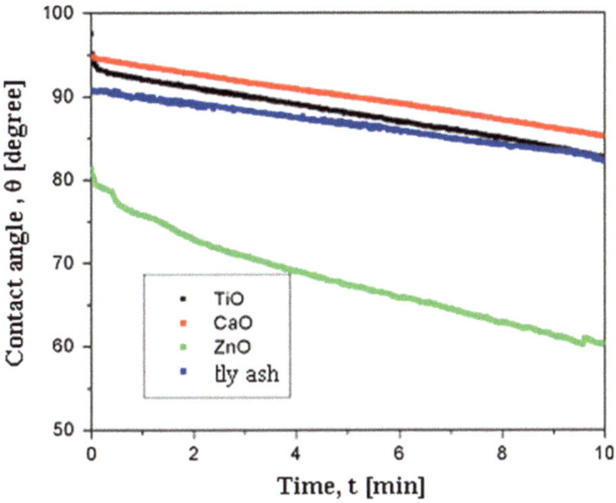


**Fig. 12** $\Theta_{water}$ angle variation with time for samples kept at high temperatures

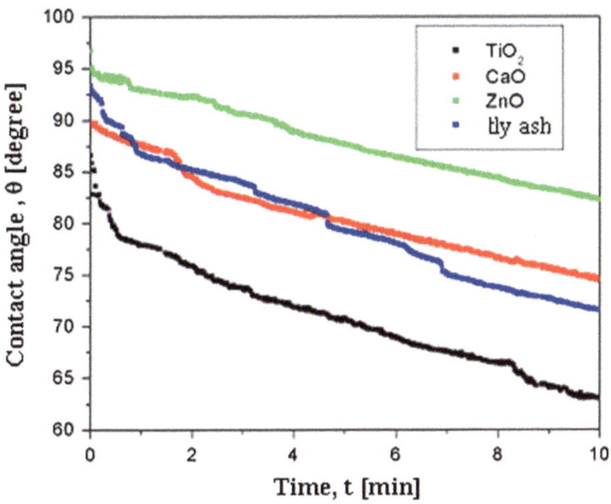


**Fig. 13** $\Theta_{salt}$ angle variation with time for samples kept at high temperatures

investigated cases, this behavior is caused, obviously, by the surface energy which is influenced by at least two factors:

1. Surface composition: the oxide powder increases the surface energy and emphasizes the wetting phenomenon, while migration/agglomeration of powders in the composite volume leaves free a hydrophobic surface with low energy.
2. Surface roughness: a rough surface, fractured, with edges and corners has a great(er) surface energy and a more hydrophilic/less hydrophobic behavior compared with a less rough surface.

**Table 5** Linear dependence of specific fitting parameters $\theta_{water}$ for the specimens kept at $T_{high}$

| Composite with | First-order pseudo-kinetics | | Second-order pseudo-kinetics | | |
|---|---|---|---|---|---|
| | k | $R^2$ | $k_2$ | $\theta_e$ | $R^2$ |
| $TiO_2$ | 0.0117 | 0.9986 | 0.0631 | 83 | 0.9995 |
| CaO | 0.0103 | 0.9982 | 0.0695 | 85 | 0.9996 |
| ZnO | 0.0273 | 0.9738 | 0.0410 | 59 | 0.9968 |
| Fly ash | 0.0094 | 0.9984 | 0.0786 | 82 | 0.9996 |

**Table 6** Surface energy values for specimens kept $T_{high}$

| Composites with | Surface tension (mN/m) | Dispersive component (mN/m) | Polar component (mN/m) |
|---|---|---|---|
| $TiO_2$ | 67.4 | 58.18 | 9.22 |
| CaO | 100.56 | 98.25 | 2.32 |
| ZnO | 78.56 | 68.51 | 10.05 |
| Fly ash | 22.48 | 0.04 | 22.44 |

The specific fitting parameters for the linear dependence of the contact angle versus time are presented in Table 5. From the slope values the rate constants are calculated and prove that, for samples containing ZnO, the adsorption/absorption phenomena are faster than for the other composites, mainly because the samples have a more porous aspect. Pseudo-second order and pseudo-first order kinetics are parallel mechanisms and the first one is faster, the contribution of first-order pseudo-kinetics mechanisms being much lower (10–15 %).

The surface energy for the samples kept at higher temperatures are presented in Table 6. It is noted that for the samples containing $TiO_2$, ZnO and CaO the dispersive component prevails, confirming the predominant hydrophobic surface, in agreement with the contact angle values. For the samples containing fly ash, a surface charge is obvious, resulting from possible interactions between the active centers of fly ash and parts of the polymeric components.

With the inorganic fillers in the composite materials, increased values of mechanical strength are observed, the oxide particles acting as nucleation agents that can lead to a size decrease of spherulite and hence the increase system packing. Samples containing CaO showed the best mechanical strength under the test conditions (impact resistance being 11.86 $kJ/m^2$). As result of their properties, these composite materials can be used as construction materials (mats, paving slabs, impact parapets, etc.) resistant to temperatures up to 50 °C, both for indoor environments and for outdoor use.

### 11.7.3 Influence of UV Radiation

Aging tests were conducted by assessing the composites behavior under UV radiation. Samples with fillers: $TiO_2$, CaO, ZnO, fly ash, were subjected to

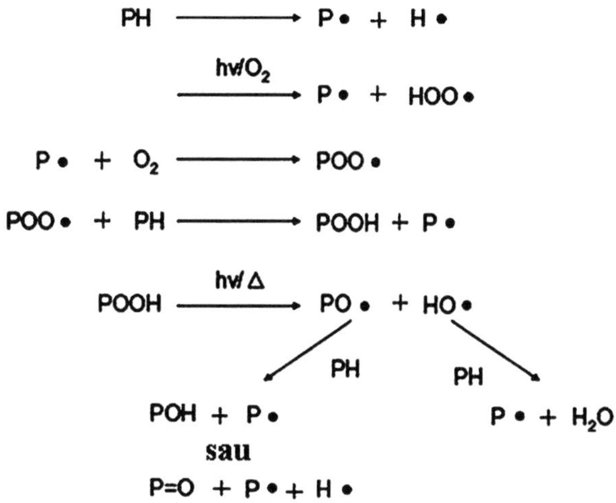


**Scheme 1** The process of polymer degradation under UV exposure

accelerated aging tests, for 500 h using a fotoreactor with three lamps (F18 W-P8 Philips) placed annular. The UV-B fluorescent lamp emits up to 365 nm. In these conditions, $TiO_2$ has a photocatalytic activity making possible the degradation of the organic/polymeric components (rubber, PET, HDPE) in systems containing traces of water. This water can result from simple humidity condensation, in regular storage conditions.

Previous studies have suggested that photo-degradation produces carbonyl and hydroxyl groups by cleavage of the polymer chain and it can be synthesized by the mechanism given in Scheme 1 by Rabek. A number of authors [79, 80] have reported that final carboxyl groups are formed during the photodegradation of PET, and they act as a catalyst to further promote degradation [81].

During photodegradation, PET molecules are subject to division on both ends of the chain, forming the final carboxyl groups. Possible reactions are shown in Scheme 2. PET strongly absorbs radiations up to 310 nm, while higher wavelengths support the generation of carboxyl marginal groups also inside the composite not only in the surface layers. However, the results indicate that degradation is more intense near the samples surface, [82].

A certain protection is given by the carbon black, existent in recycled rubber, [83], which strongly absorbs into the ultraviolet energy domain. Thus, the carbon black from rubber could protect the sample's components from UV degradation. The same role can be played by the metal oxides, including $TiO_2$ [84], the protective actions of this compound depending on the polymorph, rutile or anatase. In rutile form, the hydroxyl groups (OH) on the filler surface are relatively stable, while in the anatase form they are highly reactive and can initiate the polymers degradation. $TiO_2$ used in this work was a mixture of 75 and 25 % rutile anatase (Degussa P25), thus explaining the relatively poor performance of the samples.

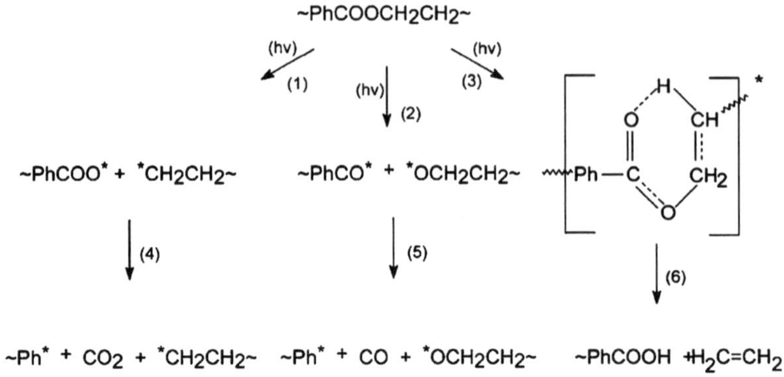


**Scheme 2** Possible reactions during UV degradation of PET

## FTIR Analysis

Interface characterization and photochemical behavior under UV irradiation of the samples were investigated by FTIR analysis, to monitor chemical changes and to identify the degradation products.

The IR spectra of the composite containing $TiO_2$ showed a band shift, from 1,473 to 1,486 $cm^{-1}$ and the formation of new bands at 3,387 and 3,207 $cm^{-1}$. The change in the distribution of the molecular conformation is also observed in the interface $TiO_2$/PET, and is confirmed in these experiments by the newly formed bands.

For samples with CaO, bands shift were recorded from 2,116 to 2,113 $cm^{-1}$, from 1,714 to 1,718 $cm^{-1}$ (band corresponding to C=O group and resulting from the oxidation of polymers; it is a measure of UV degradation) and from 1,060 to 1,041 $cm^{-1}$ (C–C valence vibration); the disappearance of the band at 1,537 $cm^{-1}$ and the development of a low-intensity band at 1,409 $cm^{-1}$ corresponding to the planar deformation of the alcohol group –OH support the assumptions of oxidative degradations in the composite.

For samples containing ZnO the following changes have been noted: the formation of low-intensity bands at the wave numbers 1,559 $cm^{-1}$ (assigned to stretching vibrations of C=C from aromatic nuclei), 1,406, 1,203, 793 $cm^{-1}$, the band from 1,990 $cm^{-1}$ is no longer shown and bands shift, from 1,714 to 1,701 $cm^{-1}$, 1,494 to 1,489 $cm^{-1}$, 1,098 to 1,091 $cm^{-1}$. It is to remember that ZnO also can act as photocatalyst in moisture conditions and (almost) neutral pH.

Also, we note that bands disappeared (1,302, 1,340, 1,141 $cm^{-1}$ –CH plane deformation, C–C valence vibration) in ash containing samples and the formation of a band at 1,720 $cm^{-1}$. Band shifts were shown from 1,793 to 1,900 $cm^{-1}$, 932 to 958 $cm^{-1}$ (corresponding to the deformation vibration of carbon-hydrogen bond at aromatic ring). This can be linked with the rather high iron oxide content in fly ash, likely to support a photo-Fenton degradation mechanism.

**Table 7** Thermophysical properties of samples containing $TiO_2$ kept UV

| Thermophysical properties | $T_m$ (°C) | $\Delta H_m$ (J/g) | $T_g$ (°C) | $\Delta C_p$ (J/g*°C) | $\chi_c$ |
|---|---|---|---|---|---|
| Composite with $TiO_2$ | 124.01 | 12,08 | −50.7 | 0.586 | 0.08 |
| | 237.33 | 10.85 | | | |

## DSC Thermal Analysis

The DSC diagram of the sample containing $TiO_2$ demonstrates that the UV exposure led to an increased crystallinity, that is confirmed by the lower glass and melting transition temperatures $T_g$ and $T_m$, Table 7.

## X-ray Diffraction

The XRD pattern of the sample containing $TiO_2$ kept for 500 h under UV radiation is presented in Fig. 14. Broad diffraction peaks are observed corresponding to very small particles and/or to a significant amount of amorphous phase. The remained crystalline phase may be the result of (re)arrangement of the polymer and/or phase transitions (amorphous-crystalline). Crystallite sizes are about 280 nm, larger than in the standard sample (157 nm).

The specific peaks of rubber and polyethylene are easily displaced. In addition there are present specific HDPE peaks at 21.43° and those of $TiO_2$. These findings confirm the results of FT-IR and DSC analysis.

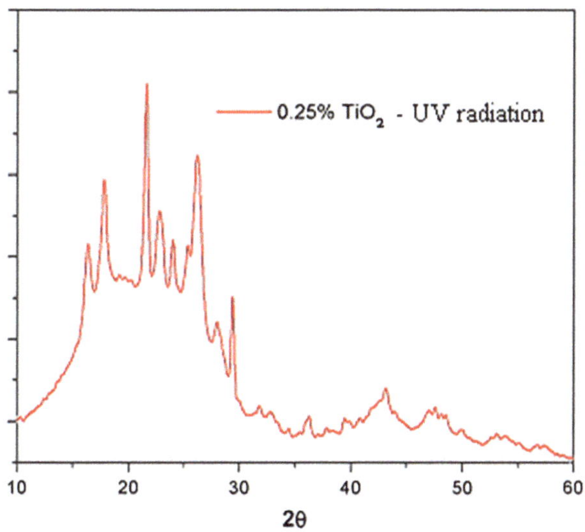


**Fig. 14** Diffractometry for samples containing $TiO_2$ kept in UV radiation

**Fig. 15** Images of atomic force microscopy for samples kept in UV radiation

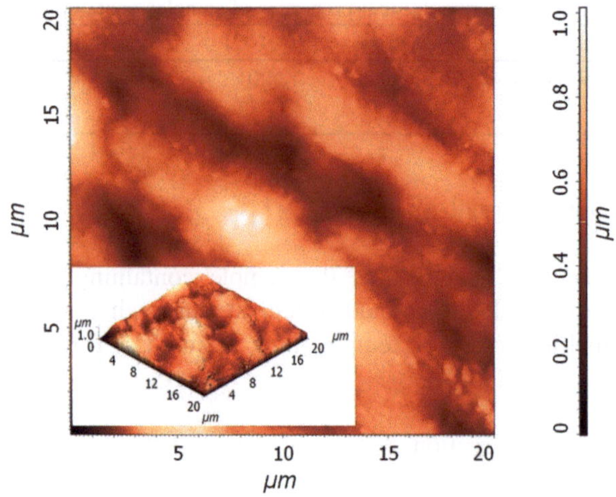


AFM Analysis

The average roughness (RMS) was calculated based on the AFM images, Fig. 15, for a scanned area of 25 × 25 μm$^2$, being 89.27 nm. There is a sharp decrease of roughness compared with the standard sample and these values can be corroborated with the expansion of the crystallites' size. One can distinguish two majority phases: rubber and PET, as well as spherical nanometric $TiO_2$ particles, clustered in the interface area.

Contact Angle Measurements

Following UV irradiation at the polymer surface, oxidation compounds result which contain C=O groups, –OH and –COOH leading to changes in surface polarity. In order to understand these mechanisms, contact angle ($\theta$) measurements were made and calculation for the surface energy, [85] were developed, using as test liquids water and 3.5 % NaCl solution; the results are presented in Figs. 16 and 17.

The contact angle measurements show higher values when 3.5 % NaCl solution is used, as an ionic solution forms large contact angles on predominantly hydrophobic surfaces. Also, the use of water as test fluid produced high values of angles, the adsorption/absorption rate in this case being smaller, which means that the samples have a more compact structure and form stronger interfaces with low wetting capacity.

The fitting parameters, specific to a linear dependence of the $\theta_{water}$ angle variation with time, for the samples UV irradiated are presented in Table 8. Adsorption/absorption mechanisms are carried out predominantly as a pseudo-second order kinetics, and the highest rate corresponds to the samples containing fly ashes.

In both types of tests, the samples which formed large angles were those containing ZnO (102° for the angle formed by water and NaCl solution and 97°)

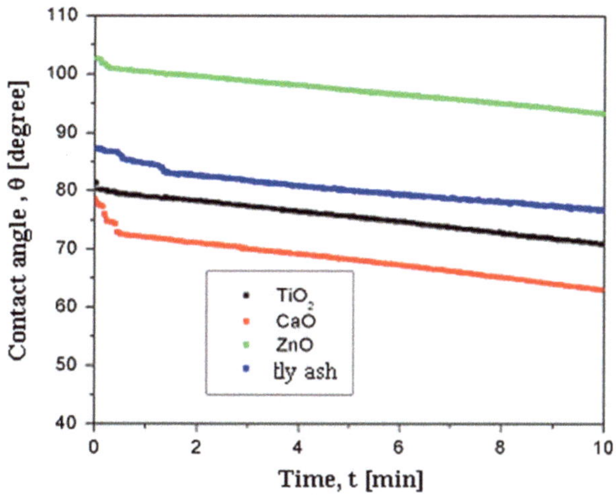


**Fig. 16** $\Theta_{water}$ angle variation with time for samples kept in UV radiations

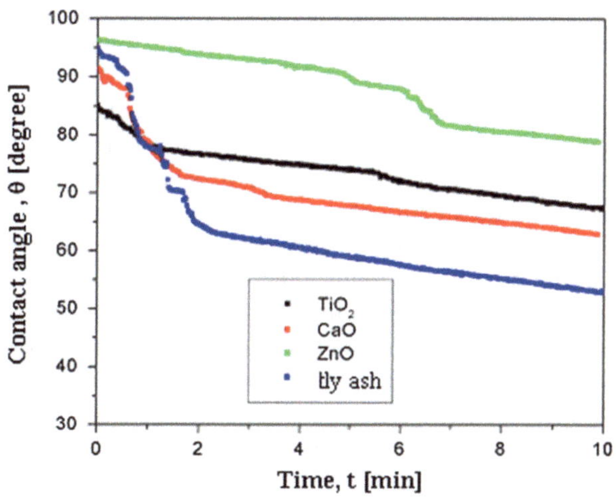


**Fig. 17** $\Theta_{salt}$ angle variation with time for samples kept in UV radiations

**Table 8** Linear dependence of specific fitting parameters $\theta_{water}$ for the specimens kept at UV radiations

| Composites with | First-order pseudo-kinetics | | Second-order pseudo-kinetics | | |
|---|---|---|---|---|---|
| | k | $R^2$ | $k_2$ | $\theta_e$ | $R^2$ |
| $TiO_2$ | 0.0121 | 0.9985 | 0.0713 | 71 | 0.9994 |
| CaO (0…0,8 min) | 0.0155 | 0.9843 | 0.0658 | 69 | 0.9991 |
| CaO (0,8…10 min) | 0.0231 | 0.9878 | 0.0756 | 64 | 0.9996 |
| ZnO | 0.0080 | 0.9974 | 0.0828 | 93 | 0.9997 |
| Fly ash (0…1,5 min) | 0.0001 | 0.9761 | 0.0872 | 86 | 0.9997 |
| Fly ash (1,5…10 min) | 0.0023 | 0.9854 | 0.0854 | 79 | 0.9993 |

**Table 9** Surface energy values for samples kept UV radiation

| Composites with | Surface tension (mN/m) | Dispersive component (mN/m) | Polar component (mN/m) |
|---|---|---|---|
| $TiO_2$ | 24.90 | 8.09 | 16.81 |
| CaO | 28.07 | 2.44 | 25.63 |
| ZnO | 14.15 | 10.25 | 3.90 |
| Fly ash | 21.85 | 3.23 | 18.61 |

and ash (85° for the angle formed by water and 95° for solution NaCl). The photochemical degradation is possible to form surfaces more or less porous, or rough, so the samples on which the UV effect was the strongest exhibit the lower contact angles (surface rich in polar/ionic groups). For example, there is a decrease of the $\theta_{salt}$ angle by 34.7 % for the samples with fly ash, due to the large pores on the composite surface, emerged from the UV degradation.

The surface energy values are presented in Table 9. Except the samples containing ZnO, all the other samples show a predominant polar surface, confirming oxidation/degradation. The overall surface energy corresponding to the sample with ZnO is low, the surface is thus predominant hydrophobic, supporting the idea that, the photocatalytic effect of ZnO is not very strong and/or the oxides particles are not predominantly stacked at the surface.

As a result of oxidation processes, on the surface of the samples new volatile chemical species may occur, which cause cracks and holes that lead to a brittle material, mechanically weaker. As degradation occurs, the composite matrix is damaged, so it reduces the quality of the filler—additive—matrix interfaces. Supplementary, inorganic nano- and micro-particles can agglomerate at the surface and/or in the bulk. In these cases the contact angle values can be modified, due to the roughness values that changes the droplet disposition on the sample surface, [86]. So, although the contact angle measurements provide an overall view of the hydrophobic/hydrophilic surface behavior, they cannot uniquely assign the cause.

Based on the values of the polar and dispersive components of the surface energy, the work of adhesion chart can be developed, allowing to estimate the sample's wettable behavior, Fig. 18. It is noted that the two characteristic points corresponding to the test liquids are not under red lines, indicating the phenomenon of wetting. The work of adhesion for water is 84.44 mN/m and for the NaCl solution it is 78.88 mN/m.

UV radiation mainly affects the rubber structure in the surface layer, releasing volatile products, [87] and forming pores. The PET is also susceptible to degradation under the UV irradiation, leading to molecular weight reduction [88]. These changes could be limited or expanded by adding the inorganic fillers.

Samples containing ZnO are more resistant to UV irradiation, comparing to the samples without any inorganic additive. The results prove that ZnO acts as a photo-stabilizer in small amounts and can effectively improve the photo-stability of the composite polymers [89–91]. A similar effect is registered for samples

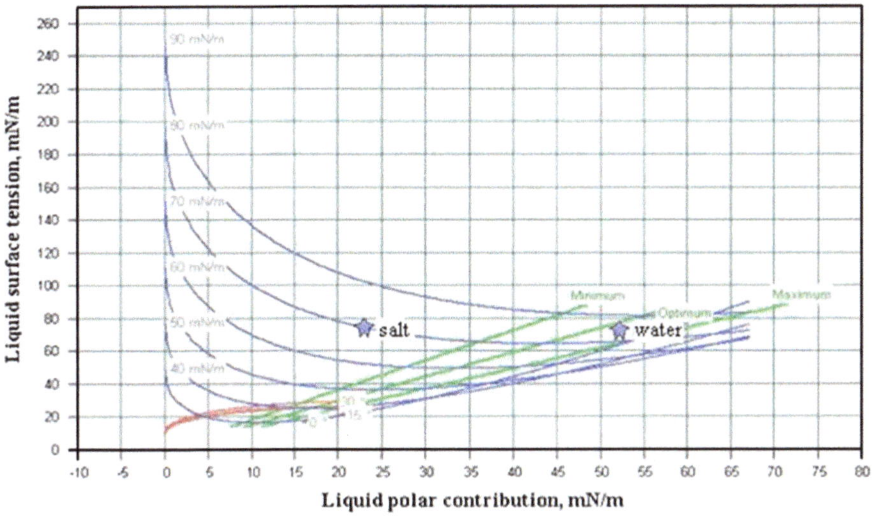


**Fig. 18** Adhesion work diagram for the sample containing $TiO_2$ kept in UV radiation

containing CaO, which showed the best UV stability (having, for example, the tensile strength diminished only by 5 % comparing to the non-irradiated samples).

On the other hand, titanium dioxide promotes UV photodegradation, forming active hydroxyl radicals, [92, 93], and accelerate the samples' degradation.

Choosing the inorganic filler must be done according to the applications. Indoor products, seldom subjected to UV irradiation, can benefit on the enhanced mechanical properties of the composites containing $TiO_2$ or fly ash, while outdoor goods are recommended to be developed using fillers that never promote photodegradation, as CaO.

## 12 Conclusion

Composite materials can be considered advanced materials with controlled properties, being a particular priority area supported worldwide. Their design, the structural and functional optimization, the development of specific products, of manufacturing and processing technologies is mostly characterized by extended interdisciplinary.

The properties of composite materials can be modified under the action of certain physical, chemical, environmental or mechanical factors, both at micro- and macroscopic level.

Tailoring a composite must target a set of properties, directly related to the applications. It is also important to remember that this set of properties must be preserved during their lifetime, under the working conditions. Being the result of a synergic mix of components, the composites may act as a new material or as an assembly of components, depending especially on the newly developed interfaces. Strong interfaces can result by new chemical bonds or by cumulative (but large) physical interactions. The PET-rubber composites, using HDPE as compatibility agent represent a good example, detailed in this chapter.

Second raw materials used for composite development raise supplementary problems, derived from the various degrees of degradation, existent in each component. Still, as the results proved, rubber—PET—HDPE composites, with good mechanical properties can be obtained. Adding low amounts of inorganic fillers represents a supplementary path to tailor the functional properties.

Not any inorganic filler is suitable for any application. The studies on the influence of environmental factors (temperature, UV radiation), [39, 40] on rubber-PET-HDPE composites with inorganic fillers ($TiO_2$, ZnO, CaO, ash) found that:

- The fillers in the polymer nanocomposites structure increase the thermal stability of the material;
- Interactions between polymer and inorganic filler materials are strong and increase molecular cohesion, thus increasing the heat required for thermal degradation;
- Low temperatures have a much stronger influence on composites degradation (aging) comparing to average and high temperatures;
- Photocatalytic materials are to be avoided for outdoor products and have no or little effect in indoor applications;
- The use of an inorganic waste, fly ash, results in composites with very good mechanical properties, and these composites are recommended especially for indoor applications.

"The results prove that the use of waste-polymers as second raw materials represent a topic where research is still expected, for increasing the quality of the resulted materials and for extending the applications, towards high-tech products. Following the increasing demands of the recycling legal framework, it is expected that the interfaces control, particularly in organic-inorganic composites will be able to provide new paths, readily up-scalable for a truly sustainable industrial development."

## References

1. Rosato, D., Plastics Processing Data Handbook, Springer, (1997)
2. Callister, D., Materials science and enginnering an introduction, 7th ed., John Willy & Sons, Inc., 2007
3. Challa, G.: Polymer chemistry: An introduction. By Ellis Horwood, London (1993)
4. Liu H.S., Richard C.P., Mead J.L., Stacer R.G., Development of novel applications for using recycled rubber in thermoplastics. Technical report 18, Chelsea Center for Recycling and Economic Development, Massachusetts, USA, 2000

5. Pramanik, P.K., Baker, W.E.: Toughening of ground rubber tire filled thermoplastic compounds using different compatibilizer systems. Plast. Rubber Compos. Process. Appl. **24**(4), 229–237 (1995)
6. Scaffaro, R.: Effect of adding new phosphazene compounds to poly(butylene terephthalate)/ polyamide blends I: preliminary study in a batch mixer. Polym. Degrad. Stab **90**(2), 234–243 (2005)
7. De, S.K., Bhowmick, A.K.: Thermoplastic Elastomers from Rubber Plastic Blends. Horwood, London (1990)
8. Walker, B.M., Rader, C.P.: Handbook of Thermoplastic Elastomers. Van Nostrand Reinhold, New York (1988)
9. Jha, A., Bhowmick, K.: Rubber Chem. Technol **70**, 798–814 (1997)
10. ASTM D 5593 – 99: "Standard Classification for Thermoplastic Elastomers-Olefinic (TEO)"
11. ASTM D 5046 – 98: "Standard specification for Fully Crosslinked Elastomeric Alloys"
12. Wang, W., Wu, Q., Qu, B.: Polym. Eng. Sci. **43**, 1798–1805 (2003)
13. Manoj, K.C., Unnikrishnan, G.: J. Appl. Polym. Sci. **105**, 908–914 (2007)
14. Sadek, E.M., El-Nashar, D.E., Motawie, A.M.: Polymer-Plastics Technology and Engineering **42**, 627–642 (2003)
15. Legge, N.R., Holden, G., Schroeder, H.E.: Thermoplastic Elastomers: A comprehensiveReview. Hanser Publishers, Munich (1987)
16. De, S.K., Bhowmick, A.K. (eds.): Thermoplastic Elastomers from Rubber- Plastic Blends. Ellis Horwood Ltd, London (1990)
17. Abdou-Sabet, S., Fifty years of thermoplastic elastomer Innovations, presented at the Symposium on Thermoplastic Elastomers, ACS Rubber Division Meeting: Cincinnati. OH, USA (2000)
18. Sadhan, K.De., Avraam, I.I., Klementina, K., Rubber recyling, New York: CRC Press, 2005
19. Xanthos, M., Dagli, S.S.: Compatibilization of polymer blends by reactive processing. Polym. Eng. Sci **31**, 929–935 (1991)
20. Deanin, R.D., Manion, M.A.: Handbook of Polyolefines, 2nd edn. Marcel Dekker Inc, New York, USA (2000)
21. Deanin, R.D., Manion, M.A.: Compatibilization of Polymer Blends. Marcel Dekker, Inc, New York (1999)
22. Alexandre, M., Dubois, P.: Polymer-layered Silicate Nanocomposites: Preparation, Properties and Uses of a New Class of Materials. Mat. Sci. Eng **12**, 1–63 (2000)
23. Abdou-Sabet, S., Patel, R.P.: Morphology of elastomeric alloys. Rubber Chem. Technol **64**, 729–769 (1991)
24. Feng, W., Isayev, A.I.: In-situ Ultrasonic Compatibilization of PP/EPDM Blends during Ultrasound Aided Extrusion. Polymer **45**, 1207–1216 (2004)
25. Xiao, H.W., Huang, ShQ, Jiang, T.: Cheng ShY. Miscibility of blends of ethylene-propylenediene terpolymer and polypropylene, J Appl Polym Sci **83**, 315–322 (2002)
26. Nakason, C., Jarnthong, M., Kaesaman, A., Kiatkamjornwong, A.: Thermoplastic elastomers based on epoxidized natural rubber and high-density polyethylene blends: Effect of blend compatibilizers on the mechanical and morphological. J. Appl. Polym. Sci **109**, 2694–2702 (2008)
27. Supri, M., Ismail, H.: Effects of Dynamic Vulcanization and Glycidyl Methacrylate on Properties of rPVC/NBR blends. Polym. Testing **25**, 318–326 (2006)
28. Hope, P.S., Bonner, J.G., Milles, A.F.: Plastics. Rubber and Composites Processing and Applications **22**, 147–158 (1994)
29. Papke, N., Karger-Kocsis, J.: Thermoplastic elastomers based on compatibilized poly(ethylene terephthalate) blends: effect of rubber type and dynamic curing. Polymer **42**, 1109–1120 (2001)
30. Papadopoulou, C.P., Kalfoglou, N.K.: Comparison of compatibilizer effectiveness for PET/ PP blends: their mechanical, thermal and morphology characterization. Polymer **41**, 2543–2555 (2000)

31. Aravind, P.A., Ranganathaiah, C., Kurian, J.V., Thomas, S.: Compatibilizing effect of EPM-g-MA in EPDM/poly(trimethylene terephthalate) incompatible blends. Polymer **45**, 4925–4937 (2004)
32. Henderson, A.M.: IEEE Electr. Insul. Mag. **9**, 30 (1993)
33. Mohamad, Z., Ismail, H., Chantara, R.: Characterization of epoxidized natural rubber/ ethylene vinyl acetate (ENR-50/EVA) blend: Effect of blend ratio. J. Appl. Polym. Sci. **99**, 1504–1515 (2006)
34. Klinklai, W., Kunyawut, C.: Preparation of thermoplastics elastomer between natural rubber and poly(lactic acid) in the presence of epoxidized natural rubber, pp. 23–29. International Conference on the Role of Universities in Hands-On Education Rajamangala University of Technology Lanna, Chiang-Mai, Thailand (2009)
35. Ismail, H., Supri, M., Yusof, A.M.: Polym. Testing **23**, 675 (2004)
36. Soares, B.G., Almeida, M.S.M., Guimaraes, P.I.C.: The reactive compatibilization of NBR/ EVA blends with oxazoline-modified nitrile rubber. Eur. Polymer J. **40**, 2185–2194 (2004)
37. Bonnin, E., Thibault, J.-F., Jansen, P., Soares, B.G.: Effect of compatibilizer and curing system on the thermal degradation of natural rubber/EVA copolymer blends. Polym. Degrad. Stab **52**, 95–99 (1996)
38. Macaubas, P.H.P., Demarquette, N.R.: Morphologies and Interfacial tension of immiscible polypropylene/polystyrene blends modified with triblock copolymer. Polymer **42**, 2543–2554 (2001)
39. Chen, J.S., Liao, M.C., Shiah, M.S.: Asphalt modified by styrene-butadiene-styrene triblock copolymer: morphology and model. J. Mater. Civ. Eng. **143**, 224–229 (2002)
40. Kim, D.H., Fasulo, P.D., Rodgers, W.R., Paul, D.R. Effect of the ratio of maleated polypropylene to organoclay on the structure and properties of TPO-based nanocomposites. Part I: Morphology and mechanical properties, Polymer 48 (2007) 5960e5978
41. Silva, S.M.L., López-Manchado, M.A., Arroyo, M.: Thermoplastic Olefin/Clay Nanocomposites. Effect of Matrix Composition, and Organoclay and Compatibilizer Structure on Morphology/Properties Relationships, Journal of Nanoscience and Nanotechnology **7**, 4456–4464 (2007)
42. Mirzadeh, A., Lafleur, P.G., Kamal, M.R., Dubois, C., Co-continuity of thermoplastic elastomer rubber-based nanocomposites, Society of Plastics Engineers (SPE), 10.1002/ spepro.002992, 2010
43. Mahallati, P., Arefazar, A., Naderi, G.: Thermoplastic elastomer nanocomposites based on PA6/NBR. Int. Polym. Proc. **25**, 132–138 (2010)
44. Ganguly, A., Bhowmick, A.K.: Quantification of surface forces of thermoplastic elastomeric nanocomposites based on poly(styrene-ethylene-co-butylene-styrene) and clay by atomic force microscopy. J. Appl. Polym. Sci. **111**, 2104–2115 (2009)
45. Nishitani, Y., Naito, T., Sekiguchi, I., Kitano, T., Effect of addition of thermoplastic elastomers on tribological anti mechanical properties of vapor grown carbon filled polybutylene terephthalate composites, World Tribology Congress 2009 - Proceedings, 2009, 421
46. Nakason, C., Nuansomsri, K., Kaesaman, A.: Dynamic vulcanization of natural rubber/high-density polyethylene blends: Effect of compatibilization, blend ratio and curing system. Polym. Testing **25**, 782–796 (2006)
47. Bhowmick, A.K.: Stephens. H.L, Handbook of Elastomers, Second Edition (2001)
48. Machado, A.V., Van Duin, M.: Dynamic vulcanization of EPDM/PE-based thermoplastic vulcanisates studied along the extruder axis, polymer **46**, 6575–6586 (2005)
49. Bhowmick, A.K., Chiba, T., Inoue, T.: Reactive processing of rubber-plastic blend: Role of chemical compatibilizer. J. Appl. Sci. **50**, 2055–2064 (1993)
50. Lachmann, M., Use of hot runner method in the injection molding of TPE, Thermoplastische Elastomere – Herausforderung an die Elastomerverarbeiter 127 (1997) 141–161
51. Fried, J.R.: Polymer science and Technology. Prentice Hall, Second Edition (2003)
52. Rodriguez, F., Cohen, C., Ober, C.: Archer. Taylor & Francis, L.A., Principles of Polymer Systems (2003)

53. Knieps, H., Extrusion of thermoplastic elastomers, Thermoplastische Elastomere – Herausforderung an die Elastomerverarbeiter 127 (1997) 103–115
54. Nagaoka, T.: Blow molding of thermoplastic elastomers (TPE). Porima Daijesuto **51**, 42–54 (1999)
55. Hull, J.L., Concise Encyclopedia of Polymer Science and Engineering, John Wiley & Sons, 1990
56. Popa, G.A., Rubber: Types, Properties and Uses, Chapter 8, ISBN: 978-1-61761-464-4, Nova Science Publishers, Inc., 2010
57. Pasch, H., Augenstein, M.: Chromatographic investigations of macromolecules in the critical range of liquid chromatography: 5- Characterization of block copolymers of decyl and methyl methacrylate. Makromol. Chem **194**, 2533–2541 (1993)
58. Pash, H.: Liquid chromatography at the critical point of adsorption – A new technique for polymer characterization. Macromol. Symp. **110**, 107–120 (1996)
59. Wang, T.L., Huang, F.J.: Preparation and characterization of novel thermoplastic elastomers by step/chain transformation polymerization. Polymer **41**, 5219–5228 (2000)
60. Otsuka, N., Yang, Y.: Saito H and Inoue T Structure and properties of PP/Hydrogenated SBR. Ann Tech Conf Soc Plast Eng **54**, 2331–2333 (1996)
61. Inoue, T., Svoboda, P.: Structure-properties of PP-EPDP thermoplastic elastomer: Origin of strain recovery, 58th Ann Tech Conf Soc. Plast. Eng. **2**, 1676–1679 (2000)
62. Huang, F.J., Wang, T.L.: Synthesis and characterization of new segmented polyurethanes with side-chain, liquid-crystalline chain extenders. J Polym Sci Part A Polym Chem Ed **42**, 290–302 (2003)
63. Li, H., White, J.L.: Preparation and characterization of biaxially oriented films from polybutylene terephthalate based thermoplastic elastomer block copolymers. Polym. Eng. Sci. **40**, 2299–2310 (2000)
64. Volegova, I.A., Godovsky, Y.K., Soliman, M.: Glass transition of undrawn and drawn copolyetherester thermoplastic elastomers. Int. J. Polym. Mater **52**, 549–564 (2003)
65. Thakkar, H., Goettler, L.A.: The effects of dynamic vulcanization on the morphology and rheology of TPV's and their nanocomposites. Tech Papers ACS Rubber Div **163**, 482–491 (2003)
66. Pham, L.P., Sung, C.: Effects of blending SIBS and SMA on morphology and mechanical properties. Mat Res Soc Symp Proceed **734**, 391–395 (2002)
67. Sauer, B.B., McLean, R.S., Brill, D.J., Londono, J.D., Morphology and orientation during deformation of segmented elastomers studied by SAXS and atomic force microscopy, Abstr 222nd ACS Meet, Chicago III, USA, PMSE–163 (2001)
68. Burford, R.P., Markotsis, M.G., Knott, R.B.: Small angle neutron scattering and transmission electron microscopy studies of interpenetrating polymers networks from thermoplastic elastomers. Nucl Instrum & Meth in Phys Res Sect B **208**, 58–65 (2003)
69. Schmalz, H., Abetz, V., Lange, R.: Thermoplastic elastomers based on semicrystalline block copolymers. Compos. Sci. Technol. **63**, 1179–1186 (2003)
70. Svodoba, P., Saito, H., Chiba, T., Inoue, T., Takemura, Y.: Morphology and elastomeric properties of isotactic polypropylene/hydrogenated polybutadiene blends. Polym. J. **32**, 915–920 (2000)
71. Duncan J. Shaw, Introduction to Colloid and Surface Chemistry, Elsevier Science Ltd., 1992
72. Bhattacharya, De Sati N., Kamal, M.R., Gupta, R.K., Polymeric nanocomposites: theory and practice, Hanser Verlag, Munich, 2008
73. Vladuta, C., Andronic, L., Visa, M., Duta, A.: Ceramic interface properties evaluation based on contact angle measurement. Surf. Coat. Technol. **202**, 2448–2452 (2008)
74. Voinea, M., Vladuta, C., Bogatu, C., Duta, A.: Surface properties of copper based cermet materials. Mater. Sci. Eng., B **152**, 76–80 (2008)
75. Rubber: Types, Properties and Uses, Nova Science Publishers, Inc., Cazan, C., Duta, A. co-authors Capitol 8: Recycled Rubber – Composite Matrix, 2010, ISBN: 978-1-61761-464-4

76. Vladuta, C., Voinea, M., Purghel, E., Duta, A.: Correlations between the structure and the morphology of PET- rubber nanocomposites with different additives. Mater. Sci. Eng., B **165–3**, 221–226 (2009)
77. Vladuta, C., Andronic, L., Duta, A.: Effect of TiO2 nanoparticles on the interfaces PET-rubber composites. J. Nanosci. Nanotechnol. **10**, 2518–2526 (2010)
78. Popovic, I.G., Katsikas, L.: The characterization of polymer composites by thermogravimetry. Materialy composite **40**(1), 7–12 (2006)
79. Allen, N.S., Edge, M., Mohammadian, M.: Polym. Degrad. Stab. **43**, 229–237 (1994)
80. Nakayama, Y., Takahashi, K., Sasamoto, T.: Surf Interf Anal **24**, 711–717 (1996)
81. Bikiaris, D.M., Karayannidis, G.P.: Polym. Degrad. Stab. **63**, 213–218 (1999)
82. Fechine, G.J.M., Rabellob, M.S.: Souto Maiora, R.M., Catalani, L.H., Surface characterization of photodegraded poly(ethylene terephthalate). The effect of ultraviolet absorbers. Polymer **45**, 2303–2308 (2004)
83. Horrocks, A.R., Mwila, J., Miraftab, M., Liu, M., Chohan, S.S.: Polym. Degrad. Stab. **65**, 25–36 (1999)
84. Allen, N.S., Edge, M., Corrales, T., Catalina, F.: Polym. Degrad. Stab. **61**, 139–149 (1998)
85. Garbassi, F., Mora, M., Occhiello, E.: Polymer Surfaces: From Physics to Technology. John Wiley & Sons, Chichester (1998)
86. Use of different surface analysis techniques for the study of the photo-degradation of a polymeric matrix composite A. Larena*, S. Jimenez de Ochoa, Applied Surface Science 238 (2004) 530–537
87. Dick, J.S., Rubber Technology, Hanser Publishers, 2001
88. Wang, H.Y., Hu, G.L., Zhou, Y., Pan, X.D., Effects of UV-raddiation on structure and properties of poly(ethylene terephthalate), Cailiao Kexue yu Gongyi/Materials Science and Technology, 16-4 (2008) 582-584 + 588
89. Lupa, L., Negrea, P., Sallai, M., Ciopec, M., Ghiga, M., Valorificarea cenuşii de zinc sub formă de oxid de zinc de înaltă puritate, Buletinul AGIR n r. 1-2/2008
90. Mijin, Dušan: D., Savic, M., Snežana, P., Smiljanić, A., Glavaški, O., Jovanović, M., Petrović, S., A study of the photocatalytic degradation of metamitron în ZnO water suspensions. Desalination **249**, 286–292 (2009)
91. Wang, X., Chen, S., Cheng, X., Tu, D.: The UV aging of nano-ZnO/polyethylene composite. Diangong Jishu Xuebao/Transactions of China Electrotechnical Society **23**, 6–10 (2008)
92. Qamar, M., Muneer, M.: A comparative photocatalytic activity of titanium dioxide and zinc oxide by investigating the degradation of vanillin. Desalination **249**, 535–540 (2009)
93. Mathews, R.W., McEvoy, S.R.: Photocatalytic degradation of phenol în the presence of near UV illuminated titanium dioxide. J. Photochem. Photobiol. A: Chem. **64**, 231–246 (1992)

# Rubber-Thermoset Blends: Micro and Nano Structured

**Jin Kuk Kim and Sanjoy Datta**

**Abstract** Highly cross-linked thermosets which are susceptible to brittle failure can be effectively toughened by blending them with rubbers. However, if the materials are already cross-linked, then blending with rubber as it is done in the conventional way with thermoplastics, is virtually impossible. Thus it asks for an altogether different method to accomplish successful blending. Initially, miscible liquid rubbers in small amounts or preformed rubber particles are incorporated in the matrix of curing agent incorporated precured thermosets resins and then the whole mass is subjected to curing. The phase separation, in case of liquid rubber toughening depends upon the formulation, processing and curing conditions and incomplete phase separation may occur resulting in unwanted lowering of glass transition temperature. The phase separation in case of liquid rubber is based upon nucleation and growth. In case of preformed rubber particles, these difficulties are not encountered and the resulting morphology can be better controlled. However, the problem of proper dispersion of these particles in the themoset resins limits the use of this method. The improvement in fracture resistance occurs in either case due to dissipation of mechanical energy by cavitation of rubber particles followed by shear yielding of the matrix. Rubber particle size plays an important role in improving toughening and very small or very large sizes are undesirable. The toughenability increases with increase in inherent ductility of the matrix.

## 1 Introduction: Concept of Rubber-Thermoset Blends

Highly crosslinked thermosets like epoxies, phenolics and cyanate esters have come to the fore with many application outlets but one disadvantage in common with these thermosets is that they all have low toughness and are susceptible to

J. K. Kim · S. Datta (✉)
School of Materials Science and Engineering, Polymer Science and Engineering, Gyeongsang National University, 501 Jinju-daero, Jinju 660-701, Korea
e-mail: rubber@gnu.ac.kr

P. M. Visakh et al. (eds.), *Advances in Elastomers I*,
Advanced Structured Materials 11, DOI: 10.1007/978-3-642-20925-3_8,

brittle failure. Thus, for applications asking for higher toughness without substantially sacrificing other desirable properties, it is mandatory to modify them. They may be modified with plastics, oils or fibres.

Another route to accomplish the same relatively easily with cheaper initial investment is by blending with rubber, though the processing may sometimes turn out to be somewhat more difficult. Progressive research works have carved routes to do away with the impediments of processing obstacles, and customary to say, rubber modified thermosets are gaining importance. However, it is noteworthy to mention that rubber/thermoplastic blends and thermoplastic vulcanizates (TPVs) are far more established over rubber/thermoset blends. It is the business of the present chapter to focus on the latter. Studies show that rubber toughening in thermosetting epoxy resin is very well established. However, this is not the case for phenolics, especially for the resol type. Some aspects of both rubber toughening of epoxies and phenolics, the two most commonly encountered thermosets will be the main area of discussion of this chapter.

## *1.1 Morphology of Rubber-Thermoset Blends*

Thermosets can be toughened by rubber particles with mainly two different morphologies which are either the use of 'core–shell rubber particles' or initially 'miscible reactive rubbers'. Emulsion polymerization is the route to the preparation of Core–shell rubber particles and exhibit alternating rubbery and glassy layers. These particles have been successfully used to modify thermosets, such as epoxies, cyanates and vinyl ester resins. The constituents of such a particle are a rubbery core and an outer shell of a glassy polymer. The rubbery core is generally based on polybutadiene. The outer glassy shell is required to prevent coalescence of rubbery particles during synthesis, and enhances a good interface with the matrix. The shell is usually based on styrene/acrylonitrile copolymers. However, the core–shell toughening does not bring about good dispersion [1]. This is why the most widely used method for the rubber toughening is the addition of initially miscible rubbers into the thermosetting matrix. The rubber may form a secondary phase during the polymerization (curing) reaction depending upon the type of rubber and thermosetting material used. The degree of this phase separation can control not only the amount of toughening obtained, but also a number of other properties such as modulus of the system along with glass transition temperature [2].

## *1.2 Toughening Mechanism of Rubber-Thermoset Blends*

Usually more than one toughening mechanism takes place before failure occurs in thermoset systems. Therefore, it is important to first consider the most effective

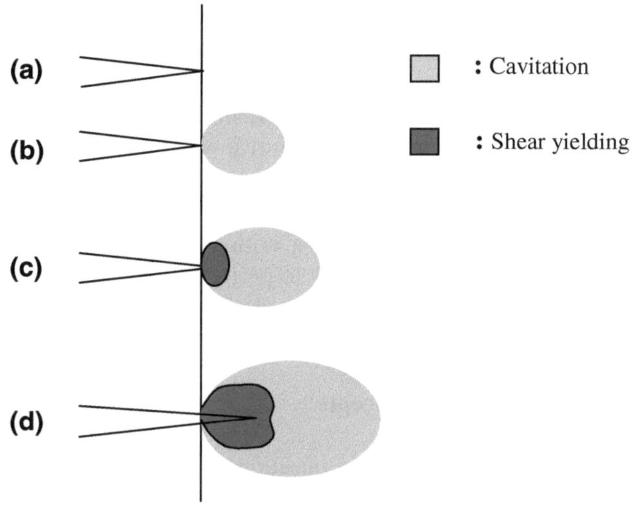


**Scheme 1** A sketched sequence of the toughening mechanism of core shell rubber modified epoxy thermoset system (a) initial starter crack; (b) formation of a cavitation zone in front of the crack tip; (c) geometry with the initial yielded plastic zone; (d) shear yielding

toughening mechanism like 'shear yielding' and then mention other effective but minor toughening mechanistic routes.

In the mechanism of 'shear yielding', shear bands or deformation zones are initiated by rubber particles leading to stress concentration at the surrounding matrix. The improvement upon toughness is proportional to the number of such particles, higher number contributing to the creation of more deformation zones before fracture occurs. The general approach to shear yielding is briefly discussed. Small rubber particles to the tune of fractions of a micrometer are quite effective in promoting extended shear yielding of the brittle thermoset matrix. The matrix undergoes plastic deformation. This type of plastic deformation is supported by the fact that the rubber particles at the crack tip elongates to the same extent as the matrix evidenced by microscopic studies compared with undeformed spherical rubber particles. The major toughening mechanism, as it appears to be is cavitation of the rubber particles followed by shear yielding of the matrix. The cavitation of the rubber particles helps relieve the plane strain constraint induced by thick specimen and sharp crack. The shear stress component around the crack tip is greatly raised and causes extended yielding of the matrix. A schematic diagram is shown in Scheme 1 [3] for understanding the phenomenon more clearly. It is, therefore, possible to toughen highly crosslinked thermosets via shear yield mechanism as long as the toughener can effectively alter the crack-tip stress state from one that favours brittle fracture to one that promotes shear yielding. In addition to shear yielding, some other toughening mechanisms such as micro-cracking, crack deflection, crack bifurcation, crack pinning, crack bridging or multilevel fracture path (enlargement of fracture surface area) can also be operative. However, these mechanisms are less effective than the 'shear yielding' mechanism and relatively low in energy absorption capacity.

In the scheme, (a) represents the initial starter crack followed by the formation of a cavitation zone in front of the crack tip when the specimen is initially loaded (b). This develops into a geometry shown in (c), when formation of initial shear-yielded plastic zone around the crack tip takes place with the relieving of the hydrostatic tension. This is due to the cavitation of the rubber particles. Once the build up of shear strain energy reaches a critical value, the material begins to undergo shear yielding allowing the crack to propagate and leaving a damage zone around the propagating crack as is shown in (d). Thereafter the crack grows again.

### 1.2.1 Review of Toughening Concepts

The first serious attempt to explain the question of how an immobile, crosslinked glassy thermoset can be toughened by rubber particles was made by Kunz and Douglass [4]. The model developed by them was based on the principle of energy dissipation during stretching and bridging of the crack surfaces by rubber particles. The particle bridging mechanism is rather a straightforward phenomenon. When the crack advances in the rubber modified epoxy system, it has a tendency to grow preferentially in the more brittle epoxy matrix phase, in effect around the rubber particles in the initial stage of the crack growth. Subsequent to this, when the crack begins to open up, the rubber particles start to span between the two separating crack planes. These rubber particles are extremely ductile and strain-harden rapidly. As a result the fracture energy required to make the crack grow is somewhat increased. In other words the brittle thermoset is toughened. The bridging particle should have the capacity to stretch between the two crack planes and so they must exhibit ductility. The size of the particle has to be much larger than that of the crack tip radius for it to function as an effective bridge. Finally, the interfacial adhesion between the particle and the matrix needs to be stronger the cohesive strength of the particle itself. If all these requirements are satisfied then the particle bridging mechanism will be effective. However, it was proved later that rubber bridging mechanism only plays a secondary role in the toughening of brittle thermosets.

Sultan and McGarry [5] highlighted the role of rubber particle size in the use of rubber toughening in a thermosetting matrix and they chose an epoxy system. In their study, they used carboxyl-terminated acrylonitrile (CTBN) liquid rubber to toughen diglycidyl ether of bisphenol A (DGEBA) epoxy resin. They set off with the concept that toughening depends on the rubber particle size and showed that 40 nm particles were not as efficient as 1 mm particles. Thus nearly five-fold increase in fracture energy values were obtained when large particles (1 mm) were used instead of small ones (40 nm). Further to this, Pearson and Yee [6] tried to investigate the rubber particle size dependence on toughening mechanisms. They prepared an epoxy system (DGEBA) with liquid reactive rubber (CTBN). From their study, they concluded that relatively large particles provided only a modest increase in fracture toughness by a particle bridging/crack deflection mechanism.

In contrast, smaller particles provided a significant increase in toughness by a shear banding mechanism.

Frounchi et al. [7] experimented with solid acrylonitrile-butadiene rubbers (NBR) added to DGEBA type epoxy resin matrix. Their study showed that increasing the acrylonitrile content of the rubber (from 19 to 33 %) caused better compatibility between NBR and epoxy resin. Thus, they obtained effective toughening and 40 % increase in impact resistance. Kaynak et al. [8–12] investigated rubber toughening of DGEBA-type epoxy resin through the incorporation of various modifiers, such as hydroxyl terminated polybutadiene (HTPB) rubber, a silane coupling agent (SCA), recycled scrap tire rubber particles and a liquid elastomer. They used these modifiers separately and with various combinations. They also studied effects of mixing order during specimen preparation, and surface modification of these rubber particles. They indicated that toughness of the brittle epoxy can be improved most effectively when these modifiers were used at certain synergistic combinations with certain surface treatments.

Chen and Jan [13] prepared bimodal distributed liquid rubber particles (CTBN) in epoxy resin (DGEBA) to study the fracture behaviour of the system. They eventually found out that a 171 % increase in fracture energy was obtained by using bimodal rubber particles over the unimodal ones. However, this was not reflected in the work of Pearson and Yee [6]. They tried out the bimodal mixtures of epoxies containing small and large particles, but did not observe any significant improvement in fracture energy.

Geisler and Kelley [14] used a combination of alumina and core–shell rubber particles with epoxy matrix in order to overcome some of the drawbacks produced by rubber toughening (decrease in modulus and high temperature performance). They observed that a cured epoxy system having rubber and alumina particles had fracture toughness values 25 % higher than those of epoxy systems having only rubber or only alumina particles. They also argued that toughness improvement does not lead to any decrease in glass transition temperature ($T_g$) and modulus. Yee and Pearson [15] also analysed the effect of matrix Tg on rubber toughening. They demonstrated that the low crosslink density epoxies were far more readily toughened than the high crosslink density ones.

From what has been discussed so far, it is customary to say that rubber toughening in thermosetting epoxy resins is very well studied and quite established. However, this is not the case for phenolics, especially for resol type phenolic resins. The reason may be due to the difficulties encountered during curing such specimens by casting. Curing in resol-type phenolic resins should be conducted at temperatures below 100 °C to avoid boiling of water which is a byproduct of the curing reaction. Water evaporation causes bubble formation and densely voided structures. To get rid of such a condition, curing is accomplished at a temperature around 80 °C leading to very long curing times of 2 or 3 days. With further lowering of the curing temperature to a measure of 40 °C bubble formation can almost be eliminated but the curing period becomes very impractical. It was observed that even after a period of 4 days enough rigid crosslinked structure was not obtained.

# 2 Various Types of Rubber-Thermoset Blends

Though many kinds of thermoset materials can be and have been experimented upon with blends of rubbers purely on research basis, yet here only two such most important blends will be considered; rubber-phenolic and rubber-epoxy.

## *2.1 Rubber-Phenolic Blends*

Phenolic resins (PR) have been widely used as coatings, adhesives, composites, etc. due to their excellent flame resistance, heat resistance, insulativity, dimensional stability and chemical resistance. However, their application has significantly been limited by inherent brittleness. Materials used as toughening agents of PR include elastomers such as natural rubber and nitrile butadiene rubber (NBR) [16], reactive liquid polymers such as liquid nitrile butadiene [17] and carboxyl terminated butadiene acrylonitrile (CTBN) [18], plastics [19, 20] such as polysulfone and polyamide, oils such as cashew nut shell liquid [21], tung oil [22] and linseed oil [23], and fibres such as glass fibres and aramid fibres [24].

The most widely used toughening agents are elastomers due to their high efficiency and low cost. However, the phenolic network is subject to deterioration in heat resistance, strengths and modulus after the incorporation of elastomers and flexible compounds.

Stiff aromatic heterocyclic structures are usually introduced into PR molecules to improve its heat resistance. However, the same stiff structures will also decrease the PR toughness. Obviously, the method for toughening and heat resistance improving is incompatible. Recently, studies on PR modification by addition of nanoparticles, including carbon nanotubes [25] and layered silicates [26, 27], were reported. These nanoparticles are able to greatly improve heat resistance and stiffness of phenolic materials, but can only slightly enhance toughness.

Modification of PR with organic nanoparticles has been reported [28]. A new kind of PR/organic nanoparticle composite, including PR/nitrile butadiene elastomeric nanoparticle (NBENP) composite and PR/carboxylic nitrile butadiene elastomeric nanoparticle (CNBENP) composite have been successfully developed.

Since phenolic blends are less openly published, it is discussed in some details in this chapter with reference to two relevant and promising works.

### 2.1.1 Rubber Toughening of Phenolic Resin Using Nitrile Rubber and Amino Silane

Kaynak and Cagatay [29] have reported the toughening of phenolic resin using nirtile rubber and aminosilane. For the phenolic resin, a resol-type liquid phenol–formaldehyde was used. Powder rubber particles used for toughening were

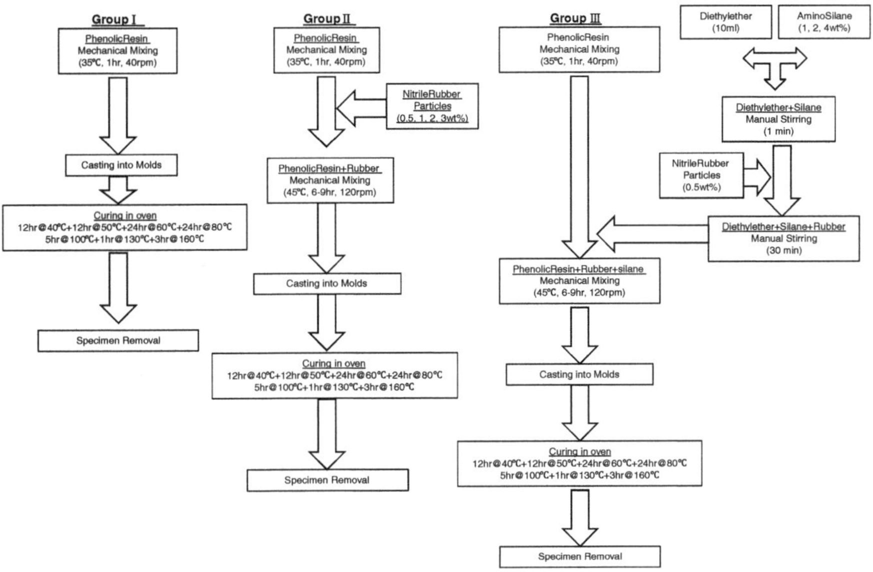


**Fig. 1** Production scheme of the three groups of specimens [29]

acrylonitrile–butadiene rubber (NBR). An amino silane (3-aminopropyltriethoxy-silane) was also used together with the nitrile rubber particles contemplated to investigate synergistic effect in rubber toughening if any. Specimens were produced in three main groups. In the first group, only neat phenolic resin was used. In the second group, phenolic matrix was modified with nitrile rubber particles. Modification was carried out both with nitrile rubber and amino silane in the third group. The production schemes of these three main groups are given in Fig. 1.

This is of utmost importance because without a proper approach to blending only poor toughening is usually observed.

Group I specimens were prepared by putting the neat phenolic resins from refrigerator onto a hotplate. Then, it was mechanically mixed at 40 rpm at 35 °C for 1 h in order to decrease the viscosity. Finally, the resin was poured into PTFE (polytetrafluoroethylene) moulds and cured in an oven. In the preparation of Group II specimens, the resin was first mechanically mixed at 35 °C for 1 h. Then, rubber particles without further purification were added to the liquid phenolic resin, and this mixture was mixed by a stirrer at 120 rpm and 45 °C to dissolve the rubber particles. In this group, since the matrix resin was modified with four different amounts of rubber at 0.5, 1.0, 2.0, and 3.0 weight percent, therefore, the mixing period was different for varying concentrations of rubber and ranged from 6 to 9 h with the increasing rubber concentration. Finally, the resin–rubber mixture was poured into PTFE molds and oven cured. The preparations of Group III specimens were somewhat more complicated with the further incorporation of the amonos-ilane. In this group, first of all, liquid amino silane was dissolved in 10 ml diethylether solvent by stirring manually for 1 min. followed by the addition of nitrile rubber particles to the silane–diethylether system. Since specimens having 0.5 wt% rubber content resulted in highest performance in Group II, so this concentration was kept fixed in Group III.

However, to investigate the effects of amino silane addition, three different silane concentrations were used at 1, 2, and 4 wt% with respect to the amount of rubber particles. Solvent–rubber–silane mixture was kept at room temperature for 30 min for evaporation of the solvent. Then, this silane–rubber mixture was added to the mechanically mixed (at 35 °C for 1 h) resin system. Next, this resin–rubber–silane mixture was mechanically mixed (120 rpm, 45 °C, 6–9 h) and, finally, poured into the mould to cure.

As explained previously, phenolic resins should be cured carefully, else the evaporation of the by-product water molecules may lead to void formation in the specimens. This study was devoted to various trials in order to obtain an efficient curing schedule. First, a high curing temperature of 160 °C was used and it was observed that specimens were cured in only 30 min. This led to the unwanted formation of a large amount of bubbles and the entire structure was rendered useless. A gradual decrease in curing temperature can be an alternative solution. However this approach reflected the formation of some voids even at a temperature of 100 °C. When the curing temperature was lowered to a measure of 40 °C then bubble formation was almost eliminated but the curing period was impractical. It was observed that even a period of 4 days was not enough to obtain a rigid crosslinked structure.

The best method is to cure the specimens, starting from low temperatures with successive increases. An optimized curing method which can be applied based on the studies is as follows: 12 h at 40 °C, another 12 h at 50 °C followed by 24 h at 60 °C and finally 24 h at 80 °C. However, this cure scheme did not furnish a concrete route to produce a specimen with sufficient mechanical properties. It was judged best to post cure the specimens at 100 °C for 5 h and then 130 °C for 1 h and finally 160 °C for 3 h. The specimen designation in this case study is shown in Table 1.

Flexural, notched Charpy impact and plane-strain fracture toughness tests were performed in order to characterize the mechanical behaviour, especially toughness, of the specimens. Figure 2 gives the mechanical properties of the resol-type

**Table 1** specimen designation used [29]

| Designation | Specimen |
|---|---|
| P | Neat phenolic resin |
| PR0.5 | Phenolic resin with 0.5 wt% nitrile rubber |
| PR1 | Phenolic resin with 1 wt% nitrile rubber |
| PR2 | Phenolic resin with 2 wt% nitrile rubber |
| PR3 | Phenolic resin with 3 wt% nitrile rubber |
| PRS1 | Phenolic resin with 0.5 wt% nitrile rubber and 1 wt% amino silane (with respect to nitrile rubber) |
| PRS2 | Phenolic resin with 0.5 wt% nitrile rubber and 2 wt% amino silane (with respect to nitrile rubber) |
| PRS4 | Phenolic resin with 0.5 wt% nitrile rubber and 4 wt% amino silane (with respect to nitrile rubber) |

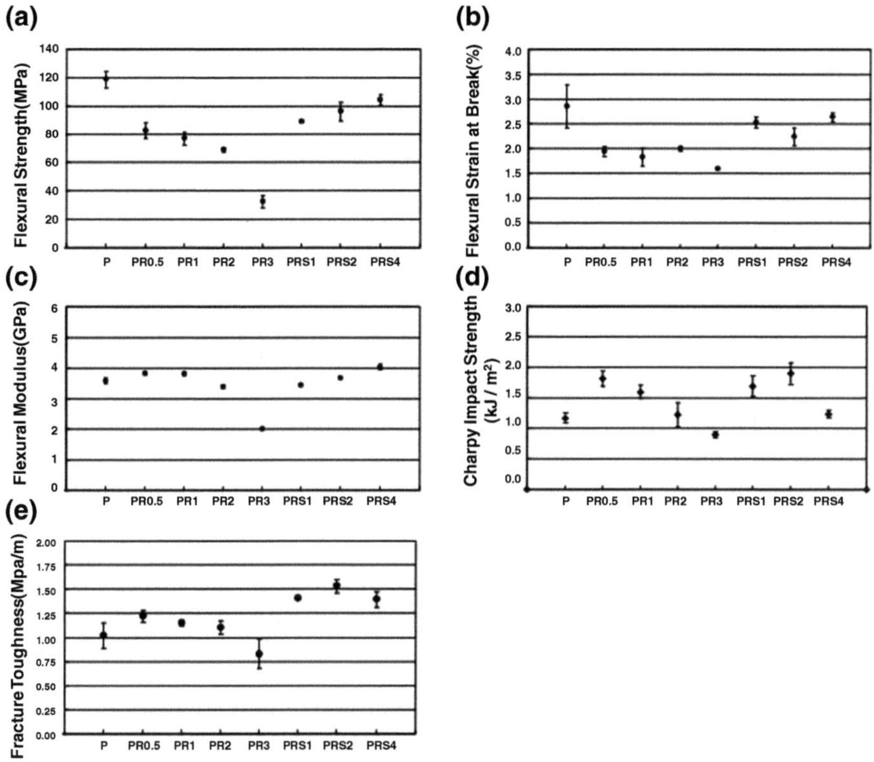


**Fig. 2** Results of flexural, impact, and fracture toughness tests: **a** flexural strength, **b** flexural strain at break, **c** flexural modulus, **d** charpy impact strength, and **e** fracture toughness [29]

**Table 2** Mechanical properties of the specimens [29]

| Specimen designation | Flexural strength (MPa) | Flexural strain at break (%) | Flexural modulus (GPa) | Charpy impact strength (KJ/$m^2$) | Fracture toughness (Mpa $m^{1/2}$) |
|---|---|---|---|---|---|
| P | 119 ± 6 | 2.85 ± 0.44 | 3.58 ± 0.10 | 1.17 ± 0.08 | 1.02 ± 0.13 |
| PR0.5 | 83 ± 6 | 1.93 ± 0.10 | 3.82 ± 0.05 | 1.82 ± 0.12 | 1.22 ± 0.06 |
| PR1 | 77 ± 5 | 1.82 ± 0.18 | 3.81 ± 0.07 | 1.60 ± 0.11 | 1.15 ± 0.03 |
| PR2 | 69 ± 1 | 1.99 ± 0.05 | 3.39 ± 0.05 | 1.22 ± 0.20 | 1.10 ± 0.07 |
| PR3 | 32 ± 4 | 1.59 ± 0.02 | 2.00 ± 0.04 | 0.89 ± 0.05 | 0.83 ± 0.15 |
| PRS1 | 89 ± 1 | 2.53 ± 0.11 | 3.43 ± 0.04 | 1.70 ± 0.17 | 1.41 ± 0.02 |
| PRS2 | 96 ± 7 | 2.24 ± 0.18 | 3.67 ± 0.03 | 1.90 ± 0.18 | 1.53 ± 0.07 |
| PRS4 | 104 ± 4 | 2.64 ± 0.09 | 4.04 ± 0.09 | 1.24 ± 0.06 | 1.39 ± 0.08 |

phenolic resin specimens modified by nitrile rubber particles and an amino silane determined by flexural, impact, and fracture toughness tests. For comparison, the data are also tabulated in Table 2.

Figure 2a shows that use of nitrile rubber particles in phenolic matrix decreased flexural strength. These decreases can be due to two main reasons: first of all, since nitrile rubbers have elastomeric behaviour with low strength, it is a reasonable argument that the flexural strength values will decrease for the phenolic matrix. The second reason may be attributed to the difficulties encountered during specimen production. For instance, formation of voids in the specimens due to the water evaporation during very long curing schedules. Another problem in specimen production was the difficulty in dissolving solid rubber particles in liquid phenolic resin, which increased more with increase in rubber content. Figure 2a also shows that use of amino silane together with nitrile rubber in the phenolic matrix increased the flexural strength.

Specimens having 0.5 % rubber and 2 % silane (PRS2) and 4 % silane (PRS4) have 16 and 25 % higher flexural strength than the specimen having 0.5 % rubber only i.e., (PR0.5). Here 'synergistic' effect of amino silane with nitrile rubber particles took place.

As seen in Fig. 2b, flexural strain at break also decreased when nitrile rubber particles were used in the phenolic matrix. Normally, since these particles are elastomeric materials, it is expected that the strain values should increase.

Unfortunately, due to the specimen production problems, this was not observed. Void formation and debonded solid rubber particles led to lower strain values at failure. However, synergistic effect of amino silane with nitrile rubber increased these strain values. It is seen that strain at failure increased by 16 and 37 % in the specimens PRS1 and PRS4, respectively, compared to PR0.5 specimen.

Figure 2c indicates that flexural modulus values did not decrease when phenolic matrix was modified with nitrile rubber particles, except for the specimen PR3 having the highest rubber content, which again can be due to the specimen production problems mentioned above. Use of amino silane with nitrile rubber increased the modulus values slightly, the highest increase being 13 % in the specimen PRS4 compared to neat phenolic specimen (P).

Figure 2d shows that Charpy impact strength value of neat phenolic specimen (P) improved when modified with nitrile rubber particles, for instance the increases being 56 and 37 % in the specimens PR0.5 and PR1, respectively. This is reasonable due to the 'rubber toughening' effect of nitrile rubber domains in the phenolic matrix. However, increasing rubber content decreased the improvement so that the specimen having highest rubber content (PR3) had lower impact strength than the neat phenolic specimen (P). This could again be due to the problems encountered during specimen production. As the rubber content increased, the time period required to dissolve these rubber particles in liquid phenolic resin also increased. These longer periods led to some pro-curing in the phenolic structure and made the mixture more viscous. Due to the difficulty in stirring the more viscous mixture, very small air bubbles were introduced into the mixture. Although many of these bubbles might have left the system from the surface, yet, some of them, especially those located at the bottom could not escape. As a result, these tiny voids at the bottom surface made the specimen more brittle by acting as stress raisers or cracks. Figure 2d also shows that the use of amino silane together with nitrile rubber increased Charpy impact strength of the neat phenolic specimen (P) significantly. This synergistic increase in rubber toughening was as much as 46 and 63 % in the specimens PRS1 and PRS2, respectively.

Figure 2e indicates that fracture toughness test results are very well correlated with the Charpy impact test results. When the neat phenolic specimen (P) was modified with nitrile rubber, its fracture toughness increased by 20, 13, and 8 % in the specimens PR0.5, PR1, and PR2, respectively. Again, increasing rubber content decreased the improvement due to the specimen production difficulties discussed earlier. Figure 2e also indicates that rubber toughening is more effective when nitrile rubber was used together with amino silane. In this synergistic case, fracture toughness of the neat phenolic specimen (P) increased by 38, 50 and 36 % in the specimens PRS1, PRS2, and PRS4, respectively.

Toughness improvement via rubber toughening was achieved by the formation of rubber domains and silane domains, which delayed and/or decreased the growth rate of main and secondary cracks propagating in the stiff but brittle phenolic matrix. Thus, phenolic resins were toughened by several energy absorption mechanisms.

Fracture surfaces obtained from flexural and fracture toughness test specimens were examined by scanning electron microscopy (Fig. 3). Fractographic studies were especially useful in determining possible rubber toughening mechanisms and distribution and interaction of rubber domains with the phenolic matrix.

Low magnification fractographs indicate that nitrile rubber domains formed in the phenolic matrix were generally uniformly distributed (Fig. 3a). They also indicate that increasing the rubber content not only increased the number of the rubber domains, but also the amount of deformation lines and the level of fracture surface roughness. Furthermore, higher magnification shows that these nitrile rubber domains have round shapes with a diameter around 10 mm (Fig. 3b), and there is a good interface between the rubber domains and the phenolic matrix with no debonding (Fig. 3c).

Due to the high viscosity and stirring difficulties mentioned before, when the rubber content was high (e.g., PR3 specimen) all the rubber particles were not dissolved in the liquid phenolic matrix. In this case, some rubber particles remained as solid particles in the crosslinked phenolic structure (Fig. 3d). Since rubber particles are not as effective as rubber domains, PR3 specimens had lower mechanical properties.

Figure 3e shows that domains of amino silanes formed in the phenolic matrix are similar to the nitrile rubber domains, but they are more spherical and have a lighter colour. To differentiate amino silane domains from the nitrile rubber domains, EDX analysis was also carried out.

Figure 3f reflects domains of both nitrile rubber and amino silane together, which resulted in synergistic improvement in rubber toughening of the brittle phenolic structure. All the fractographs indicated that the main rubber toughening mechanism was shear yielding observed as deformation lines, especially initiated at the domains of nitrile rubber and amino silane.

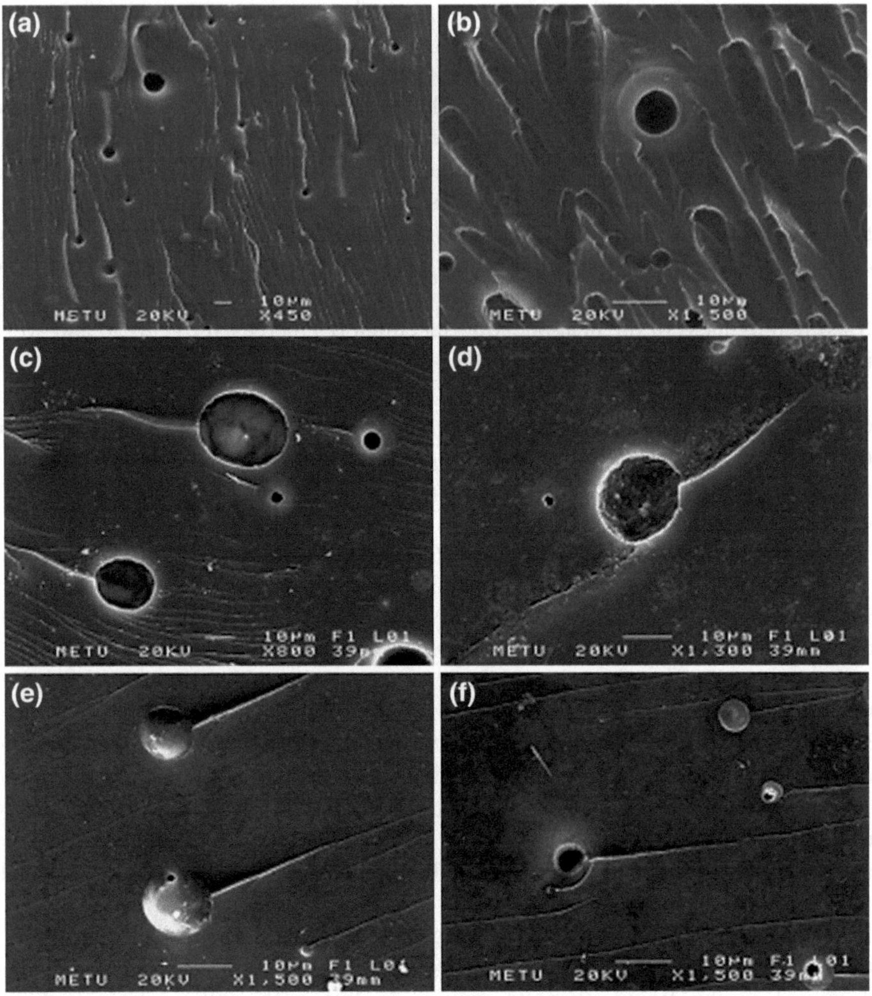


**Fig. 3** Sem fractographs of the specimen fracture surfaces: **a** uniform distribution of the nitrile rubber domains, **b** round shaped nitrile rubber domains and deformation lines, **c** proper interface between nitrile rubber domains and phenolic matrix, **d** undissolved nitrile rubber particles left in the matrix, **e** amino silane domains, **f** domains of both nitrile rubber and amino silane [29]

Dynamic mechanical analysis over a temperature range of 25–250 °C was done in order to obtain storage modulus ($E_0$) and glass transition temperature ($T_g$), and the results are shown in Fig. 4.

Figure 4b shows that Tg (peaks of tan d curves) of neat phenolic specimen (P) increased slightly when modified by nitrile rubber particles alone or together with amino silane. This might be due to certain interactions between the phenolic matrix and domains of nitrile rubber and amino silane. As shown in Fig. 4c, storage modulus values (at 50 °C) of neat phenolic specimens (P) increasd slightly

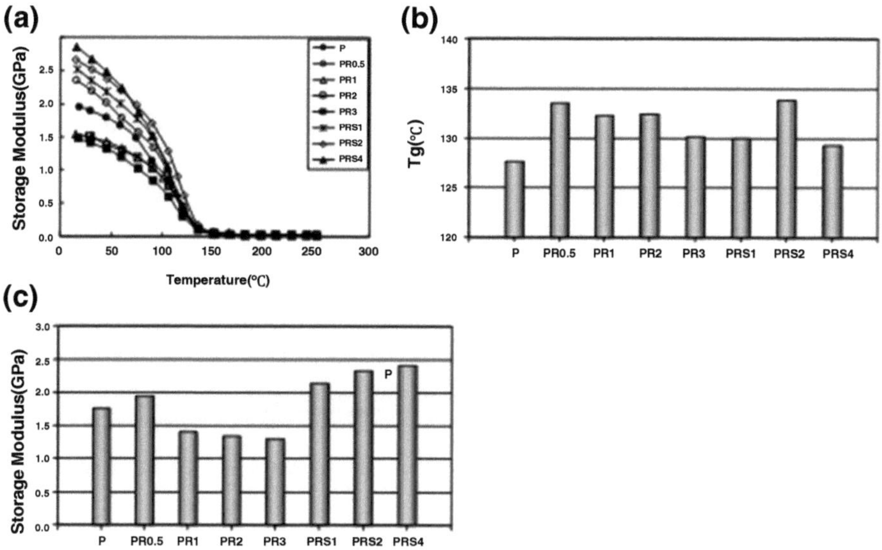


Fig. 4 DMA results of the specimens: **a** storage modulus versus temperature curves, **b** glass transition temperature values (tan d Peaks), **c** storage modulus values (at 50 °C) [29]

when modified with very low amount of nitrile rubber (PR0.5). Higher rubber contents decreasd the modulus value, possibly due to the specimen production difficulties discussed before. However, when the neat phenolic specimen was modified with nitrile rubber and amino silane together, significant increase in the storage modulus values was obtained.

From this elaborate case study it can well be concluded that a seemingly difficult development becomes easy if the underlying scientific principles can be deciphered and explained. Resol-type phenol–formaldehyde resin can be toughened by a judicious choice of nitrile rubber along with the use of an aminosilane.

### 2.1.2 Effect of Elastomeric Nanoparticles on Properties of Phenolic Resin

Ma et al. [28] reported the use of elastomeric nanoparticles (ENP) to modify the impact strength of phenolic resins (PR) and achieved the simultaneous modification of flexural strength and heat resistance. They used two sets of blends in their studies, namely, PR/nitrile butadiene elastomeric nanoparticle (NBENP) composite and PR/carboxylic nitrile butadiene elastomeric nanoparticle (CNBENP) composite. The elastomeric nanoparticles (ENP) studied were special ultra-fine full vulcanized powdered rubbers, prepared by a special irradiation technique [30, 31].

The ENP were uniformly dispersed in phenolic matrix (Fig. 5) with a diameter of about 100 nm. The size of dispersion phase was much smaller than that of

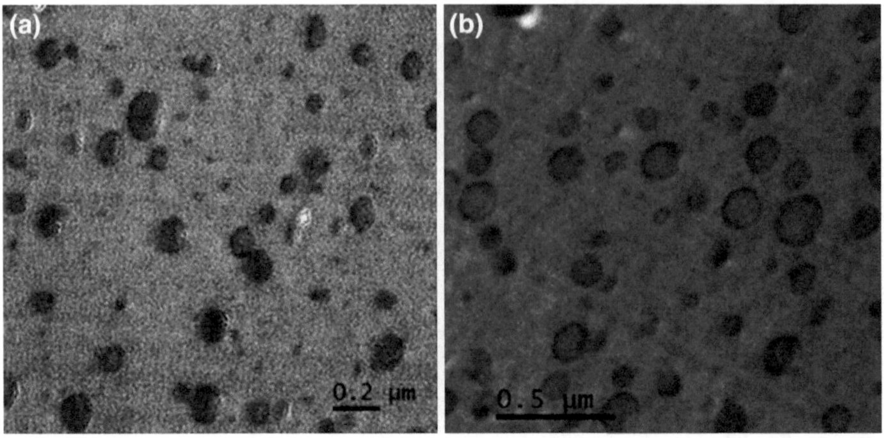


**Fig. 5** TEM photographs of (**a**) PRC5 wt% NBENP and (**b**) PRC5 wt% CNBENP blends [28]

conventional rubbers [32]. The positive outcome of such class of composites are their high toughening effect, low addition amount (less than 5 wt%) and simple addition process. ENPs have already found application in PR industry [33].

However, the problem of aggregation was encountered when the rubber concentration mounted above 5 weight percent. The new type of PR composite modified by ENP exhibited large interface because of the large specific surface area of nanoparticles. Furthermore, it was also found that the uniform dispersion of rubbers can be ensured by adequate blending time and appropriate addition amount.

SEM photographs of impact fracture surface of ENP modified PR are shown in Fig. 6.

It was observed that the average distance between microcracks was over 5 mm on the fracture surface of pure phenolic network, as shown in Fig. 6a, b; however, the average distance between microcracks initiated by ENP wasonly about 0.5 mm on the fracture surface of PR/ENP blends, as shown in Fig. 6c, d. Therefore, there were much more microcracks on the fracture surface of PR/ENP blends than that on the fracture surface of pure phenolic network due to smaller distance between microcracks. It is well known that plastic deformation of the matrix is limited in thermoset plastics with high crosslink density and microcrack is a main toughening mechanism [34, 35]. The amount of microcracks, initiated by rubber particles, depends on the particle size of rubber in PR/rubber composite if same amount of rubber is used; therefore, it is explanatory that the ENP toughened PR system, where rubber particles are nanostructured (about 100 nm), has much more microcracks and much higher impact strength than the conventional rubber toughened PR system.

The reaction between PR and ENP was supported through FTIR studies which are shown in Fig. 7.

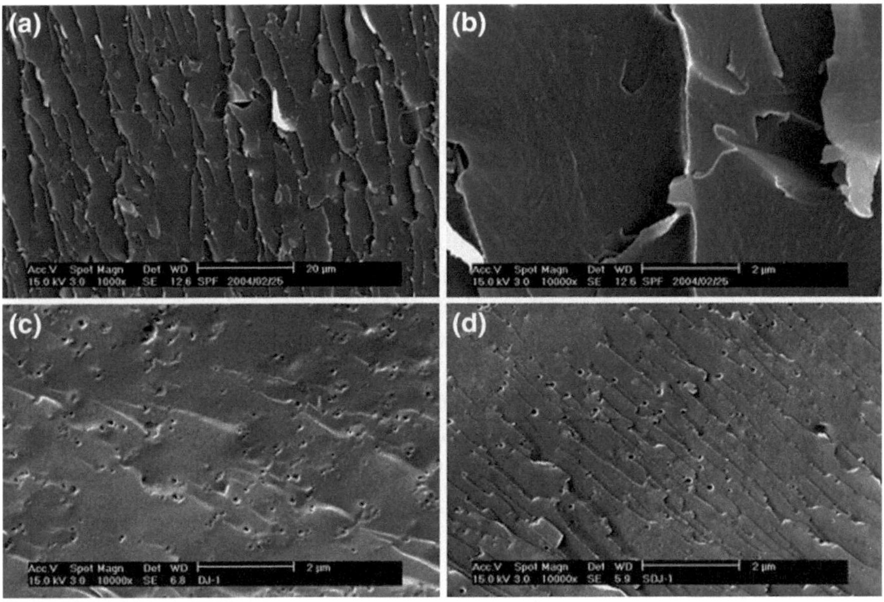


**Fig. 6** SEM photographs of fracture surface of toughened phenolic network (**a**) and (**b**) un-toughened; **c** 5 wt% NBENP toughened; **d** 5 wt% CNBENP toughened [28]

The characteristic peaks of phenolic hydroxyl group were changed after melt blending. Before melt blending, the stretching vibration of hydroxyl group showed a broad peak at 3,352 $cm^{-1}$ (hydrogen bond) with a shoulder peak at 3,490 $cm^{-1}$ (free hydroxyl group). After melt blending, only one symmetric peak at

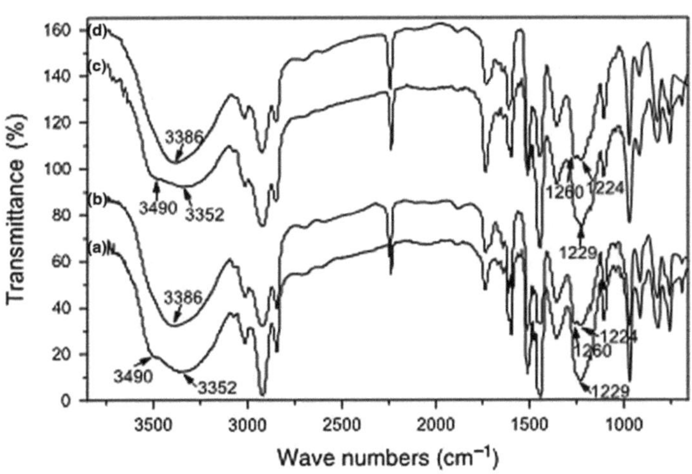


**Fig. 7** FTIR spectra of: *a* NBENP/PR and *c* CNBENP/PR before melt blending; *b* NBENP/PR and *d* CNBENP/PR after melt blending [28]

3,394 $cm^{-1}$ was noticeable, which indicated that some hydrogen bonds were broken by ENP. Moreover, the peak at 1,230 $cm^{-1}$, which is attributed to stretching vibration of phenolic C–OH, splited and shifted to higher wave numbers after melt blending, which shows the influence of ENP on some phenolic hydroxyl groups. Since the hydrogen atoms at ortho or para position of phenolic hydroxyl group are very active, they are subject to substitution by some active functional groups on ENP surface during melt blending. As a result of substituent reaction, crosslinked ENP, acting as a large substituent group, was introduced onto the benzene ring of PR. Therefore, the induction effect of large substituent groups causeed a shift of peak at 1,230 $cm^{-1}$ to higher wave numbers.

Since only active groups on rubber surface can react with PR so other weak changes were not observed in the FTIR spectra.

However, FTIR analysis conclusively indicated chemical reaction between rubbers and PR during melt blending of the components as a result of which enhancement of interfacial adhesion between ENP and PR took place.

## *2.2 Rubber-Epoxy Blends*

One of the most important classes of thermosetting polymers is the epoxy resins as their applications can cover a wide spectrum. They are extensively used in various fields of coating, high performance adhesives and engineering applications. Cured epoxy polymers are characterized by high chemical and corrosion resistance simultaneously having good mechanical and thermal properties. Epoxy resins are reactive monomers, which are commonly cured with amine to form thermosetting polymers. When cured with an aromatic amine of sufficient functionality, it results in a highly cross-linked network with relatively high stiffness, glass transition temperature ($Tg$) and chemical resistance. However, the inherent toughness of tightly cross-linked polymer networks is relatively low. It is therefore desirable to enhance toughness without adversely affecting the other useful properties of the polymer. Aside from inorganic reinforcement, elastomer modification is one of the most frequently used methods of toughening rigid network polymers [36–39]. The principle mechanism of rubber toughening for network polymers is thought [36, 37, 39] to be the enhancement of shear yielding at the crack tip through a change in stress state in the region around a rubber particle. Stress fields around the particles must overlap to bring about optimal shear banding. For determining the toughness of the blended and processed material the particle diameter and inter-particle spacing are important.

The most common methods of rubber toughening are by the use of liquid rubbers or preformed rubber particles. In the former method, the rubber is initially dissolved into the epoxy resin [40, 41], but during cure the rubber phase separates as a discrete particulate phase. Studies [42–45] show that the phase separation process is a result of the decrease in configurational entropy due to the increase in molecular weight as the epoxy cures. As a consequence the free energy of mixing changes, leading to a decrease in the solubility of the rubber. This is then the driving force for phase separation. Thus the functionality of the matrix monomers, which control the development of the network and the cross-link density of the epoxy matrix, has an effect on the phase precipitation process. The particle size and concentration of the precipitated rubber also depends on the curing process and the interaction between the rubber and the epoxy resin.

With butadiene-based rubbers the solubility may be increased by forming a copolymer with the more polar acrylonitrile monomer. Another way to modify the liquid rubber is to alter their interaction with the matrix by functionalising the chain ends with carboxyl, amine or epoxide groups that may couple with the reacting matrix [39, 43, 46]. The mechanism of reinforcement has been discussed in the introductory part along with some reiteration in this section.

### 2.2.1 Epoxy Toughening by Liquid Rubber

It is well over 40 years since rubber toughened epoxy was first patented. McGarry and Willner [47] reported the use of low molecular weight carboxyl terminated copolymer of butadiene and acrylonitrile as the liquid rubber in their effort to toughen the epoxide. Various diglycidyl ether of bisphenol A (DGBEA) epoxides were cured with piperidine (PIP) and CTBN as the toughening agent. Since then much efforts have been put by various researchers to decipher the morphology development during curing, morphology and fracture property relationship and the mechanism of toughening. These researches also paved the way to various new types of toughening agents as potential replacements for CTBN.

A comprehensive idea of rubber toughened epoxy is somewhat difficult to write in a nutshell because of complications arising out of several factors. Literature study reveals the use of different types of curing agents (amines, anhydrides) and a spectrum within each type. The target of these researches was of course the same, i.e., to improve the resistance to mechanical and thermal shocks. The modified epoxy resins were characterized with respect to their fracture behaviour and adhesive bond strengths. The frequently conducted tests were peel strength, resistance to peel force, resistance to crack propagation, fracture toughness, impact strength and critical strain energy rate. Usually difunctional epoxy resins were used for experimentation. The chemical structures of a typical epoxy resin (DGBEA) and some of the more important amine curing agents are shown in Fig. 8 [48].

### 2.2.2 Chemistry

The liquid rubber should have to be chemically bonded to the epoxy matrix to bring about effective toughening [49, 50]. If there is weak bonding then failure of the toughened system may take place through debonding of the particles. Free

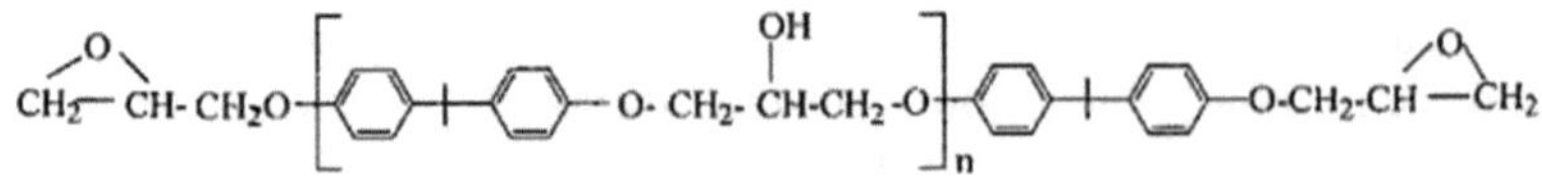

Diglycidyl ether of bis phenol-A (DGEBA)

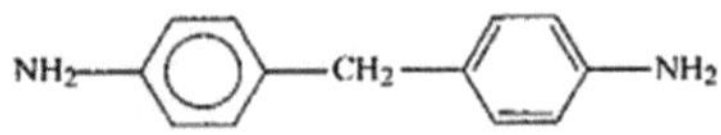

Diamino diphenyl methane (DDM)

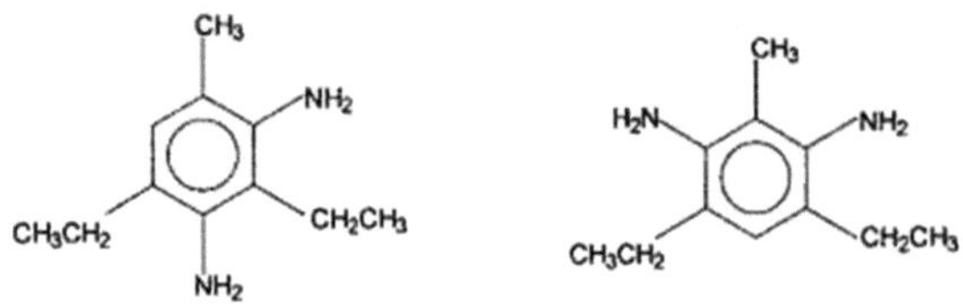

Diethyl toluene diamine (DETDA)

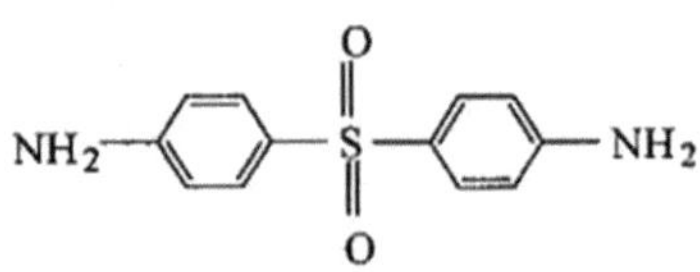

Diamino diphenyl sulfone (DDS)

$$NH_2\text{—}C_2H_4\text{—}NH\text{—}C_2H_4\text{—}NH\text{—}C_2H_4\text{—}NH_2$$

**Fig. 8** Chemical structure of epoxy and amine curing system [48]

liquid rubber is also undesirable because the collection of these molecules at the metal interface will create weak boundary layers, thus decreasing the joint strength [51, 52]. Sibert and Riew [53] proposed the formation an epoxy-CTBN-epoxy adduct which is chain extended and crosslinked with more epoxy resin. The dispersed rubber phase and the resin matrix are chemically bonded through the progression. PIP is a selective catalyst for carboxyl-epoxy reaction. However, most of the curing agents favour epoxy-amine reactions suppressing the epoxy-carboxyl one. The problem has been resolved [54, 55] by precipitating the liquid rubber and the epoxy resin in an alkyl hydroxyl esterification reaction in presence of triphenyl phosphene catalyst. The acid adducts so formed can then be cured with any curing agent as they contain only epoxy groups similar to the epoxy resins. The reaction scheme is furnished in Fig. 9.

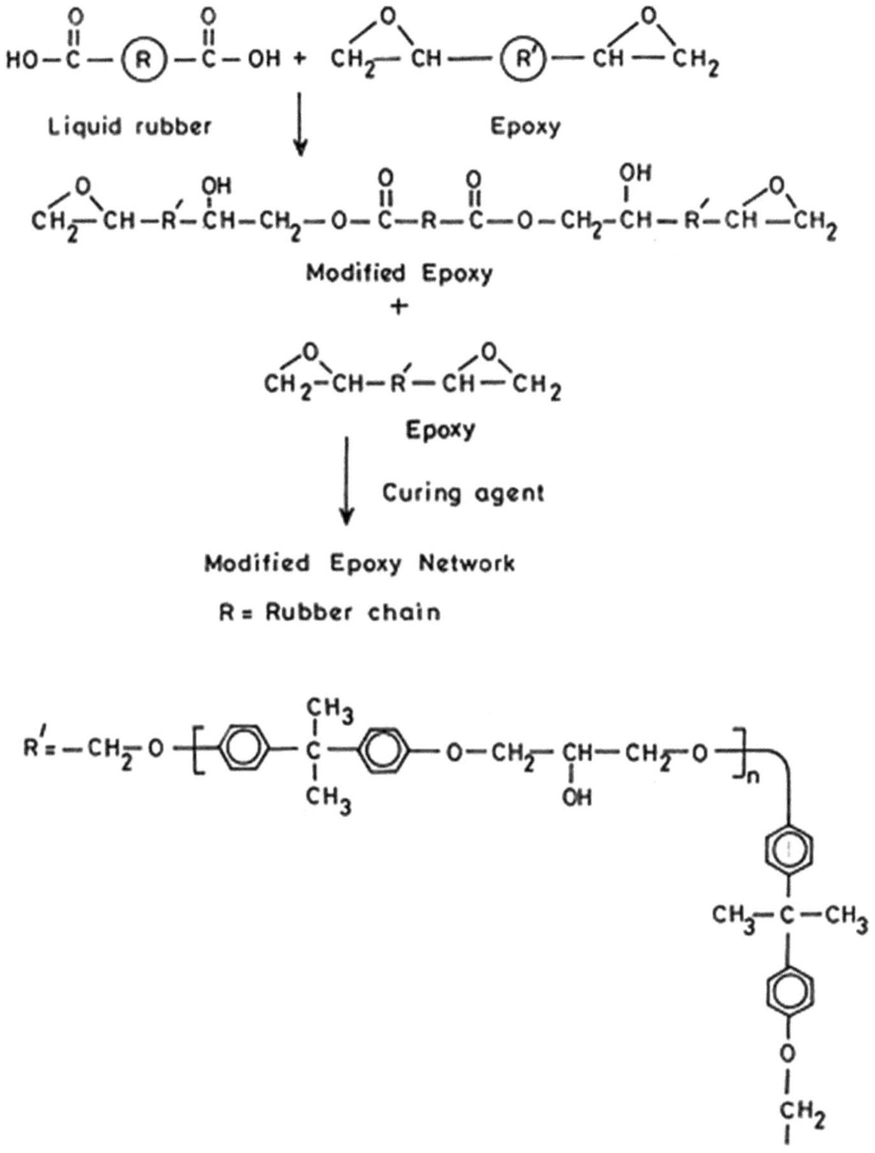


**Fig. 9** Prereaction of liquid rubber with epoxy [48]

### 2.2.3 Thermodynamic Consideration

There should be thermodynamic compatibility between the liquid rubber and the epoxy resin before curing at the curing temperature [56]. Thus, the change in Gibb's free energy should have to be negative [56, 57].

$$(\Delta G_m)_{P,T} < 0 \tag{1}$$

The combined Flory-Huggin's equation and the Hildebrand equation give the expression of the free energy of mixing as follows:

$$(\Delta G_m)/V = \phi_e \phi_r (\delta_e - \delta_r) + RT(\frac{\phi_e}{V_e} \cdot \ln \phi_e + \frac{\phi_r}{V_r} \cdot \ln \phi_r) \tag{2}$$

where $\phi_e$, $\phi_r$ are the volume fractions, $\delta_e$, $\delta_r$ are the solubility parameters and $V_e$ and $V_r$ are the molar volumes of the epoxy resin and the rubber respectively. The second term of the right hand side of the equation is always negative since both $\phi_e$ and $\phi_r$ is a fraction.

For a fixed rubber/epoxy weight composition, negative free energy change is favoured if the value of $V_r$ is low and simultaneously the values of $\delta_e$ and $\delta_r$ are quite close. This means that the rubber should have low molecular weight and the solubility parameter values of the rubber and the epoxy resin should be close. If these two parameters are controlled and $\Delta G_m$ is slightly negative the rubber will be compatible with the epoxy matrix. However, as the curing advances on the time scale, both $V_e$ and $V_r$ increase due to increase in molecular weight of rubber as well as the epoxy resin. With this advancement a time comes when the free energy just becomes positive and the rubber starts to phase separate out at the cloud point.

Hence the phase separation process is explained [58–60] and is pictorially shown in Fig. 10.

An initially homogeneous monophasic phase ($p = 0$) passes through a cloud point conversion ($p_{cp}$) and the final morphology is arrested at gelation ($p_{gel}$) [61, 62]. Actually due to manifold increase in viscosity of the reacting system, the final morphology is arrested well before gelation. However, there are reports of changes in phase composition even after gelation [58, 63]. For complete phase separation, the phase separation time should necessarily be greater than the time required for the diffusion of rubber particles from the epoxy matrix and here the role of temperature

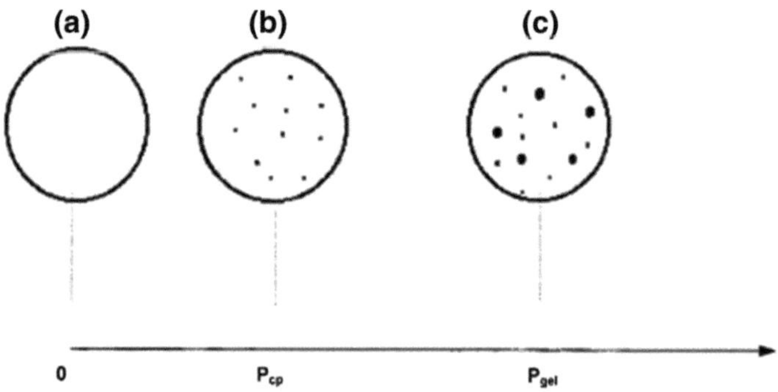


**Fig. 10** Phase separation of rubber in a typical epoxy matrix: **a** homogeneous solution at the start of the reaction ($p = 0$), **b** phase separation at cloud point conversion ($p_{cp}$), **c** morphology at gelation ($p_{gel}$) [48]

becomes important. The initial cure temperature greatly affects the resulting morphology but post curing temperature does not because phase separation is arrested at gelation.

CTBN with higher acrylonitrile content is shown to have greater compatibility with epoxy matrix and undergoes phase separation at a much advanced curing state [64, 65]. With acrylonitrile content above 30 % no phase separation occurs resulting in a single phase morphology.

The morphology for rubber modified epoxy system is a discrete one as reported by many authors and consists of spherical particles of the rubber dispersed in the epoxy matrix [64–66]. A model was developed by William et al. [67] to calculate the fraction, composition and average particle diameter of the dispersed phase formed during the thermoset polymerisation based on the principle of nucleation, coalescence and growth. The thermodynamic consideration as discussed above was applied in their approach to predict the same. The morphological variation with cure temperature as was observed experimentally was presumably a result of the effect of temperature on the rate of nucleation and subsequent growth of the dispersed rubbery phase [67–70]. Increase in cure temperature also increase rate of epoxy reaction with a decrease in the viscosity of the system. Since the activation energy of curing reaction is higher than that for prepolymer viscosity, the rate of reaction increases more than the rate of diffusion with an increase in the growth of rubber particles. However for a significant increase in temperature the situation may be different [71, 72]. Here, $\Delta G_m$ may attain such a high value that no phase separation takes place making the system completely miscible. Another possible explanation may be that the demixed particles cannot attain larger size due to the high system viscosity arising as a result of high levels of conversion. Therefore, the plot of average particle diameter versus temperature goes through a maximum as has been experimentally observed [70].

Some other authors stressed upon a spinodal decomposition as the origin of phase separation assigned to textures exhibiting a cocotinuous structure. Yamanaka et al. [73, 74] and Kims ruled out the possibility of the nucleation-growth mechanism based on the fact that nucleation is supposed to be a very slow process. They argued that the spherical domain structures evolved from an initial cocontinuous structure although they were incapable of producing direct evidence of this model.

### 2.2.4 Liquid Rubbers Other than CTBN

From the foregoing discussion, it is generalized that it is the presence of rubbery particles that toughens the epoxy matrix and it is not mandatory that the particles be necessarily made up of CTBN. Other kinds of nitrile rubbers having some other end groups like amine, mercaptan and hydroxyl have also been studied [75]. However it is the carboxyl-terminated acrylonitrile (CTBN) gives the best performance [76]. The presence of the carboxyl group in CTBN provides better adhesive strength with the substrate in CTBN modified epoxy.

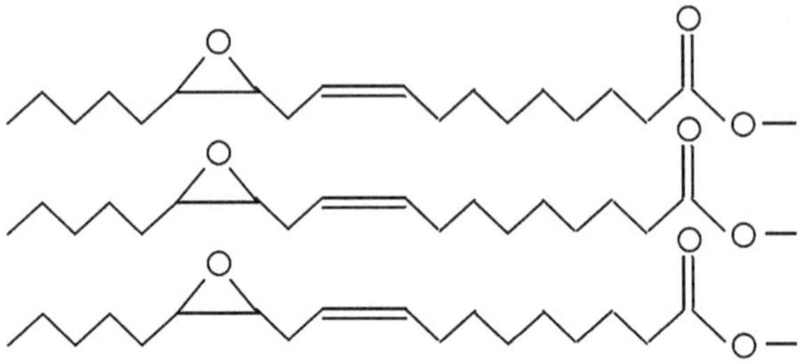

**Fig. 11** Chemical structure of vernonia (epoxy group containing triglyceride) oil

Significant enhancement in toughness can be brought about by using poly(methyl methacrylate) (PMMA) grafted natural rubber in place of CTBN as has been reported by Rezaiford et al. [77] Carboxyl terminated polyisobutylene [78] and polysulphide [79] rubbers are also reported as effective toughening agents for DGEBA resin using different types of amine hardener. Mizutani [80] has reported the use of hydroxyl terminated liquid polychloroprene rubber as a toughening agent containing up to a maximum of 10 volume percent of this liquid rubber, completely phase separated. However, due to similar refractive index with the epoxy, the modified networks are transparent.

Epoxy group containing modified triglyceride oils have been tried as modifiers of epoxy resin. Amongst these, the use of vernonia oil, epoxidised soybean oil and castor oil are reported in the literature [81, 82]. Figure 11 shows the structure of vernoria oil which is a plant product. Vernonia oil is extracted from the seeds of Vernonia galamensis (ironweed), a plant native to eastern Africa. The seeds contain about 40–42 % oil of which 73–80 % is vernolic acid. Products that can be made from vernonia oil include epoxies for manufacturing adhesives, varnishes and paints, and industrial coatings. Its low viscosity recommends its use as a nonvolatile solvent and thus it is eco-friendly.

If the epoxidised oil is directly mixed with the resin then it results in a two-phase microstructure with a plasticizing effect. Two-phase microstructure was however also observed when the prepolymer of epoxidised soybean oil (ESO) and amine hardener was used instead of pure ESO. Since the prepolymer has a high molecular weight, so compatibility with the epoxy matrix is reduced resulting in phase separation. The effect of ESO content on impact properties of modified epoxy network was studied by Ratna et al. [83] Both one stage and two stage routes were studied and the impact strengths were reported. It was found that impact strength increased as a function of modifier in both the cases, but it was more significant in the two stage process. This is illustrated in Fig. 12. It is evident from the figure that at 20 phr concentration of the toughener in the two stage process the maximum impact strength was obtained. Above 30 phr, phase inversion occurs [82, 83].

A novel route has been tried by Kim, Moon and Boonkerd to synthesise and couple poly(butadiene) with ESO. Butadiene monomer was reacted in a high pressure glass reactor in a solvent (hexane) and modifier (tetrahydrofuran) purged with nitrogen gas. The temperature was maintained at 30 °C and n-butyllithium initiator was injected into the system. After the temperature was stabilised, ESO

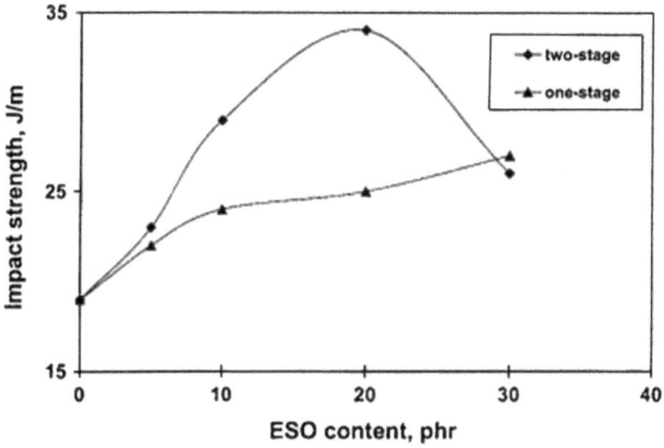


**Fig. 12** Effect of ESO content on impact strength of toughened epoxy network [83]

was injected into the reaction system. The reaction was terminated by methanol containing BTH as the antioxidant. Precipitation of the product using excess of methanol followed by removal of unreacted ESO by acetone was done to get the pure product. Finally drying was done using a vacuum oven to remove the excess of acetone. The reaction variables were the temperature and catalyst concentration. Molecular weight distribution was determined using gel permeation chromatography (GPC). This product can be tried as an alternative for CTBN to improve the impact resistance of epoxy thermosets. The reaction scheme is shown in Fig. 13.

### 2.2.5 Saturated Liquid Rubbers

The outstanding modification in the fracture properties of epoxy resins exhibited with the toughening by liquid rubber like carboxyl terminated copolymer of acrylonitrile butadiene (CTBN) in the field of technology and engineering of adhesives [84] is not altogether an unmixed blessing. Since the butadiene component of the elastomer contains unsaturation, it is a site for premature thermal and/or oxidative stability. Thus, even such modified resins are not suitable for application at high temperature [85]. Excessive crosslinking may take place with time which would detract from otherwise improvements brought about in their structures. Another problem encountered is related to some unreacted acrylonitrile remaining in the system which may be carcinogenic [86]. These are the reasons which asked for an alternative to CTBN and considerable efforts have been made to come up with potential substitutes.

Siloxane rubbers were thought to be attractive in this direction because of some of the more important features like high chain flexibility, very low glass transition temperature ($T_g \sim -100$ °C) low surface tension and surface energy along with hydrophobic nature which are rewarding properties for toughening. However, poly (dimethyl siloxane) oligomer cannot be used as such, because it is extremely

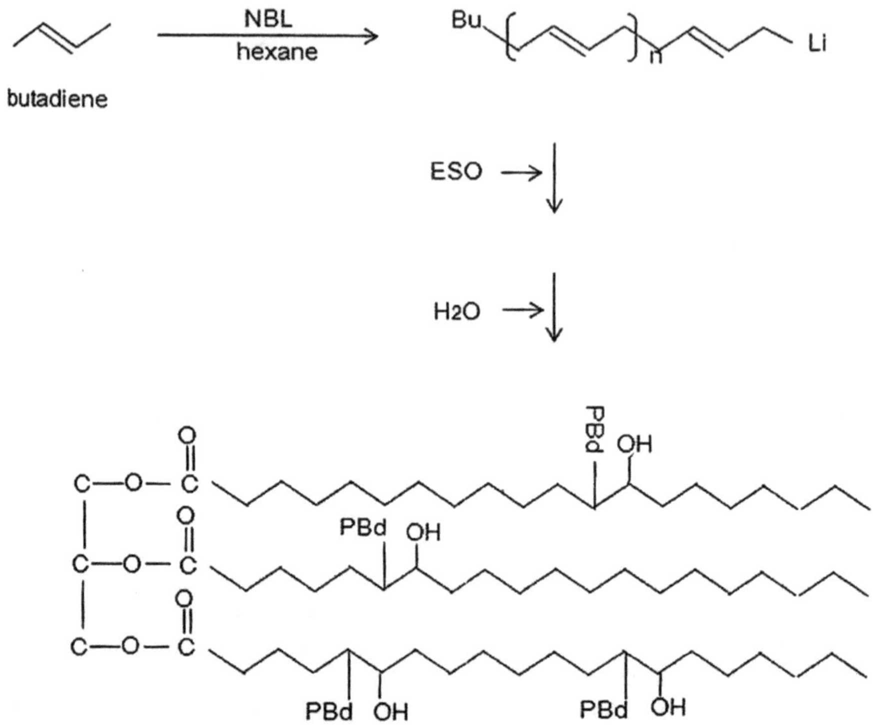


**Fig. 13** Coupling reaction of polybutadiene with ESO

incompatible with epoxy. The compatibility is increased by copolymerizing dimethyl siloxane with diphenyl siloxane or dimethyl fluoropropylsiloxane [87]. Such a copolymer having controlled structure can be used as a toughening agent [87–89]. Riffele et al. [90, 91] developed an epoxy-terminated siloxane oligomer and it was successfully used to toughen epoxy resin. The epoxy-terminated oligomer was reacted with piperazine to produce secondary amine group end capped oligomer which has been used as toughening agent. Silicone-epoxy block copolymers have also been reported [92] as toughening agent for epoxy. Laurienzo et al. [93] have modified epoxy resin by hydroxyl terminated block copolymers of poly(dimethylsiloxane) (PDMS) and polyoxyethylene (PDMS-co-POE) elastomer and functionalized saturated polybutene. Oxidative stability of these networks showed improvement but the mechanical properties were inferior to CTBN.

Polyepichlorohydrin (PECH) has been successfully incorporated [94] as toughening agent. The degree of toughening was somewhat proportional to the molecular weight of PECH, It was found that matrix properties like Tg, modulus and hot/wet properties were comparable with CTBN.

Many other polymers were tried and some of the more important ones with their roles are stated in the following discussion. Carboxyl terminated poly(propylene glycol) was developed for an ambient temperature curing of epoxy adhesives [95]. Recently polyurethane (PU) elastomer has been investigated as a modifier by Wang et al. [96]. They developed hydroxyl, amine and anhydride terminated PU prepolymer and found that the hydroxyl terminated oligomers at 20 wt% gave the best result with a fivefold increase in fracture energy over the unmodified epoxy.

Many other modifiers have been used in systems where the curing system is not a diamine one. Laeuger et al. [97] have reported the toughening of anhydride cured DGEBA resin using diol and bis(4-hydroxy benzoate) terminated poly(tetrahydrofuran) liquid rubber. In their work they have supported the morphology development via spinodal decomposition. The bis(4-hydroxy benzoate) terminated poly(tetrahydrofuran) gives better performance over the diol terminated one as far as the mechanical and the fracture properties are concerned. Wang et al. have synthesized polyfunctional poly(n-butyl acrylates) (containing epoxy and carboxyl groups) by photo polymerization, and used it as a modifier to toughen epoxy resin cured with diamino diphenyl methane (DDM). The concept of optimum functionality to get maximum impact resistance epoxy and carboxyl functionalized and epoxy functionalized liquid rubber modified systems was introduced in the work. Similar results were obtained by Lee and coworkers [98] using n-butyl acrylate/acrylic acid copolymers. They reported improvement of adhesion strength in a DGEBA epoxy matrix and showed that an optimum functionality existed to impart the highest interfacial tension.

A new class of reactive dendritic hyperbranched polymers (HBPs) has been used as liquid rubbers as a modifier for epoxy resin [99–105]. HBPs offer much lower prepolymer viscosity owing to their spherical structure. They also provide better internal bonding and stronger adhesion with the matrix due to the presence of highly dense surface functional groups.

### 2.2.6 Toughening by Preformed Particles

Preformed, insoluble particles are often directly used in place of liquid rubbers as toughening agents for the epoxy matrix. This is to minimize some of the difficulties encountered while using liquid rubbers as such [106–109]. The phase separation, in case of liquid rubber toughening depends upon the formulation, processing and curing conditions. Incomplete phase separation lowers the glass transition temperature ($T_g$) significantly. Moreover, the difficulty in controlling the rubber phase which separates during curing may result in uneven particle size. The differences in morphology and also the volume of the separated phase affect the mechanical properties of the product. It is very difficult to study the effect of individual parameter such as morphology, particle size and composition as they are independent.

In case of preformed particle, the size, morphology and composition, shell thickness and crosslink density of the rubbery core can be controlled separately by employing emulsion polymerization and so the effects of various parameters on the toughening of the epoxies can be investigated [110–114]. Interfacial architecture can be controlled by changing the following parameters: (1) shell thickness which depends on the ratio of core–shell material and the mechanism of polymerization; (2) chemical bonding and physical interaction between particles and matrix enhanced by the introduction of functional groups on the surface of the shells; (3) grafting between the shell and the core and (4) molecular weight of the shell material. Core shell, occluded and multilayer morphologies of the composite material can be obtained by controlling the emulsion polymerization [115–117]. The incorporation of the preformed rubber particles in epoxy matrix is enhanced through mechanical mixing. Dispersibility can be increased either by introducing crosslinks into the shell or by using comonomer like acrylonitrile or glycidyl metacrylate (GMA) which increase the interfacial adhesion by polar or chemical interactions [118, 119].

### 2.2.7 Some Miscellaneous Studies with Rubber-Epoxy Blends

Jansen et al. [120] reported the preparation of thermoset rubbery epoxy particles as novel toughening modifiers for glassy epoxy resins. Two types of liquid epoxy resin with an aromatic backbone were used in this work; a diglycidyl ether of bisphenol A (DGEBA) and a diglycidyl ether of bisphenol F (DGEBF). The aliphatic epoxy resin was diglycidyl ether of polypropyleneoxide (DGEPPO). They showed that curing of water dispersed droplets of an aromatic or aliphatic epoxy resin, respectively, produces glassy or rubbery thermosetting epoxy spheres which can be prepared in a relatively easy way. The rubbery epoxy particles can successfully be applied as a toughening agent for glassy epoxy matrices. The advantage of these preformed modifiers is the control over the final morphology as the size and concentration of the dispersed rubber phase can be chosen independently. The improvement in fracture toughness and the morphological features of the fractured surfaces are identical to standard. The use of this new class of rubber modifiers can be used to produce impact modified composites with predetermined particle size and rubber content. The authors claimed this study to be the first attempt to prepare preformed rubbery aliphatic epoxy particles in dispersion and to apply these particles as toughening agents for glassy aromatic epoxy matrices.

Boyonton and Lee [121] in their work used a synergistic combination of a liquid carboxyl-terminated butadiene acrylonitrile rubber and solid rubber particles of different sizes, the latter obtained from recycled automobile tires. They found no significant improvement in the fracture toughness of the composite when used with solid rubber alone. However, when used in combination with the liquid rubber modifier, it was observed that the fracture toughness of these hybrid epoxies was higher than that of those toughened with liquid rubber alone. This synergistic effect is explained in terms of crack deflection and localized shear yielding. Furthermore, they observed a slight improvement in the fracture toughness as the size of the solid rubber particles increased. Although using a combination of both reactive liquid rubber and solid rubber particles as toughening agents had been investigated previously, here, the solid rubber particles used were from recycled rubber tires. Therefore this work demonstrated an application of producing high quality engineering epoxy systems using toughening modifiers that are relatively low in cost and created higher-value products for recycled solid rubber.

A very similar work has been reported by Bagheri et al. [122] The objective of the research was to investigate the feasibility of using surface treated recycled rubber particles for toughening of epoxy polymers. The particles were obtained through grinding of scrap tires followed by oxidizing the surface of the particles in a reactive gas atmosphere. Surface treated recycled rubber particles with a nominal particle size of approximately 75 μm and a commonly used reactive liquid elastomer, CTBN, was incorporated in a DGEBA epoxy resin. Microscopic studies reveals that, when used alone, recycled rubber particles simply act as large stress concentrators and modestly contribute to toughening via crack deflection and microcracking. In the presence of micron size CTBN particles, which cavitate and induce massive shear yielding in the matrix, however, the recycled particles "stretch" the plastic deformation to distances far from the crack tip. This mechanism cause plastic zone branching and provides an unexpectedly high fracture toughness value. This study, therefore, provides a practical approach for manufacturing engineering polymer blends utilizing the surface modified recycled rubber particles.

## 2.3 Rubber Toughening of Cyanate Esters

Cyanate ester (CE) monomers undergo polycyclotrimerization to form trifunctional triazine rings, which exhibit excellent dielectrical, thermal and adhesive properties [123]. Consequently, CE resins are currently important thermosetting materials for the encapsulation of electronic devices, high-temperature adhesives and structural aerospace materials. However, low toughness is a major drawback with most crosslinked thermosetting materials, including CE resins. Although a cured cyanate resin has a relatively higher toughness than a cured bismaleimide (BMI) or a cured epoxy resin, it still requires suitable modification to improve toughness without reducing the intrinsic physical strength for structural applications [124]. Accordingly, a number of studies were carried out to improve the toughness of cyanate resin, such as the preparation of flexibilised cyanate resins, the incorporation of monocyanates, the utilization of rubber-toughening technologies, the use of organoclay toughening, and the preparation of semi-interpenetrating networks [124]. Among them, using carboxyl-terminated butadiene-acrylonitrile rubber (CTBN) and other similar functionalized liquid nitrile rubber was proven to be one of the most successful techniques [125–132]. For example, Hillermeier et al. [126] used epoxy-terminated butadiene-acrylonitrile rubber (ETBN), and Hayes and Seferis [127] used CTBN and amine-terminated butadiene-acrylonitrile rubber (ATBN). It was found that incorporation of all of these terminal-functionalized liquid butadieneacrylonitrile rubber provided increased toughness over CE resin. Of the various types of functional groups investigated, carboxylic acid end-groups proved to be most effective [133]. CTBN greatly improves the impact strength of CE resins, but reduces the tensile strength and modulus of CE resins due to its rubbery characteristic. In comparison, Liang et al. [134] found that epoxy resin (EP) could improve the toughness of CE without significant sacrifice of tensile strength or modulus.

Feng et al. [135] used, epoxy resin (EP), which could react with both the cyanate group of cyanate ester (CE) and the carboxyl group of carboxyl-terminated butadiene-acrylonitrile rubber (CTBN). This EP was used together with CTBN as a synergistic toughening component. Its effect on the thermostability of the blends was also evaluated. They found that the addition of an appropriate amount of EP to CE/CTBN improved the mechanical properties and thermostability of the resulting blends. This is attributed to the decrease in the mobility and increase in the stability of CTBN caused by the reaction between CTBN and EP via terminal carboxyl (–COOH) and hydroxyl (–OH) groups. However, very high EP concentration decreases the crosslinking density of CE, consequently reducing the mechanical properties and thermostability of the blends.

## 3 Concluding Remarks

Brittle thermoset materials can be toughened successfully by blending with proper liquid rubbers in small amounts or by incorporating preformed rubber particles directly. Both of the mentioned methods are in vogue and the choice of using one depends on the feasibility of a system of thermoset. The phase separation, in case of liquid rubber toughening depends upon the formulation, processing and curing conditions and incomplete phase separation may occur resulting in unwanted lowering of glass transition temperature. Difficulty in controlling the rubber phase which separates during curing may result in uneven particle size. It is very difficult to study the effect of individual parameter such as morphology, particle size and composition as they are independent. In case of preformed rubber particles, these difficulties are not encountered and the resulting morphology can be better controlled. However, the problem of proper dispersion of these particles in the themoset resin limits the use of this method. The improvement in fracture resistance occurs in either case due to dissipation of mechanical energy by cavitation of rubber particles followed by shear yielding of the matrix. The toughenability increases with increase in inherent ductility of the matrix. Rubber particle size plays an important role in the direction of toughening and very small or very large sizes are undesirable. The phase separation in case of liquid rubber is based upon nucleation and growth. However, some researchers opposed this theory on the fact that nucleation is a very slow process. They stressed upon spinodal decomposition as the origin of phase separation assigned to textures exhibiting a cocotinuous structure although they were incapable of producing direct evidence of this model.

## References


1. Pascault, J.P.: Yielding and Fracture of Toughened Networks. Thermosetting Polymers. Marcel Dekker, New York (2002). (Chapter 13)
2. Burton, B. L., Bertram, J. L., Arends, C.B. (eds.): Design of Tough Epoxy Thermosets. Marcel Dekker, New York, p. 340 (1996)
3. Nicholas, P. Cheremisinoff (Ed.), (1993) Elastomer Technology Handbook" edited by by CRC Press, Inc.Chapter 18 "Toughening concept in rubber-modified high performance epoxies" by Hung-Jue Sue, Eddy I. Garcia-Meitin, and Dale M. Pickelman P.680
4. Kunz, S., Douglass. (1980) Low temperature behavior of epoxy-rubber particulate composites, Jounal of Material Science **5**, 1109
5. Sultan, J.N., McGarry, F.J.: Effect of rubber particle size on deformation mechanisms in glassy epoxy. J. Polym. Sci. **13**, 29–35 (1973)
6. Pearson, R.A., Yee, A.F.: Influence of particle size and particle size distribution on toughening mechanisms in rubber-modified epoxies. J. Mater. Sci. **26**, 3828–3844 (1991)
7. Frounchi, M., Mehrabzadeh, M., Parvary, M.: Toughening epoxy resins with solid acrylonitrile-butadiene rubber. Polym. Int. **49**, 163–169 (2000)
8. Kaynak, C., Ozturk, A., Tincer, T.: Flexibility improvement of epoxy resin by liquid rubber modification. Polym. Int. **51**, 749–756 (2002)
9. Ozturk, A., Kaynak, C., Tincer, T.: Effects of liquid rubber modification on the behavior of epoxy resin. Eur. Polym. J. **37**, 2353–2363 (2001)
10. Kaynak, C., Sipahi-Saglam, E., Akovali, G.: A Fractographic study on toughening of epoxy resin using ground tyre rubber. Polymer **2**, 4393–4399 (2001)
11. Sipahi-Saglam, E., Kaynak, C., Akovali, G.: Studies on epoxy modified with recycled rubber. Polym. Eng. Sci. **41**(3), 514–521 (2004)
12. Kaynak, C., Elikbilek, C.C., Akovali, G.: Use of silane coupling agents to improve epoxy-rubber interfaces. Eur. Polym. J. **39**, 1125–1132 (2003)
13. Chen, T.K., Jan, Y.H.: Fracture mechanism of toughened epoxy resin with bimodal rubber particle size distribution. J. Mater. Sci. **27**, 111–121 (1992)
14. Geisler, B., Kelley, F.N.: Rubbery and rigid particle toughening of epoxy resins. J. Appl. Polym. Sci. **54**, 177–189 (1994)
15. Pearson, R.A., Yee, A.F.: Toughening mechanisms in elastomer-modified epoxies. J. Mater. Sci. **24**, 2571–2580 (1989)
16. Sasidharan, P.A., Ramaswamy, R.: Reactive compatibilization of a nitrile rubber/phenolic resin blend: Effect on adhesive and composite properties. J. Appl. Polym. Sci. **69**, 1187–1201 (1998)
17. Camino, G., Alba, E., Buonfico, P., Vikoulov, K.J.: Thermal behavior of elastomer-resin mixtures used as friction materials. Appl. Polym. Sci. **82**(6), 1346–1351 (2001)
18. Gouri, C., Nair, C.P., Ramaswamy, R.J.: Effect of elastomer modification on the adhesive characteristics of maleimide-functional phenolic resins. Appl. Polym. Sci. **74**(9), 2321–2332 (1999)
19. Carter, J.T.: The development of thermoplastic toughened phenol-formaldehyde molding compounds for under the bonnet applications. Plast. Rubber Compos. Process. Appl. **16**(3), 157–170 (1991)
20. Dong, R., Yang W., Qi, S.: Modern Plast. Process Appl. **13**(5), 16–18 (2001)
21. Sathiyalekshmi, K.: Bull. Mater. Sci. **16**(2), 137–150 (1993)
22. Yu, G.: Polym. Mater. Sci. Eng. **11**(2), 121–126 (1995)
23. Chakrawarti, P.B., Mehta, V.: Preparation and properties of linseed oil-modified CNSL (cashew nut shell liquid)- formaldehyde resin and m-cresol novolac modified by epoxy resin. Indian J. Technol. **25**(3), 109–113 (1987)
24. Qi, S., Zeng, Z., Li, Y: China Plast. Ind. **27**(6), 15–16 (1999)
25. Su, Z., Liu, Y., Wei, Q.: Carbon Tech. **1**, 8–11 (2002)

26. Choi, M.H., Chung, I.J., Lee, J.D.: Morphology and curing behaviors of phenolic resin-layered silicate nanocomposites prepared by melt intercalation. Chem. Mater. **12**, 2977–2983 (2000)
27. Byun, H.Y., Choi, M.H., Chung, I.J.: Synthesis and characterization of resol type phenolic resin/layered silicate nanocomposites. Chem. Mater. **13**, 4221–4226 (2001)
28. Ma, H., Wei, G., Liu, Y., Zhang, X., Gao, J., Huang, F., Tan, B., Song, Z., Qiao, J.: Effect of elastomeric nanoparticles on properties of phenolic resin. Polymer **46**, 10568–10573 (2005)
29. Kaynak, C., Cagatay, O.: Rubber toughening of phenolic resin by using nitrile rubber and amino silane. Polym. Testing **25**, 296–305 (2006)
30. Liu, Y., Zhang, X., Wei, G., Gao, J., Huang, F., Zhang, M.: Special effect of ultra-fine rubber particles on plastic toughening. Chin. J. Polym. Sci. **20**(2), 93–98 (2002)
31. Huang, F., Liu, Y., Zhang, X., Wei, G.: Effect of elastomeric nanoparticles on toughness and heat resistance of epoxy resins. Macromol. Rapid Commun. **23**, 786–790 (2002)
32. Dutta, A., Mauman, E.B.: Role of elastomers in the toughening of fiber-reinforced phenolics. Polym. Mater. Sci. Eng. **67**, 482–483 (1992)
33. Fan, Z., Tan, Y., Ma, H., Xia, Y., Ren, C., Zhu, D.: New Chem Mater. **32**(2), 35–37 (2004)
34. Lu, F., Plummer, C.J.G., Cantwell, W.J., Kausch, H.H.: Toughening mechanisms in modified epoxy resins with different crosslink densities. Polym. Bull. **37**, 399–406 (1996)
35. Shiann, H.L., Nauman, E.B.: Effect of crosslinking density on the toughening mechanisms of rubber-modified thermosets. J. Mater. Sci. **26**, 6581–6590 (1991)
36. Riew, C.K., Gillham, J.K.: editors. American Chemical Society, Rubber modified thermoset resins Washington, DC (1984)
37. Pearson, R. A., Yee, A. F.: Toughening mechanisms in elastomer-modified epoxies. Part 2. Microscopy studies. J. Mater. Sci. **21**, 2475 (1986)
38. Riew, C.K.: Rubber Toughened Plastics. American Chemical Society, DC (1984)
39. Huang, Y., Kinloch, A.J., Bertsch, R.J., Seibert, A.R.: In: Riew, C.K., Kinloch, A.J. (eds.) Toughened Plastics I, Science and Engineering, p. 189. American Chemical Society, Washington, DC (1993)
40. Sultan, J. N., Laible, R. C., McGarry, F. J.: Appl. Polym. Symp. **16**, 127 (1971)
41. Sultan, J.N., McGarry, F.J.: Polym. Eng. Sci. **13**, 29 (1973)
42. Williams, R.J.J., Borrajo, J., Adabbo, H.E., Rojas, A.J.: Rubber Modified Thermoset Resins, p. 13. American Chemical Society, chap, DC (1984)
43. Verchere, D., Sautereau, H., Pascault, J.P., Moschiar, S.M., Riccardi, C.C., Williams, R.J.J.J.: Rubber-modified epoxies. I. Influence of carboxyl-terminated butadiene-acrylonitrile random copolymers (CTBN) on the polymerization and phase separation processes. Appl. Polym. Sci. **41**, 467 (1990)
44. Verchere, D., Pascault, J.P., Sautereau, H., Moschiar, S.M., Riccardi, C.C., Williams, R.J.J.: Rubber-modified epoxies. II. Influence of the cure schedule and rubber concentration on the generated morphology. J. Appl. Polym. Sci. **42**, 701 (1991)
45. Moschiar, S.M., Riccardi, C.C., Williams, R.J.J., Verchere, D., Sautereau, H., Pascault, J.P.: Rubber-modified epoxies. III. Analysis of experimental trends through a phaseseparation model. J. Appl. Polym. Sci. **42**, 717 (1991)
46. Chen, D., Pascault, J.P., Sautereau, H.: Rubber-modified epoxies. I. Influence of presence of a low level of rubber on the polymerization. Polym. Int. **32**, 361 (1993)
47. McGarry, F.J., Willner, A.M.: March) Toughening of an Epoxy Resin by an Elastomer Second Phase, R 68–8. MIT, Cambridge (1968)
48. Ratna, D., Banthia, A.K.: Rubber toughened epoxy. Macromol. Res. **12**(1), 11–21 (2004)
49. Sohn, J.E.: Morphology of solid uncured rubber-modified epoxy resins. Org. Coat. Plast. Chem. **44**, 38 (1981)
50. Daly, J., Pethrick, R.A., Fuller, P., Cunliffe, A.V., Dutta, P.K.: Rubber-modified epoxy resins. 1. Equilibrium physical properties. Polymer **22**, 32 (1981)
51. Xuzong, N., Lijuan, W., Rulian, X., Yiming, L., Yunchao, Y: In: Lee, L. H. (ed.) Adhesive Chemistry: Developments and Trends, p. 659. Plenum, New York (1984)

52. Dodiuk, H., Kenig, S., Liran, I.: Room-temperature-curing epoxy adhesives for elevated temperature service. J. Adhes. **22**, 227 (1987)
53. Siebert, A.R., Riew, C.K.: Chemistry of rubber toughened epoxy resins. I. Org. Coat. Plast. Chem. **31**, 552 (1971)
54. Bascom, W.D., Moulton, R.J., Rowe, E.J., Siebert, A.R.: The fracture behavior of an epoxy polymer having a bimodal distribution of elastomeric inclusions. Org. Coat. Plast. Chem. **39**, 1964 (1978)
55. Romanchick, W. A., Sohn, J. E., Gaibel, J. F.: In: Bauer, R.S. (ed.) Epoxy Resins Chemistry II, p. 93. Adv. Chem. Ser. 221, American Chemical Society, DC (1983)
56. Bucknall, C.B., Partridge, I.K.: Phase separation in crosslinked resins containing polymeric modifiers. Polym. Eng. Sci. **26**, 45 (1986)
57. Olabisi, O., Robeson, L. M., Shaw, M. T.: Polymer- Polymer Miscibility, Academic Press, London (1979) (Chaps. 2 & 3)
58. Visconti, S., Marchessault, R.H.: Small angle light scattering by elastomer-reinforced epoxy resins. Macromolecules **7**, 913 (1974)
59. Bitner, J.L., Rushford, J.L., Rose, W.S., Hunston, D.L., Riew, C.K.: Viscoelastic fracture of structural adhesives. J. Adhes. **13**, 3 (1981)
60. Nae, H.N.: Phase separation in rubber-modified thermoset resins: Optical microscopy and laser light scattering. J. Appl. Polym. Sci. **31**, 15 (1986)
61. Bartlet, P., Pascault, J.P., Sautereau, H.: Relationships between structure and mechanical properties of rubber-modified epoxy networks cure with dicyandiamide hardener. J. Appl. Polym. Sci. **30**, 2955 (1985)
62. Gillham, J.K., Glandt, C.A., McPherson, C.A.: Am. Chem. Soc. Div. Org. Coat. Plast. Chem. **37**, 195 (1977)
63. Verchere, D., Pascault, J.P., Sautereau, H., Moschair, S.M., Riccardi, C.C., Williams, R.J.J.: Rubber-modified epoxies. II. Influence of the cure schedule and rubber concentration on the generated morphology. J. Appl. Polym. Sci. **42**, 701 (1991)
64. Manzione, L.T., Gillham, J.K.: Rubber-modified epoxies. II. Morphology and mechanical properties. J. Appl. Polym. Sci. **26**, 906 (1981)
65. Hwang, J.F., Manson, J.A., Hertzberg, R.W., Miller, G.A., Sperling, H.L.: Structure-property relationships in rubber-toughened epoxies. Polym. Eng. Sci. **29**, 1466 (1989)
66. Manternal, S., Pascault, J.P., Sautereau, H.: in *Rubber Toughened Plastics*, C. K. Riew, Ed, Adv. Chem. Ser., 222, p. 193. American Chemical Society, Washington, DC (1989)
67. Williams, R. J. J., Adabbo, H. E., Rojas, A. J., Borrajo, J.: In: Martuscelli, E., Marchetta, C. (eds.) New Polymeric Materials. VNU Science Press, Utrecht (1987)
68. Vazquez, A., Rojas, A.J., Adabbo, H.E., Borrajo, J., Williams, R.J.J.: Rubber-modified thermosets: Prediction of the particle size distribution of dispersed domains. Polymer **28**, 1156 (1987)
69. Villiani, R.V., Bianchi, N., Braglia, R.: Relationship between gelation and morphology of rubber-modified epoxy resins. Polym. Int. **26**, 69 (1991)
70. Verchere, D., Sautereau, H., Pascault, J.P., Moschiar, S.M., Riccardi, C.C., Williams, R.J.J.: Rubber-modified epoxies. I. Influence of carboxyl-terminated butadiene-acrylonitrile random copolymers (CTBN) on the polymerization and phase separation processes. J. Appl. Polym. Sci. **41**, 467 (1990)
71. Verchere, D., Sautereau, H., Pascault, J.P., Moschair, S.M., Riccardi, C.C., Williams, R.J.J.: Buildup of epoxycycloaliphatic amine networks. Kinetics, vitrification, and gelation. Macromolecules **23**, 725 (1990)
72. Roginskya, G.F., Volkov, V.P., Vogdanova, L.M., Chalykh, A.Y., Rosenberg, B.A.: Polym. Sci. USSR **25**, 2305 (1983)
73. Yamanka, K., Takagi, Y., Inoue, T.: Reaction-induced phase separation in rubber-modified epoxy resins. Polymer **30**, 1839 (1989)
74. Kim, D.S., Kim, S.C.: Rubber modified epoxy resin. II: Phase separation behavior. Polym. Eng. Sci. **34**, 1598 (1994)

75. Poicus, A.V.: Elastomer modification of structural adhesives. Rubber Chem. Technol. **58**, 622 (1985)
76. Frigione, M.E., Mascia, L., Acierno, D.: Oligomeric and polymeric modifiers for toughening of epoxy resins. Eur. Polym. J. **31**, 1021 (1995)
77. Rezaiford, A.H., Hodd, K.A., Tod, D.A., Barton, J.: Toughening epoxy resins with poly(methyl methacrylate)-grafted natural rubber, and its use in adhesive formulations. M Int. J. Adhes. Adhes. **14**, 153 (1994)
78. Slysh, R.: in *Epoxy Resins*, Advances in Chemistry Series, 92, p. 108. American Chemical Society, Washington, DC (1970)
79. Kemp, T.J., Wilford, A., Howarth, O.W., Lee, T.C.P.: Structural and materials properties of a polysulfide-modified epoxide resin. Polymer **33**, 1860 (1992)
80. Mizutani, K.: Transparency and toughness characterization of epoxy resins modified with liquid chloroprene rubber. J. Mater. Sci. **28**, 2178 (1993)
81. Frischinger, I., Dirlikov, S.K.: Two-phase epoxy interpenetrating thermosets. Polym. Mater. Sci. Eng. **65**, 178 (1991)
82. Frischinger, I., Dirlikov, S.K.: Two-phase epoxy thermosets that contain epoxidized triglyceride oils. V. Phase separation. Polym. Mater. Sci. Eng. **70**, 1 (1993)
83. Ratna, D., Banthia, A.K.: Epoxidized soybean oil toughened epoxy adhesive. J. Adhes. Sci. Technol. **14**, 15 (2000)
84. Kinloch, A.J.: Adhesion and Adhesive. Science and Technology, Chapman Hall, London (1987)
85. Okamoto, Y.: Thermal aging study of carboxyl-terminated polybutadiene and poly(butadieneacrylonitrile)-reactive liquid polymers. Polym. Eng. Sci. **23**, 222 (1983)
86. Duseck, K., Lendnicky, F., Lunak, S., Mach, M., Duskova, D.: In: Riew, C. K. (ed.) Rubber Modified Thermoset Resins, p. 28. Adv. Chem. Ser. 208, American Chemical Society, DC (1984)
87. Jin, S.J., Yee, A.F.: Effects of mechanical properties of the toughener phase on toughening efficiency—a mechanistic study using a microgel-toughened epoxy. Polym. Mater. Sci. Eng. **63**, 107 (1990)
88. Hedric, J.C., MaGrath, J.E.: Polym. Mater. Sci. Eng. **63**, 107 (1990)
89. Shao, J.Z., Li, S.: Physical aging studies of polysiloxane-modified epoxy resin. Polymer **36**, 3479 (1995)
90. Riffle, J., Yilgor, I., Banthia, A.K., Wilkes, G.L., McGrath, J.E.: Org. Coat. Plast. Chem. **42**, 122 (1982)
91. Riffle, J., Yilgor, I., Trans, C., Wilkes, G.L., McGrath, J.E., Banthia, A.K.: Epoxy resin chemistry II. ACS Symp. Ser. **221**, 21 (1983)
92. Ochi, M., Yamana, H.: Kobunsi Ronbunghu **49**, 499 (1992)
93. Laurienzo, N.P., Malinconico, M., Martuscelli, E., Rongosta, G., Valpe, M.G.: Formulation, curing, morphology and impact behavior of epoxy matrixes modified with saturated rubbers. J. Mater. Sci. **27**, 786 (1992)
94. Jackson, M.B., Edmond, L.N., Varley, R.J., Warden, P.G.: Toughening epoxy resins with polyepichlorohydrin. J. Appl. Polym. Sci. **48**, 1259 (1993)
95. Sasidharan, Achary P., Gouri, C., Ramamurty, R.: Carboxyl-terminated poly(propylene glycol) adipate-modified room temperature-curing epoxy adhesive for elevated temperature service environment. J. Appl. Polym. Sci. **42**, 743 (1991)
96. Wang, H.B., Li, S.J., Ye, J.Y.: J. Appl. Polym. Sci. **44**, 789 (1989)
97. Laeuger, J., Albert, P., Kressler, J., Grouski, W., Muelhaupt, R.: Anhydride-cured epoxy resins toughened with diol- and bis(4-hydroxybenzoate)- terminated poly(tetrahydrofuran) liquid rubbers. Part I. Morphology and blend properties. Acta Polym. **46**, 68 (1995)
98. Lee, Yu-Der, Wang, S.K., Chin, W.K.: Liquid-rubber-modified epoxy adhesives cured with dicyandiamide. I. Preparation and characterization. J. Appl. Polym. Sci. **32**, 6317 (1986)
99. Boogh, L., Pettersson, B., Manson, E.J.: Dendritic hyperbranched polymers as tougheners for epoxy resins. Polymer **40**, 2249 (1999)

100. Wu, H., Xu, J., Liu, Y., Heiden, P.A.: Investigation of readily processable thermoplastic-toughened thermosets. V. Epoxy resin toughened with hyperbranched polyester. J. Appl. Polym. Sci. **72**, 151 (1999)
101. Ratna, D., Simon, G.P.: Dendritic hyperbranched polymers and epoxy blends. J. Polym. Mater. **19**, 349 (2002)
102. Ratna, D., Simon, G.P.: Thermal and mechanical properties of a hydroxyl-functional dendritic hyperbranched polymer and trifunctional epoxy resin blends. Polym. Eng. Sci. **41**, 1815 (2001)
103. Ratna, D., Simon, G.P.: Thermomechanical properties and morphology of blends of a hydroxy-functionalized hyperbranched polymer and epoxy resin. Polymer **42**, 8833 (2001)
104. Ratna, D., Varley, R., Singh, R.K., Simon, G.P.: Studies on blends of epoxy-functionalized hyperbranched polymer and epoxy resin. J. Mater. Sci. **38**, 147 (2003)
105. Mezzenga, R., Boogh, L., Manson, E.J.: Evaluation of solubility parameters during polymerisation of amine-cured epoxy resins. J. Polym. Sci. Polym. Phys. **38**, 1883 (2000)
106. Sue, H.J.: Study of rubber-modified brittle epoxy systems. Part I. Fracture toughness measurements using the double-notch four-point-bend method. Polym. Eng. Sci. **31**, 270 (1991)
107. Sue, H.J.: Crazelike damage in a core-shell rubber-modified epoxy system. J. Mater. Sci. **27**(11), 3098–3107 (1992)
108. Maazouz, A., Santerean, H., Genard, J.F.: Toughening of epoxy networks using preformed core-shell particles or reactive rubbers. Polym. Bull. **33**, 67 (1994)
109. Nakamura, Y., Tabata, H., Sujuki, H., Iko, K., Ocubo, M., Matsumoto, T.: Internal stress of epoxy resin modified with acrylic core-shell particles prepared by seeded emulsion polymerization. J. Appl. Polym. Sci. **32**, 4865 (1986)
110. Vanderhoff, J.W., El-Aasser, M.S., Micale, F.J., Sudol, E.D., Tseng, C.M., Silwanowicz, A., Shew, H.R., Kornfeld, D.M.: Polym. Mater. Sci. Eng. Prep. **54**, 587 (1986)
111. Ugelstad, J., Mork, P.C., Kaggurud, K.H., Ellingsen, T., Berge, A.: Adv. Colloid Interface Sci. **13**, 101 (1980)
112. Barret, K.E.J.: Disperson Polymerization in Organic Media. Wiley, New York (1975)
113. Shen, S., Sudol, E. D., El-Aasser, M. S.: J. Polym. Sci., Part A: Polym. Chem. **32**, 1393
114. Markel, M. P., Dimonie, V. L., El-Aasser, M. S., Vanderhoff J. W.: Morphology and grafting reactions in core/shell latexes. J. Polym. Sci., Part A: Polym. Chem. **25**, 1219
115. Lee, D.L.: ACS. Sym. Ser. **165**, 893 (1981)
116. Jhonson, J.L., Hassander, H., Jansson, L.H., Tornwell, B.: Morphology of two-phase polystyrene/poly(methyl methacrylate) latex particles prepared under different polymerization conditions. Macromolecules **24**, 126 (1991)
117. Lee S. and Rudin, A.: In: Daniels, E. S. Sudol, E. D., El-Aasser, M. S. (eds.) Polymer Latex Preparation, Characterization and Applications. ACS Symp. Series 492, 234
118. Lin, K.F., Shieh, Y.D.: Toughening of epoxy resins with designed core-shell particles. Polym. Mater. Sci. Eng. **76**, 358 (1997)
119. Sue, H. J., Garcia- Meitin, E. L., Orchard, N. A.: J. Polym. Sci., Polym. Phys. Ed. **31**, 595
120. Jansen, B.J.P., Tamminga, K.Y., Meijer, H.E.H., Lemstra, P.J.: Preparation of thermoset rubbery epoxy particles as novel toughening modifiers for glassy epoxy resins. Polymer **40**, 5601–5607 (1999)
121. Boynton, M.J., Lee, A.: Fracture of an epoxy polymer containing recycled elastomeric particles. J. Appl. Polym. Sci. **66**, 271–277 (1997)
122. Bagheri, R., Williams, M.A., Pearson, R.A.: Use of surface modified recycled rubber particles for toughening of epoxy polymers. Polym. Eng. Sci. **37**(2), 245–251 (1997)
123. Hamerton, I. (ed.): Chemistry and Technology of Cyanate Ester Resins, Blackie Academic and Professional, Glasgow (1994)
124. Sabyasachi, G., Derrick, D.: Mechanical properties of intercalated cyanate ester–layered silicate nanocomposites. Polymer **44**, 1315–1319 (2003)
125. Cao, Z.Q., Mechin, F., Pascault, J.P.: Effects of rubbers and thermoplastics as additives on cyanate polymerization. Polym. Int. **34**, 41–48 (1994)

126. Hillermeier, R.W., Hayes, B.S., Seferis, J.C.: Processing of highly elastomeric toughened cyanate esters through a modified resin transfer molding technique. Polym. Compos. **20**, 155–165 (1999)
127. Hayes, B.S., Seferis, J.C., Parker, G.A.: Rubber modification of low temperature cure cyanate ester matrices and the performance in glass fabric composites. Polym. Eng. Sci. **40**, 1344–1349 (2000)
128. Borrajo, J., Riccardi, C.C., Williams, R.J.J., Cao, Z.Q., Pascault, J.P.: Rubber-modified cyanate esters: thermodynamic analysis of phase separation. Polymer **36**, 3541–3547 (1995)
129. Hayes, B. S., Seferis, J. C.: Rubber toughened epoxy-cyanate ester liquid resins: Effect of rubber functionality. In: International SAMPE Symposium and Exhibition, 46 I, pp. 1072–1078
130. Wise, C.W., Cook, W.D., Goodwin, A.A.: CTBN rubber phase precipitation in model epoxy resins. Polymer **41**, 4625–4633 (2000)
131. Auad, M.L., Frontini, P.M., Borrajo, J., Aranguren, M.I.: Liquid rubber modified vinyl ester resins: fracture and mechanical behavior. Polymer **42**, 3723–3730 (2001)
132. Partridge, I.K., Maistros, G.M.: Effect of brominated bisphenol A on the cure, morphology and properties of epoxy/CTBN blends. Plast. Rubber Compos. Process. Appl. **23**, 325–330 (1995)
133. Riew, C. K., Rowe, E. H., Seibert, A. R.: In: Deanin, R. D., Crugnola, A. M. (eds.) Toughness and Brittleness of Plastics. Advances in Chemistry Series, No 154, p 326. American Chemical Society, DC (1976)
134. Liang, et al.: Research of cyanate ester resin modified by epoxy. Mech Sci Technol. **19**, 137–139 (2000)
135. Feng, Y., Fang, Z., Gu, A.: Structure and properties of CE/CTBN/EP blends: II. Effect of EP on the mechanical properties and thermostability of the CE/CTBN system. Polym. Int. **54**, 369–373 (2005)

# Interphase Modification and Compatibilization of Rubber Based Blends

**Bağdagül Karaağaç and Veli Deniz**

**Abstract** Blending of two or more elastomers is carried out for several purposes. The properties of an elastomer blend depend strongly on its state of compatibility and miscibility. In this chapter, recent advances on development of interphase modification and compatibilization of rubber-based blends are summarized. Current trends in compatibilization of rubber/rubber blends, TPEs and other rubber/thermoplastic blends, natural polymer blends, rubber-based blends with and without filler modification are discussed in detail. Finally, new challenges and opportunities of rubber-based blends are given.

## 1 Introduction to Phase Modification and Compatibilization

Elastomeric compounds are often consist of more than one type of rubber as polymer matrix. Blending of two or more elastomers is carried out for several purposes, such as improving the physical and mechanical properties of the first elastomer, obtaining good processing characteristics of rubber compound and/or decreasing the compound cost [1]. For example, a tire compound has to be soft and elastic in order to conform to the road surface. At the same time, it has to be stiff and strong in order to bear load, and be resistant to abrasion to provide long service life. Similarly, various rubber based goods, such as seals and 'O' rings, have to be soft and conform to the contour of the equipment; but at the same time, they have to undergo minimum compression set in order to perform under high stress over a

B. Karaağaç (✉) · V. Deniz
Department of Chemical Engineering, Kocaeli University, Kocaeli, Turkey
e-mail: bkaraagac@kocaeli.edu.tr

V. Deniz
e-mail: vdeniz41@gmail.com

P. M. Visakh et al. (eds.), *Advances in Elastomers I*,
Advanced Structured Materials 11, DOI: 10.1007/978-3-642-20925-3_9,

long period of time. Since different elastomers have different type of responses to external effects, blending of different rubbers has been practiced to meet all the needed properties [2]. However, the mechanical, thermal, rheological and other properties of an elastomer blend depend strongly on its state of compatibility. Most of elastomer blends have cure rate differences due to discrepancies in polarity and unsaturation. Also, different dispersion quality of fillers and other compound additives in two or more rubbers causes deterioration of mechanical properties of the blend [1, 3–5].

Also, there are some contradictory requirements in some plastic products where rubber is added to plastics to provide toughness and prevent failure. Toughening of thermoplastics has been practiced over a long period of time through blending of plastics with rubber. In fact, the recent development of thermoplastic elastomers (TPE) can be traced to fulfill this expectation. Many TPEs are made either in the reactor or in mixing equipment. TPEs can be produced either by ionic and condensation polymerization or can be construed as blends of different blocks of polymers, covalently bonded to each other. Adequate interphase morphology is required and very important for both TPEs and rubber blends.

However, prediction of miscibility or compatibility of polymers is considerably complicated because the polymer molecules generally have large molecular weight and because the segments which are the interacting entities and are connected at each end are constrained by their neighboring segments. So, they cannot be moved to fill any available site in a lattice model, often used for estimating thermodynamic parameters. This is only one example of the complicated differences between polymer molecules and small molecules that must be worked out in achieving successful prediction of polymer-polymer miscibility or compatibility. The other factors include the small entropy change and volume change during mixing, the heterogeneity of molecular composition, the high polydispersity, the complex morphology, and the influence of processing parameters on compatibility [2].

In elastomeric blends, dispersion of compounding ingredients is one of the most important parameters which provide appropriate conditions for compatible blends. A variety of ingredients including one or more type of rubber, fillers, plasticizers, processing aids, antioxidants, and vulcanizing agents are mixed together to conceive effective vulcanization and to provide the required physical and mechanical properties. In contrast, a thermoplastic is generally processed with only a few ingredients such as fillers, stabilizers and processing aids. Besides, while most thermoplastics are processed much above their melting point or glass transition temperature at comparatively low viscosity, rubber compounds are mixed in highly viscous state. It is therefore much more difficult to get fair dispersion of ingredients in a rubber compound than in a thermoplastic compound. This situation is further complicated for elastomeric blends because there may be partition or preferential dispersion of one or more ingredients in one component compared to the other one, depending on the polarity or surface characteristics of the additive. This may influence the morphology and physical properties of the vulcanized rubber blend appreciably. It is therefore, important to discuss the potential of uneven distribution of compounding ingredients in elastomeric blends and its net effect on physical properties of the compound [2].

It is possible to classify the compounding ingredients in two general categories, namely, reactive and inert ingredients. Vulcanizing ingredients such as sulphur, accelerators, zinc oxide and stearic acid take place in the cross linking reaction. The distribution in the different elastomers in a rubber component is influenced not only by mechanical mixing, but also by their migration to potential active sites. The latter is also controlled by the reactivity of the additives with rubber. In other words, the negative free energy of chemical reaction largely overwhelms the small free energy of physical mixing. Since rubber vulcanization is carried over a long period, compared to less than a minute in thermoplastic processing time is small, for example, in injection molding, the likelihood of uneven distribution and uneven cure may exist. The non-reactive ingredients include fillers, plasticizers, processing aids and antioxidants. The last two are added in small quantities, much below their saturation level. Hence, their distribution ordinarily does not pose any problem. Plasticizers are often liquid. Hence, it is likely that they get a fair opportunity to mix well in the rubber compound. There may be some degree of partitioning of these compounds in the different components of the blends, but the effect of uneven distribution will be minimal [2].

Since most blended polymers are immiscible, in many cases additional compatibilization process is required to obtain maximum synergy. There are several excellent techniques on the compatibilization of polymer blends exist. However, compatibilization of rubber based blends is more difficult because of crowd and complex compound matrix [6]. Our approach is to present some of these compatibilization applications as a kind of toolbox: what can be added to an incompatible blend, or how should we modify the blend components to obtain a compatible blend?

Elastomers are generally immiscible with each other and their blends undergo phase separation with poor adhesion between the main matrix and dispersed phase. The properties of such blends are often poorer than the weight average property of individual components. In fact, the main objectives of blending are to combine the performance characteristics of two or more polymers, to develop high performance products. These can be accomplished by compatibilizing the blend, either by adding a third component, called compatibilizer, or by enhancing the interaction of the two component polymers, chemically or mechanically. The roles of compatibilizer are reducing interfacial energy and improve adhesion between two or more polymer phases, achieving finer dispersion during mixing and stabilizing the fine dispersion against agglomeration during processing and through out the service life.

In blending process, the first step is mixing the compound ingredients. Among extruders, particularly twin screw extruders are used for blending thermoplastics in their melt phase. In contrary, elastomers are blended by Banbury type mixers or by open two roll mills in solid form. In both cases, the materials are exposed to shear stress. The size of the dispersed phase is determined by the balance between drop break up and coalescence process, which is controlled by the type and severity of the stress, interfacial tension between the two phases and the rheological characteristics of the components [7].

If the interfacial tension is small, a less severe agglomeration process is initiated due to the need for reducing potential energy. In this case, addition of a small amount of compatibilizer into the solution acts like a solid emulsifier and stabilizes the droplets, so that decreasing the dispersed phase size. The component with major surface area acts as matrix. In case of blends with equimolar components, the component with lower viscosity tries to encapsulate the one with higher viscosity. It has been observed and theoretically established that better dispersion is achieved when the viscosity of both phases are close to each other. The necessity of condition for forming and keeping a co-continuous phase is given as:

$$\mu_1 \varphi_2 / \mu_2 \varphi_1 = 1$$

where $\mu_1$ and $\mu_2$ are the viscosities and $\varphi_1$ and $\varphi_2$ are the weight fraction of each component in the blend. Co-continuous phase provides the special morphology to the blends, where the two phases act in tandem and the blend exhibits the best performance of the two components. On the other hand, if the viscosity of the minor phase is very high, the phase does not get fractured and broken down into small dispersed particles. Therefore, component with high viscosity can be better dispersed, if it is premixed with a plasticizer in order to bring the viscosity closer to that of the law viscosity component [8].

Even if plasticizers are added to reduce the viscosity and processing aid to improve flow, many elastomeric blends separate into different phases after the mixing is completed, possibly due to re-agglomeration. Hence, compatibilization is essential to reduce the size of the dispersed phase, and to provide a blend with co-continuous phase [2].

The ultimate objective is to land on a morphology that will allow smooth stress transfer from one phase to the other and allow the product to resist failure under multiple stress. In case of elastomeric blends, compatibilization may be necessary to aid uniform distribution of fillers, curatives and plasticizers to obtain a morphologically and mechanically sound product [9].

Compatibilization of rubber blends can be carried out by both reactive and non-reactive methods. In non-reactive methods, an external polymeric material is added into elastomeric blend, such as a block or graft co-polymer to improve compatibilization. The essential function of a compatibilizer is to wet the interface between the two phases. Block and graft co-polymers achieve this by spreading at the interface and mixing with both phases through their component parts, which are similar to one phase or the other. In reactive methods, block and graft polymers are formed in situ, during mixing of the two or more components [2].

# 2 Current Trends in Interphase Modification and Compatibilization

## 2.1 Compatibilization of Rubber/Rubber Blends

Blends of elastomers are often used as a basis for compound formulations. For instance, tire tread formulations may contain Natural Rubber (NR), Butadiene Rubber (BR) and Styrene-Butadiene Rubber (SBR) in order to provide cost-effective products with good mechanical properties, dynamic properties and abrasion resistance. Blends of dissimilar Ethylene Propylene Diene Monomer Rubber (EPDM) grades are often used when formulating EPDM compounds. In many cases such blends prepared with different EPDM types blends are used to obtain synergy, for example properties of the parent EPDM grades are not averaged, but favorable properties of the blend components are combined or unfavorable properties suppressed [10]. The constituent polymers in the blend are generally immiscible due to thermodynamic barriers and structural dissimilarities. Miscibility is a term used for blends having a homogeneous phase, showing additive properties of individual polymers resulting from interaction of chemical constituents through secondary forces of attractions such as hydrogen bonding, dipole-dipole interaction, dispersion forces, etc. The miscible polymer blends show definite thermodynamic properties, which alter the physico-mechanical behavior of the blend. Several miscible blends have been studied in the past decades [11].

Scanning electron microscopy (SEM) can be utilized to evaluate phase morphology of the blends. Subsequently, these studies can also be extrapolated to characterize compatibility of different rubbers such as NBR and EPDM rubber by many other analytical techniques, viz., FTIR, ultrasonic interferometry, atomic force microscopy (AFM) and thermal techniques, e.g., DMA, MDSC, TMA, etc. [12].

Attention was focused on compatibilization of the rubbers which have different unsaturation and different polarity beside their chemical dissimilarity. Flow visualization of mixing and phase morphology of binary and ternary rubber blends of NBR with other rubbers were of large interest. Efficacy of different compatibilizers, e.g. chlorinated polyethylene (CM) or chlorosulponated polyethylene (CSM) to impart blend homogenization is interested, too. Addition of a suitable compatibilizer was found to result in development of finer scale of dimension of the dispersed phase into the matrix as well as enhancement of physico-mechanical properties of the NBR based blend vulcanizates [12].

In the last three decades, ionizing radiation is drawing attention as a powerful source of energy for chemical processing applications. Thus, it can be applied in different industrial areas. The application of radiation on polymers can be employed in various industrial sectors, i.e. biomedical, coatings, electrical applications, textile, membrane, rubber goods, automotive tires, foam, footwear, aerospace and pharmaceutical industries. Radiation-based processes have many advantages over other conventional methods. In radiation processing, no catalyst or additives are required to initiate the reaction. With this technique, adsorption of free energy by the backbone, polymer initiates a free radical process. Free radicals are formed by the decomposition of the initiator into segments and these segments then attack the base polymer. Due to this, radiation-induced processes are free from initiator contamination and it is preferable to use radiation-induced polymerization and reactive compatibilization processes for producing bio-compatible and medical materials [13]. As an example; N-vinyl pyrolidone-grafted silicone rubber material is widely used in biomedical applications. However, DMMA and N,N-dimethyl amino ethylacrylate grafted NR tubes are better bio-compatible than silicone rubber. Blood compatibility of the material is improved with the increasing degree of grafting of the monomers [14, 15].

The conventional vulcanization process occurs heating and adding sulphur and/or other chemicals to form cross links between the characteristic long chains of elastomer molecules. This process started long time ago and is still being used. The properties of the polymer depend on the amount of sulphur used and withstand higher temperature, pressure and mechanical challenges to its integrity. But sulphur vulcanization has severe drawbacks with respect to human health and environment though some economic advantages. However, it needs high temperature to start the chemical reaction and emits odorous and toxic gases as well as producing numerous hazardous chemical residues that have to be removed from the final product. Radiation cross linking is the well-proven method that bypasses all these negative effects of the vulcanization process. It is a room temperature method having in itself an important cost advantage. It is easily controlled and the desired properties of the polymer are obtained simply by changing the dose (irradiation time). The transformed materials are in no way inferior to those produced by sulphur vulcanization [13].

Adhesion of an elastomer to itself (self-adhesion or auto adhesion) or to another elastomer is an important technological problem about interfacial action. Most published studies considered mainly symmetric joints that means that same formulation and same degree of cross linking when two sheets of the same elastomer are brought into contact. This is opposed to asymmetric joints where the formulation and/or the degree of cross linking of the two sheets of the same elastomer are brought into contact. Different stiffness of the two sheets of the assembly increases the difficulty of interpretation of the experimental data of the peel test usually used to evaluate the interfacial strength. Three mechanisms of adhesion are liable for the interfacial strength: adsorption, diffusion of the polymer chains through the interface (especially in the case of auto adhesion or adhesion of compatible polymers) and cross linking of the polymer chains at the interface. There is a direct relationship between the number of covalent bonds created at the interface and the interfacial strength. The contribution of chemical bonding can be estimated by comparing the results of peel tests in air and in liquid media. This approach has been used in a study of symmetric joints of SBR and Isoprene Rubber (IR). In the case of auto adhesion of SBR, the obtained results are in agreement with this approach; whereas, for the IR joints strain-induced crystallization causes additional adhesion. Also, when considering asymmetric joints, for peroxide cured samples of SBR, the threshold strength of auto adhesion is still proportional to the number of interfacial links and equal to one-half of that for symmetric joints. For sulfur-cured carbon black-filled BR/SBR blends, the tear strength of the blend is reached when about half of the complete interlinking is reached. However, in asymmetric joints as mentioned, the mechanism of interlinking is complex because of possible concentration gradients of vulcanizing components and their different solubility behavior when two partially-compatible or incompatible elastomers are considered [16].

## 2.2 Compatibilization of TPEs and Other Rubber/ Thermoplastic Blends

Besides fully elastomeric blends, another type of polymer blends which have similar properties is called thermoplastic elastomers (TPEs). These types of materials exhibit functional properties of conventional elastomeric materials and they also can be processed using thermoplastic processing machines. Such materials have heterophase morphology and can be produced either as block copolymers or as blends. The hard domains of the TPEs undergo dissociation at elevated temperatures. This allows the material to flow. The hard domains are again solidified when lowering the temperature. So, the strength properties of the material at service temperatures can be improved. The application fields of TPEs based on rubber-plastic blends have grown large, giving two distinctly different classes. One class consists of a simple blend and is commonly called a thermoplastic elastomer polyolefin (TEO) or thermoplastic polyolefin (TPO). In the other class, the rubber phase is dynamically vulcanized, giving rise to thermoplastic vulcanizates (TPVs) [17].

A large volume of work has been published on the TPEs generated by rubber-plastic blending. Even though blending has many attractive features, it suffers from certain drawbacks. Matching viscosities and identical molecular structures are some of the requirements to get a compatible blend. However, the use of compatibilizer has improved results by locating at the interface of an immiscible blend. As a result the interfacial tension is reduced, coalescence is suppressed and improved adhesion between phases is achieved. Compatibilization can be achieved by several ways. Dynamic vulcanization and cross linking in several ways can be considered as an efficient way for compatibilization. Various attempts were done about compatibilization by addition of appropriate block or graft copolymers that act as interfacial agent. Reactive compatibilization is possible when both constituent polymers have reactive functional groups. Reactive conpatibilization occurs when both polymers have reactive functional groups. The effectiveness of reactive compatibilization of polysulfone/polyamide (PSF/PA) blends by addition of modified PSF is emphasized in various studies. In this case, PSF-PA block copolymer is formed and it effectively reduced the size of the dispersed phase [18].

In dynamic vulcanization process, cross linking occurs in the elastomeric phase during melt mixing with the thermoplastic phase. The slightly cross linked rubber particles can withstand more stress but at the same time plastics phase can maintain its flow ability. Cross linking increases the viscosity of elastomeric phase and phase morphology of the blend face to some changes. There are a lot of studies that indicate the effect of blend ratio, reactive compatibilization and dynamic vulcanization on the properties of TPE blends e.g. PP/NBR blends, polycarbonate/acrylonitrile–butadiene–styrene copolymer (PC/ABS) blends, NR/PP blends, EPDM/PP blends and HDPE/NBR blends. HDPE/NBR blend is a good example for TPVs for hot and oil-resistant industrial applications. But the incompatibility between the two polymers is an important problem. The effects of blend ratio, compatibilization, dynamic vulcanization and filler incorporation were investigated in detail. Maleic anhydride modified polyethylene (MAPE) was selected as the compatibilizer and MAPE was prepared by melt-mixing of HDPE (100 parts) with maleic anhydride (5 parts) and dicumyl peroxide (DCP) (0.5 parts) in a Brabender Plasticorder. Then it was sheeted and cut into small pieces for further use as compatibilizer. Melt mixing of two phases was carried out by using traditional dynamic vulcanization system. Dynamic vulcanization was carried out using additional DCP (5 phr). Compatibilization using MAPE, and by dynamic vulcanization significantly reduce the damping peak values indicating enhanced interaction between the consistituents. Dynamic vulcanization is more effective when compared with MAPE in bringing about compatibility. Filler (carbon black) incorporation results in significant increase in dynamic modulus values, especially in rubber rich blends. This is because of that reinforcing action of the filler is predominant when the rubber is continuous or dominant phase [18].

The blending of special elastomers and commodity polymers are shown great interest for several decades. For example, the use of EPDM-PP TPEs is continuously growing in various industrial domains recently. Since PP is used in commodity as well as in engineering application, due to good mechanical properties, low density and low cost. The one the good way out to improve low temperature impact resistance of PP, is blending it with EPDM. However, this type of thermoplastic elastomer (Olefinic, TPO) has an unstable morphology because of coalescence of dispersed rubber particles and low compatibility between rubber phase and thermoplastic matrix. The incompatibility between PP and EPDM may also be attributed to differences of two polymers in their crystallinities. As it is possible to mix EPDM and PP in any ratio, there is theoretically a wide spectrum of materials from elastified PP to EPDM rubber reinforced with thermoplastics. Investigations were performed especially on different types of blend systems prepared with different quantity of EPDM and with or without compatibilizer and dynamic vulcanization. There are two common compatibilizers that can be considered as examples: a Maleic anhydride grafted PP one (MagPP, MFI50), and an Ionomer called Surlyn (MFI16), chemically known as, ethylene-co-methacrylic acid neutralized by sodium ion. Different combinations can be prepared by using a co-rotating intermeshing twin screw extruder. It is well known that the adherence between particles (PP) and matrix (EPDM) plays a crucial role in the mechanical behavior of the blends, even if two phases have been blended in melt phase together. It is therefore essential to quantitatively evaluate the role of compatibilizers by introducing them. When each compatibilizer and a combination by mixture of both together in the PP/EPDM blends are tested, the experimental results show that PP particles (10, 25, and 30 % w/w) effect the fraction, significantly. Also, introduction of a compatibilizer increases the rigidity of the blends and affects notably their macroscopic behavior. These observations can be interpreted as a consequence of the modification at micro level of adherence between particles and matrix phases. More specifically, Ma-g-PP is a chemical compatibilizer that attacks the unsaturated locations in the chain; this is provided by norbornene ring that is attached to EPDM chain structure. However, for the compatibilizer Surlyn, only physical interactions take place through ion interactions. It may be expected that with an adequate concentration of compatibilizer a perfect contact might be reached. But, the rate of Surlyn or Ma-g-PP beyond 7 % wt. is found to provide no supplementary benefit.

Also, TPE blends of NBR with either neat HDPE or phenolic resin modified HDPE coupled with dynamic vulcanization of the blend by peroxide. This material was developed as a substitute for the NBR-polyvinyl chloride (PVC) blend particularly for oil and fuel resistance applications. Morphological analysis of NBR and neat HDPE blend show phase separated morphology. Treatment of HDPE with phenolic resin and dynamic vulcanization results in an improvement of the miscibility characteristics which can be ascribed to generation of compatibility between these two phase components. Thermal conductivity and topographic images reveal that NBR and PVC is a miscible blend [12].

NR based TPEs are called as thermoplastic NRs (TPNRs). NR has an inherent affinity with some olefin thermoplastic that permits the formation of TPVs characterized by sub-micron scale morphologies and excellent physical properties. Various types of thermoplastics are used to prepare the TPNRs. As well as classical virgin NR, modified forms of NR can also been used to prepare the TPNRs. Epoxidized NR (ENR) is one of the modified forms currently used to prepare TPNRs. This is done through blending of ENR with thermoplastics such as PVC and PMMA. Graft copolymers of NR with PMMA are used to prepare TPNRs. NR and HDPE based TPNRs can also be prepared via a dynamic vulcanization process. Various compositions and types of compatibilizers for these blends were studied [17]. Two types of phenolic resin are generally used as blend compatibilizer. These are dimethylol phenolic resin or octylphenol-formaldehyde resin and phenolic resin with active hydroxmethyl (methylol) groups. Compatibilizers can be added in masticating stage of the process with a two-roll mill at almost 80 °C for 40–45 min in the presence of a peptizer. Phenolic modified PE can also be prepared by melt-mixing of HDPE (100 parts) with dimethylol phenolic resin (4 parts) and stannous chloride (0.8 parts) using an extruder. The other phenolic modified PE can be prepared with using the same procedure. NR and HDPE blends can be prepared in an appropriate ratio (60/40) via a melt mixing process. Mixing can be performed for 2 min at a rotor speed of 60 rpm at 180 °C. The compatibilizer can be then incorporated into the mixing chamber and mixing is continued for another 1 min. And then, NR is added and mixed for a few minutes. If necessary, other additives can be incorporated into the mixing chamber. In blending

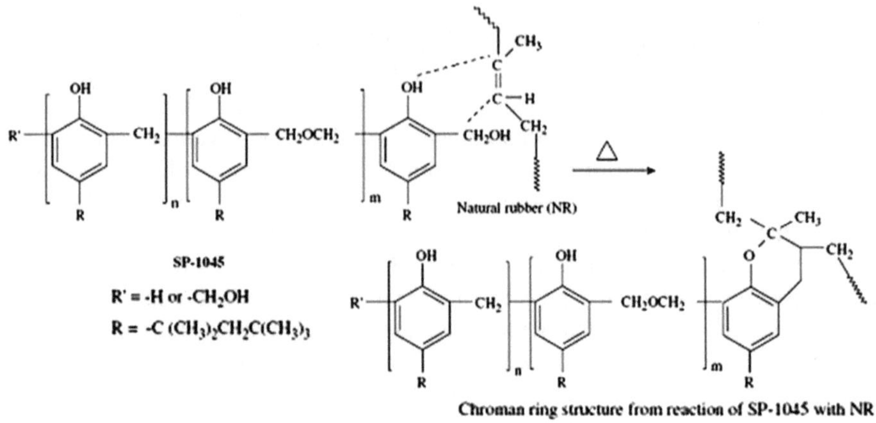


**Scheme 1** Possible reaction between phenolic resin (SP-1045) and natural rubber [17]

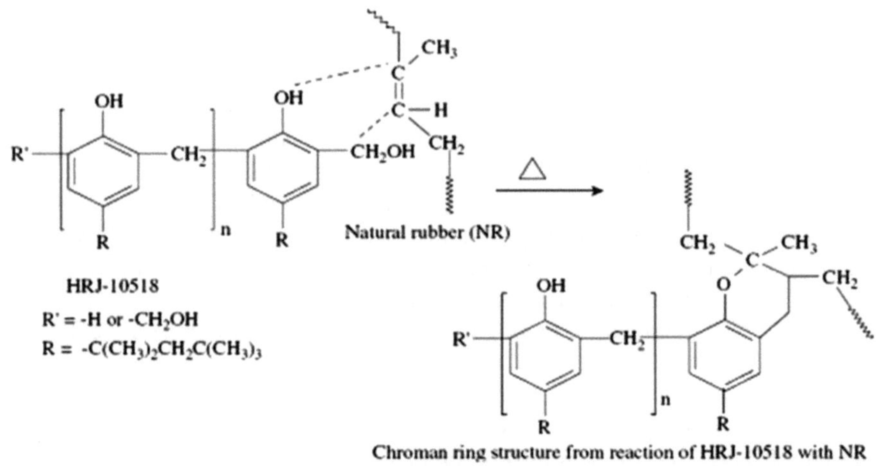


**Scheme 2** Possible reaction between phenolic resin (HRJ-10518) and natural rubber [17]

systems, with phenolic resins, the methylol groups are capable of reacting with the unsaturated sites in the NR molecules to produce a Chroman ring structure, as shown in Scheme 1, 2, respectively. Similarly, another reactive functional group (–$CH_2OH$) in the molecules can react with double bonds in the rubber molecules. Therefore, the phenolic resins behave as a bridge connecting the thermoplastic and rubber molecules. This leads to higher mechanical strength of the TPVs prepared using these two types of compatibilizers [17].

Since the beginning of 1980, there has been a lot of research on the blends of a thermoplastic polymer and a minor liquid crystalline polymer (LCP). It is known that thermo tropic liquid crystalline polymers (TLCP), especially aromatic TLCPs can be used as a processing aid by reducing the viscosity of the main matrix material. Moreover, the LCP phase domains in a thermoplastic/TLCP blend can be deformed into elongated fibrils under shear flow to act as in situ composites to reinforce the thermoplastic matrix. However, mechanical properties of the thermoplastic/LCP blends depend strictly on the morphology of the LCP phase which is controlled by several factors, especially interfacial properties of the blend components and blend composition. A major part of thermoplastic/LCP blends are incompatible and there has been poor adhesion between two phases due to difference in their chemical structures. This disadvantageous can be eliminated by using reactive or non-reactive compatibilization techniques with appropriate compatibilizer during melt mixing. Introducing a second thermo tropic LCP such as PET/PHB60 as a compatibilizer can improve both adhesion and dispersion between incompatible thermoplastic and LCP. These compatibilizers, acting as polymeric surfactants, reduce the interfacial tension, which promotes interfacial adhesion, and improve dispersion. In some cases, it can be shown that compatibilizers reduce the polydispersity of the dispersed phase particle size by emulsifying of blends. For instance, when LCP content is 20–25 % (w %) (gives maximum fibril aspect ratio) in poly(ether imide)/LCP (PEI/LCP) blends, the position of characteristic absorption for chemical groups in LCP, which indicates no chemical changes or specific interaction occurred. While with the incorporation of a small amount of silicone rubber, some characteristic absorption shifts to lower frequency, which implies the specific interactions of these functional groups may be involved. These interactions possibly include hydrogen bonds formed between silicone rubber and amino end-groups or hydroxyl end-groups in LCP, such as O–H–O bonding. The addition of silicone rubber to the PEI and LCP leads to significant mechanical property improvements over the PEI/LCP binary blend owing to better adhesion and fibrillation of LCP in PEI. Also, silicone rubber decreases crystallinity of the tribulent system as compatibilizer [19].

## 2.3 *Compatibilization of Natural Polymer Blends*

The use of polymer blends from renewable resources (PFRR) is an environmentally advantageous alternative to synthetic polymers in some applications. Generally, PFRR can be classified into three groups: (1) natural polymers, such as starch, protein and cellulose; (2) synthetic polymers from natural monomers, such as polylactic acid (PLA); and (3) polymers from microbial fermentation, such as polyhydroxybutyrate (PHB) [14].

One of the main disadvantages for biodegradable polymers obtained from renewable sources is their hydrophilic character, fast degradation rate and, in some cases, unsatisfactory mechanical properties, particularly under wet environments. In principle, the properties of natural polymers can be improved by blending with synthetic polymers, significantly. Polymer blending is a well-used technique whenever modification of properties is required [14]. The progress of blends from three kinds of polymers from renewable resources—(1) natural polymers, such as starch, protein and cellulose; (2) synthetic polymers from natural monomers, such as poly(lactic acid) (PLA); and (3) polymers from microbial fermentation, such as polyhydroxybutyrate (PHB)—are described with an emphasis on potential application [14].

Many polysaccharide biopolymers have been examined in this context, with the long-term aim of value-adding to waste agricultural byproducts such as sugar beet pulp and rice husks. Pure starch is a good model for such biopolymers, and also is useful as a substrate in its own right [20, 21].

Starch, a hydrophilic renewable polymer, and has been used as a filler for environmentally friendly plastics for about two decades. To obtain useful materials from starch, the native properties must be enhanced, because of starch's higher water sensitivity and poor mechanical properties compared to those of sorbitol have been investigated to decrease the brittleness of these materials [22, 23]. Vegetable fibers and mineral fillers can be chemically modified or blended with synthetic polymers to improve mechanical strength and water resistance [24–30].

A suitable polymer for use in starch modification, natural rubber latex (NRL) is very attractive. NR has been blended with starch for a number of different applications by using different techniques. In latex form, which facilitates blending with a starch solution and indeed with any particulate substance such as sawdust or pulp; it is a renewable resource, which should help compatibilization with starch. The polymer is essentially 100 % *cis*-1,4-polyisoprene, and this conformation leads to a number of useful mechanical properties such as improved mechanical strength on stretching. It is also inexpensive [20]. Carvalho et al. [31] studied the blending of starch with NR. Thermoplastic starch/NR polymer blends were prepared using NRL and cornstarch. The blends were prepared in an intensive batch mixer at 150 C, with the NR content varying from 2.5 to 20 %.

A novel method has been developed to modify NRL by grafting a 'hairy layer' of hydrophilic polymers [30]. Various hydrophilic monomers can be used for this purpose. One of the important materials studied is dimethylaminoethyl methacrylate (DMAEMA). The amine functionality of DMAEMA confers considerably enhanced colloidal stability on NRL: while NRL coagulates when the pH is below ~8.5, the polyDMAEMA-modified NRL is stable to pH values as low as 2. DMAEMA is biodegradable, although has some ecotoxicity [20].

NRL and NRL modified by polyDMAEMA can directly be used in the wet casting of starch films. In grafting procedure, the weight of DMAEMA can be chosen as 5, 10 and 15 % w/w with respect to dry content of NRL. NRL, DMAEMA, cumene hydro peroxide (CHP), 10 % aqueous solution of tetraeyhylene pentamine (TEPA), and 2.5 % ammonia solution are the content of the grafting reaction. Reaction mixture can be agitated with a low-shear impeller. 10 % aqueous solution of TEPA is then added shot-wise over 1 h at 5 min intervals, and the system is cooled in an ice/water bath during the first 8 h of reaction before gradually warming to room temperature, and allowed to continue react for another 16 h. After reaction, the latex is filtered through glass wool and neutralized to pH 7 with 1 M hydrochloric acid solution [20].

Films formed from the addition of two types of rubber latexes (NRL, and NRL grafted with hairs of a cationic hydrophillic polymer, polyDMAEMA) to a glycerol/starch solution show a range of physical properties which can be explained by the competing effects of the hydrophilicity of the starch and the polyDMAEMA, the hydrophobicity of the NRL, and hydrogen bonding between the grafted polyDMAEMA and the starch. NRL acts essentially as inert filler: the tensile strength and modulus decrease and the elongation slightly increases. The adhesion between the phases is imperfect, and there is an increase in the hydrophobicity of the starch film.

The DMAEMA-modified NRL significantly changes the properties of the starch/latex films. In the cationic form, the polDMAEMA chains enhance greatly the compatibility between the rubber and starch phases, with an optimum compatibility obtained for about 10 % (w/w) DMAEMA relative to rubber. This leads to a large increase in film elongation and thougness, but also in an increase of the water sensitivity of the composite material. These results have potential applications in improvement of material properties of materials made from judicious addition of modified and unmodified NRL to polysaccharide-based agricultural waste material surface [20].

Nowadays, tissue and biomedical engineering that natural polymers are preferred are two rapidly expanding and important fields of research as a consequence of their potential and already significant impacts in the medical field. In many actual potential applications, either a vector or a scaffolding material is required to achieve or realize the purpose of the medical procedure. This is a challenging task since numerous properties and constraints must generally be satisfied in order to obtain the final desired material in terms of mechanical properties and processability, porosity (average pore size, distribution and interconnection), surface texture, biocompatibility and host response, and response to external stimuli. Multiphase polymer blends have been explored as a route to fabricate scaffolds with complex architectures, controlled porosities and pore size distributions [32].

Current trends for tissue engineering focus on PLA. It is used to be preparing novel blends to overcome some disadvantages of PLA. Polystyrene (PS), polycaprolactone (PCL) and EPDM can be used for preparing satisfactory blends with PLA, especially after some modifications. In a relevant study, the activity of polystyrene-block-poly(L-lactite) (PS-b-PLLA) and polystyrene-block-poly (methyl methacrylate) (PS-b-PMMA) diblock copolymer brushes located at a PS/PLLA interface were employed as a route to control the final microstructure of 95 % void volume, ultraporous PLLA scaffolds. The latter were initially prepared from melt-processed quaternary blends of EPDM/poly(e-caprolactone)/polystyrene/poly(L-lactite) (EPDM/PCL/PS/PLLA) 45/45/5/5 (vol %) modified with the diblock copolymers. The blends display a layer comprised of the PS and PLLA phases located at the interphase of the co-continuous EPDM and PCL phases. When the PS-b-PLLA copolymer is added, sub-micrometric PLLA droplets are encapsulated within the PS continuous layer phase. In comparison, both the PS and PLLA phases compete for the encapsulation process when the PS-b_PMMA is used, indicating that the microstructure of the PLLA phase can be fine-tuned with an adequate choice of interfacial modifier. These effects were investigated by analyzing the microstructure of ternary high density polyethylene (HDPE)/PS/PMMA 89/10/10 (vol %) blends displaying PS/PMMA shell/core composite droplets in a HDPE matrix [32].

For the EPDM/PCL/PS/PLLA quaternary systems modified with the PS-b-PMMA, the PLLA homopolymer phase significantly penetrates and swells the PMMA blocks due to their mutual high affinity, as compared to the classical like-prefers-like compatibilization approach. The swelling of the blocks will tend to bend the interface towards the PS phase in order to minimize the lateral compression of the PS/PMMA shell/core structure in the HDPE/PS/PMMA ternary system. This level of control ultimately leads to quite significant differences in microstructures and surface textures for the PLLA scaffolds.

Blends of EPDM/PCL/PS/PLLA copolymer and HDPE/PS/PMMA copolymer can be prepared with the respective volume compositions (1) 45/45/5/5 and (2) 80/10/10. Both systems were prepared by simultaneously mixing all the components in a one-step process using a plasticorder at almost 200 °C for 8 min under a constant nitrogen flow. The blends were then quenched in cold water to freeze-in the morphology. Quiescent annealing was subsequently performed during 15, 30 and 60 min at 200 °C in an air athmosphere, followed by quenching in cold water. After 60 min of quiescent annealing time at 200 °C, the PLLA average molecular weight decreases by approximately 40 % [32]. In the same study, it is demonstrated that how 95 % vol ultraporous PLLA scaffolds can be prepared from melt-processed EPDM/PCL/PS/PLLA blends modified with PS-b-PLLA and PS-b-PMMA block copolymers. The modified quaternary systems present themselves as co-continuous EPDM/PCL blends with PS/PLLA/PS-b-PMMA or PS-b-PLLA sub-blend layers located at the EPDM/PCL interface: a blend is located at the interface of another one. Using a PS-b-PMMA copolymer leads to an increased percolation of the PLLA phase within the layered phase as compared to the blend modified with PS-b-PLLA copolymer which shows sub-micron PLLA droplets forming a tight array within the layer. This is explained by the copolymer enthalpic brush activity at the interface. As a result of compatibilization of the PS and PLLA phases within the layer not only decreases the interfacial tension and stabilizes the morphology over long annealing periods of times, it also strongly influences the final microstructure of the scaffolds, which is an important parameter to control when designing these materials [32].

## *2.4 Compatibilization of Rubber Based Blends by Filler Modification*

Nanoclays are an attractive alternative to traditional compatibilizers because they can be compounded quite easily. They also stabilize different crystalline phases of polymers, and improve mechanical and thermal properties. Blending of rubbers is a very common practice and incorporation of nanoclays to reinforce the rubber blend is obviously a common approach to get a rubber compounds with superior physical properties [33].

Polymer nanocomposites are widely used in various applications. But, in spite of their other superior mechanical properties, especially low toughness is still an important problem which should be solved. This is a reason for increasing amount of papers dealing with application of elastomeric tougheners in nanocomposites, but this approach leads usually to a certain compromise with an increase at the expense of stiffness and strength. In some cases, addition and dispersion of clay into rubber phase may provide synergistic effects leading to very fair balance of mechanical behavior. This seems to be a consequence of complex influencing the multiphase system by clay such as modification of components and interface is improved by the dynamic phase behavior. Compatibilization with clay is effective in many polymer combinations including toughened nanocomposites with, e.g., PA, PET and PP matrixes. When applying this compatibilization/reinforcement concept to an elastomer (EPR, EMA)—toughened PA6 nanocomposite, there is a significant size reduction and modification of dispersed phase morphology. As in the binary PA6/EPR matrix, the application of clay to PA6/PS/EPR matrix leads to a decrease in particle size. However, the presence of the clay in ternary matrix causes predominantly opposite changes in mechanical behavior to these in binary blends. The differences include a decrease in toughness with increasing clay content and a less effective toughening effect of core–shell (elastomer/clay) particles. At the same time, in contrast to the PA/PS system, combination of core–shell particles formed by PS/C15 (Cloisite 15A, modified with dialkyl dimethylammonium chloride 95 meq/100 g) preblending with EPR or EPR/C15 preblend leads to fair mechanical behavior including enhanced toughness. In summary, proper combination of rigid and elastomeric inclusions can lead to nanocomposite with balanced enhanced mechanical behavior [34].

There are lots of applications on rubber/clay blends. ENR is used as a compatibilizer between NR and clay to get intercalated and exfoliated clay particles in NR matrix. Due to the polar nature of ENR, the exfoliation of the clay layers is becoming easier and the exfoliated clay particles are located at the NR/ENR interface. Similar particular amount of organically modified clay can be used in different NR/ENR blends with varying proportions of ENR. Besides reinforcing, the vulcanization acceleration effect of organoclays was found to be very prominent in this NR/ENR system. The success of ENR as a compatibilizer is also proved in nanoclay filled NBR and nanoclay filled SBR. Also, extra amount of stearic acid in curing recipe improves the reinforcing efficiency of nanoclay in NBR [33].

Polychloroprene (CR) is a special synthetic rubber with more than 75 years of proven performance in a broad industrial spectrum due to its unique combination of properties: ozone resistance, toughness, dynamic flex life, good adhesion to other materials and heat resistance. These properties can also be satisfied by blending with polyolefin elastomers such as EPR or EPDM which have better heat, ozone, and cut growth resistance. However, these CR/EPR or EPDM blends are incompatible. Quaternary ammonium modified montmorillonite can be taken as an alternative means to control the phase compatibility between CR and EPDM. For this, the clay is treated with stearic acid in a mortar and placed in the oven at almost 100 °C for 30 min. This compound is used as filler. The EPDM rubber is mixed with ZnO and the stearic acid modified clay is incorporated into it in two-roll mill. This mix can be added with pre-masticated CR and finally the curatives are incorporated [35].

Also, rubber blend-layered silicate nanocomposites have received significant attention, recently. Nanocomposites are interesting for academic and industrial point of view because of the remarkable improvements of some mechanical and thermo mechanical properties, enhanced barrier properties or reduced flammability. However, the efficiency of nanoclay with regard to improving the solid-state properties of rubber blends is extremely dependent on the degree of clay dispersion, i.e. intercalation and exfoliation as well as the distribution of the exfoliated platelets and the phase specific clay distribution, respectively. Regarding the phase specific distribution of clay in heterogeneous polymer blends several works showed that clay preferentially resided in that blend phase having the better affinity to clay. Clay can form in situ grafts by adsorbing large amounts of polymer, which in turn are very effective for reducing the interfacial tension and inducing compatibilization in blends from highly immiscible components. This is attributed this phenomenon to the fact that the two immiscible polymer chains can co-exist between the intercalated clay platelets. These two chains then play the role of a block copolymer, which acts as a compatibilizer for thermoplastic blends (e.g. PS/PP blend). During mixing of pure HNBR with NR/clay master batch, the apparent clay migration from the NR phase to the HNBR phase was ascribed to the different interactions between the polymers and clay. Because in the clay galleries, both NR and HNBR molecules can intercalate the bi-intercalated clay tactoids act as compatibilizing agent like a block copolymer. Therefore, the refinement of blend morphology is caused by the compatibilizing effect of clay [35].

The adhesion of tire cords between rubber compounds is an important objective of the tire industry, because this adhesion influences the service performance of the tire, directly. Nylon 66 cords have shown outstanding adhesion in NR or NR/SBR blends in tires. However, the adhesion of RFL dipped polymer cord to rubber is adversely affected by exposure to ozone, UV, humidity, heat and certain chemicals (e.g. $NO_2$, $SO_2$) before cured with rubber. At present, the main studies to solve these problems focus on modifying RFL formulation, adding adhesion promoters into rubber compounds and modifying RFL dipping technology. In a novel method, addition of maleic anhydride modified carbon black (by solid state grafting) can be added into rubber compound and this promote the adhesion between RFL dipped nylon 66 tire cords and NR. The main reason that MAH modified carbon black enhances RFL dipped nylon 66 cords to NR adhesion is that active reaction groups introduced onto carbon black surface in MAH modifying process improve the formation of integration effect between RFL dipped nylon 66 cords and NR vulcanizate interface. At the same time, MAH modified carbon black can enhance mechanical properties of the vulcanizate, especially for aged samples. The reason of which is probably the formation of whole network effect in the vulcanizate because of the application of MAH modified carbon black. Also, DMA results indicate that the modification of carbon black may effectively lower the rolling resistance of NR vulcanizates [36].

## 3 Future Prospects

In parallel to emerging technologies, smart polymeric blends, nanofiller application in blends are growing interests recently. Controlling the migration of additives will get some application opportunities in food packaging and biomedical industries. Nanaoclays and nanosilvers are the materials to be used in these studies. In this context, some attempts on the novel compatibilizers and/or compatibilization processes are in progress. Nowadays, shape-memory polymers are present between emerging issues for academy and also industry. However, at present, the industrial application of these materials is limited, due to the manufacturing and scale-up difficulties. However, these difficulties tend to be reduced by using smart polymer blends and new compatibilizers. Self-adjusting smart materials are promised for rapid growing in electronic industry.

The properties of polymers can be improved by using multilayered smart pellet additives in the blends. But, there is still some need to improve the adhesion between filler and polymer interphase. This and some further improvements can simply be achieved by surface modification of these fillers using appropriate compatibilizers.

## References

1. Karaağaç, B., Kaner, D., Deniz, V.: The effects of compatibility on the mechanical properties and fatigue resistance of IIR/EPDM rubber blends. Polym. Compos. **31**(11), 1869 (2010)
2. Mangaraj, D.: Elastomer blends. Rubber Chem. Technol. **75**(3), 365 (2002)
3. Sebenik, U., Zupancic-Valant, A., Krajnc, M.: Investigation of rubber-rubber blends miscibility. Polym. Eng. Sci. **46**, 1649 (2006)
4. Scares, B.G., Sirqueira, A.S., Oliveira, M.G., Almeida, M.S.M.: Compatibilization of elastomer-based blends. Macromol. Symposia **189**, 45 (2002)
5. Mathew, M., Ninan, K.N., Thomas, S.: Compatibility studies of polymer-polymer systems by viscometric techniques: nitrile rubber-based polymer blends. Polymer **39**(25), 6235 (1998)
6. Koning, C., Van Duin, M., Pagnoulle, C., Jerome, R.: Strategies for compatibilization of polymer blends. Prog. Polym. Sci. **23**, 707 (1998)
7. Paul, D.R., Barlow, J.W.: Polymer blends(or alloys). J Macromol Sci: Reviews in Macromolecular Chemistry and Physics **C18**, 109 (1980)
8. Multicomponent polymeric materials. In: Paul, D.R., Sperling, L.H. (eds.) Advanced Chemistry Series, no. 211, ch. 2
9. Heggs,R.P., Marcus, J.L., Markham, D., Mangaraj,D.: Viscosity modifier for improved PC/ Nylon blends. Plast. Eng. **4**, 29 (1988)

10. Dikland, H.G., Van Duin, M.: Miscibility of EPM-EPDM blends. Rubber Chem. Technol. **76**(2), 495, (2003)
11. Abdul Kader, M., Bhowmick, A.K.: New miscible elastomer blends from acrylate rubber and fluorocarbon rubber. Rubber Chem. Technol. **73**(5), 889 (2000)
12. Setua, D.K., Gupta, Y.N.: One the use of micro thermal analysis to characterize compatibility of nitrile rubber blends. Thermochim. Acta **462**, 32 (2007)
13. Bhattacharya, A.: Radiation and industrial polymers. Prog. Polym. Sci. **25**, 371 (2000)
14. Yu, L., Dean, K., Li, L.: Polymer blends and composites from renewable resources. Prog. Polym. Sci. **31**, 576 (2006)
15. Razzak, M.T., Otsuhata, K., Tabata, Y., Onashi, F., Takeuchi, A.: Blood compatibility assessment of graft copolymer (NR-g-DMAA) tubes. Radiat. Phys. Chem. **39**(6), 547 (1992)
16. Vallat, M.F., Giami, S., Coupard, A.: Elastomer-elastomer autoadhesion-Interphase gradient of elastic modulus. Rubber Chem. Technol. **72**(4), 701 (1999)
17. Nakason, C., Nuansomsri, K., Kaesaman, A., Kiatkamjornwong, S.: Dynamic vulcanization of natural rubber/high density polyethylene blends: Effect of compatibilization, blend ratio and curing system. Polym. Testing **25**, 782 (2006)
18. George, J., Neelkantamn, N.R., Varughese, K.T., Thomas, S.: Dynamic mechanical properties of high density polyethylene and nitrile rubber blends: Effect of blend ratio, compatibilization and filler incorporation. Rubber Chem. Technol. **78**(2), 286 (2005)
19. Rath, T., Kumar, S., Mahaling, R.N., Khatua, B.B., Das, C.K., Yadaw, S.B.: Mechanical, morphological and thermal properties of in situ ternary composites based on poly(ether imide), silicone rubber and liquid crystalline polymer. Mater. Sci. Eng., A **490**, 198 (2008)
20. Rouilly, A., Rigal, L., Gilbert, R.G.: Synthesis and properties of starch and chemically modified natural rubber. Polymer **45**, 7813 (2004)
21. Rouilly, A., Rigal, L.J.: J. Macromol. Sci.: Agro-materials: A bibliographic review. Polym. Rev. **C42**, 441 (2002)
22. Tomka, I., Sala, R.: In: Blanshard, J.M.V., Lillford, P.J. (eds.) The Glassy State in Foods, p. 475. Nottingham University Press, England (1993)
23. Gaudin, S., Lourdin, D., Le Botlan, D., Ilari, J.L., Colonna, P.J.: Plasticisation of mobility in starch-sorbitol films. J. Cereal Sci. **29**, 273 (1999)
24. Dufresne, A., Vignon, M.R.: Improvement of starch film performances using cellulose microfibrils. Macromolecules **31**, 2693 (1998)
25. de Carvalho,A.J.F., Curvelo,A.A.S., Agnelli, J.A.M.: A first insight on composites of thermoplastic starch and kaolin. Carbohydr. Polym. **45**, 189 (2001)
26. Wilhelm, H.M., Sierakowski, M.R., Souza, G.P., Wypych, F.: Starch films reinforced with mineral clay. Carbohydr. Polym. **52**, 101 (2003)
27. Onteniente, J.P., Etienne, F., Bureau, G., Prudhomme, J.C.: Fully biodegradable lubricated thermoplastic starches: Water desorption on extruded samples. Starch/Starke **48**, 10 (1996)
28. de Graaf, R.A., Janssen, L.P.B.M.: The production of a new partially biodegradable starch plastic by reactive extrusion. Polym. Eng. Sci. **40**, 2086 (2000)
29. Mohanty, A.K., Misra, M., Hinrichsen, G.: Biofibres, biodegradable polymers and biocomposites: An overview. Macromol. Mater. Eng. **276**, 1 (2000)
30. Halley, P., Rtgers, R., Coombs, S., Kettels, J., Gralton, J., Christie, G., Jenkins, M., Beh, H., Griffin, K., Jayasekara, R., Lonergan, G.: Developing biodegradable mulch films from starch-based polymers. Starch/Starke **53**, 362 (2001)
31. de Carvalho, A.J.F., Job, A., Alves, N., Curvelo, A., Gandini, A.: Carbohydr. Polym. **53**, 95 (2003)
32. Virgilio, N., Sarazin, P., Favis, B D.: Ultraporous poly(L-lactide) scaffolds prepared with quaternary immiscible polymer blends modified by copolymer brushes at the interface. Polymer **52**(7), 1483 (2011)
33. Das, A., Mahaling, R.N., Stöckelhuber, K.W., Heinrich, G.: Reinforcement and migration of nanoclay in polychloroprene/ethylene-propylene-diene rubber blends. Compos. Sci. Technol. **71**, 275 (2011)

34. Kelnar, I., Rotrekl, J., Kotek, J., Kapralkova, L., Hromadkova, J.: Effect of montmorillonite on structure and properties of nanocomposite with PA6/PS/elastomer matrix. Eur. Polymer J. **45**, 2760 (2009)
35. Ali, Z., Le Hong, H., Illish, S., Turn-Albrecht, T., Radush, H.J.: Morphology development and compatibilization effect in nanoclay filled rubber blends. Polymer **51**, 4580 (2010)
36. Jia, D., Zhang, X.: Effect of MAH modified carbon black prepared by solid state grafting in situ on the adhesion between nylon 66 cords and natural rubber and dynamic mechanic properties of the vulcanizates. Rubber Chem. Technol. **75**(4), 669, (2002)

# Interpenetrating Polymer Networks: Processing, Properties and Applications

**Aji. P. Mathew**

**Abstract** Interpenetrating polymer networks (IPNs) are defined as combination of two or more polymers in network form with at least one of which is polymerised and/or crosslinked in the immediate presence of the others. IPNs are based on combinations of two or more polymers and are younger cousins to polymer blends, blocks and grafts. All these are members of a larger class of multicomponent polymeric systems, where as in IPNs, the polymers are crosslinked, thus providing a mechanism for controlling the domain sizes and reducing creep and flow. Though the idea behind IPN synthesis is to effect molecular level interpenetration of the polymer networks, most IPNs form immiscible systems with phase separation during some stage of synthesis. Aylsworth, in 1914 invented the first known IPN, but the term IPN was coined much later in 1960, by Millar who developed PS/PS IPNs to be used as ion exchange resin matrices (Aylsworth, US Patent 1, 111, 284, 1914), (Millar, J. Chem. Soc. 1311, 1960). The literature review shows that Sperling and coworkers at Lehigh university, USA followed by Frisch from University of Detroit and Frisch from Suny, Albany have made the most contributions to this research area. The current review on IPNs summarises the processing, properties and applications of IPNs, with special focus on some recent developments and trends.

## 1 Introduction

Interpenetrating polymer networks (IPNs) are a special class of polymer blends.

Polymer blends can be divided broadly into six subclasses based on the structure of component polymers. The blends of two polymers without any primary bonds are the simplest form of blends and are called mechanical blends [3]. Introduction of some chemical bonds gives rise to graft copolymers [4] and end-to

Aji. P. Mathew (✉)
Division of Materials Science, Luleå University of Technology, Luleå 97187, Sweden
e-mail: aji.mathew@ltu.se

P. M. Visakh et al. (eds.), *Advances in Elastomers I*,
Advanced Structured Materials 11, DOI: 10.1007/978-3-642-20925-3_10,

end joining of polymer components give rise to block copolymers [5]. When one of the polymeric phase is crosslinked a semi-IPN is formed where as crosslinking of both the components results in a full-IPN [6, 7].

Figure 1 shows the schematic representation of IPNs in comparison with different types of polymer blends discussed above.

According to IUPAC Compendium of chemical terminology IPNs are a polymer comprising two or more networks which are at least partially interlaced on a molecular scale but not covalently bonded to each other and cannot be separated unless chemical bonds are broken. It is also stated that a mixture of two or more preformed polymer networks is not an IPN.

IPNs involve the polymerization of one monomer in the immediate presence of the other, and the crosslinking of one or both the polymers. The interesting and unique properties of IPNs emerge when the deliberately introduced crosslinks out number the accidently introduced grafts in the polymerization stage. However, some amount of graft is usually present in all IPNs and contributes to the IPN performance in a favorable manner. The two characteristics features of IPNs that distinguishes itself from the other types of polymeric blends are (1) IPNs swells but does not dissolve in solvents (2) creep and flow are suppressed [8].

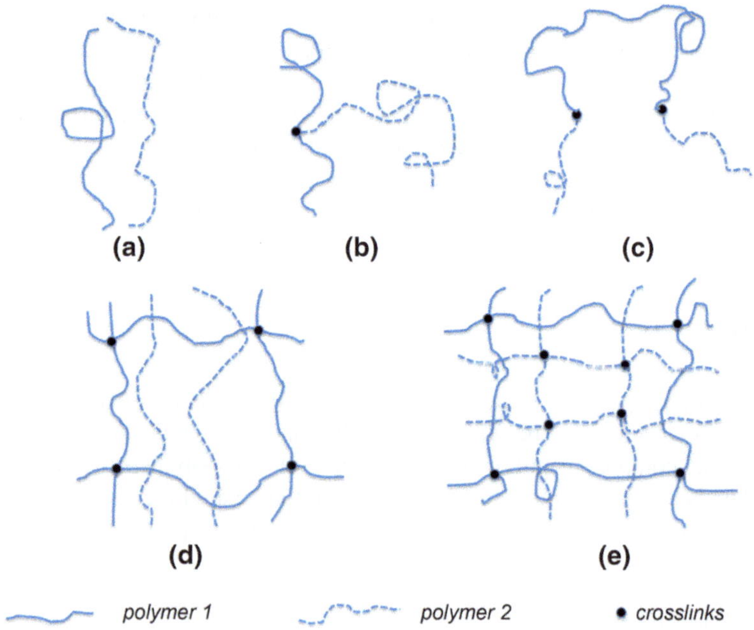


**Fig. 1** Schematic representation of (**a**) mechanical blend, (**b**) grafts, (**c**) blocks, (**d**) semi-IPNs, and (**e**) full-IPNs (based on Ref. [8])

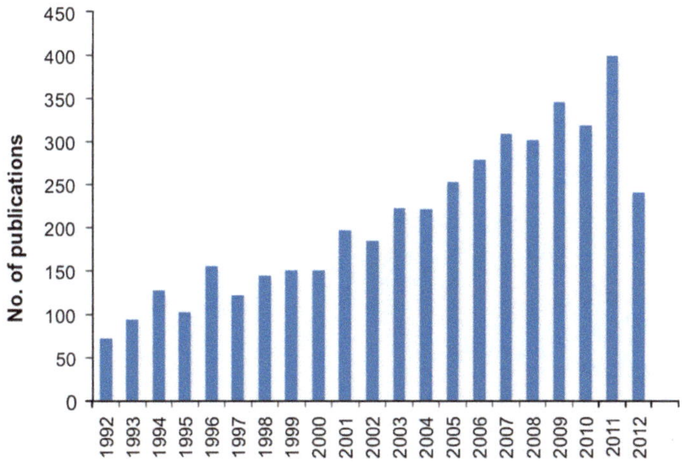


**Fig. 2** Publications on IPNs during the last 20 years (based on ISI web of science)

Though all these are termed IPNs, true interpenetration occurs only at phase boundaries and most IPNs are phase separated to some extent. Molecular level interpenetration occurs in the case of mutual solubility only.

Figures 2 and 3 shows the publications during the last 20 years and also the countries that are most active in this research area, respectively. The data from ISI web of science indicate that China, USA and India are the leaders in this research area. Since 1960 [1, 2], there was a slow but steady growth for this research field. About 30–50 publications per year can be found until 1990, from literature search. After 1990 the growth was more rapid and currently about 400 papers per year is being published in this research field.

## *1.1 Types of IPNs Based on Synthesis Mode*

According to the mode of synthesis, the IPNs are of five different types viz. sequential IPNs, simultaneous IPNs, latex IPNs, gradient IPNs and thermoplastic IPNs.

1. Sequential IPNs where polymer I is crosslinked initially followed by the swelling of polymer network I with monomer II and crosslinker and initiator and subsequent polymerization of monomer II, in situ [9].
2. Simultaneous IPNs, where a mutual solution of monomer I and monomer II, crosslinker I and crosslinker II are taken together and then polymerized simultaneously by non-interfering routes [10].
3. Latex-IPNs where two lattices of linear polymers are mixed and coagulated and are crosslinked simultaneously. They are also called interpenetrating elastomeric networks (IENs), especially when both polymers are above the glass transition temperature [11].

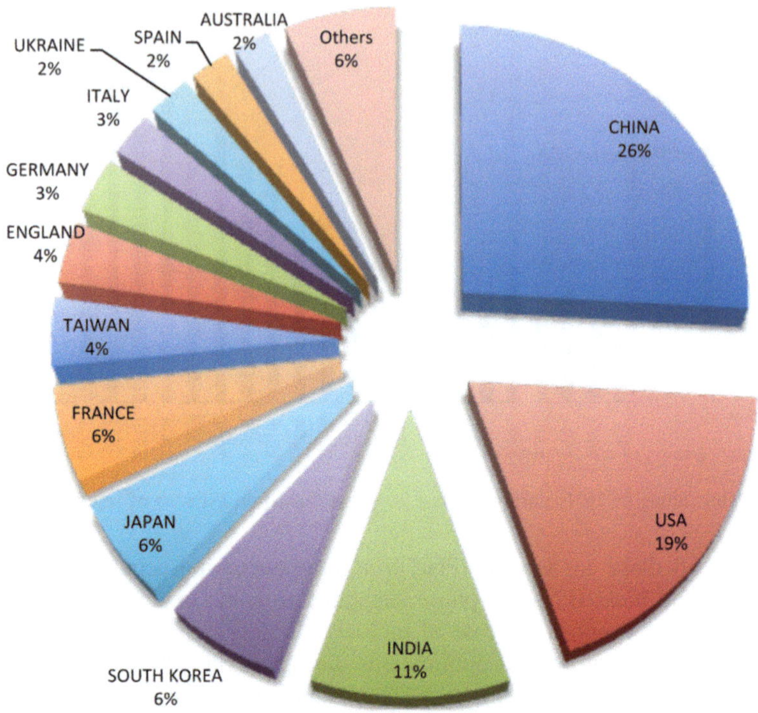


**Fig. 3** Countries contributing to IPN research (based on ISI web of science)

4. Thermoplastic IPNs where physical crosslinks are present rather than chemical crosslinks. In this case some degree of phase continuity exists and behaves as thermoset in ambient conditions and as thermoplastic at elevated temperature [12].
5. Gradient IPNs, where the composition is varied within the sample at the macroscopic level. This is carried out by swelling the network in the monomer for a given amount of time and polymerizing rapidly before equilibrium sets in. This methodology allows a higher concentration of polymer 1 in one surface and polymer 2 in the other and a gradient through out the bulk [13].
6. Semi-IPNs (SIPNs) or psuedo IPNs are those materials in which one phase is crosslinked where as the second phase is linear or branched [7].

## *1.2 Unique Properties of IPNs*

IPNs are composed of two or more chemically distinct networks held together predominantly by trapped mutual entanglements rather than by covalent bond grafting. From the investigation it is known that IPNs do not interpenetrate on a monomer scale, but have a microheterogeneous morphology with regions enriched by segments of one of the components [14, 15].

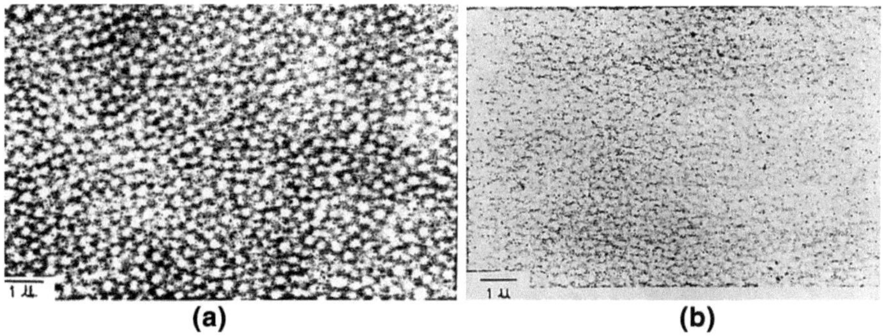

**Fig. 4** TEM image of (**a**) PU/PS linear blend with 1000 Å domains, and (**b**) PU/PS semi-IPNS with 500 Å domains. The PU/PS ratio is 25/75 in both cases (based on Ref. [19])

The properties of IPNs depend on (1) properties of component polymers (2) phase morphology (3) compatibility/interaction between the phases. In many cases the properties of IPNs are approximately the simple average of the properties of the component polymers but in some cases synergism is obtained [16, 17]. Unique properties as improved mechanical properties, thermal stability, chemical resistance etc., compared to conventional blends can be expected from IPNs. In PC/PU semi-IPNs reported by Shyu and Chen [18], two glass transition temperatures is found to be shifted inwardly indicating the interpenetrating network of PU increased the mutual miscibility of PC and PU. The average domain size of PC was 500 Å in semi-IPNs while the blends of the corresponding polymers has phase domains of 1000–6000 Å (Fig. 4).

## *1.3 Miscibility and Phase Separation*

In IPNs the two separate glass transitions for each polymer, with or without inward shift can occur or one broad Tg intermediate to the Tgs of the individual polymers or one sharp Tg intermediate to the Tgs of the component polymers can be obtained, depending of the extent of compatibility or phase mixing. An inward shift or merging of Tgs can be considered as evidence for interpenetration of the phases and is the case in many IPNs. This is interpreted as increase in miscibility of the phases due to crosslinking [19].

Kong and Narine studies sequential IPNs of PU using canola oil based polyol and PMMA and compared it with those castor oil based PU/PMMA IPNs [20]. The DMTA study of these materials showed are given in Fig. 5. One sharp peak corresponding to the glass transition was observed for both pure PUR samples where as two peaks (at 31 and 134 °C) were observed for pure PMMA. In the case of canola oil based IPNs, the following features like a) inward shift of second and third transitions for CN-PUR15 %-PMMA,; combining and broadening of peaks

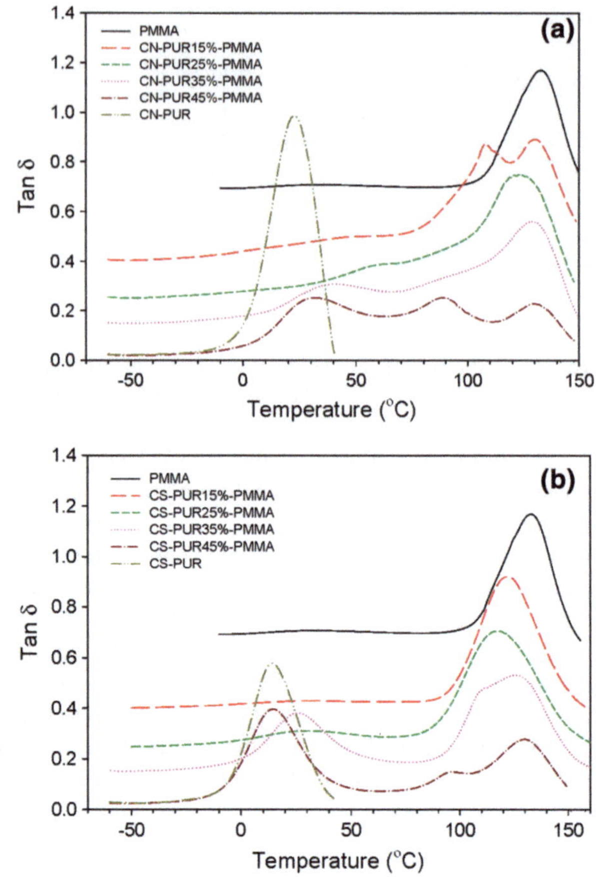


**Fig. 5** Effect of temperature on the Tan delta peak of PU/PMMA IPNs with different compositions (**a**) PU based on canola oil, and (**b**) PU based on castor oil polyols (reproduced with permission from Ref. [21])

or appearance of additional peak at lower temperature and (3) three distinct transition peaks in castor oil based IPNs: only one transition peak at higher temperature was present for CS-PUR15 %-PMMA; splitting of peaks at PUR >35 wt %, and well-resolved transition peak at lower temperature was also observed. The highest degree of phase mixing was obtained for the sample with 25 wt % of PUR content in both series of IPNs. By comparing the tan $\delta$ curves, presented in Fig. 5, it is obvious that the transitions of canola oil based IPNs are broader than those of castor oil based IPNs, implying again that a modest degree of molecular mixing exists in the former IPNs.

Also the domain sizes found in interpenetrating networks are smaller than corresponding blends, which supports this concept of forced miscibility due to crosslinking. Most IPNs and related materials show phase separation. It was pointed out by Sperling that the morphology of semi and full IPNs depends on many factors of which (1) degree of crosslinking and (2) mechanism of phase separation are of great importance. During IPN synthesis a certain degree of compatibility between the polymers are introduced by the formation of two polymer networks interlocked in a 3-dimensional structure. The phase domains are smaller in high compatibility systems and even phenomenon like nanostructured morphology, dual phase continuity can be observed. In PEA/PMMA IPNs the dispersed domain sizes were about 100 Å, which was attributed to the chemical compatibility between PEA and PMMA, which is isomeric. In the case of PEA/PS where the compatibility is lower, much bigger dispersed domains are obtained [21]. Mathew et al. have obtained domain sizes in the range of 15–25 nm were obtained in NR/PS sequential IPNs [13]. The materials showed intimate mixing of the polymers with phase separation as well as dual phase continuity.

Further more, Gennes [22] points out that crosslinked polymers, may be more miscible than corresponding blends because the chains must be extended to enable phase separation. The case for full IPNs can be inferred to be more like that of the double cross-linked polymers. Both network chains must be extended (lower entropy) to enable phase separation. It is expected that phase separation occur after gelation, which explains why the kinetics of phase separation are much slower than the kinetics of polymerization. Suthar et al. gave a good review with respect to the kinetics studies on the formation of interpenetrating polymer networks [23]. They concluded that the faster the rates of the respective chain extension and crosslinking reactions are and the closer they are to simultaneity, the more homogeneous are the IPNs. In addition, the individual components sometimes can polymerize more rapidly in the IPNs than alone, due to a "solvent effect" of the IPNs.

# 2 Processing of IPNs and Resultant Properties

## *2.1 Sequential IPNs*

Sequential IPNs are synthesised by a two-step process as shown by route A in Fig. 6 Polymer network 1 is first formed which is then swollen by monomer 2 and its activator and crosslinker. Then polymer 2 is polymerized and crosslinked in situ in network 1

The morphology of the IPNs are developed during processing stage and depends on the degree of crosslinking of phase 1 and phase 2 and the miscibility between the phases. If the polymers gel before phase separation occurs, the crosslinking of polymer chains of the first polymer will restrict phase separation of the second phase, which results in smaller domain sizes. On the contrary, if phase separation precedes gelation, the domains will be large and the crosslinks will stabilize the phase separated morphology. In. M.L sequential IPNs resulted in domain sizes as low as 20 nm when PU is polymerized first, where as when P2EHAs polymerised first the domain size of P2EHA was about 100 nm [24].

Usually, if the miscibility and crosslinking density of two constituents are greater, then the phase domain of the IPNs is smaller. In case of IPNs with immiscible polymers, at low degree of crosslinking, phase separation occurs, but

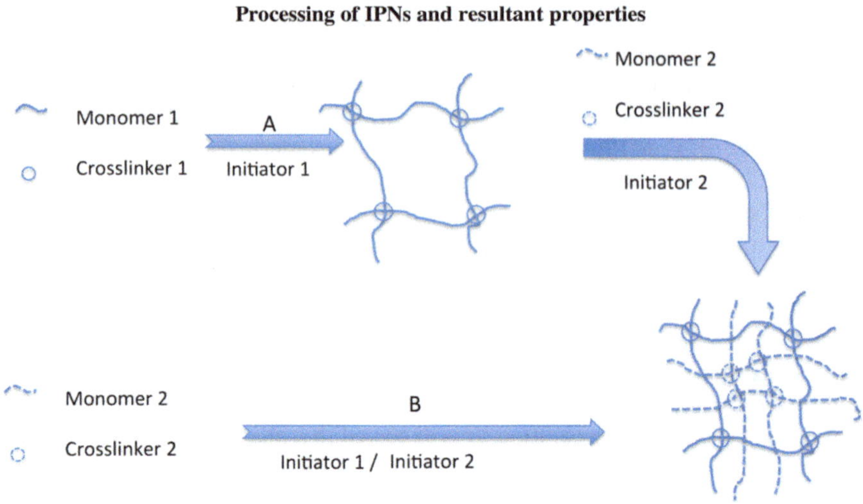


**Fig. 6** Schematic representation of the synthesis of IPNs by (*A*) sequential, and (*B*) simultaneous routes

with increasing crosslinking density the domain sizes become smaller which promotes intermixing of phases, leading to partial or complete miscibility. For example PnBA/PBMA, PVAc/PMAc where the miscibility increased by increasing the crosslink density.

In the case of sequential full-IPNs, the morphology is very much dependent on the network structure that is first formed [25]. Donatelli et al. developed an equation to show the inverse proportionality of the domain size of second phase to the crosslink density of first network for semi-IPNs [26]. Yeo et al. also developed a theory which explained the relation of domain size with network I and II [27]. From these studies it was clear that morphology and properties of IPNs may be manipulated by controlling the crosslink level of either components.

In sequential IPNs the following generalizations are applicable:

1. Polymer I tends to be continuous in space for all compositions. Mid-range and high concentrations of polymer II tend to have dual phase continuity that is most likely interconnected cylinders for polymer II.
2. Domain sizes vary from about 100 nm for highly immiscible systems to about 10 nm for more miscible systems.
3. The domain size decreases with increasing cross-link density. For many compositions, the effect is about 10 times as large for cross-links in network I as for network II.
4. For domains in the size range of 50–100 nm, a potential application is for tough or impact-resistant plastics. For domain sizes of about 10 nm, sound and vibration damping materials should be considered.

### 2.1.1 Case Study: NR/PS Sequential IPNs

NR/PS IPNs were prepared by the sequential technique [13, 28]. The NR is crosslinked (vulcanized) first using DCP to form network 1. This network is then swollen in styrene monomer followed by the polymerisation and crosslinking to PS phase. The vulcanised NR sheets were weighed and kept immersed in inhibitor free styrene monomer containing 1 % BPO as initiator and 2, 4 or 6 % of DVB which acts as crosslinker for PS phase. The NR sheets were swollen to different time intervals to obtain varying weight percentages of PS. The swollen samples were kept at 0 °C for few hours to achieve equilibrium distribution of styrene monomer in the matrix. The swollen networks were heated at 80 °C for 6 h and at 100 °C for 2 h to complete the polymerisation and crosslinking of styrene monomer. The hardened sheets were then kept in a vacuum air oven to make it free of unreacted styrene. The final weight of the IPN was taken and the composition of the sample was determined. In this method, DCP, AIBN is used as the initiator or and 2, 4 or 6 % of DVB which acts as crosslinker for PS phase. The different concentrations of DVB resulted in different degree of crosslinking of the PS phase. The hardened sheets, in all cases, were then kept in a vacuum air oven to make it free of unreacted styrene. The final weight of the sample was taken and the composition of the sample was determined. In all the four series, NR/PS semi-IPNs with PS content up to 70 % were prepared.

The TEM studies showed that the system is nanostructured (see Fig. 7). In all the cases, dual phase continuity was observed as both the phases are crosslinked. The nano-structured morphology is due to intimate mixing of the phases with micro level phase separation.

It is well known that the combination of a glassy polymer with another polymer that is rubbery at room temperature can produce IPNs possessing a range of properties, from reinforced elastomers to high-impact plastics depending on the composition. Miscibility can be determined by dynamic mechanical analysis, that

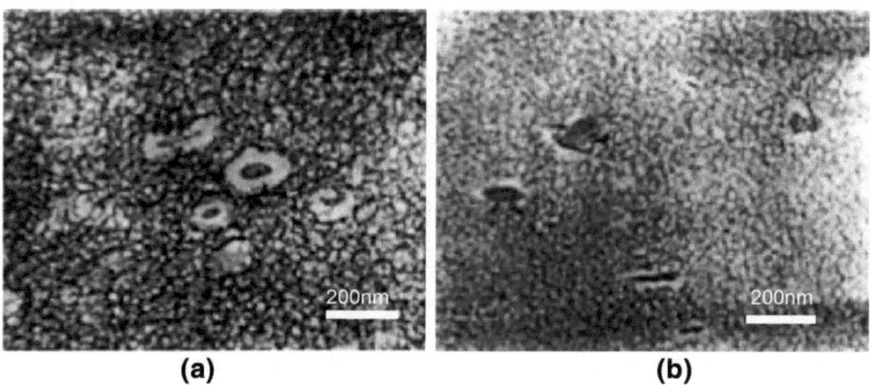


**Fig. 7** Transmission electron micrographs showing the nano- structured morphology: NR-PS/30–70, sequential INS using (**a**) DCP, and (**b**) AIBN as initiator (based on Ref. [13])

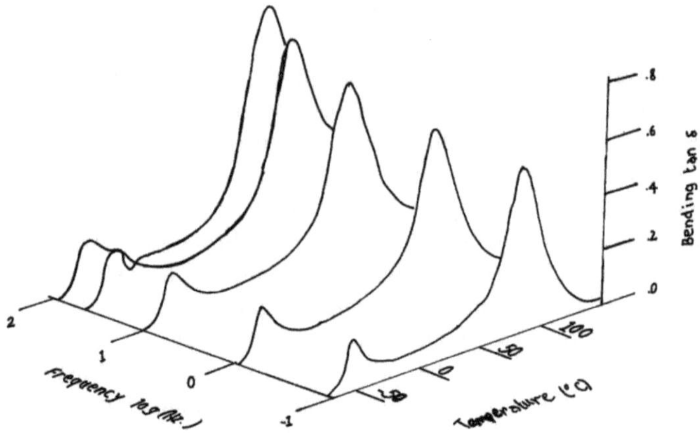


**Fig. 8** Effect of temperature and frequency on the Tan delta peak of NR/PS sequential IPNs (reproduced from Ref. [28])

is, the analysis of the glass-transition temperatures (Tg's) of virgin polymers and their IPNs. A single Tg for a composite is an indication of homogeneity, and two separate Tg's are indications of immiscibility.

In all cases of NR/PS IPNs, two Tg's were observed, corresponding to the transitions of NR and PS phases (See Fig. 8). For IPNs with 70 % PS, the PS peak was at 110 °C and was quite prominent, whereas the peak due to NR was at −35 °C and was very weak. In the 50/50 IPN sample, two distinct peaks also were observed at 101 and −37 °C, corresponding to PS and NR transitions. For samples with 30 % PS sample, the NR peak at −37.5 was quite prominent and sharp. The PS peak was at 99 °C and was very weak. Tg of the PS phase shifted to the low-temperature side, and that of NR shifted to a high temperature. This inward shift indicates phase mixing and supports the morphology observed using TEM.

With an increase in the NR content, tan delta max increased. The tan delta peak height is a measure of damping. As the rubber content increased, the damping increased because the introduction of rubber reduced the brittleness of PS and the system as a whole reduced its plastic nature. Above 50 % NR content, tan delta max increased drastically. The chain flexibility of the rubber phase was restricted by the interpenetration with the PS phase, and as a result, the damping decreased. As the PS content increased, the number of crosslinks increased, and this made the damping property lower.

In this study it was found that incase where only NR phase is crosslinked (semi-IPNs) the co-continuity is observed only when the PS concentration is above 50 %. However in the case of full-IPNs, all the studied composition viz, 30, 50 and 70 % of PS showed a co-continuous morphology. This demonstrates the effect of crosslinking on phase separation and co-continuity of phases.

## 2.2 Simultaneous IPNs

Simultaneous IPNs are synthesised by a two-step process as shown by route B in Fig. 5. Monomers 1 and 2 together with their respective activators and crosslinkers are mixed together and crosslinked simultaneously, but by independent non-interfering routes.

Among the two modes of synthesis, the simultaneous IPN (SIN) is generally the best one to have a high degree of intermixing compared with the sequential PN (SIPN) processes because of the compatibility of the monomeric mixture, which is much higher than that of a polymeric mixture [29]. During polymerization, two competing processes take place simultaneously. Phase separation of the forming polymer chains proceeds by diffusion through an increasingly viscous medium to form phase domains. The formation of cross-links restricts diffusion, and at gelation, then the present situation is frozen in [30]. With highly incompatible polymers, the phase separation is so serious that the gross phase separation occurs before gelation. The SIN process can demonstrate very fine micro heterogeneous morphology.

Polyurethane-based SINs are very popular model materials [31, 32]. The other polymer is most frequently poly(methyl methacrylate), but sometimes polystyrene or another monomer or co-monomer mix [33, 34]. In dynamic mechanical experiments, the SINs show either two well-defined glass transitions one broad transition or transitions shifted inward all from substantially the same starting materials. A possible explanation is that the time order of the three events viz. Gelation of polymer I, gelation of polymer 2 and phase separation of polymer I from polymer 2, be different for these otherwise very similar chemistries.

### 2.2.1 Case Study: Polyurethane (PU)/Polystyrene (PS) [33, 35]

A series of simultaneous interpenetrating polymer networks (SINs) based on penta-erythritol modified castor oil polyurethane and polystyrene were prepared. Polyurethane prepolymer was thoroughly mixed with the styrene. The initiator BPO and the DVB (1.4 and 2 % by weight of styrene, respectively) as well as the catalyst dibutyl tin dilaurate (DBTL; 0.5 % by weight of polyol) were added. As the mixture became viscous, it was poured into the mold and kept under pressure and the rest of the reactions were allowed to proceed in the mold. It was kept at room temperature for 24 h and at 120 °C for 8 h. A series of IPNs of different compositions were obtained following the same procedure (Table 1). The finished sheets were cut in desired shapes and sizes for further study.

The position of the glass transition temperature of a polymer mixture is often considered as a criterion to establish its miscibility. The glass transition temperatures of the PU-modified CO and PS homopolymers were 35 and 100 °C, respectively. The IPNs showed single Tg which shifted progressively towards high temperatures with increasing PS content. Figure 9 shows graphs of the glass

**Table 1** Some commercially used IPNs [47]

| Manufacturer | Trade name | Components | Application |
|---|---|---|---|
| Shell Chemical company | Kraton IPN | SEBS-polyester | Automotive parts |
| Petrarch Systems Inc | Rimplast | Silicone rubber-PU | Medical gears |
| ICI Americas Inc. | ITP | PU-polyester-polystyrene | Sheet molding compounds |
| DSM N.V. | Kelburon | PP-EP rubber-PE | Automotive parts |
| Monsanto | Santoprene | EPDM-PP | Tires, hoses, belts and gaskets |
| Dupont | Somel | EPDM-PP | Outdoor applications |
| B F Goodrich | Telcar | EPDM-PP or PE | Tubing, liners, wire, cables |
| Exxon | Vistalon | EPDM-PP | Paintable automotive parts |
| Dentsply International | Trubyte Bioform | Acrylic based | Artificial teeth |
| Hitachi Chemicals | – | Vinyl-phenolics | Damping compounds |

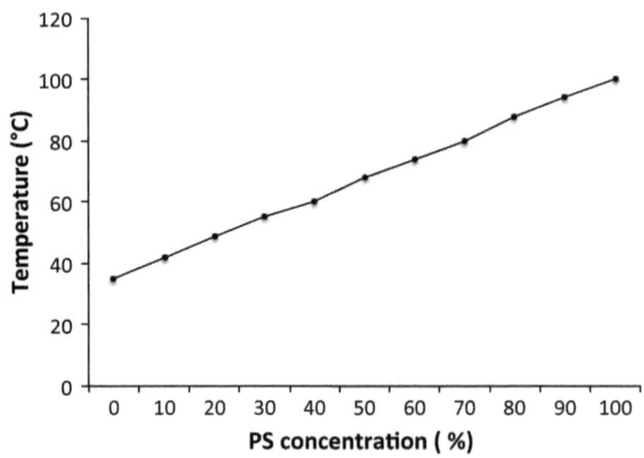


**Fig. 9** Glass transition temperature of simultaneous IPNs based on modified castor oil polyurethane and polystyrene versus polystyrene composition (experimental data and Tg calculated using Wood equation (based on Ref. [36])

transition temperature versus composition, the shifting of Tg's is solely attributed to the interpenetration in these cases, since there is little or no possibility for chemical interaction between components networks. The Tg analysis showed that the synthesized SIPNs had a semi-miscible behavior.

The mechanical properties of SINs obtained with varying amounts of styrene shows that the pentaerythritol modified castor oil polyurethane and polystyrene inter-penetrating polymer networks (exhibited better mechanical properties as compared to unmodified castor oil polyurethane/polystyrene IPN (CO-IPNs), due to more crosslinking density. The tensile strength increase up to 20 % styrene, and starts to decrease thereafter indicating that there is little intermixing between the two polymers, and that styrene is dispersed in the polyurethane continuous phase. The steep rate of decrease in the properties beyond styrene concentration of 25 % implies phase inversion. These drops in the properties suggest that phase inversion

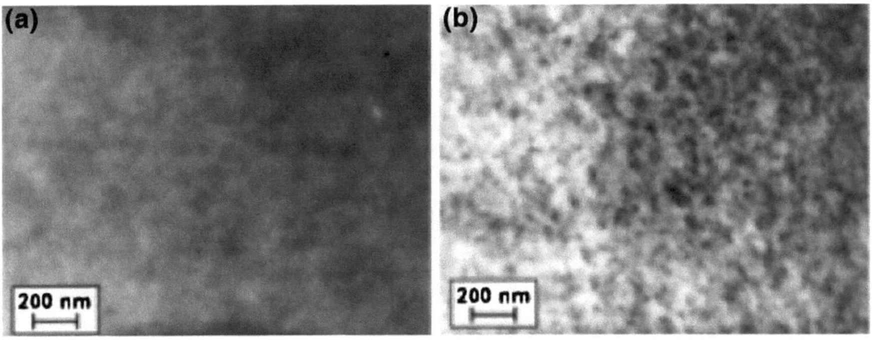


**Fig. 10** TEM photos of IPNs with 80 wt% PMMA with (**a**) an inhibitor and TDI, (**b**) no inhibitor and TDI, (based on Ref. [37])

occurs at about 75 % PU/25 % PS, where the continuous matrix changes from the elastomeric PU to the rigid PS.

## *2.3 Comparison of Simultaneous and Sequential IPNs*

Bird et al. have reported the processing of PU/PMMA interpenetrating polymer networks by a simultaneous process route (i.e. polymerization of monomers occurring at the same time) and sequential route (polymerisation occurring at different temperatures) by controlling the polymerisation process [36]. The use of an inhibitor during polymerisation resulted in a sequential IPN, where as in the absence of inhibitors, simultaneous IPNs resulted. The effect of the two processing routes on the properties and the morphology of the prepared IPNs were evaluated.

Figure 10a shows that the TEM image of sequential PU/PMMA IPN prepared and a fine dispersion of the component polymer domains is observed throughout the entire sample. This observation is an indication of the interpenetration process that is produced at the molecular level. When a sequential IPN was formed, one of the monomers, in this case the MMA, TRIM, and the initiator, polymerized after the first polymer, PU. The continuous phase of the PU network was filled with the MMA and TRIM monomers and the PMMA polymerized in situ inside the crosslinked PU network. Less phase separated materials were created in this manner. Figure 10b shows the electron microscopy analysis IPNs prepared without inhibitors. In the absence of an inhibitor in the PMMA phase, the two polymers polymerized at the same time, thus producing a simultaneous IPN. This generated clear spherical domains (dark areas in the TEM photos), possibly due to incompatibility of the growing species during the polymerization process.

Further more samples without inhibitor in the monomeric MMA phase (simultaneous IPN) resulted in materials with lower transparency than the counterpart produced with inhibitor (sequential IPNs see Fig. 11). This results in

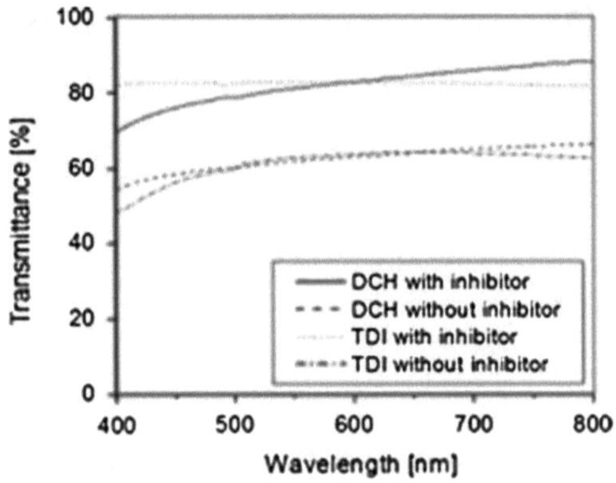


**Fig. 11** Comparison of transmittance through IPNs with and with out inhibitors, resulting in sequential and simultaneous IPNs, respectively (reproduced with permission based on Ref. [37])

disagreement with the usual observation that simultaneous IPNs results in better compatibility and phase mixing.

## 3 Recent Trends in IPNs

More recently, there is an increasing trend to have a third phase like a nanoreinrforcement in the IPN system. These type of materials have shown interesting properties and applications as catalysts, hydrogels, reinforced composites etc. [37–41].

Dongyan et al. reported the synthesis and applications of castor oil based PU/PMMA simultaneous IPNs containing $BaTiO_3$ fibers. The synthesis resulted in graft IPNs and the TEM analysis showed that the domain sizes were in nanometric scale. The mechanical studies of the nanocomposites showed a change from elastomeric behavior to brittle nature, due to the addition of $BaTiO_3$ [38].

Organically modified palygorskite was used as reinforcing agent in PU/Epoxy IPNs to prepare nanocomposites. These materials showed superior mechanical and yermal properties compared to unreinforced PU/EP IPNs [39].

Use of copper nanoparticles with 10–20 nm diameter in Polyvinyl alcohol (PVA)/polyacrylamide (PAAm) IPNs resulted in hydrogels which can be tuned for drug release or tissue engineering application [40]. The study showed that the complexation of PVA and PAAm with Cu2+ resulted in low aggregation of Cu nanoparticles and thereby controlling and stabilizing the dispersion and distribution of nanoparticles in the polymer network.

Zhan et al. reported the use of nanoporous IPNs with Pd nanoparticles and its use as catalyst for Heck reactions [41]. The encaging of Pd nanoparticles in the IPN structure with mesh sizes in the order of magnitudes smaller than the average size of the nanoparticles resulting in their stabilization.

Peterson et al. has reported a new sequential in situ IPN with self-healing characteristics, consisting of a PACM-cured DGEBA cross-linked polymer and a pMPGE linear polymer. Soxhlet extraction shows that pMPGE is mobile within the IPN, which is a necessary condition for healing via linear polymer diffusion [42].

IPN hydrogels based on poly(PEGDA) and PMAA were synthesized by sequential interpenetrating technology showed potential in heavy metal removal processes. The metal ion adsorption capacity of the IPN hydrogels increased with the pH values of the feed solution between 3 and 5 and PMAA content in the IPN. Regeneration studies on these IPN hydrogels suggested that the IPN hydrogel may be reused several times without significant loss of its metal ion capacity [42].

IPN scaffolds using biopolymers as sodium alginate and sodium hyaluronate was developed by Chung et al. for chondrocyte culture. The crosslinking of the two phases were crosslinked with polyethyl glycol diglycidyl ether (PEGDG) and calcium chloride and freeze dried to obtain porous structures [43].

IPN based responsive hydrogel systems were developed using end-linked poly ethylene glycol and loosely crosslinked polyacrylic acid with hydrogen bond reinforced strain hardening behavior in presence of water [44]. The mechanism of double network formation is shown in Fig. 12. The template polymerisation of second network in the presence of preexisting first network is found to be the mechanism of IPN formation in these hydrogels.

The authors suggest that at least two prerequisite processing elements may also be necessary: (1) that the pre-stress must be gradually built into the IPN through

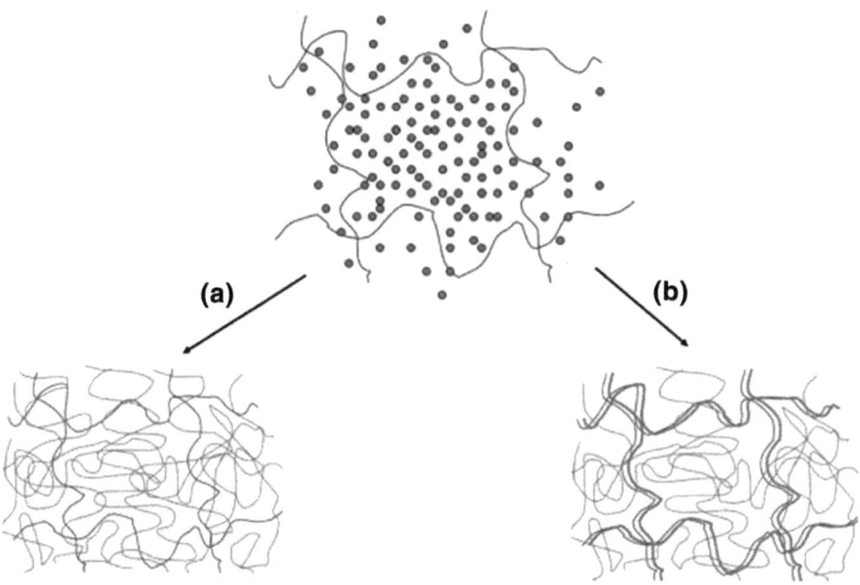


**Fig. 12** Schematic illustrating the differences in IPN structure after (**a**) standard polymerization and (**b**) template polymerization of a second network in the presence of a preexisting first network, at excess concentrations of acrylic acid relative to ethylene glycol monomer units. (reproduced with permission based on Ref. [44])

buffer-induced ionization of the PAA network after polymerization, and (2) that the second network is formed by a template polymerization process along the preformed first network. Mechanical enhancements was observed in this "double network" IPN systems despite being composed of otherwise weak polymers. Further more an excess of ionizable second network material in the presence of a neutral first network is necessary to produce an increase in fracture properties. Thermally responsive polyacrylamide/poly(acrylic acid) interpenetrating polymer network nanoparticles were synthesized by Owens et al. [45]. The IPN nanoparticles had spherical shape and had diameters of about 250–300 nm and exhibited a unique rapid sigmoidal swelling transition as a function of temperature. These systems showed larger relative swelling volume compared to random copolymer and homopolymer particles comprised of acrylamide and acrylic acid which decreased with increased cross-linker density.

A new polymeric silicone hydrogel prepared by sequential synthesis has shown potential as contact lenses. These materials has high mechanical properties, transparency, hydroåhilicity and O2 and water permeability [46].

## 4 Applications of IPNs

The patent literature reveals that there are many products like optically smooth surfaces, toughened plastics, adhesives, damping materials, ion exchange resins, impact modifiers etc. based on IPNs or related materials [8]. See Table 1 for a list of few of the commercialized IPN materials.

Research publications also suggest applications like ateriovenous shunts, piezodialysis membranes and adhesives [48–50], IPN structures are used to develop porous structures for different applications [51]. Table 2 shows some potential applications of IPNs, found in research literature.

More recently IPNs based materials have shown potential as self healing materials, hydrogels and catalyst, thermosensitive materials, contact lenses etc. [39–45].

**Table 2** Potential application of IPN and related materials [8]

| Components | Suggested application |
|---|---|
| Natural leather/rubber | Improved leather |
| Anionin/cationic | Peizo dialysis membranes, Ion exchange resins |
| Plastic/rubber | Noise damping, tough plastic |
| Plastic/plastic | Optically smooth surfaces, denture base materials |
| Rubber/rubber | Pressure sensitive adhesives |
| Rubber/plastic | Impact modifier |
| Rubber/crystalline plastic | Thermoplastic elastomer |
| Rubber/water swellable phase | Arteriovenous shunt |

# 5 Conclusions

Interpenetrating polymer networks (IPNs) have been studied extensively since their advent in the 1960s. In most cases IPNs provide intimate mixing of phases, smaller domain sizes, better mechanical properties and damping properties than the corresponding blends, which are physical mixtures of two polymers. The mechanism of phase separation and morphology of simultaneous and sequential IPNs, are different and in most cases simultaneous IPNs leads to better phase mixing, especially when gelation point is reached at the same point of time.

IPNs have different applications and is a commercially successful form of polymer blends, probably owing to the crosslinked structure that provides better thermal stability, mechanical properties, chemical resistance etc. Though IPNs are traditionally used as damping materials, impact resistant materials, adhesives, etc., more recently they are used in combination with nanoparticles to produce nanocomposites for use as responsive hydrogels, medical implants, porous scaffolds or catalyst supports. The number of publications in this research area has shown a regular increase and is about 400 publications per year, currently.

**Acknowledgments** The collaboration support by Swedish Research Links, SIDA under project No. 348-2008-6040 is acknowledged

# References

1. Aylsworth, J.W.: US Patent 1, **111**, 284 (1914)
2. Millar, J.R.: Interpenetrating polymer networks. Styrene–divinylbenzene copolymers with two and three interpenetrating networks, and their sulphonates. J. Chem. Soc. 1311 (1960)
3. Work, J.L.: Solid-state structure of melt blended incompatible polymeric mixtures involving poly(vinyl chloride). Polym. Eng. Sci. **13**, 46 (1973)
4. Molau, G.E.: Colloidal and morphological behavior of block and graft copolymers. Plenum, New york (1971)
5. Szwarc, M.: Block and graft polymers their synthesis, especially by living polymer technique, and their properties. Polym. Eng. Sci. **13**, 1 (1973)
6. Sperling, L.H., Mishra, V.: The current status of interpenetrating polymer networks. Polym. Adv. Technol. **7**, 197 (1996)
7. Mathew, A.P., Packirisamy, S., Thomas, S.: Morphology, mechanical properties failure topography of semi-interpentrating polymer networks based on natural rubber polystyrene. J. Appl. Polym. Sci. **78**, 2327–2344 (2000)
8. Sperling, L.H.: IPN and relatedmaterials. Plenum press, Newyork (1981)
9. Chakraborty, D., Das, B., Roy, S.: Epoxy resin–poly(ethyl methacrylate) interpenetrating polymer networks: Morphology, mechanical, and thermal properties. J. Appl. Polym. Sci. **67**, 1051 (1998)
10. Tan, S., Zhang, D., Zhou, E.: Dynamic mechanical properties of interpenetrating polymer networks based on polyacrylates and epoxy. Acta Polymerca **47**, 507 (1996)
11. Frisch, H.L., Klempner, D.: Topological Isomerism and Macromolecules. In: Pasika, W.M. (ed.) Advances in Macromolecular Chemistry, vol. 2, Academic Press(1970)
12. Coran, A.Y., Patel, R.P.: In: Holden, G., Legge, N.R., Quirk, R., Schroeder, H.E. (eds.) Thermoplastic elastomers. Hanser, Munich (1996)

13. Lipatov, Y.S., Karanova, L.V., Gorbach, L.A., Lutsyk, E.D., Sergeeva, L.M.: Temperature transitions and compatibility in gradient interpenetrating polymer networks. Polym. Int. **28**(2), 99 (1992)
14. Touhsaent, R.E., Thomas, D.A., Sperling, L.H.: Epoxy/acrylic simultaneous interpenetrating networks. J Polym. Sci. **46C**, 175 (1974)
15. Touhsaent, R.E., Thomas, D.A., Sperling, L.H.: In: Deanin, R.D., Crugnola, A.M. (eds.) Toughness and brittleness of plastics. Advances in Chemistry, Ser. 154. American Chemical Society, Washington (1976)
16. Wang, S.H., Zawadzki, S., Akcelrud, L.: Morphology and Damping Behavior of Polyurethane/PMMA Simultaneous Interpenetrating Networks. Mater. Res. **4**(1), 27–33 (2001)
17. Mathew, A.P., Packirisamy, S., Thomas Eur, S.: Effect of initiating system, blend ratio crosslink density on the mechaical properties failure topography of nano-structured full-interpenetrating polymer networks from natural rubber polystyrene. Polym. J. **37**, 1921 (2001)
18. Shyu, S.S., Cen, D.S.: Polycarbonate-polyurethane semi-interpenetrating polymer networks: *Tg* behavior and morphology. J. Appl. Polym. Sci. **34**, 2151 (1987)
19. Kim, S.C., Klempner, D., Frisch, K.C., Frisch, H.L.: Polyurethane interpenetrating Polymer Networks. II. Density and Glass Transition Behavior of Polyurethane-Poly(methyl methacrylate) and Polyurethane-Polystyrene IPN's . Macromolecules **9**(2), 263 (1976)
20. Kong, X., Narine, S.S.: Sequential interpenetrating polymer networks produced from vegetable oil based polyurethane and poly(methyl methacrylate). Biomacromolecules **9**(8), 2221 (2008)
21. Heulck, V., Thomas, D.A., Sperling, L.H.: Interpenetrating Polymer Networks of Poly(ethyl acrylate) and Poly(styrene-co-methly methacrylate). I. Morphology via Electron Microscopy. Macromolecules **5**, 340 (1972)
22. De Gennes, P–.P.: Sliding gels. Phys. A. **271**, 231 (1999)
23. Suthar, B., Xiao, H.X., Klempner, D., Frisch, K.C.: A review of kinetic studies on the formation of interpenetrating polymer networks. Polym. Adv. Technol. **7**, 221 (1996)
24. Robeson, L.M.: Polymer Blends: A comprehensive review, Hanser (2007)
25. Donatelli, A.A., Sperling, L.H., Thomas, D.A.: Interpenetrating Polymer Networks Based on SBR/PS. 1. Control of Morphology by Level of Cross-Linking. Macromolecules **9**(4), 676 (1976)
26. Donatelli, A.A., Sperling, L.H., Thomas, D.A.: Interpenetrating Polymer Networks Based on SBR/PS. 2. Influence of Synthetic Detail and Morphology on Mechanical Behavior. Macromolecules **9**(4), 671 (1976)
27. Yeo, J.K., Sperling, L.H., Thomas, D.A.: Theoretical prediction of domain sizes in IPN's and related materials. Polymer **24**, 307 (1983)
28. Mathew, A.P., Groeninckx, G., Radhusch, H.J., Michler, G.H., Thomas, S.: Viscoelastic properties of nanostructured natural rubber/polystyrene interpenetrating polymer networks. J. Polym. Sci. Polym. Phys. **41**, 1680 (2003)
29. Chen, C.H., Chen, W.J., Chen, M.H., Li, Y.M.: Simultaneous full-interpenetrating polymer networks of blocked polyurethane and vinyl esterPart I. Synthesis, swelling ratio, thermal properties and morphology. Polymer **41**, 7961 (2000)
30. Hourston, D.J., Schafer, F.U.: Poly(ether urethane)/poly(ethyl methacrylate) interpenetrating polymer networks: Morphology, phase continuity and mechanical properties as a function of composition. Polymer **37**, 3521 (1996)
31. Klempner, D., Berkowski, L.: Encyclopedia of polymer science and engineering, vol. 8. Wiley, New York (1988)
32. Sun, Y-Y., Chen, C-H: Interpenetrating polymer network of blocked polyurethane and phenolic resin. I. synthesis, morphology, and mechanical properties. Polym. Eng. Sci. **51**, 285–293 (2011)
33. Roha, M., Dong, F., Appl, J.: The effects of functional azo initiator on PMMA and polyurethane IPN systems. III. Tear resistance and crack growth of PBD(1,2)-PU/PMMA (50%) blends. Polym. Sci. **45**, 1397–1409 (1992)

34. Valero, M.F.: Polyurethane–Polystyrene simultaneous interpenetrating networks from modified castor oil. J. Elast. Plast. **42**, 255–265 (2010)
35. Valero, M.F., Pudino, J.E., Ramirez, A., Cheng, Z.: Simultaneous interpenetrating polymer networks from pentaerythritol-modified castor oil and polystyrene: Structure-property relationship. J. Am. O Chem. Soc. **86** (4), 383–392 (2009)
36. Bird, S.A., Clary, D., Jajam, K.C., Tippur, H.V., Auad, M.L.: Synthesis and characterization of high performance, transparent interpenetrating polymer networks with polyurethane and poly(methyl methacrylate). Polym. Eng. Sci. (2012). doi:10.1002/pen.23305
37. Dongyan, T., Hong, L., Weimin, C.: Synthesis and application studies of castor oil PU/PMMA IPNs with BaTiO3 fiber nanocomposites. Ferroelectrics **265**, 259 (2002)
38. Lei, Z., Yang, Q., Wu, S., Song, X.: Reinforcement of polyurethane/epoxy interpenetrating network nanocomposites with an organically modified palygorskite. J. Appl. Polym. Sci. **111**, 3150 (2009)
39. Luo, Y.-L., Feng, Q.S., Xu, F.: Preparation and Properties of PVA/PAAm IPN Hydrogels-Copper Nanoparticles Nanocomposites. Adv. Mater. Res. **2397**, 284–286 (2011)
40. Zhan, K., You, H., Liu, W., Lu, J., Lu, P., Dong, J.: Pd nanoparticles encaged in nanoporous interpenetrating polymer networks: A robust recyclable catalyst for Heck reactions. React. Func. Polym **71**, 756 (2011)
41. Peterson, A.M., Kotthapalli H., Pahmathullah, M.A.M., Palmsese, G.R.: Investigation of interpenetrating polymer networks for self-healing applications. Comp. Sci. Tech. **72**(2), 330 (2012)
42. Wang, J., Liu, F., Wei, J.: Enhanced adsorption properties of interpenetrating polymer network hydrogels for heavy metal ion removal. Polym. Bull. **67**(8), 1709 (2011)
43. Chung, C_W., Kang, J.Y., Yoon, I-S., Hwang, H-D., Balakrishnan, P., Cho, H-J., Chung, K-D., Kang, D-H., Kim, D–D.: Interpenetrating polymer network (IPN) scaffolds of sodium hyaluronate and sodium alginate for chondrocyte culture. Colloids Surf. B. **88**, 711 (2011)
44. Myung, D., Waters, D., Wiseman, M., Duhamel, P-E., Noolandi, J., Ta, C.N., Frank, C.W.: Progress in the development of interpenetrating polymer network hydrogels. Polym. Adv. Technol. **19**, 647 (2008)
45. Owens, D.E., Jian, Y., Fang, J-E., Slaughter, B.V., Chen, Y-H., Peppas, N.A.: Thermally responsive swelling properties of polyacrylamide/poly (acrylic acid) interpenetrating polymer network nanoparticles. Macromolecules **40**, 7306 (2007)
46. Chekina, N.A., Pavlyunchenko, V.N., Danilichev, V.F., Ushakov, N.A., Novikov, S.A., Ivanchev, S.S.: A new polymeric silicone hydrogel for medical applications: synthesis and properties. Poym. Adv. Technol. **17**, 872 (2006)
47. Sperling L.H.: Multicomponent polymeric materials. In: Paul, D.R., Sperling, L.H. (eds.) Advances in chemistry 211, American Chemical Society, Washington (1986)
48. Predecki, P.: A method for hydron impregnation of silicone rubber. J. Biomed. Mater. Res. **8**, 487 (1974)
49. Odian, G., Bernstein, B.S.: Monomers improve radiation crosslinking in polymers. Nucleonics **21**, 80 (1963)
50. Sperling, L.H., Florenza, V.A., Manson, J.A.: Interpenetrating polymer networks as piezodialysis membranes. J. Poly, Sci Polym. Lett. Ed. **13**, 713 (1975)
51. Balaji, R., Loileau, S., Guerin, P., Grande, D.: Design of Porous Polymeric Materials from Miscellaneous Macromolecular Architectures: An Overview. Polym. News **29**, 205 (2004)

# Micro and Nanofillers in Rubbers

**Mehmet Kodal and Guralp Ozkoc**

**Abstract** As a most general definition, filler is a finely divided solid that is used to modify the properties of a material in which it is dispersed. From the inception of the rubber industry, fillers have a crucial role in either providing durability and performance or in reducing the price by decreasing the rubber partition in the compound. The fillers used in rubbers can be divided into two main groups such as black and non-black fillers. Besides the conventional micron size fillers, nanofillers recently have gained both academic and industrial importance. In this chapter, it is aimed to introduce the fillers used in rubbers. Their characteristics and their impact on properties of rubbers are discussed by giving examples from the recently published literature. In addition to the conventional ones, the new emerging nanofillers and their added value to the rubbers are given in detail by highlighting some selected studies.

## 1 Introduction

The reinforcement of elastomers by mineral or carbon particulate fillers has been widely investigated in the past. Besides the reinforcement, particulate fillers are added to rubbers for reducing the cost and improving the mechanical properties and for contributing the process-ability. Fillers are generally introduced to rubber matrices between 50 and 120 phr, or sometimes even higher. The size of the filler is probably one of the most important properties for reinforcement. The particulate fillers obtained by grinding minerals or by coarse precipitation are usually non-

M. Kodal · G. Ozkoc (✉)
Department of Chemical Engineering, Kocaeli University, Kocaeli, Turkey
e-mail: guralp.ozkoc@kocaeli.edu.tr

M. Kodal
e-mail: mehmet.kodal@kocaeli.edu.tr

P. M. Visakh et al. (eds.), *Advances in Elastomers I*,
Advanced Structured Materials 11, DOI: 10.1007/978-3-642-20925-3_11,

reinforcing fillers because of their too big size. These fillers can be used in rubbers to decrease the expensive rubber partition in the mixture. They sometimes slightly increase the modulus and hardness, but usually a very significant drop in break properties occurs. Small particle sized materials participated in rubbers as a dry powder define as active with strengthening effect (reinforcing agent). Active or reinforcing fillers strengthen physical and mechanical properties (tensile strength, modulus, tear and abrasion resistance, etc.) by interacting with polymer molecules. Effectiveness of mineral fillers depends on particle shape, particle size (micrometer to nanometer), particle specific surface area and particle-matrix compatibility. The fillers most commonly used in rubbers are carbon black, silica, aluminum trihydrate, aluminum oxide, clay, calcium carbonate, talc, mica, zinc oxide, magnesium oxide, calcium, magnesium silicate and etc. In the rubber industry, carbon blacks are consumed more than three times with respect to other fillers. Kaolin, calcium carbonate and types of silica follow carbon black respectively. Carbon black (size of 10–30 nm) and precipitated silica (size of 30–100 nm) still remain the conventional fillers due to difficulty in dispersing them at nanometer level. However, recently several other nanofillers have pulled attention for reinforcement of rubbers, such as organoclays, nanosilica, carbon nanotubes and nanocalcium carbonate.

In this chapter, it is aimed to introduce the fillers used to reinforce the rubbers. Their characteristics and their impact on properties of rubbers are discussed by giving the examples from recently published literature. Beside the conventional filler systems, the new emerging nanofillers and their added value to the rubbers are given in detail by highlighting some selected studies.

## 2 Fillers Used in Rubbers

### *2.1 Carbon Black*

Carbon black is the most widely used reinforcing agent in the rubber industry. They were discovered by Mote and Mathews et al. in England in 1904. Since 1910, carbon blacks have been used in tire industry as the one of the well-known ingredients of a tire [1].

Approximately 95 % of carbon black produced in the world is being handled in rubber industry and nearly 5 million metric tons of carbon blacks are consumed annually. Carbon blacks are also used as a coloring pigment, printing ink ingredient and adsorbent material. They are obtained from combustion or thermal degradation of natural gas, light and heavy crude oils and aromatic hydrocarbons under controlled oxygen conditions. Carbon blacks with approximately 90–99 % elemental carbon combined with oxygen and hydrogen complexes, i.e., carboxylic, quinonic, lactonic, phenolic, ketonic and others, are located on the surface of blacks. They include agglomerates comprising sphere-shaped particles called

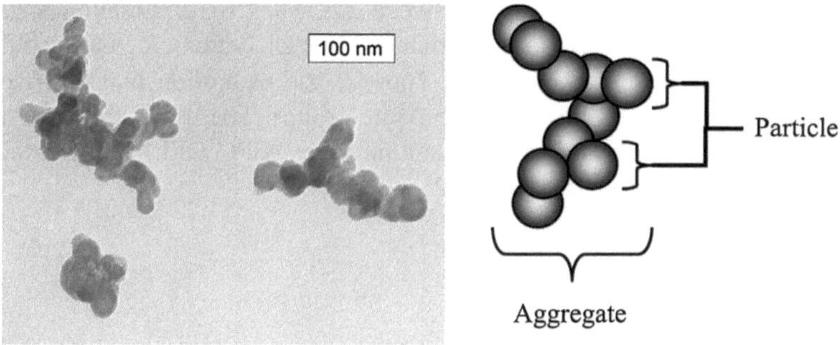


**Fig. 1** (*Left*) *Carbon black* aggregate as viewed by transmission electron microscopy and (*right*) a schematic showing the distinction between carbon black particles and the aggregate [4]

aggregates. Aggregates show differences in size from 100 to 800 A°. Particle size of a carbon black ranges from 10 to 40 nm and can be measured with an electron microscope. Figure 1 shows the distinction between a particle and an aggregate in carbon black. There are a variety of carbon blacks, e.g., super abrasion furnace (SAF), intermediate super abrasion furnace (ISAF), high abrasion furnace (HAF), general purpose furnace (GPF), high modulus furnace (HMF), semi-reinforcing furnace (SRF), fast extrusion furnace (FEF), super conducting furnace (SCF), conductive furnace (CF), easy processing channel (EPC), medium processing channel (MPC), high processing channel (HPC), fine thermal (FT) and medium thermal (MT) [1–5].

### 2.1.1 Furnace Blacks

Furnace blacks with particle size between 18 and 85 nm are obtained by pre-heated hydrocarbons burning in the furnaces containing partial oxygen at 1200–1600 °C. In the first stage, the product is filtered, then it is passed through a cyclone separator and finally it is dried after sweeping volatile gases. Their pH value ranges from 6.5 to 10. Furnace blacks are the basic carbon blacks used in the rubber industry. The most widely known types of the furnace blacks are ISAF, HAF, FEF, GPF and SRF [1, 6].

### 2.1.2 Channel Blacks

The raw material for production process of channel blacks is natural gas. In this process, natural gas is burned on an iron plate under limiting oxygen conditions. Channel blacks are not produced as much as furnace blacks because of low efficiency and leading to environmental pollution. Channel blacks have acidic properties (their pH value is around 5.5). The particle size of them is between 15 and

40 nm. Their oxygen content varies between 2.5 and 3 %, or sometimes rising to 11 %. When the oxygen content of channel blacks increases, the particle size decreases and the pore size increases. These kinds of carbon blacks dissolve in water due to oxygen in the form of carboxylic groups. Therefore, they are suitable to be used in printing inks and water-soluble dyes. EPC and MPC are the most important types of channel blacks [1, 2].

### 2.1.3 Thermal Blacks

Thermal blacks are obtained by thermal degradation of hydrocarbons. They are used in obtaining medium particle sized carbon blacks. PT and MT are examples of this type. Acetylene black is also a kind of thermal black. It is obtained by exothermic decomposition of acetylene under high temperature (800–1000 °C).

Carbon black particles create clusters with connecting each others in the form of chains. They resemble bunch of grapes. The bigger the aggregation, the higher the structure is. Surface activity of a carbon black depends on the groups including oxygen located on the surface of blacks. Channel blacks have higher surface activity due to higher oxygen content on their surface compared to furnace and thermal blacks. Table 1 shows the standard classification system for carbon blacks used in rubber products.

As the particle size gets smaller, the surface area becomes higher, as well. Therefore, the most reinforcing carbon black in rubbers is N110 SAF, although its usage is very limited in rubber industry due to the difficulties in processing and dispersion problems. N220 ISAF and N330 HAF carbon blacks can be used in tire manufacturing, because they provide higher abrasion and tear strength. N550 FEF is developed for extrusion process. N660 GPF and N762 SRF are utilized manufacturing of many rubber goods due to their easy processability. Processability, cost and desired tire properties should take into consideration when selecting the carbon black type [1, 2, 7].

## *2.2 Silica*

Silica with the general formula $SiO_2$ or $SiO_2.xH_2O$ is a naturally occurring mineral such as sand, quartz, quartzite, perlite, tripoli and diatomaceous [2, 8, 9]. They can be amorphous or crystalline forms. Silica is used in many applications such as plastics, rubbers, coatings, cosmetics, adhesives, sealants, inks and toners. Its white color is a significant advantage in competition with carbon blacks [2]. Silica with a specific surface area in the range of 125–250 $m^2/g$ is designated as "reinforcing", while products with a specific surface area in the range of 35–100 $m^2/g$ are "semi-reinforcing".

**Table 1** Carbon black properties [7]

| ASTM classification | Pour density D1513 kg/m$^3$ | DBP no. D2414 $10^{-5}$ m$^3$/kg | DBP No. compressed sample D3493 $10^{-5}$ m$^3$/kg | Iodine adsorption no. D1510 g/kg | Tint strength D3265 |
|---|---|---|---|---|---|
| N110 | 345 | 113 | 97 | 145 | 123 |
| N115 | 345 | 113 | 97 | 160 | 123 |
| N120 | 345 | 114 | 99 | 122 | 129 |
| N121 | 320 | 132 | 111 | 121 | 119 |
| N125 | 370 | 104 | 89 | 117 | 125 |
| N134 | 320 | 127 | 103 | 142 | 131 |
| N135 | 320 | 135 | 117 | 151 | 119 |
| N220 | 355 | 114 | 98 | 121 | 116 |
| N231 | 400 | 92 | 86 | 121 | 120 |
| N234 | 320 | 125 | 102 | 120 | 123 |
| N293 | 380 | 100 | 88 | 145 | 120 |
| N299 | 335 | 124 | 104 | 108 | 113 |
| N326 | 455 | 72 | 68 | 82 | 111 |
| N330 | 380 | 102 | 88 | 82 | 104 |
| N335 | 345 | 110 | 94 | 92 | 110 |
| N339 | 345 | 120 | 99 | 90 | 111 |
| N343 | 320 | 130 | 104 | 92 | 112 |
| N347 | 335 | 124 | 99 | 90 | 105 |
| N351 | 345 | 120 | 95 | 68 | 100 |
| N356 | – | 154 | 112 | 92 | 106 |
| N358 | 305 | 150 | 108 | 84 | 98 |
| N375 | 345 | 114 | 96 | 90 | 114 |
| N539 | 385 | 111 | 81 | 43 | – |
| N550 | 360 | 121 | 85 | 43 | – |
| N582 | – | 180 | 114 | 100 | 67 |
| N630 | 500 | 78 | 62 | 36 | – |
| N642 | – | 64 | 62 | 36 | – |
| N650 | 370 | 122 | 84 | 36 | – |
| N660 | 440 | 90 | 74 | 36 | – |
| N754 | – | 58 | 57 | 24 | – |
| N762 | 515 | 65 | 59 | 27 | – |
| N765 | 370 | 115 | 81 | 31 | – |
| N772 | 520 | 65 | 59 | 30 | – |
| N774 | 490 | 72 | 63 | 29 | – |
| N787 | 440 | 80 | 70 | 30 | – |
| N990 | 640 | 43 | 37 | – | – |
| N991 | 355 | 35 | 37 | – | – |

After carbon blacks, they show the best reinforcing effect in rubbers because of their small particle size and high surface activity. Increasing in surface area results in higher tensile, abrasion and tear strength [1]. Anti blocking effect, promotional effect on adhesion of rubber to brass-coated wires and textiles, improving the thermal and electrical properties of plastics are some of the other advantages of them [2]. Ground mineral silica, precipitated silica and fumed or pyrogenic silica are of three specific types used in rubber industry [3]. Physical properties of various silicas and their functions in rubber compounds are shown in Tables 2 and 3.

Ground silica with 300 mesh in size is used as a cheap heat resistant filler. It does not improve the rate or state of cure [3]. Precipitated silica is obtained from the reaction of sodium silicate and acids such as sulfuric, hydrochloric and carbonic acid. The hydrogen bonds among the silica particles result in the formation of aggregates like carbon black. Fumed silica with 5–10 nm particle size includes less than 2 % combined water. It generally finds application with silicone rubber. Fumed silica has smaller particle size and higher surface area than precipitated silica. It contains fewer surface hydroxyls and lower moisture because of its pyrogenic manufacture [3, 10]. Precipitated silica contains about 6 % adsorbed water. This water quantity is very important in rubber compounds. It can affect the dispersion of filler in the matrix and vulcanization rate. Hydroxyl groups of silica lead to strong filler–filler interactions, which results in poor dispersion in the matrix, delay of the scorch time and reduction of the delta torque [3, 4, 11, 12]. To overcome these problems and to obtain a good dispersion of particles in the matrix, a surface treatment is necessary to reduce interaction force between particles [13–15]. For this purpose, surface modifiers such as fatty acids, silane coupling agents and titanate coupling agents are used [16]. Silane coupling agents (silanes) are the most widely used surface modifiers.

Typically, silanes have a general formula of R–Si(OY)$_3$ with four-functional molecules. R and Y are ethoxy or methoxy groups, which can be hydrolyzed. Besides, there are also non-hydrolyzed organo-functional groups (amino, methacrylate, or vinyl mercapto groups), which are able to provide interaction between polymer and fillers [16]. Three methoxy or ethoxy groups are generally attached to central silicon atom and these groups hydrolyze to form Si–OH groups. These Si–OH groups can be adsorbed to silicate or silica surface. R linked to silicate sometimes constitutes a primary chemical bond with a polymer molecule. Thus, the interaction between polymer matrix and reinforcement are increased [8].

As shown in Eqs. (1), (2), and (3), surface treatment of fillers with silane coupling agents was carried through hydrolysis, condensation and coupling reactions as follows [16, 18]:

*Hydrolysis reaction* (1):

$$RSi(O\text{-}Y)_3 + 3H_2O \longrightarrow RSi(OH)_3 + 3YOH \qquad (1)$$

**Table 2** Physical properties of various silicas [9]

| Characteristics | | Pyrogenic silicas | | Silicas made by wet methods | | |
|---|---|---|---|---|---|---|
| | | Fumed silica | Arc silica | Precipitated silica | Xerogels | Aerogels |
| Spesific BET area | $m^2/g$ | 50–600 | 25–300 | 30–800 | 250–1,000 | 250–400 |
| Size primary particles | nm | 5–50 | 5–500 | 5–100 | 3–20 | 3–20 |
| Size aggregation/agglomeration | μm | [a] | 2–15 | 1–40 | 1–20 | 1–15 |
| Density | $g/cm^3$ | 2.2 | 2.2 | 1.9–2.1 | 2.0 | 2.0 |
| Volume | ml/100 g | 1,000–2,000 | 500–1,000 | 200–2,000 | 100–200 | 800–2,000 |
| Mean pore diameter | nm | Non porous till ca. 300 $m^2/g$ | Non porous | 30 | 2–20 | 25 |
| Pore diameter distribution | | [a] | [a] | Very broad | Narrow | Narrow |
| Shape of interior surface | | 0 | 0 | Poor | Very much | Much |
| Aggregation and agglomeration structure | | Chain-like agglomeration (open surface) | Dense spherical aggregates/particle non-agglomerated | Slightly aggregated nearly spherical particles | Highly porous agglomerated particles | Macroporous agglomerated particles |

[a] not applicable

**Table 3** Forms and properties of silica used in rubber compounding [17]

| | Primary size, μm | Function in rubber |
|---|---|---|
| *Natural (crystalline)* | | |
| Ground quartz | 1–10 | Extending |
| Diatomite | 1–5 | Processing; extending |
| Neuberg silica | 1–5 | Extending |
| *Synthetic (amorphous)* | | |
| Fumed | 0.005–0.02 | Reinforcing |
| Precipitated | 0.01–0.03 | Reinforcing |
| Precipitated | 0.04 | Semi-reinforcing |
| Precipitated | 0.08 | Processing; color |
| Ferro-silicon by-product | 0.10 | Extending |

*Condensation reaction* (2):

(2)

*Coupling reaction* (3):

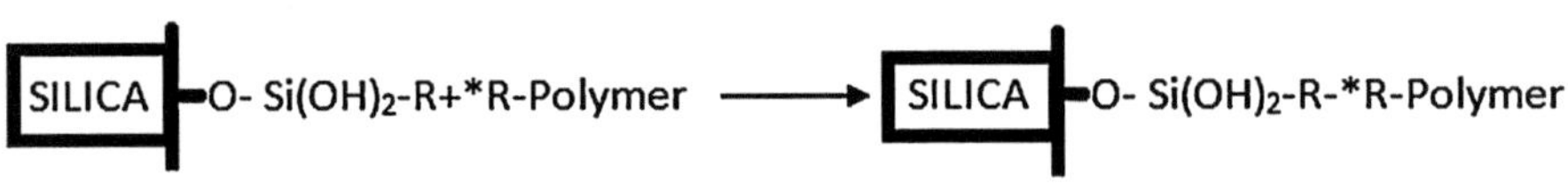


(3)

Silanol groups formed with hydrolysis reaction react with hydroxyl groups that are bonded to filler surface with siloxane bonds during condensation reactions. As a result of these reactions, silane-coupling agents modify the interface by creating a bond between components [3, 16, 19, 20]. Better tear strength, tensile strength, abrasion resistance, and low hysteresis can be obtained by using silane-coupling agents [4].

## *2.3 Clays*

Clays are composed of plate/layered-shaped hydrous silicates of Al, Mg, Fe and other less abundant elements. The individual platelets (layers) are aggregated together to form clay particles. The forces binding them together are mainly Van der Waals forces. Isomeric substitution (for example tetrahedral $Si^{+4}$ by $Al^{+3}$ or octahedral $Al^{+3}$ by $Mg^{+2}$ or $Fe^{+2}$) within the layers generates negative charges that are counterbalanced by alkali and alkaline earth cations (typically $Na^{+1}$ or $Ca^{+2}$) situated inside the galleries (gaps between the layers). This type of layered silicate

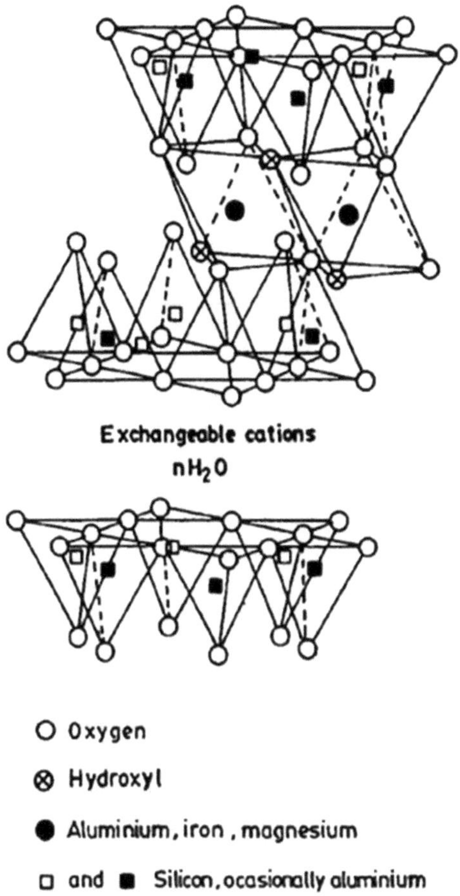


**Fig. 2** Structure of tetrahedral-octahedral-tetrahedral layered structure of smectite clay [21]

is characterized by a moderate surface charge known as the cation exchange capacity (CEC). Details regarding the structure and chemical formula of the layered silicates are provided in Fig. 2 [21].

Particle sizes of clay minerals generally range from 0.5 to 5 μm [3, 22]. Surface treatment of clays can be carried out to improve the interfacial interaction with polymers. For this purpose, stearic acid, fatty acids and a cationic scattering agent (i.e., quaternary ammonium compound) can be used [13]. Kaolin which can be found in four different forms called hard, soft, calcined and treated is the most well known clay mineral. Because of their structure, clays have the ability to absorb water. Therefore, clays are always moist. Clays are in plastic form when they are swollen with water.

There are five main groups of clays; the first is the kaolinite group containing kaolinite and halloysite; the second is the illite groups containing illite; the third is the smectite group containing montmorillonite and hectorite; the

**Table 4** Characteristic of some clay minerals [22]

| | Kaolinite | Attapulgite | Montmorillonite | Vermiculite |
|---|---|---|---|---|
| Chemical composition | $Al_4(Si_4O_{10})(OH)_8$ | $K_{1-1.5}Al_4(Si_{7-6.5}$ $Al_{1-1.5}O_{25})(OH)_4$ | $Al_2Si_4O_{10}$ $(OH)_2XH_2O$ | $Mg_3Si_4O_{10}(OH)_2$ $XH_2O$ |
| Structure | Triclinic or monoclinic | Monoclinic | Monoclinic | Monoclinic |
| Specific density g/cm$^3$ | 2.65 | 2.4 | 2–3 | 2.3 |
| Refractive index | 1.56 | 1.57 | 1.48–1.64 | 1.52 |
| Mohs' hardness | 2–2.5 | 1–2 | 1.2 | 1.5 |

fourth one is the palygorskite group including sepiolite and attapulgite and last one is the vermiculite. In these groups there are approximately thirty different types of clays [3, 22]. Table 4 shows the characteristic of some selected clay minerals.

### 2.3.1 Kaolinite

The other known names of kaolinite are china clay and porcelain earth. Kaolins with hexagonal structure are hydrated aluminum silicates such as dickite, nacrite and halloysite. Kaolins comprise to two sheet; alumina octahedral sheet and silica tetrahedral sheet (1:1 non-expandable structure). This alumina tetrahedral sheet chemically bonds on one side to the silica tetrahedral sheet. The molecular formula of kaolins is $Al_2O_3.SiO_2.2H_2O$ including approximately 46.3 % $SiO_2$, 39.8 % $Al_2O_3$ and 13.9 % $H_2O$. Kaolins are highly resistant to chemicals usually have high degree of whiteness. They do not conduct electricity [2, 8]. If the finished product has high modulus and at least 75 % of all particles are less than 2 μm, it is the hard kaolin. However, soft kaolins have lower modulus, tensile strength and abrasion resistance and are coarse grained [3, 8]. Calcined kaolins are heat treated to remove water. They are harder compared to soft and hard kaolin. Their electrical resistance is too good to use in cable insulating material. Kaolins can be modified with silane coupling agents to obtain a good dispersion in polymers and to improve interfacial interactions. Silane treated clays can improve the mechanical properties of compounds.

### 2.3.2 Montmorillonite

The clay known as montmorillonite (MMT), a member of smectite clays, consists of platelets with an inner octahedral layer sandwiched between two silicate tetrahedral layers as illustrated in Fig. 3 [23]. The octahedral layer can be an aluminum oxide sheet where the aluminum atoms have been replaced with magnesium by leaving a negative charge due to the difference in valences of Al

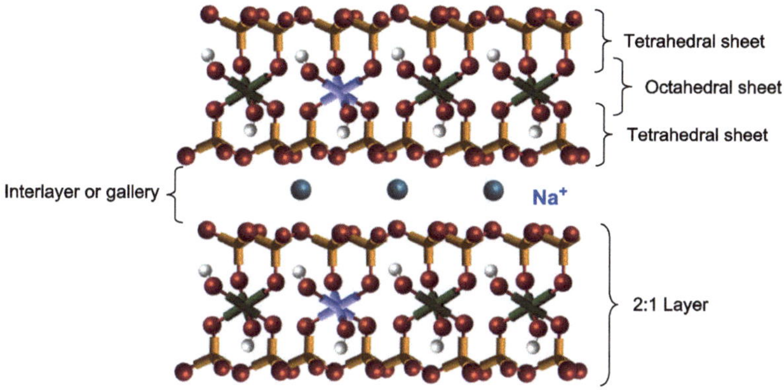


**Fig. 3** The structure of MMT [23]

and Mg distributed within the platelets that are balanced by positive counter ions, typically sodium ions, located between the platelets or in the galleries as shown in Fig. 3. The clay exists in the nature as stacks of many platelets. Hydration of the sodium ions results in the galleries to expand (swelling of clay). In the presence of enough water, the platelets can be fully separated from each other. The sodium ions can be exchanged with other cations. The extent of the negative charge of the clay is characterized by the cation exchange capacity, i.e., CEC. The layer thickness is around 1 nm, and the lateral dimensions of these layers may vary from 30 nm to several microns. The distance between the layers varies from 0.98 to 1.8 nm while it is dry [21]. This distance increases while it is wet. If it is possible to disperse these layers in a polymer matrix, one could obtain a nanocomposite meaning that the filler and the matrix have a nanoscaled interaction. The difficulty resisting to this is the high attraction forces between the layers of MMT.

The distance between the organoclay layers is called “basal spacing” or “d-spacing”. This is very important to produce a nanocomposite, because, the dispersion of clay layers inside a polymer matrix can be started by the insertion of polymer chains into the galleries of the clay. However, due to the polarity difference between the polymer matrix and the clay surface, the diffusion of the polymer chains into these galleries is not possible. It is possible to increase the basal spacing and at the same time organophilically modify the surface is possible with ion exchanging with $Na^+$ cation and alkyl ammonium cations [21]. The schematic of the ion exchange is shown in Fig. 4.

The increasing gallery gap helps the polymer chains easily penetrate into the basal spacing. Thus, wettability of the filler by matrix is improved and clay dispersion is increased by formation of an interface. Due to hydrophilic character of $Na^+$ or $Ca^{+2}$ MMT, MMTs are incompatible with hydrophobic rubber matrices that results in poor mechanical properties. In such a case, the chemical structure of the organic modifier plays an important role [24].

Three different types of nanocomposites can be obtained by the use of MMT. If the MMT layers are still in the form of stacks and the average size is few microns,

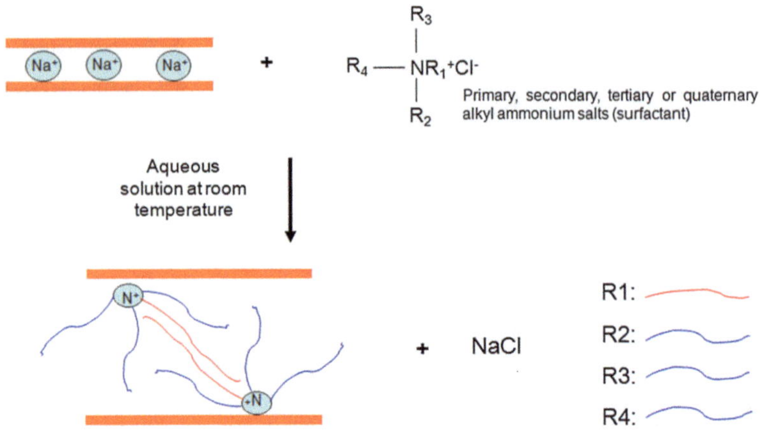


**Fig. 4** Surface modification of montmorillonite with ion exchange

then the resulting structure is called conventional composite. If the interlayer distance (basal spacing) is increased due to the diffusion of polymer chains but not separated from each other, the resulting structure is called intercalated nanocomposite. Intercalated nanocomposites can be characterized from XRD patterns indicating a shift to lower angles in basal spicing peak of the MMT. If all the platelets are separated from each other and dispersed in the polymer matrix, the resulting composite structure is called exfoliated (delaminated) nanocomposite. Exfoliated nanocomposites can be characterized by disappearing of basal spacing peak in XRD pattern of MMT. These possible structures are shown in Fig. 5.

Nanocomposites can be prepared by applying different techniques:

**In situ polymerization technique**: In this technique, the MMT is swollen in the monomer. By the help of shear mixing, the swollen layers are separated from each other, and simultaneously polymerization is carried out. A schematic representation of this technique is shown in Fig. 6.

**Solution intercalation process**: In this method, MMT in the dilute polymer solution is swollen by solvent. Polymer penetrates between the galleries of swollen MMT. At this stage shearing helps the dispersion of the clay layers from each other in solution phase. At the final stage the solvent is evaporated to solidify the nanocomposite (see Fig. 7).

**Melt intercalation**: Molten polymer is mixed with MMT under high shear stresses and shear rates. The resulting structure of the nanocomposites depends on the intensity of the shear, process parameters of the shearing device (speed, temperature and residence time) and the compatibility between polymer and the MMT (Fig. 8).

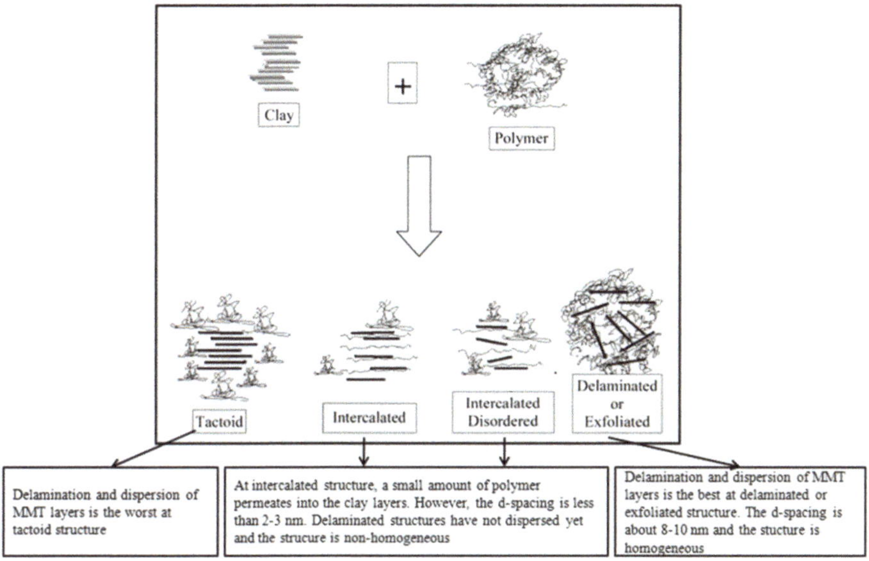


**Fig. 5** Schematic illustration of terminology used to describe nanocomposites formed from organoclays [25]

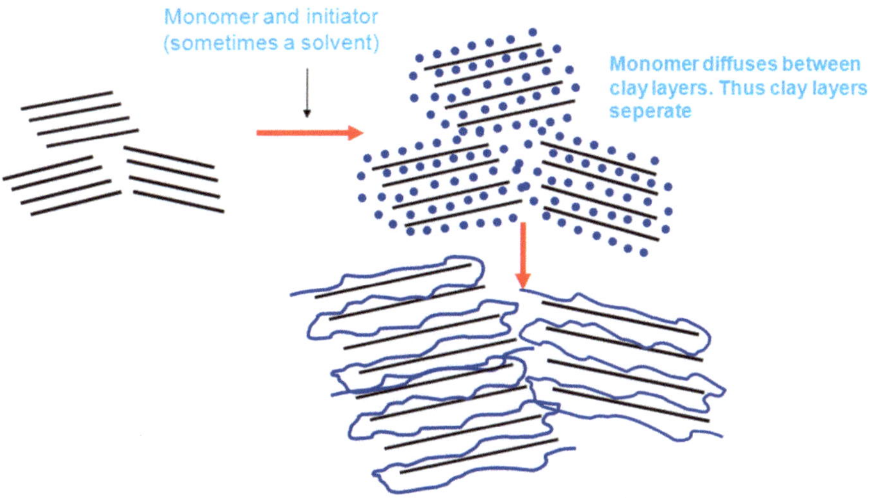


**Fig. 6** The mechanism of in situ polymerization technique

### 2.3.3 Other Silicates

**Sodium aluminum silicate**: They are derived from kaolin. As compared to kaolin, they are fine-grained. Therefore, they have better reinforcing effect in rubber compounds than kaolin. Sodium aluminum silicate with easier mixable property

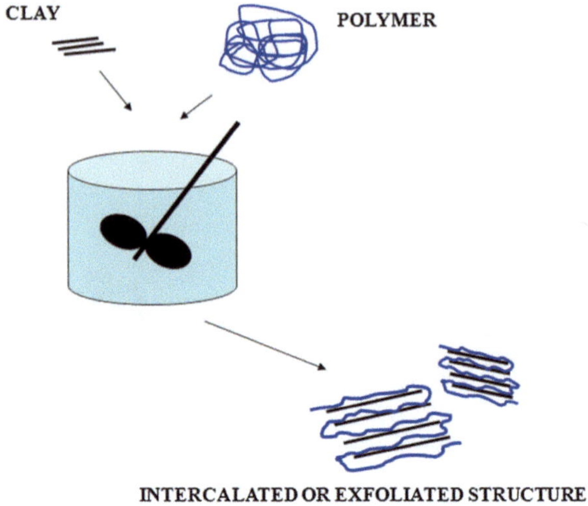


**Fig. 7** The mechanism of solvent-assisted process (solution intercalation) technique

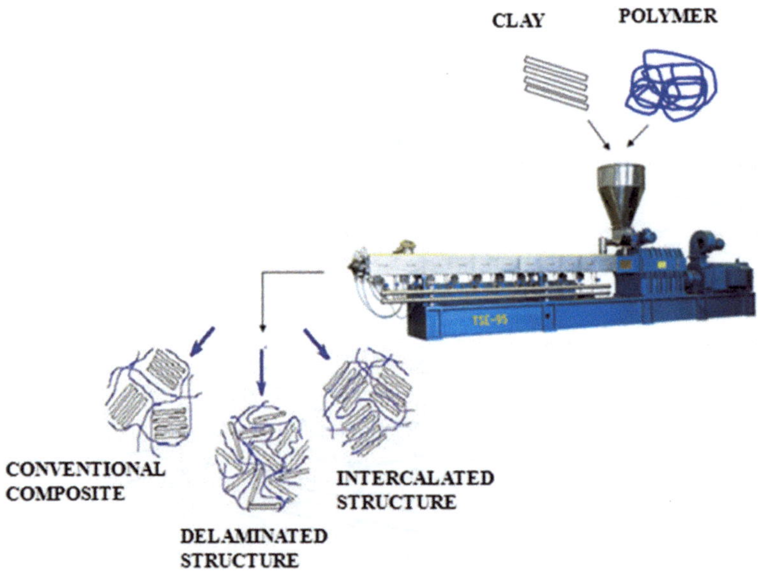


**Fig. 8** Melt blending technique

gives a good tear strength and elasticity properties to rubbers [1]. They are less reactive or reinforcing than calcium silicate. Vulcanization rate may decrease with using sodium aluminum silicate. They generally use in high electrical resistivity applications in spite of more expensive compared to silica and kaolin [3, 26].

**Calcium silicate**: Calcium silicate known as hydrated calcium silicate or hydrous calcium silicate with a relative density of 2.26–2.40 and the molecular formula of $CaO.3SiO_2.xH_2O$ is obtained by mixing sodium silicate solution with calcium chloride solution [1, 26]. It may decrease the vulcanization rate. Therefore, accelerators may be used in order to increase vulcanization rate [3, 26].

**Aluminum potassium silicate (Mica)**: Muscovite $[KAl_2(AlSi_3O_{10})(OH)_2]$, phlogopite $[KMg_3(AlSi_3O_{10})(OH)_2]$ and biotite $[K(Mg,Fe)_3(AlSi_3O_{10})(OH)_2]$ are the three classes which have found commercial use [22]. They have a monoclinic structure, a density of nearly 2.8 g/cm$^3$, Mohs hardness of 2.5–3, refractive index of about 1.56 and oil absorption values of 48–500 g per 100 g powder [8, 22]. It is an abundant mineral throughout the world. It provides polymers with excellent electrical insulation, thermal stability, water resistance, chemical resistance, flexibility and high-spin off properties [27, 28]. To enhance the reinforcing effect of mica in polymers, surface treatment and ultrasonic delamination are used [22].

**Magnesium silicate (Talc)**: Talc with the molecular formula $3MgO.4SiO_2.H_2O$ is the major constituent of rocks known as soapstone and steatite. Talc is the softest industrial mineral with a Mohs hardness of 1, which provides talc-filled polymers with easy delamination and low abrasiveness. Talc has a specific density of 2.5–2.8 g/cm$^3$, refractive index of 1.54–1.57, particle size of 1–8 μm, high aspect ratio and platy structure [8, 22]. This platy structure plays an important role in reduction permeability of fluids and gas due to diffusion path is so complicated [22, 29]. Besides, talc is used in wire and cable because it is electrically non-conductive. Thermal performance of elastomers can be improved by the partial replacement of carbon black with talc. Compound and finished product costs are reduced with this approach [29, 30]. Talc might improve the dispersion of both carbon black and precipitated silica without influence the cure characteristics of thermoset rubbers and this reduces the mixing time and cost of finished products [31].

## *2.4 Calcium Carbonate*

Calcium carbonate is abundant, largely inert, low cost, white filler with cubic, blocked-shaped or irregular particles of very low aspect ratio [2, 22]. Calcium carbonate in the form of ground chalk, limestone or marble is the most common deposits formed in sedimentary rocks. It consists of following properties: 98–99.5 % $CaCO_3$, up to 0.5 % $MgCO_3$, up to 0.2 % $Fe_2O_3$, up to 1.0 % Al–silicate (colloidal) and up to 0.3 % moisture [2, 8]. Wet or dry natural limestone with particle sizes between 700 and 5,000 nm and precipitated calcium carbonate with average particle sizes as low as 40 nm are the two types of calcium carbonate used in the rubber industry [32]. Calcium carbonates range from 20 to 300 phr in rubber compounds find applications in electrical wire and cable insulation at rubber industry due to its low moisture content and natural insulating properties

[33]. Loading rate and average particle size of calcium carbonate affect the modulus, tensile strength, hardness, mooney viscosity, tear strength and abrasion resistance. Precipitated calcium carbonates have better effects on mechanical properties due to much higher surface areas. To improve dispersability, to control water absorption and to promote better wetting of the calcium carbonate by rubbers, a surface treatment with stearic acid can be applied [32, 34].

## *2.5 Polyhedral Oligomeric Silsesquioxane*

The polyhedral oligomeric silsesquioxane (POSS) nanoparticles have a high potential since they have flexible chemical and physical hybrid properties and their lower cost. POSS are usually produced by hydrolytic condensation of trifunctional monomers of $RSiX_3$, where X is a highly reactive substituent, such as Cl or alkoxy [35–37]. POSS has an empirical formula of $(RSiO_{1.5})_n$ (n 4, generally it is 8), where R is hydrogen or any alkyl, alkylene, aryl, arylene, or organo-functional derivative of alkyl, alkylene, aryl, arylene groups. It has a diameter of approximately 1–3 nm. It is a reinforcing Nanofiller [36, 38]. There are potentially an unlimited number of POSS variants due to variety of R groups. The structure of POSS contains 8 silicon atoms linked together with oxygen atoms as shown in Fig. 9 [39].

Organic substituents of POSS on its outer surface make POSS nanostructure compatible with polymers, biological systems, or surfaces. Besides, these groups can be specially designed to be nonreactive or reactive. They are scentless and

**Fig. 9** Molecular structure of POSS [39]

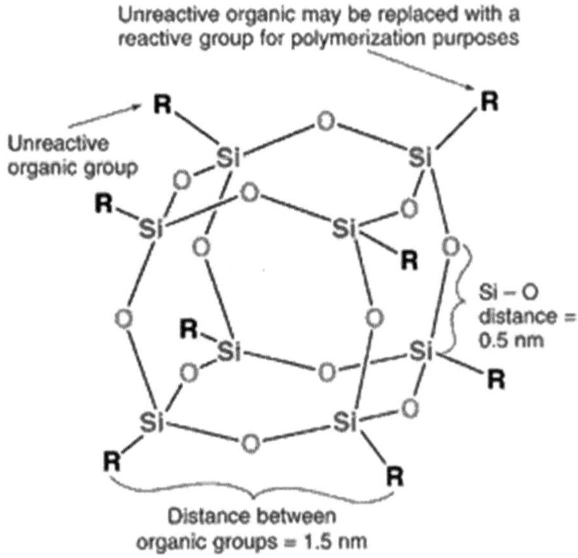

environmentally friendly chemicals since they do not release any volatile organic components [36]. Unlike the other fillers, POSS dissolves in polymers if there is a proper compatibility between POSS and polymer. Lots of POSS can be dissolved in common solvents such as THF, acetone, methanol and etc. They do not cause an abrasion problem because Mohs hardness of POSS is approximately 1 like talc. Most POSS are solid white powders that are micron sized agglomerates typically in the 1–100 $\mu$ range and some are colorless liquids. The nanoscopic dimensions of the POSS are only obtained once it is dispersed in the polymers [39].

## 2.6 Miscellaneous Inorganic Fillers

### 2.6.1 Barium Sulfate

Barium sulfate known as baryte with a specific gravity of 4.5 g/cm$^3$, a refractive index of 1.636, Mohs hardness of 2.5–3.5 and orthorhombic structure can be obtained from naturally occurring heavy spar or from the precipitation of barium salts [3, 8, 22]. It is used in lots of applications such as foams, floor coverings, protection against high energy radiation, acid resistant compounds, sound deadening, X-ray detectable materials due to its high apparent density, unique chemical resistance, inertness and high absorptivity of light [8, 10, 22]. It does not affect the cure rate in rubber compounds [26].

### 2.6.2 Calcium Sulfate

Anhydrate with the molecular formula $CaSO_4$ and Gypsum with the molecular formula $CaSO_4.2H_2O$ are the others sulfate minerals used in rubber compounds. Gypsum with the monoclinic structure, a specific gravity of 2.3 g/cm$^3$, a refractive index of 1.520, Mohs hardness of 2 is found to a high concentration in sea water. Anhydrate calcium sulfate has orthorhombic structure, a specific density of 3 g/cm$^3$, a refractive index of 1.570 and Mohs hardness of 3–3.5 [22]. The dihydrate is the most widely used form for polymer applications [40].

### 2.6.3 Magnesium Carbonate

The most common magnesium carbonate is magnesite ($MgCO_3$). It has a cryptocrystalline structure. Its specific gravity is 3–3.5 g/cm$^3$, refractive index is 1.70 and Mohs hardness is 3.5–5. It is a fine reinforcing powder which gives good physical properties [3, 22]. In polar elastomers such as nitrile butadiene rubber (NBR), sulfur is treated with magnesium carbonate to improve dispersion of sulfur on matrix [4].

### 2.6.4 Aluminum Hydroxide

Aluminum hydroxide or aluminum trihydrate with the chemical formula $Al(OH)_3$ has a monoclinic structure, with a specific gravity of 2.4 g/cm$^3$, a refractive index of 1.570 and Mohs hardness of 2.5–3.5. It is a nonflammable white powder and insoluble in water [8, 22]. Aluminum hydroxide is obtained from bauxite an aluminum ore. It is used mainly to impart flame-resistant properties to rubber [34]. Its use is restricted to plasticized PVC, unsaturated polyesters, epoxy resins, rubbers and polyethylene, because it decomposes from about 200 °C; therefore, it can only be used when the processing temperature of the polymer is below this value [41].

### 2.6.5 Metal Oxides

Titanium dioxide ($TiO_2$) is used as a filler for white and colored rubber articles. It has the ability of scattering light provides rubber with high whiteness and opacity. Anatase and rutile are the two crystalline forms of titanium dioxide used in rubber compounding. The anatase form is softer. The other form is harder than anatase and chalk resistant [33, 42].

Zinc oxide the first non-black filler used for reinforcement of rubber compounds gives high tensile strength, resilience and moderate hardness. Their main role in rubber compounding has been as activators in sulfur cure systems in the several decades [3, 33].

## *2.7 Carbon Nanotubes*

Carbon nanotubes (CNTs) with unique structural and electronic properties were discovered by Iijima in 1991 [2, 43]. Composite materials, battery electrode materials, field emitters, nanoelectronics, nanoscale sensors are some of the potential application fields of CNTs [44]. The structure of CNTs consists of rolled graphitic layers. They have small size of nm in diameter and μm length, high aspect ratio (up to $10^4$), specific surface area, Young's modulus (103 GPa) close to that of diamond, tensile strength about 50 GPa and density of about 1.3 g/cm$^3$ [45]. Single wall carbon nanotubes (SWNTs), single hexagonal layer of carbon atoms (a graphene sheet) and multi wall carbon nanotube (MWCNT), a stack of graphene sheets rolled up into concentric cylinders are the two main structures of carbon nanotubes [2]. As shown in Fig. 10, there are three types of SWNTs with differing chirality.

One dimensional unit cell has a circumference given by the chiral vector $C = na + mb$, where n and m are integers, and a and b are unit vectors of the hexagonal lattice [2]. These different chiralities determine the morphological

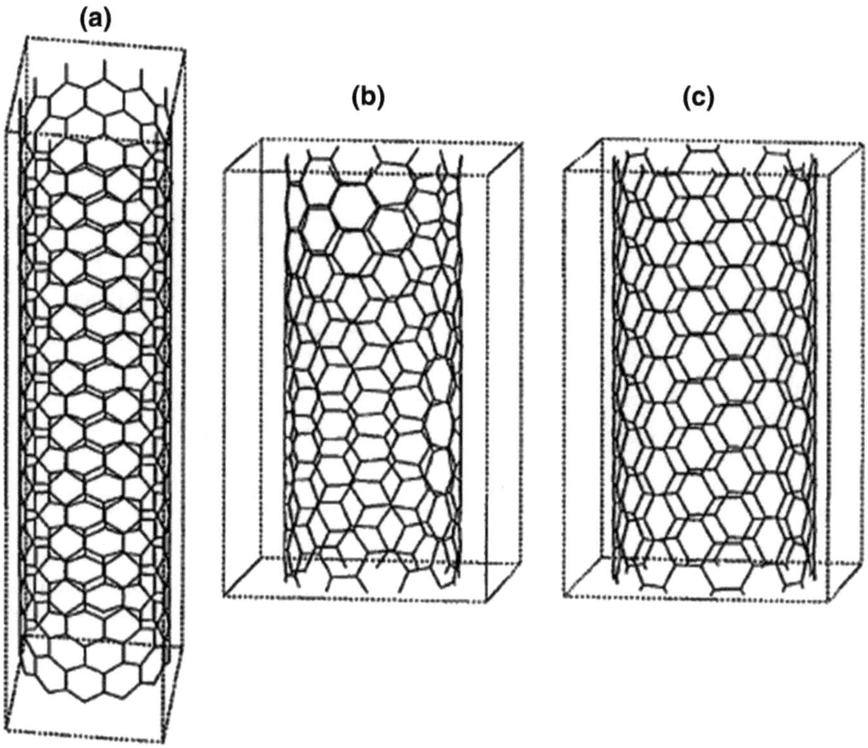


**Fig. 10** Lattice structure for **a** ideal zigzag (n, m = 12, 0) tube, **b** the ideal chiral (8, 6) tube, and **c** the ideal armchair (8, 8) tube [46]

properties. For instance, the electronic properties of an armchair nanotube are metallic; however, the electronic properties of zigzag and chiral nanotubes are semiconducting.

Multi-wall carbon nanotubes (MWCNTs) consisting of a stack of graphene sheets rolled up into concentric tubes have a diameter of 2–25 nm, whereas that of is 1–2 nm as shown in Fig. 11. The spacing between the layers of graphene in MWCNTs is 0.36 nm [2, 47].

One of the disadvantages of CNTs is the exhibition of weak interactions with the polymer matrix. To overcome this problem, CNTs can be functionalized during the purification process by exposing them to strong oxidizing acids to introduce hydroxyls and carboxyl groups on the CNT surface. Thus, compatibility can be obtained between CNT and polymer matrix [48, 49]. The acid treatment also generated ether-type oxygen groups between the graphitic layers. For example, Kuanping et al. attached the synthetic triptycene orthoquinone (TOQ) onto SWNTs surface as shown in Fig. 12.

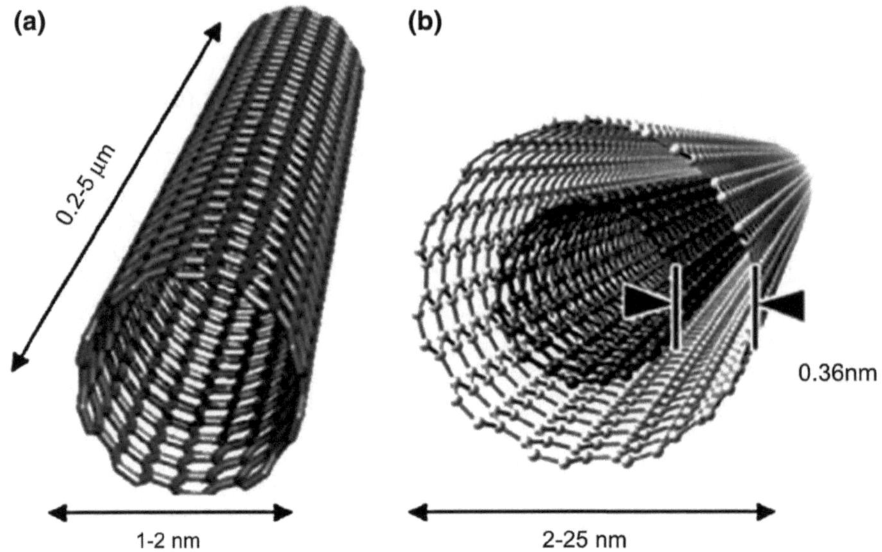


**Fig. 11** Conceptual diagram of single-walled carbon nanotube (SWCNT) **a** and multiwalled carbon nanotube (MWCNT), **b** delivery systems showing typical dimensions of length, width, and separation distance between graphene layers in MWCNTs [47]

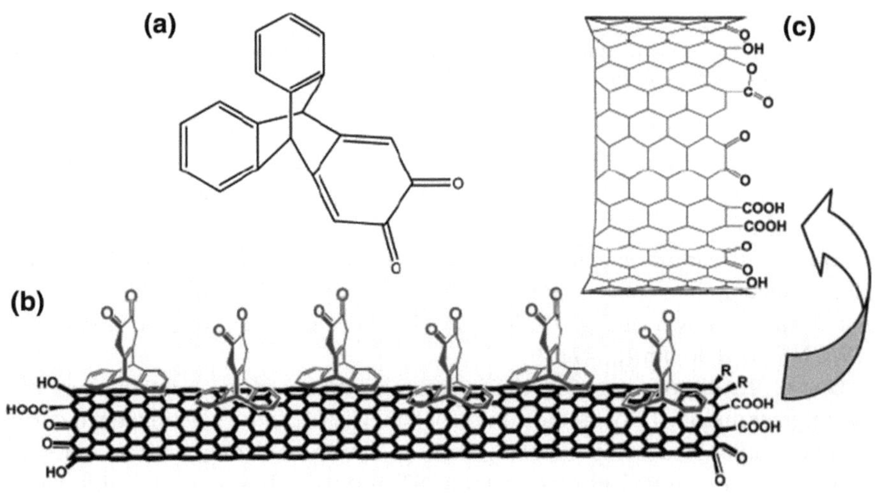


**Fig. 12** Structure of TOQ (**a**), Schematic illustration of attachment of TOQ onto carbon nanotubes (**b**), and enlarged tube ends with oxygen-containing moieties (**c**) [50]

# 3 Recent Studies Related to Micro Filler-Rubber Systems

Mechanical properties play a major role in determining the processing methods and application areas of a rubber compound. There are many research papers and several reviews recently published on the mechanical properties of filled rubber systems. The mechanical properties of particulate-polymer composites are affected by particle size, particle-matrix interface adhesion and particle loading level [51].

When the particle size gets smaller, the surface area/volume becomes higher. As mentioned before, carbon blacks are the most widely used fillers in rubber industry. Surface area of a carbon black is one of the factors affecting the mechanical properties of carbon black-rubber composites [52, 53]. Li et al. [52] stated that when the filler loading is the same, the larger surface area/volume has more irregularity and hence more bound rubber that results in loss of segmental mobility of polymer chains and consequently decreases the flexibility of the rubber matrix. They reported that elongation at break decreased with increasing surface area of carbon black as shown in Table 5 [52].

A high specific surface area can extremely increase the mechanical properties such as tensile strength. Saatchi and Shojaei [54] used XE2B, a carbon black whose mean particle size was 30 nm, with SBR. They found that XE2B led to extremely high tensile strength of SBR/XE2B that could create higher interfacial area. This behavior could be explained by taking into consideration of strong interfacial interaction between the elastomer macromolecules and XE2B surfaces through both physical and chemical interactions towards elastomer macromolecules [54–57]. Vinod et al. [58] studied the gradual replacement of high abrasion carbon black, general purpose furnace black, acetylene black, china clay and precipitated silica by aluminum powder. They found that some mechanical properties like tensile strength, tear strength, rebound resilience, hardness and compression set, etc. were comparatively better for these composites [58]. The effects of alumina particle size (average diameter: 0.5, 5, 10 and 30 μm) on the mechanical properties of silicone rubber studied by Zhou et al. [59]. They remarked that when $V_S$ (relative volume fraction) was 0.3, the tensile strength of large particles filled rubber was lower and the elongation at break values were higher than that of rubber filled with small particles due to stronger interactions between small particles and matrix [59–61]. Similar results were obtained elsewhere [62]. The effect of particle on the elongation at break, tear strength and

**Table 5** Mechanical properties of EPDM/carbon black composites [52]

| | SCB | N770 | N550 | N330 | N472 |
|---|---|---|---|---|---|
| Hardness (shore A) | 57 | 52 | 60 | 59 | 79 |
| 300 % modulus (MPa) | 4.67 | 2.91 | 5.17 | 5.23 | 15.12 |
| Tensile strength (MPa) | 8.44 | 12.12 | 15.04 | 18.38 | 17.76 |
| Elongation at break (%) | 510 | 640 | 526 | 571 | 345 |
| Permanent set (%) | 12 | 17 | 15 | 15 | 17 |

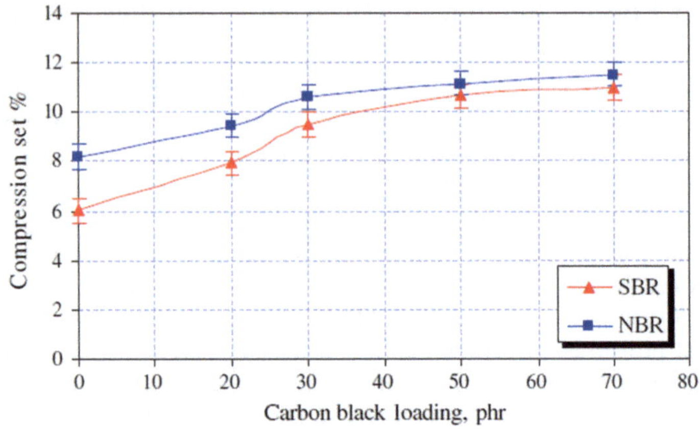


**Fig. 13** The effect of carbon black loading on the compression set of SBR and NBR filled compounds [66]

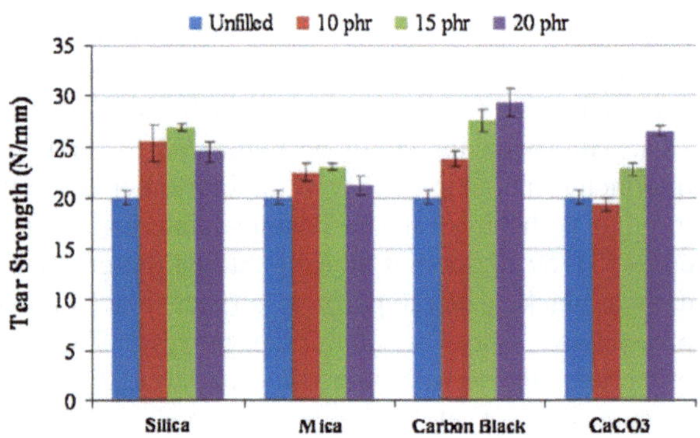


**Fig. 14** The effect of different types of fillers and filler loadings on the tear strength of the carboxylated acrylonitrile-butadine (XNBR) latex films [67]

tensile strength of $Mg(OH)_2$/EPDM composite was also studied [63]. Results showed that elongation at break, tear strength and tensile strength increased with increasing particle size. Adversely, tear strength, elongation at break, tensile strength and modulus of epoxidized natural rubber composites decreased with increasing particle size of oil palm wood flour (OPWF). Smaller particle size of OPWF gave better mechanical properties [64].

Filler loading is the other factor affecting the mechanical properties of rubbers. Mostafa et al. [65] used five different compositions in SBR and NBR as 0, 20, 30, 50 and 70 phr of carbon black (CB). It was found that the increase in carbon black loading gave rise to the compression set value as shown in Fig. 13. At similar CB loading, CB filled NBR vulcanizates showed a slightly higher compression set value than CB filled SBR vulcanizates because of better crosslinking density of NBR filled compounds than SBR filled compounds [66].

Figure 14 showed the effect of different types of fillers and filler loadings on the tear strength of the carboxylated acrylonitrile-butadine (XNBR) latex films [67].

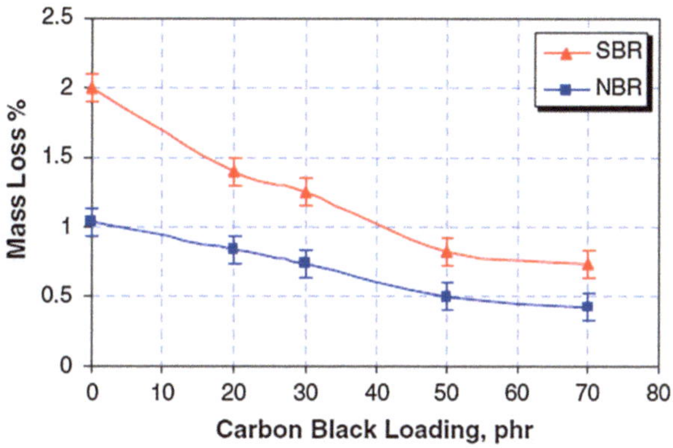


**Fig. 15** The percentage of mass loss of filled SBR and NBR vulcanizates with CB content under constant load of 20 N [69]

With the addition of various types of fillers in the XNBR latex films, the value of tear strength increased. Researchers also stated that silica and mica fillers showed the reinforcing effect at lower filler loadings, whereas CB showed the reinforcing effect at higher filler loadings. Besides, $CaCO_3$ fillers did not increase the mechanical properties.

Wang et al. [27] studied loading effect of natural and synthesized micas on the mechanical properties of EPDM. Both EPDM/mica composites showed much higher 300 %-tensile stress and tensile strength than unfilled EPDM. Natural mica filled ones had much higher elongation at break values than unfilled EPDM. However, synthesized micas had no effect on the elongation at break of EPDM. Similar results about tensile strength were obtained by Daniele et al. [68]. They found that tensile strength of NR vulcanizates increased as the amount of mica increased. The effect of particle loading (0, 20, 30, 50 and 70 phr) on the abrasion resistance expressed as mass loss of CB/SBR and CB/NBR composites is given in Fig. 15. In this research, filled SBR and NBR vulcanizates were found to have a high abrasion resistance compared with unfilled vulcanizates. Besides, abrasion resistance tended to increase when CB content increased up to 50 phr due to crosslinking density came close to maximum limit [69].

Particle/filler interfacial adhesion is the other factor that influences the mechanical properties of filled rubbers. Mechanical properties of silane treated (3-mercaptopropyl methyldimethoxysilane-MrPDMS and 3-mercaptopropyl trimethoxysilane-MrPTMS) silica particle filled polyisoprene rubber composites were studied by Nakamura et al. [70]. They found that stress at the same strain increased with the silane treatment and it was higher in the dialkoxy structure than in the trialkoxy structure above 300 % strain. There was no significant influence of the loading amount on the stress for the trialkoxy silane structure. Silica filled SBR vulcanizates with different rubber/filler interactions were examined by Suzuki et al. [71]. In this research, mono-functional decyltrimethoxy and bi-functional bis-(triethoxysilylpropyl)-tetrasulfide were used as the coupling agents. The combination of electron spin resonance results and stress–strain data revealed that at a given strain, the tensile stress increased with increasing the interfacial

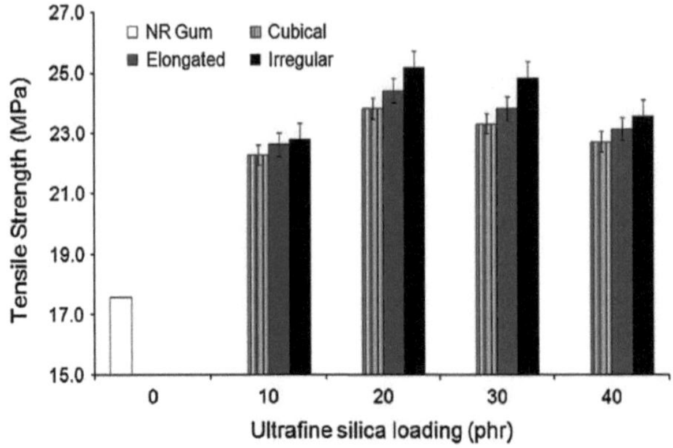


**Fig. 16** Tensile strength of ultrafine silica filled natural rubber vulcanizates [74]

interactions between rubber molecules and silica surface [71]. Effect of a coupling agent, bis-(3-triethoxysilylpropyl) tetrasulfide (Si69), on the properties of hemp-hurd-powder (HP) filled SBR was studied by Wang et al. [72]. They concluded that the addition of Si69 improved the interfacial interaction between HP and the SBR matrix, which led to an increase in the mechanical properties.

Cross-linking density, sulfur content and particle shape are the other factors influencing the mechanical properties of rubber compounds. In a chapter, it was reported that modulus and tensile strength of N472 CB filled EPDM composites exhibited higher values than the other carbon black filled composites because of the highest sulfur content and cross-linking density of N472 [52]. Park and Choe [73] studied the influence of silane coupling agents treated silica on crosslink density of rubber composites. $\gamma$-aminopropyl triethoxysilane (APS), $\gamma$-chloropropyl trimethoxysilane (CPS), and $\gamma$-methacryloxypropyl trimethoxysilane (MPS) were used as a silane coupling agent. They stated that the surface characteristics of silicas after silane treatments lead to an increase of the cross-linking density of the silica/rubber composites compared to that of untreated silica/rubber composites, resulting an increased tearing energy of composites. Idrus et al. explained the effect of different shapes of silica as fillers in natural rubber compounds. Three shapes of filler (cubical, elongated and irregular) were used [74]. As seen in Fig. 16, the irregular shaped silica showed the highest tensile strength. Irregular shaped silica also had highest elongation at break, tensile modulus and hardness values, followed by the elongated and cubical ultrafine silica.

Rheological properties give information about deformation and flow of polymer composites. The use of fillers in rubber compounds influences the rheological properties when compared with unfilled elastomers, for instance, disappearance of any linear viscoelastic region, wall slippage effects, lower extrudate swell with increasing filler content, smoother melt fracture defects with increasing filler content, anisotropic effects in flow, etc. [75]. Montes et al. [76] studied the effect of filler loading level and structure on the shear viscosity function of natural rubber as seen in Fig. 17. As can be seen, when using of 20 phr reinforcing carbon black (N326), the Newtonian plateau is no longer observed. Obtained curves suggested

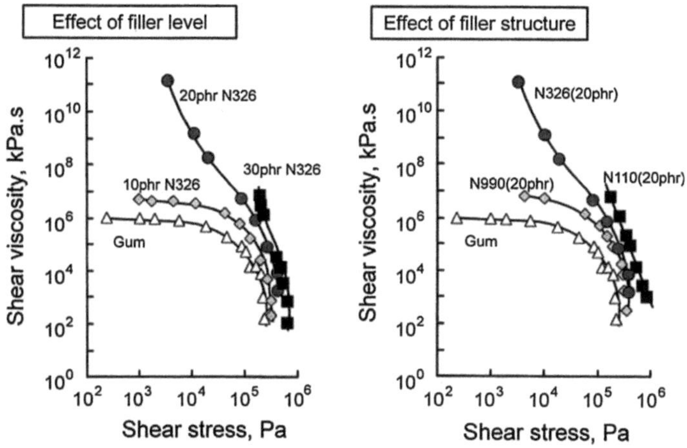


**Fig. 17** Effect of filler level and structure on the shear viscosity function of rubber compound [75, 76]

**Fig. 18** Effect of carbon black level on post-extrusion swelling [75, 77]

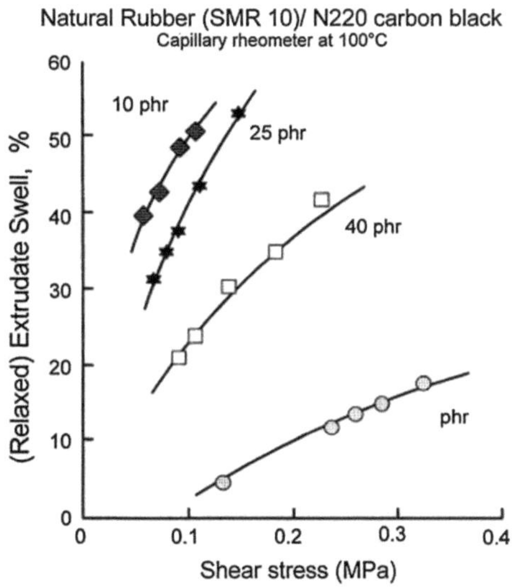


the occurrence of yield stress. No flow was observed below the yield stress. A similar behavior was seen, when substituting a non-reinforcing grade CB (N990) by a reinforcing one. Besides, extrudate swell can be decreased with increasing carbon black content as seen in Fig. 18 [75, 77].

Non-newtonian flow and high viscosity are observed by adding carbon blacks to rubbers. The increase in viscosity by the addition of carbon black can be resulted from hydrodynamic effect of carbon black loading, the structure of carbon black which can be associated with anisometry of the filler aggregates, the occlusion of rubber within the aggregates and adsorption of rubber molecules on the filler

Fig. 19 Viscous heating of gum, SBR and SBR filled with different types of carbon black and silica at 5 min, strain of 900 %, frequency of 30 cpm and 60 °C [53]

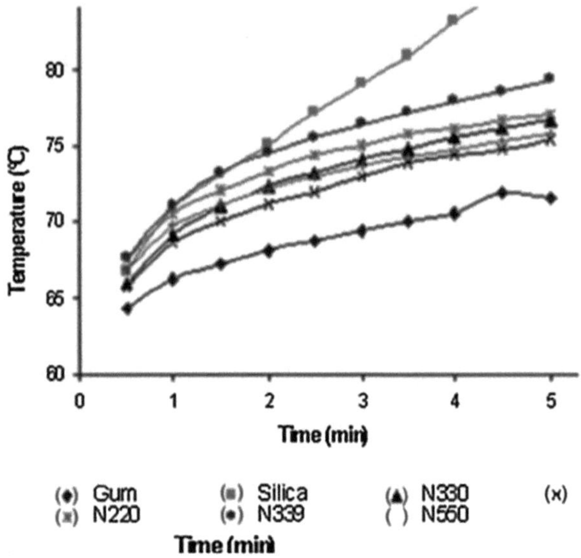


surface [53]. Soltani and Sourki [53] stated that unfilled SBR showed the lowest temperature increment. Temperature increment increased with loading carbon blacks and silica and this increment reduced in the order of the highest structure carbon blacks to the lowest structure one as shown in Fig. 19.

Saad et al. [78] found that the use of carbon black (type N347) in 80/20 NBR/CIIR blends resulted in increasing viscosity and cross-linking density and decreasing scorch time because of the restriction of the mobility and deformability of the matrix with the introduction of mechanical restraints.

Rubbers exhibit viscoelastic behavior, which can be measured by stress relaxation tests provide a measure of viscoelastic response. Soltani and Sourki stated that compounds containing larger particle size carbon blacks showed shorter relaxation time. Increasing the temperature weakened interaggregate interaction and diminish the modulus levels of the compound [53].

The ability of the thermal conduction of a rubber is low, therefore heat builds up under dynamic conditions and rubber is subjected to thermal fatigue. Particle size of the fillers affects the thermal conductivity of filled rubbers. Kemaloglu et al. [79] studied the properties of thermally conductive micro and nano size boron nitride (BN) reinforced silicon rubber composites. As seen in Fig. 20, the incorporation of BN particle into the silicone elastomer results in a reduction in coefficient of thermal expansion (CTE) of the composites. As the particle size of the BN is decreased, the CTE decreased at any given loading level. They obtained the lowest values for nano size fillers filled composites. They also stated that regardless of BN type, the addition of thermally conductive particles to the silicone matrix enhances the thermal conductivity.

Soltani and Sourki expressed that silica filler gave significantly poorer thermal conductivity than carbon blacks in cured compounds, due to filler–filler

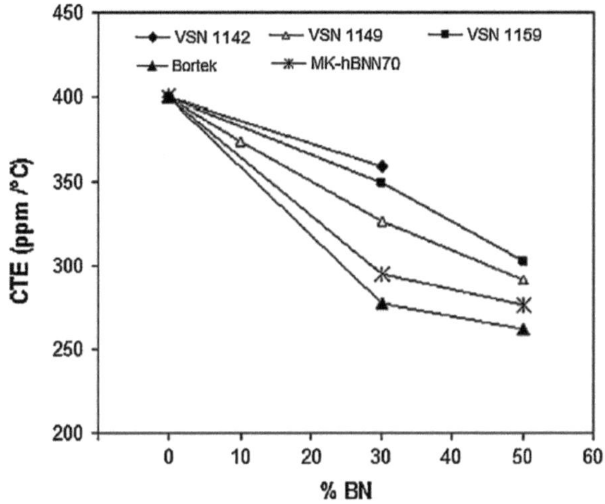


**Fig. 20** Effect of BN type and amount on the CTE of silicone rubber composites [79]

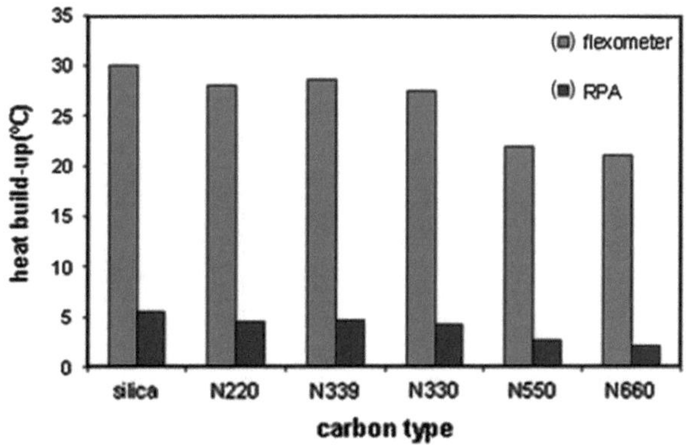


**Fig. 21** Heat build-up for SBR cured compound having different types of carbon black measured by Goodrich flexometer and RPA [53]

interactions. They also explained that with increasing the particle size, as well as, decreasing the size of carbon black, the amount of generated heat could be reduced as shown in Fig. 21 [53].

Zhou et al. [59] improved the thermal conductivity of silicone rubbers by adding different size of alumina in silicon rubbers. They found that the thermal conductivity of the larger filler particle filled rubber was higher than that of the smaller particle size filled system, namely, 1.48, 1.28 and 1.14 W/m K for the 30, 10 and 5 μm alumina particles filled system. Thermal conductivity increased with increasing particle size. They also stated that dielectric constant of composite rubber filled with large particles was slightly higher than the others [59]. Effect of particle loading and content of coupling agent (3-methacryloyloxypropyltrimethoxysilane) on the thermal conductivity of the $Al_2O_3$ filled silicon rubbers are shown in Figs. 22 and 23, respectively [62].

As seen in Fig. 22, thermal conductivity increases with increasing $Al_2O_3$ concentration up to 80 vol %. They explained that heat-conductive particles

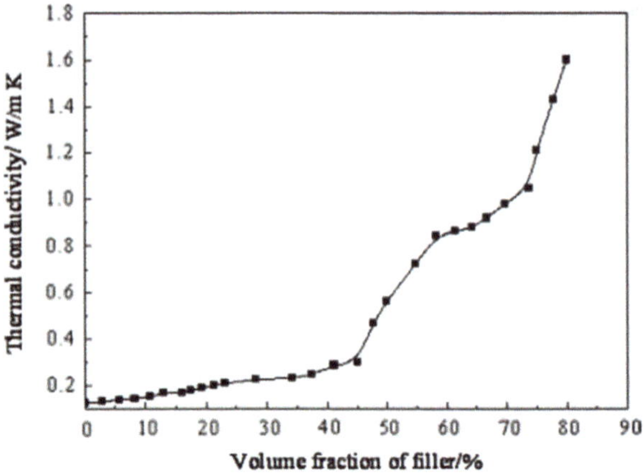


**Fig. 22** Thermal conductivity of silicone rubbers with different amounts of filler particles (25 μm) [62]

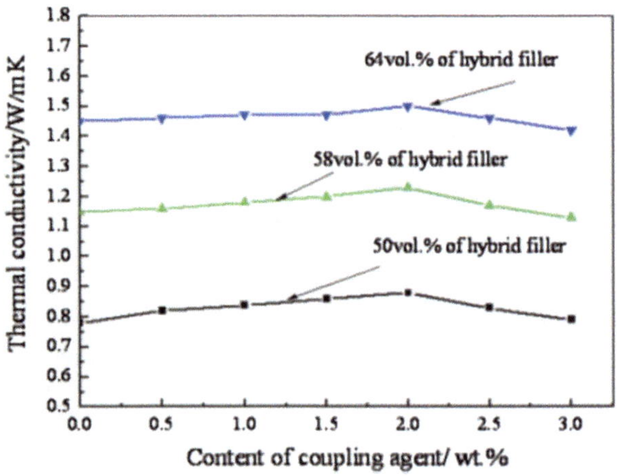


**Fig. 23** Thermal conductivity of silicone rubber versus the concentration of the coupling agent [62]

surrounded by a rubber matrix could not touch one another at low loadings. The use silane coupling agent also improved the thermal conductivity due to the decreased thermal contact resistance at the filler-matrix interface [62].

It was shown in another study that gradual replacement of carbon blacks with aluminum powder improved the thermal conductivity that resulted in decreasing vulcanization time and uniform curing throughout the material led to energy saving in the vulcanization of thick rubber articles [58].

Carbon black surfaces can easily transport electrons. Therefore, carbon blacks not only reinforce the composites but also provide electrical conductivity to filled rubbers. Conduction path theory, electron tunneling theory and electrical field radiation theory are the proposed mechanisms of the electrical conduction in the polymers filled with carbon blacks [52, 80]. These electrical conduction mechanisms are significantly affected by percolation limit, which is a loading limit where the conduction paths start to form. If the loadings of carbon black are lower than

**Table 6** Volume resistivity of EPDM/carbon black composites [52]

| N770[a] | SCB[a] | N550[a] | N330[a] | N472[b] |
|---|---|---|---|---|
| $2.3 \times 10^{14}$ Ω·m | $2.0 \times 10^{14}$Ω·m | $1.8 \times 10^{11}$ Ω·m | $1.2 \times 10^{10}$ Ω·m | 0.18 Ω·m |

*Note* [a] the sample was characterized by the high resistance meter
[b] the sample was characterized by the conventional four-point technique

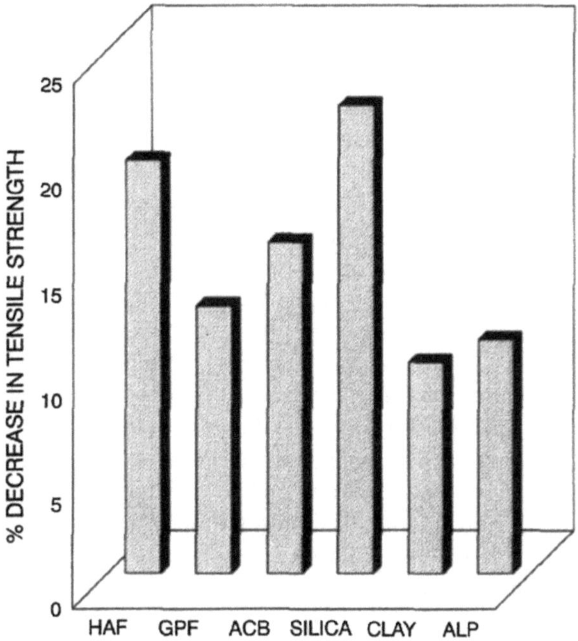


**Fig. 24** Aging characteristics of various filler incorporated (40 phr) natural rubber composites (aging at 70 °C for 14 days) [58]

the percolation limit, the rubbers filled with these kinds of carbon blacks act as insulator. On the other hand, they are electrically conductive. Zhang and Chen indicated that N472 with a large surface area which affects the percolation limit exhibited low resistivity indicating that EPDM with N472 was suitable to be used in conductive applications as shown in Table 6 [52].

El-Wakil and El-Megeed [81] prepared $CaCO_3$, SN85 and SRF filled EPDM composites and aged them at 90 °C for 2, 4, 6, 8 and 10 days then measured electrically. They observed that all of the fillers improved electrical conductivity of EPDM and dielectric constant slightly decreased with increasing the aging time. Besides, sillitin N85 and calcium carbonate decreased the dielectric constant of the EPDM vulcanizates.

Aluminum powder incorporated vulcanizates showed good resistance against thermal aging and oxidative degradation as shown in Fig. 24 [58]. Similar results were obtained by El-Wakil and El-Megeed [81] that retained tensile strength and cross-linking density of filled EPDM were better than pure one. Mostafa et al. [65] studied the influence of carbon black loading on thermal aging resistance of SBR and NBR rubber compounds under different aging temperature. It is evident from the obtained results that gum vulcanizates had more resistance (300 % modulus) towards aging as compared with loaded samples. As carbon black (CB) loading increased, the resistance was decreased. Rapid degradation of rubber occurred because CB accelerated the oxygen uptake of sulfur-cured rubber. CB worked as a catalyst for the direct oxidation.

Saad et al. [78] stated that solvent uptake of a NBR/CIIR blend could be decreased by using carbon black; because cross-linking density increased and this increment restricted the swelling of the rubber [78]. The swelling percentage decreased with increasing carbon black loading level for both SBR and NBR filled compounds. NBR filler compound exhibited lower swelling percentage than SBR ones because of having better crosslinking density of NBR. Moreover, the swelling percentage increased while increasing the exposure time and temperature due to the ease of penetration of oil into the rubber chains [66].

Zhang et al. investigated the effect of particle size on the fire resistance of $Mg(OH)_2$/EPDM composites with cone calorimetry and limiting oxygen index analysis (LOI). They found that the peak value of heat release rate decreased and $T_{ign}$ (ignition temperature) increased with decreasing particle size [63].

Surface area, surface reactivity, particle size, moisture content and metal oxide content of the fillers determine the cure rate of rubbers [64]. Having low surface area, high moisture and metal oxide content result in a faster cure rate [64, 82, 83]. Ismail et al. [64] found out that shorter $t_{90}$ and scorch times for oil palm wood flour filled epoxidized natural rubber were obtained by using larger particle sizes of filler. Besides, the maximum torque increased with increasing filler content. The effect of chitosan loading level (ranges from 0 to 40 phr) on the curing characteristic of NR, ENR and SBR was studied by Ismail [84]. A significant decrease was observed in scorch ($t_{S2}$) and cure times ($t_{90}$) with the addition of chitosan in all three types of rubber, and the values further decreased as chitosan loading increased. The lower the scorch and cure times, the faster the cure rate was. They attributed this decrease to the formation of bonds due to the interaction between the chitosan functional groups (–OH and $–NH_2$ groups in chitosan) and the compounding additives. Adversely, the maximum torque increased with chitosan loading for all three types of rubber compound [84].

The micro fillers also influence the gas barrier properties of rubber. Zhang et al. [85] remarked that the nitrogen permeability of natural rubber (NR)/silane modified kaolin (SMK) composites was decreased by 20–40 % after the addition kaolin; because, parallel kaolinite particles restricted the free movement of rubber chains and retarded the diffusion of the gas molecules.

## 4 Recent Studies Related to Nanofiller-Rubber Systems

In the last few years, numbers of articles have been published on rubber-based nanocomposites, in which potential Nanofillers such as nanoclays, talc, silica, carbon nanotubes, nano-ceramic particles, and nano-biofillers have been used. The researchers have mainly focused on the improvement of physical properties of the rubber compounds, i.e., mechanical, thermal, rheological, barrier properties, and so on, by the incorporation of Nanofillers at a very low loading level in comparison to conventional fillers. In most of the studies, the filler kind, filler-polymer compatibility, filler loading level, preparation technique and processing conditions were selected as the experimental parameters.

In this part, it was aimed to highlight some of the informative studies selected from the literature focusing on the parameters stated above. A special attention was paid to cite the most recent ones in order to figure out the direction of the field.

One of the most important properties of rubber composites is sure its mechanical property. Nanoparticles play an important role in enhancing mechanical properties in composite application including mostly thermoplastics and thermosetting resins. However, recently researchers from rubber science and technology have also been focusing on mechanical property enhancement by incorporating nanoparticles into the rubber matrices.

Sadhu and Bhowmick [86] investigated the preparation and properties of different nanoclays based on sodium montmorillonite, bentonite and potassium montmorillonite (K-MMT), and organic amines of varying chain lengths in the SBR matrix. The amines were octadecylamine (ODA), hexadecylamine (HDA), dodecylamine (DDA), and decylamine (DA). XRD studies showed that the intergallery distance was increased by the incorporation of the amines depending on their chemical structure. It was mentioned in the study that the clay structures broke down due to shear mixing and exfoliation was observed. Mechanical properties such as tensile strength, elongation at break and modulus improved. On the other hand, the properties also depend on the kind of amine compound, for instance the tensile strength increased with increasing chain length of the amine. The gallery distance and the mechanical properties were connected. Na-MMt having the highest intergallery space exhibited the best mechanical properties. As compared to neat SBR, all the nanocomposites displayed better mechanical properties.

In another SBR/Clay nanocomposite study, the effect of modified clay content was investigated with respect to styrene content of SBR (15, 23, and 40 %) [87]. TEM pictures showed uniform distribution of the organoclays in the SBR matrix with a thickness of 10–15 nm. The most important result of the work was the improved mechanical properties in nanocomposites even at a very low clay loading in comparison to the neat SBR.

Kojima et al. [88] focused on the properties of NBR based MMT nanocomposites. They modified MMT by cation exchanging with amine terminated butadiene oligomer. This organoclay was then blended with NBR by roll milling and finally sulfur cured. The results showed that only 10 phr of organoclay was enough to achieve tensile strength comparable to an NBR compound loaded with 40 phr of carbon black.

As mentioned before, the similarity in polarity has a crucial importance in dispersing clays into the matrix. In the study of Vu et al. [89], the conditions for dispersing modified and unmodified clay into nonpolar cis-1,4-polyisoprene (IR)

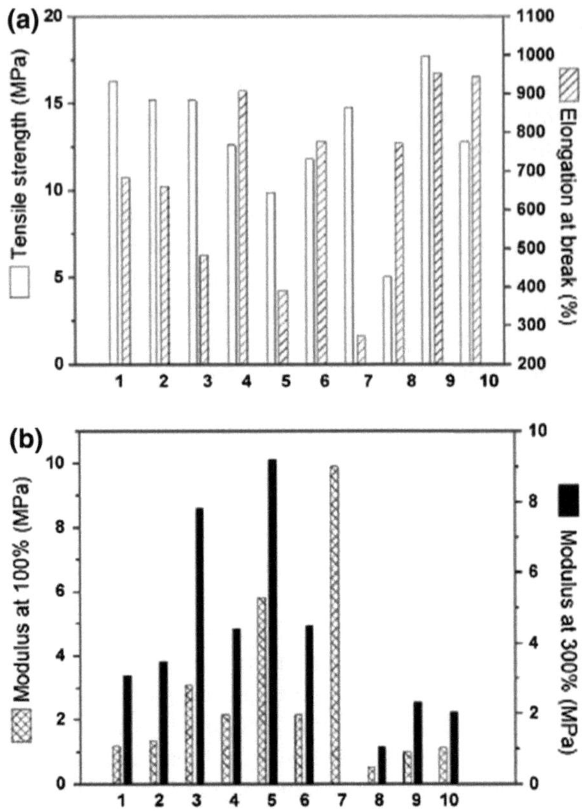


**Fig. 25** **a** Tensile strength and elongation at break of the various vulcanizates: (*1*) ENR50 with 10 phr Hi-Sil, (*2*) ENR50 with 10 phr Na-MMT, (*3*) ENR50 with 30 phr Na-MMT, (4) ENR50 with 10 phr C1, (5) ENR50 with 30 phr C1, (6) ENR50 with 10 phr C2, (*7*) ENR50 with 30 phr C2, (*8*) IR50 with 10 phr Hi-Sil, (*9*) IR50 with 10 phr C2 and (*10*) IR50 with 10 phr C3. **b** Modulus values of the vulcanizates. (Sample designations as in Fig. 5a) [90]

and more polar epoxidized natural rubber (ENR) was focused. The organically modified clays used in this work were bis-(2-hydroxyethyl) methyl tallow ammonium montmorillonite (Mod1), dimethyl dihydrogenated tallow ammonium montmorillonite (Mod2), and dimethyl hydrogenated tallow (2-ethylhexyl) ammonium montmorillonite (Mod3). Precipitated silica (trade name of Hi-Sil 233) was used as the reference filler material. The loadings level was changed as 10, 20, and 30 phr. An internal mixer was used to mix the ingredients. XRD experiments indicated intercalation of clays, followed by exfoliation (delamination) of the silicate layers. The tensile properties of the vulcanizates are compared in Fig. 25a, b [90]. It was observed that the reinforcing effect of the clays was strongly dependent on the extent of dispersion of silicate layers in the matrix.

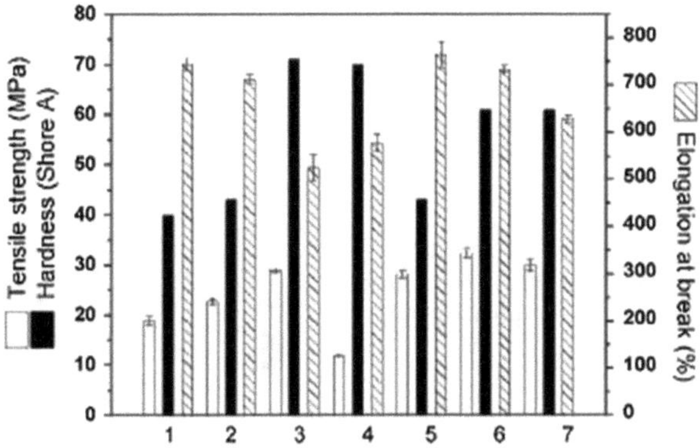


**Fig. 26** Mechanical properties of the various conventional rubber vulcanizates: (*1*) unfilled NR, (*2*) 7 phr Na-MMT in NR, (*3*) 20 phr N110 in NR, (*4*) 20 phr Hi-Sil 927 in NR, (*5*) 6.1 phr Hi-Sil 927 in NR, (*6*) 7 phr ODA-MMT in NR, and (*7*) 7 phr Octadecyltrimethy lammonium chloride-MMT in NR [90]

In another work the importance of compatibility was demonstrated. Researchers found that organically modified galleries of MMT were easily penetrated by natural rubber (NR) chains and led to intercalated structures along with partial exfoliation [91]. As a result, at 10 wt% modified MMT, modulus increased comparable to that achieved by high loadings of conventional micrometer sized fillers. Moreover, it was shown that MMT clays modified with long primary amines resulted in much more improved mechanical properties in comparison to the quaternary amines (having the same carbon numbers) in a NR matrix [92]. The authors explained this observation by the help of XRD data, which showed a greater degree of intercalation for the long chain primary amines compared to the quaternary ammonium salts. It was also observed that only 7 wt% modified clay was good enough to achieve mechanical properties exhibited by highly silica and carbon black filled NR vulcanizates (see Fig. 26).

Lim et al. [93] used solvent-casting method in order to prepare rubbery poly-epichlorohydrin (PECH)/organoclay (Cloisite 25A) nanocomposites. The XRD data showed an intercalated structure of Cloisite 25A in rubber matrix. The mechanical properties of the nanocomposites were observed to be much better than those of the pristine polymer.

The blending technique has also an impact on the properties of the resulting nanocomposites. Wang et al. [94] made a comparison on the mechanical properties of fractionated bentonite/SBR nanocomposites with respect to preparation method, i.e., the solution or latex blending techniques. The nanocomposites prepared by the latex route were better than those prepared by the solution blending technique at equivalent clay loadings.

In another study, SBR based rectorite (a layered silicate of 1:1 type of layer structure arranged regularly with an alternating dioctahedral mica layer and di-octahedral smectite layer nanocomposites) were prepared by co-coagulating SBR latex and rectorite/water suspension [95]. The aspect ratio of the rectorite layers was found to be higher than that of MMT in the nanocomposites by the TEM. At

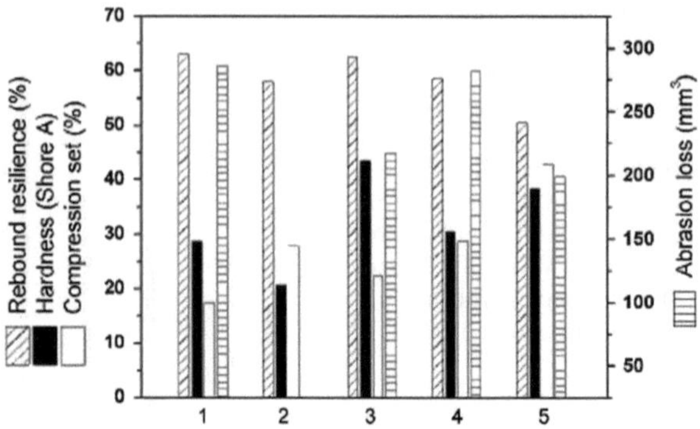


**Fig. 27** Comparison of rebound resilience, hardness (*Shore A*), compression set and abrasion loss for the various mixes: Mixes: 1-NR (*unfilled*), 2-NR + 10 phr unmodified clay, 3-NR + 10 phr organoclay, 4-NR + 10 phr *carbon black* and 5-NR + 40 phr *carbon black* [90]

equivalent filler loading (20 phr), the rectorite/SBR nanocomposites exhibited better mechanical properties than SBR filled with carbon black or calcium carbonate. Due to good mechanical properties, authors proposed tire tube and inner liner applications for rectorite/SBR nanocomposites.

Melt or direct intercalation technique is the most cost effective and environmentally friendly method since no solvent is used. However, due to the compatibility problem or the high viscosity and high molecular weight of the polymer chains, intercalation and exfoliation is very difficult to achieve. The melt intercalation of the polymer chains inside the silicate layers is primarily driven by enthalpic interaction; however, in solution technique the driving force is the entropic [90, 96]. In order to decrease the surface energy and to improve the wettability of silicate surfaces, the alkyl ammonium groups are used. The most important outcome of the polymer-clay interaction is the improvement in the strength of the resulting composite.

It is reported that compounds of epoxidized NR prepared by mixing pristine sodium MMT using direct intercalation technique results in tensile properties close to or even higher than those of compounds with precipitated silica [89]. The effect of dynamic strain amplitude on the storage modulus ($G_0$) revealed that $G_0$ of the compound with Na-MMT was higher in comparison to that of the compound with silica possibly due to a greater hydrodynamic reinforcement as a result of melt intercalation of the rubber in the Na-MMT (clay) galleries. In the similar study, it was showed that with an identical Na-MMT loading, $G_0$ of Na with 50 % epoxidization was greater than that of NR with 25 % epoxidization. This can be attributed to the stronger rubber-filler interactions in the higher epoxidized NR.

In a different study, NR was reinforced with 10 phr Na-MMT (unmodified clay) and ODA-modified Na-MMT (organonoclay) and compared the results with NR loaded with 10 and 40 phr carbon black [97]. An open two roll mil was used as the processing tool. The mechanical behavior of NR with 10 phr organoclay was comparable with the compound with 40 phr carbon black. Some of the additional mechanical properties like rebound resilience, hardness, compression set, and abrasion loss reported by Sengupta et al. are shown in Fig. 27.

Teh et al. [98] compounded octadecyl trimethylamine modified MMT with NR using an internal mixer. The compatibilization was done by using 10 phr ENR50. Two commercial fillers, carbon black (N330) and precipitated silica were also used in this study for comparison purposes. Filler loadings were 50 phr for carbon black, 30 phr for silica, and 2 phr for organoclay. The improvements of tensile strength, elongation at break and tear properties in organoclay filled vulcanizate were significantly higher compared with carbon black and silica filled vulcanizates. SEM micrographs did not reveal any significant difference in the morphological appearance between the organoclay filled vulcanizate compared with the silica and carbon black filled vulcanizates. However, the larger specific surface area and smaller particle size of the organoclay was believed to provide a better rubber-filler interaction resulting a better mechanical response than the conventional fillers. Further evidence of the rubber-filler interaction was obtained from the relaxation behavior of the filled vulcanizates from dynamic mechanical analysis (DMA) [90, 98].

Attapulgite (a natural fibrillar silicate) (AT) reinforced SBR rubber compounds were prepared by Tian et al. [99] by direct mill mixing. The dispersion of the AT particles was observed with the aid of TEM, SEM, and a Rubber Process Analyzer (RPA). It was reported that most of the AT units was separated into dispersed units with diameters less than 100 nm in SBR but a few dispersion units as large as 0.2–0.5 mm were also noticed. Moreover, researchers used a silane coupling agent in order to improve the dispersion of AT in the rubber matrix. Silane modification resulted in an enhancement of the interfacial adhesion between the AT particles and the rubber chains. At a constant filler loading of 40 phr, silane modified AT exhibited better reinforcing effect on SBR compared with a furnace carbon black (particle size range 60–100 nm), but was inferior when compared with SBR reinforced with N330 having particle size varying from 26 to 30 nm.

EPDM/Clay nanocomposites were prepared by mixing modified clay with EPDM [100] in molten state with an extruder. The silicate layers of the organoclay were exfoliated and uniformly dispersed in the EPDM matrix. The tensile strength of the EPDM/clay hybrids was higher than that of neat EPDM. The 100 % tensile stress of the EPDM/clay hybrids also increased compared to that of EPDM. Moreover, the storage moduli of the EPDM-clay hybrids were higher than that of neat EPDM.

The processing conditions and the kind processing tool have also influence on the resulting properties of nanocomposites. Gatos et al. [101] investigated the characteristics of EPDM/organoclay nanocomposites as a function of processing conditions and formulations. The EPDM was sulfur cured and was containing 10 phr organoclay. Two different mixing methods, two roll mixing mill with a friction ratio of 1.3 and Brabender plasti-Corder at 60 rpm, were used. Increasing the processing temperature and the high shear mixing (internal mixer instead of open mill) improved the mechanical performance of the rubber nanocomposites (see Fig. 28). The change in the rubber polarity was provided by replacing half of the EPDM rubber with maleic anhydride (MA) or glycidyl methacrylate (GMA) grafted EPDM. The incorporation of the grafted EPDMs strongly improved the

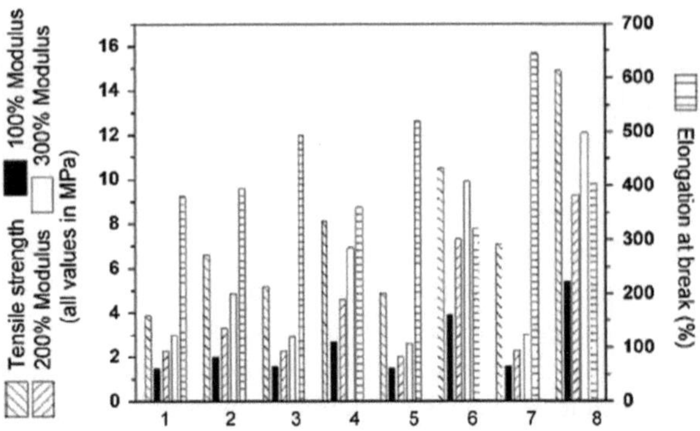


**Fig. 28** Mechanical properties of EPDM/organoclay (10 phr) compounds prepared under different conditions with and without MA grafted EPDM compatibilizer: (*1*) open mill mixing at room temperature (*RT*) of neat EPDM, (*2*) open mill mixing at RT of EPDM-MA50 %, (*3*) open mill mixing at 100C of neat EPDM, (*4*) open mill mixing at 100C of EPDM-MA50 %, (*5*) mixing in Brabender at RT of neat EPDM, (*6*) mixing in Brabender at RT of EPDM-MA50 %, (*7*) mixing in Brabender at 100C of neat EPDM and (*8*) mixing in Brabender at 100C of EPDM-MA50 % [90]

strength and stiffness of the nanocomposites due to the improved interactions at the interface. The microstructure of the rubber/organoclay systems studied by XRD, TEM, and SEM revealed that organoclay intercalation/exfoliation was accompanied by its more or less severe reaggregation and/or deintercalation. This was traced to partial (monolayer arrangement) and complete removal (cation exchange) of the original intercalant (ODA) from the clay galleries via the formation of a zinc complex in which amine groups of the ODA and sulfur participated [90, 101].

Since nanoclays are one of the first Nanofillers that have been used to reinforce the rubbers, there are numerous studies related to nanoclays. On the other hand, carbon nanotubes (CNTs) exhibit unique mechanical, electronic and magnetic properties and have been the subject of many studies [102–104]. The effectiveness of carbon nanotubes in composite applications depends on the ability to disperse the tubes uniformly and without agglomeration into the polymeric matrix. In addition, the aspect ratio of the CNT should not be reduced in order to not lose the reinforcing effect and good interfacial bonding is required to achieve load transfer across the interface.

The reported results in the literature on the rubber/CNT nanocomposites demonstrated that a large increase in effective modulus and strength was obtained with the addition of small amounts of carbon nanotubes [105–107]. In a recent study, the effects of –COOH group on the CNT added to SBR matrix on the mechanical and thermal properties of the nanocomposite were investigated [108]. Multi-walled carbon nanotubes (MWCNTs) were functionalized with a carboxylic acid group by using nitric acid at 120 °C for 48 h. The modified carbon nanotubes were incorporated to SBR by using solution casting technique. The researchers characterize the CNTs, and CNT reinforced polymer nanocomposites. TEM and

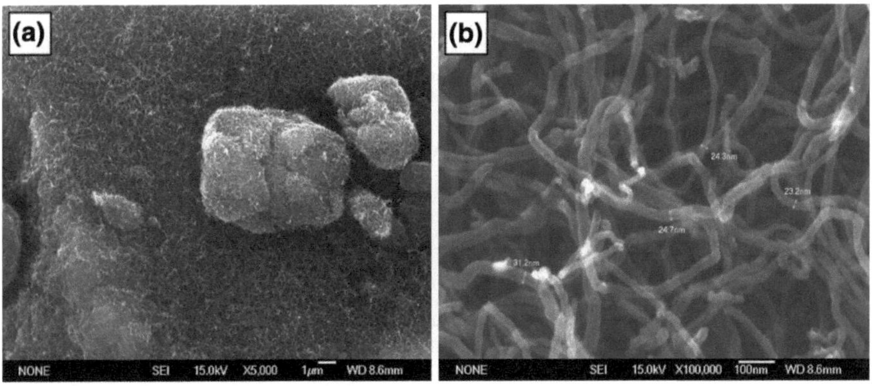


**Fig. 29** SEM images of carbon nanotubes at **a** at low resolution, **b** at high resolution [108]

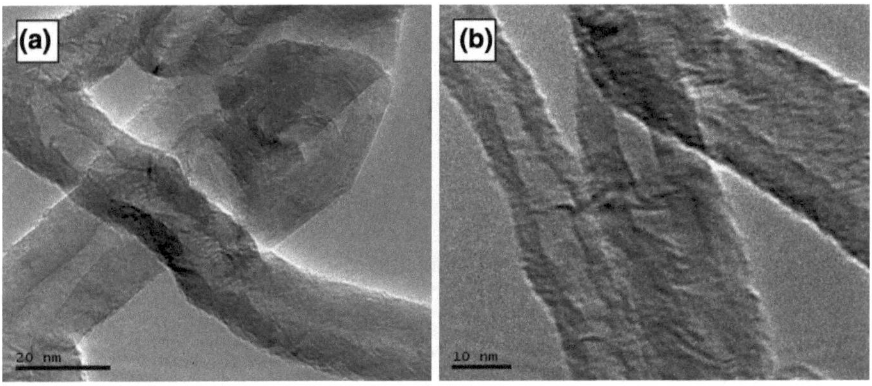


**Fig. 30** TEM images of carbon nanotubes **a** at low resolution **b** at high resolution [108]

SEM pictures showed that the nanotubes were of high purity, with uniform diameter distribution and contained no deformity in the structure (Figs. 29 and 30).

Stress–strain curves indicated that the incorporation of CNT in SBR leads to an enhancement of elastic modulus and the rupture properties, such as the stress and elongation at break. An increased loading of carbon nanotubes in the SBR matrix leads to an increase in tensile strength and modulus. With regard to the pure polymer, the stress–strain curve indicates that the strength of the rubber nanocomposites at 10 wt% of CNTs was approximately 4 times (273 %) higher than that of SBR due to well-dispersed MWCNTs and a good interface between the MWCNT and SBR matrix. As compared to the pure polymer, the Young's Modulus of the rubber nanocomposites at 10 wt% of CNTs is approximately 6 times higher than that of SBR due to well-dispersed MWNTs and good interface between the MWNT and SBR matrix [108] (Fig. 31).

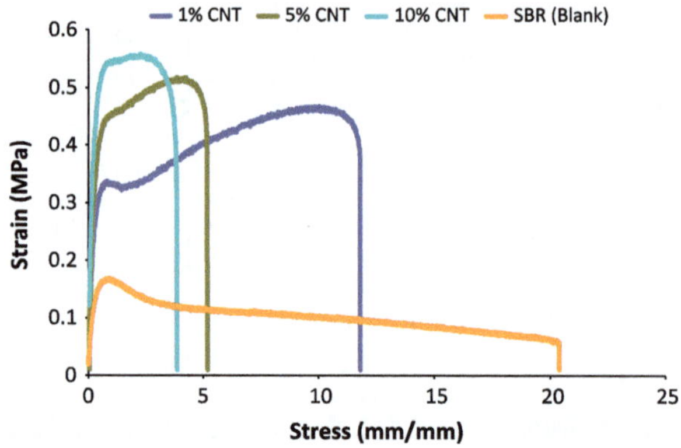


**Fig. 31** Stress-strain *curve* of functionalize CNTs/SBR nanocomposite matrix [108]

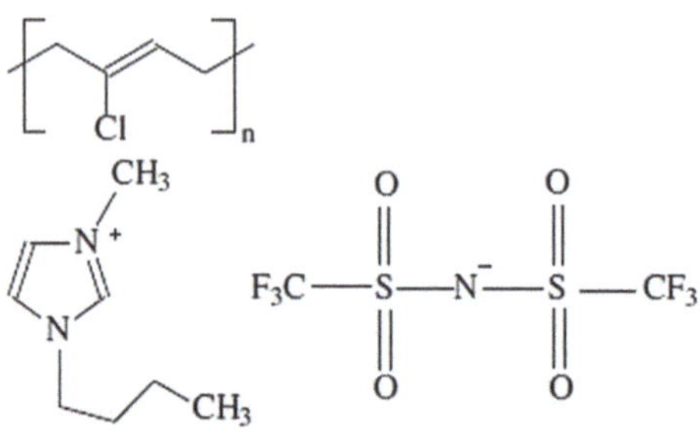


**Fig. 32** Chemical structure of BMI [109]

In a very recent work, a simplified and eco-friendly approach to develop polychloroprene rubber (CR)/MWCNT composites was reported [109]. The researchers used a room temperature ionic liquid, 1-butyl 3-methyl imidazolium bis (trifluoromethylsulphonyl) imide (BMI) (see Fig. 32) to improve the dispersion. The MWCNT content was 5 phr. TEM images exhibited an improved dispersion of the BMI modified tubes in matrix (see Fig. 33). The presence of agglomerates was found to be minimum and the dispersion of the tubes was well-marked. Hence, the researchers believed that the bundles of MWCNTs was disturbed due to the reduction in the intertubular attraction between CNTs in presence of BMI and filler–filler networks are found to be improved with well dispersed MWCNTs by increasing the proportion of BMI. A schematic representation of this action was also represented (Fig. 34). The improvement of the dispersion of CNTs in the matrix was also supported by the mechanical properties. The tensile modulus of 3 phr loaded modified MWCNT/CR nanocomposite was increased by 50 % with regard to that of the unmodified MWCNT/CR nanocomposite [109].

Other potential Nanofiller to be used in rubber compounding is the nanosilica. Aso et al. [110] studied the properties of thermoplastic elastomer/nanosilica compounds. The rubber was Hytrel 5,556 from Dupont. The samples were prepared in a twin-screw extruder. In order to obtain the standard test bars, the nanocomposites were injection molded. It was found that the dispersion of silica nanoparticles was very uniform; the typical clusters had only 2–3 particles and

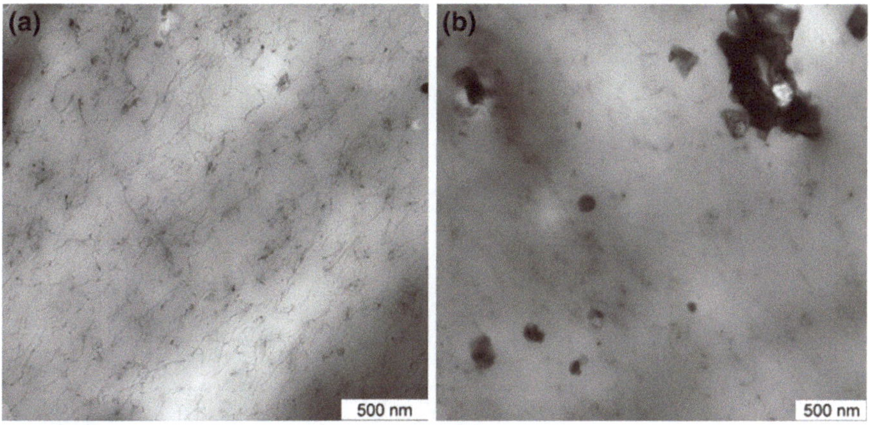


**Fig. 33** The TEM micrographs of CR composites (**a** and **b**) $CT_3BMI_O$ [109]

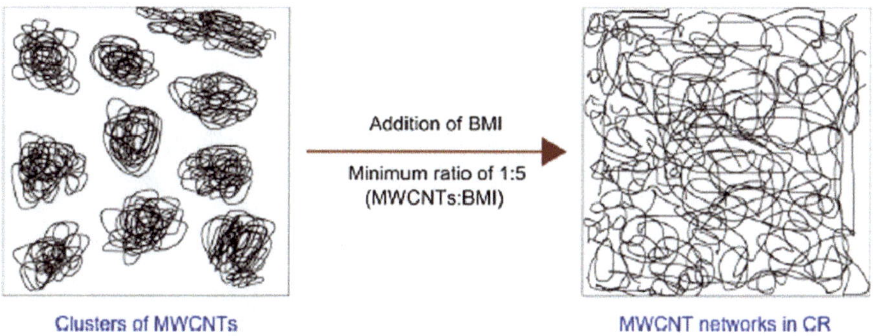


**Fig. 34** Schematic illustration of the formation of MWCNT networks in presence of BMI [109]

many individual particles were observed in SEM micrographs. The dispersion obtained was attributed to the energetic processing used and to the increase in the organophilic nature of the $SiO_2$ surface after silane modification. The former promotes better particle dispersion, while the latter hinders ulterior agglomeration. The mechanical properties were also investigated. The addition of Nanofillers into rubber brings a simultaneous improvement in modulus, elongation at break and creep resistance.

Inclusion of nanoparticles in the rubber matrices has an effect on the thermal properties of the resulting composites. Many researchers have focused on the thermal transitions, thermal degradation behavior, thermal conductivity and thermo-mechanical properties of rubber nanocomposites.

Peng et al. [111] prepared a novel natural rubber/silica (NR/$SiO_2$) nanocomposite by combining self-assembly and latex-compounding techniques (For the details of the technique, one should refer to Ref. [111]). Authors showed that when the silica content was less than 4 wt%, a homogenous dispersion of silica in rubber

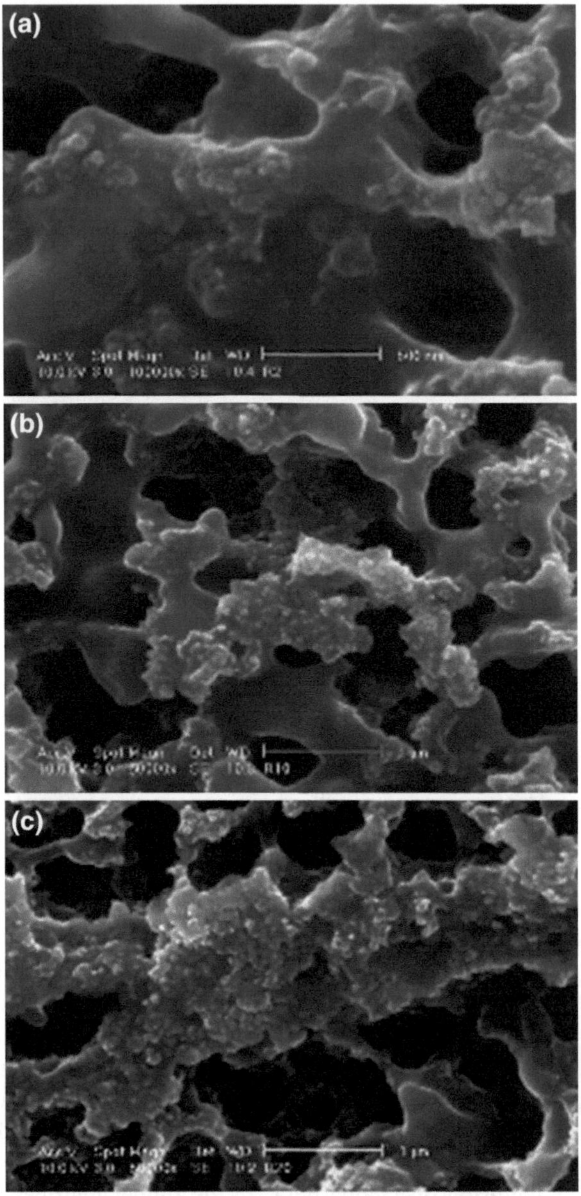


**Fig. 35** SEM micrographs of NR/$SiO_2$ nanocomposites: **a** 1.0 wt% $SiO_2$, **b** 4.0 wt% $SiO_2$, and **c** 8.5 wt% $SiO_2$ [111]

matrix was obtained as individual spherical nanoclusters, but further increase in the $SiO_2$ content lead to intensive aggregations. This can be observed from Fig. 35, which shows the morphology of the NR/$SiO_2$ nanocomposites with different $SiO_2$ contents. On the other hand, it was also shown that when only a very small amount of $SiO_2$ was added to the NR matrix (less than 1.0 wt%), most of the spherical

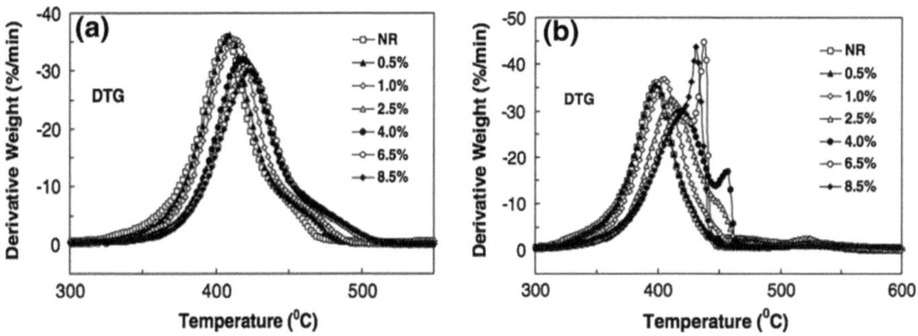


**Fig. 36** The derivative thermo gravimetric curves of NR/$SiO_2$ with respect to silica content: **a** in nitrogen, **b** in air [111]

clusters of nanoparticles were individually distributed in the NR matrixes (Fig. 35a). When more $SiO_2$ nanoparticles are loaded, the density of the $SiO_2$ clusters becomes higher. Researchers studied the thermal and thermooxidative ageing resistance of NR/$SiO_2$ nanocomposite, respectively, from the investigation of thermal and thermooxidative decomposition. It was found only one obvious thermal decomposition step of NR molecular chains, primarily initiated by thermal scissions of C–C chain bonds accompanying a transfer of hydrogen at the site of scission (see Fig. 36a, the derivative thermogravimetric curves). On the other hand, it was shown that the thermooxidative decomposition was obviously different from the thermal decomposition as shown in Fig. 36b, which are the TG and DTG curves of the pure NR and NR/$SiO_2$ nanocomposites in air, respectively. The first large decomposition peak was attributed to the oxidative dehydrogenation accompanying hydrogen abstraction, while the second weak peak was attributed to the oxidative reaction of residual carbon. For the nanocomposites, the main decomposition peak shifted to a higher temperature and divided into two peaks. As more $SiO_2$ was added (>6.5 wt%), the side peak sharpened. The most interesting result of the study is the fact that the thermal and thermooxidative ageing resistance of prepared nanocomposite was highly dependent on the distribution of $SiO_2$ nanoparticles. Since the $SiO_2$ nanoparticles were homogenously distributed in NR matrix as nano-clusters when it was less than 4.0 wt%, the $SiO_2$ and NR molecular chains strongly interacted through various effects such as the branching effect, nucleation effect, size effect, and surface effect. Therefore, the diffusion of decomposition products from the bulk polymer to gas phase was slowed down. As a result, the nanocomposite had a more complex decomposition and a pronounced improvement of ageing resistance compared to the pure NR. It was also mentioned that the other reason for the improvement in ageing resistance of NR/$SiO_2$ nanocomposites was that inorganic nanoparticles could migrate to the surface of the composites at elevated temperatures because of its relatively low surface potential energy [112]. This migration could result in the formation of a $SiO_2$/NR char, which acted as a heating barrier to protect the NR inside.

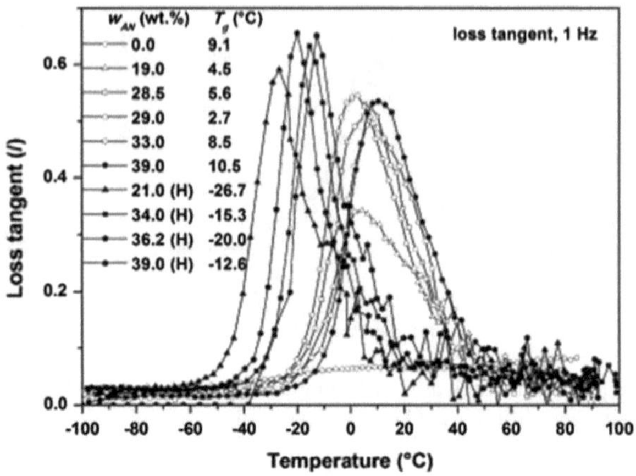


**Fig. 37** Characterization of mechanical properties–dynamic mechanical properties [loss tangent at 1 Hz vs. temperature (−100 to 100 °C)] of (H)NBR/MWCNT nanocomposites with different AN contents [113]

Likozar and Major [113] prepared nanocomposites by mixing multi-walled carbon nanotubes (MWCNT) with nitrile and hydrogenated nitrile elastomers (NBR and HNBR). MWCNT were mixed with rubbers for 0.5 h at 100 °C using melt compounding in Brabender Plasti-Corder PLD-Type 651 (Germany) instrument. The compounds were peroxide crosslinked. The homogeneity of distribution of nanotubes was found to be well from the SEM micrographs. The authors emphasized that usually, the homogeneity of the MWCNT-reinforced polymer nanocomposites may be achieved by solution mixing, but in this case a great excess of solvent would be needed per rubber amount. In their case, the high shear and temperature during compounding proved to grant excellent homogeneity. The nanotube aggregates were found to be quite small (tens of nanometers) with a few tens of micrometers scale length. The researchers also investigated the thermo mechanical properties (Fig. 37). The glass transition was strongly shifted to lower temperature as AN content increases up to 29.0 wt% when it started increasing again. $T_g$ of HNBR/MWCNT composites were higher when compared to the silica-reinforced hydrogenated nitrile rubber nanocomposites reported elsewhere previously. This was attributed to the highly reduced mobility of the HNBR chains around the MWCNT. Also, the area of the tan(delta) peak was greatly enlarged with increasing AN content up to 29.0 wt% when it started decreasing again, indicating significantly increased damping. It was also stated that for HNBR/MWCNT composites, $T_g$ changed without a trend with increasing AN content.

In a recent study, the thermal degradation of hydrogenated nitrile-butadiene rubber (HNBR)/clay and HNBR/clay/carbon nanotubes (CNTs) nanocomposites was investigated with thermo gravimetric analysis (TGA) by using several kinetic models [114]. The activation energies evaluated had a sequence of HNBR and its nanocomposites as HNBR/clay/CNTs > HNBR/clay > HNBR. It was found that HNBR/clay/CNTs nanocomposites had higher char yield at 600 °C than HNBR/clay, which was attributed to the interaction of network between clay and CNTs. The researchers also investigated the gases involved during thermal degradation in nitrogen atmosphere by Fourier transform infrared spectroscopy coupled with TGA. It was reported that the HNBR/clay/CNTs nanocomposites had lower thermal degradation rate than HNBR/clay, which could be attributed to that the clay-CNTs filler network reduced the diffusion speed of degradation products. It was mentioned that the coexistence of clay and CNTs could form compact char layers with better barrier property than clay and thus improved the thermal stability of HNBR.

It is possible to obtain Nanofillers from renewable resources, i.e., cellulose whiskers that are the most abundant polymers on the earth. It was shown in the literature that nanocellulose mostly embrittles and reinforces the matrix. Nanocellulose whiskers can also be used to reinforce the rubbers. In one of the studies from the literature, cellulose whiskers were isolated from bleached sugar cane bagasse kraft pulp [115]. The length of the isolated whiskers was in the range 84–102 nm while the width was in the range 4–12 nm. They were used as reinforcing elements in natural rubber (NR) matrix. Researchers investigated the thermal stability of rubber/cellulose whiskers nanocomposites by thermo gravimetric analysis (TGA) and compared to that of neat rubber matrix. It was found that thermal decomposition of rubber in nitrogen atmosphere started at about 380 °C and completed about 550 °C. NR/cellulose whiskers nanocomposites containing 10 wt% cellulose whiskers showed slightly lower onset degradation temperature (265 °C) than neat natural rubber. The lower onset degradation in the case of rubber/cellulose whiskers nanocomposites was attributed to the lower onset degradation temperature of bagasse whiskers than rubber. The effect of addition of cellulose whiskers on the glass transition temperature ($T_g$) of NR was investigated by differential scanning calorimetry (DSC) The results showed no significant effect of the whiskers on $T_g$ of rubber. $T_g$ was found to be located around −64 °C for both of the neat rubber and rubber filled with 2.5 and 7.5 wt% bagasse whiskers.

From rheological point of view, addition of Nanofillers to the rubber compound strongly influences the steady shear and viscoelastic properties of the nanocomposites. For example, the effect of strain-dependence of dynamic viscoelastic property of filled polymers, often referred as Payne effect, has been well known in elastomers for many years. The level of filler dispersion, filler-polymer adhesion/interaction and the intrinsic properties of fillers have play important role in defining the rheological properties of uncured and cured rubbers; therefore it is still under the scientific focus.

In a recent study, the effect of an epoxidized natural rubber (ENR, with 25 and 50 % epoxidization) and organoclay surface treatment (no treatment, hydrophobic treatment and hydrophilic treatment) on the morphology and behavior of natural rubber (NR) nanocomposites were investigated [116]. The nature and extent of the clay dispersions in the filled samples were evaluated by X-ray diffraction. In the presence of ENR, an exfoliated structure was obtained with hydrophilic clay, which suggested that rubbery polymer was incorporated into the interlayer spacing due to match-up in polarity. The effect of clay in rubber compounds was analyzed through rheological measurements, i.e., Mooney viscosity and Moving Die

Rheometer. It was reported by the authors that the Mooney viscosity has been shown to be directly proportional to the true shear viscosity at very low shear rates (1–2 $s^{-1}$). The observed shear thinning behavior was explained by the disentanglement and orientation phenomena in the direction of shear. It was shown that the Mooney viscosity gradually decreased as ENR content in the blend increased, which was attributed to the relatively low molecular weight of the ENR. It was observed that the viscosity of NR compound was strictly dependent on the inclusion of organoclay, different than observed in other conventional fillers, such as carbon black or silica. These conventional fillers normally give rise to an increase in the viscosity and modify the rheological properties of the elastomer [117, 118]. The authors also pulled attention to the fact that for the organoclays, the short modifier groups on the clay platelets could act as a plasticiser. It was reported by the authors that the incorporation of the Nanofillers hardly affect to the NR/ENR viscosity peak. They observed a slight increase in viscosity after shearing due to both the hydrodynamic effect and the formation of rubber-filler interactions. This effect was found to be more pronounced in the presence of the organoclays and as both ENR content and epoxy units in the ENR increased. The improvement in the filler-rubber compatibility by the addition of ENR, not only causes a better dispersion of the clay particles in the rubber matrix with an increase of interphase but also increases the filler-rubber interactions and decreases the agglomeration of filler particles. According to all these factors, an increment in the final viscosity is expected.

Sadhu and Bhowmick [119] investigated the rheological properties of montmorillonite clay based nanocomposites using three different grades of acrylonitrile butadiene rubber (NBR) (19, 34, and 50 % acrylonitrile contents), styrene butadiene rubber (SBR), and polybutadiene rubber (BR). Rheological study was carried out on these nanocomposites at three different temperatures (110, 120, and 130 °C) over a range of shear rates for comparison. The authors pulled following general conclusions, which are also valid for many clay based rubber nanocomposites:

1. The nanocomposites exhibited shear thinning behavior. The shear viscosity and the die swell decreased with addition of the modified nanoclay.
2. The polarity of the rubber had a very pronounced effect on the processing behavior of the nanocomposites. The higher the polarity, the higher was the decrease in shear viscosity with the incorporation of the nanoclay. The die swell of the unmodified and the modified clay filled matrix was lower than that of the gum rubber.
3. The nature of the rubber, which controls the extent of intercalation as well as the dispersion level of the Nanofiller, controlled the rheological behavior. Therefore, the trend of changes in rheology with the addition of the nanoclay was different for the different rubber systems.
4. Up to an optimum filler loading level, the viscosity decreased depending on the nature of the rubber, beyond this point; it increased due to agglomeration of the

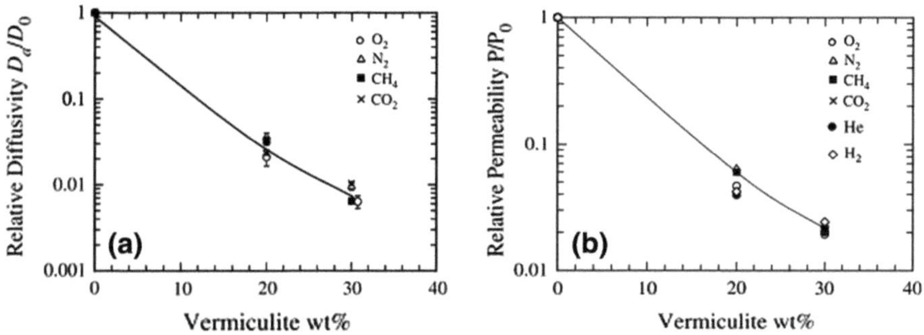


**Fig. 38** Relative diffusivity and relative permeability of butyl rubber/vermiculity nanocomposites for various gases measured at 30C and 2 atm [120]

clay layers. On the other hand, the die swell always decreased with increasing loading, as expected.

The nanoparticles in the rubber matrix may affect the gas permeation properties of rubber by creating a physical nanoscale barrier inside the rubber matrix creating a tortuous pathway for diffusion. One of the most informative studies from the literature deals with the gas permeation properties of nanocomposites based on butyl rubber with high loadings of vermiculite, i.e., 20 and 30 wt% [120]. It was reported that the gas permeability was reduced by 20–30-fold by the vermiculite (see Fig. 38). Researchers obtained solubility coefficients by dividing the experimental permeability by the diffusivity obtained from the time lag observation. Interestingly they found that these values increased significantly with vermiculite content in contradiction to that expected by theories. To be sure, researchers performed independent gas sorption isotherm experiments for $CO_2$ on butyl rubber compositions with 0 and 30 wt% vermiculite. The results also showed larger solubility in the nanocomposite than in the neat butyl rubber.

## 5 Applications of Rubber Nanocomposites

It was already shown in the literature that nanocomposites have improved performance in comparison to matrices containing conventional, micron-sized fillers due to their high surface area and aspect ratios. These improved properties are achieved at much lower additive concentrations, <10 % by weight, compared to conventional systems. Because of these properties, rubber based nanocomposites have pulled attention from rubber industry.

One of the potential applications of rubber-based nanocomposites is the tire component industry. The utilization of nanoparticles in rubber matrices for tire industry is driven by high tire performance (improved traction, less heat built-up and better wear properties), desired fuel efficiency by reducing the weight (decreasing the amount of filler for a specific property achievement) and easier processing due to the reduction of the viscosity.

In the summer of 2003, Nanoproducts Corporation had reported the successful use of its nanoscale materials, silicon carbide, to produce currently commercially available tires with improved skid resistance and nearly 50 % reduced abrasion [121]. The inventors claimed in the patent (US Patent 6,469,089) that they achieved a successful dispersion of nanoscale materials in elastomers. Improved wet skid resistance resulted in superior safety, and reduced abrasion provided extended life. Cabot, one of the world's leading tire-rubber producers, successfully tested "PureNano" silicon carbide nanoparticles designed by Nanoproducts Corporation. Because of the reduced abrasion, tires last up to twice as long and thereby significantly reduce the need for new tire-rubber. At present, 16.5 million tires are retread every year in the US alone. Presumably that number would shrink by almost half [122].

The other application is to produce airtight surfaces by the help of the nanoparticles in the rubber. Companies such as Inmat and Nanocor produce nanoparticles of clay that can be mixed with plastics and synthetic rubber to create an airtight surface. Inmat's nanoclay has already been used as a sealant for "double core" tennis balls produced by Wilson. The Double Core balls have twice the bounce-capacity because the nano-particles lock in air more effectively. Inmat, which was originally set up in co-operation with Michelin, the world's leading tire manufacturer, believes the same technology may be used to seal the inside of tires, reducing the amount of butyl rubber required and making tires lighter, cheaper and cooler running. For example, isobutylene-based rubbers are known to exhibit good barrier for air molecules; therefore they find application in tire inner tubes and inner liners. Inclusion of usually alkyl-based organo-montmorillonite clay can further enhance the permeation barrier creating a tortuous diffusion pathway. ExxonMobil Company developed a brominated isobutyl methyl styrene (BIMS) rubber blend nanocomposite (using Exxpro) with montmorillonite, which reduces air permeation by 38 % [123]. In addition, it was claimed that this technique was less expensive and offered weight savings of up to 1 kg for large truck tires which could result in fuel efficiency. Moreover, it was reported that tire curing was faster due to the higher thermal conductivity of the clay.

The patent of Inmat encloses the information about special tire liners containing smectic silicates making tires to hold air longer with reduced weight and improved milage [124]. In this technology, butyl rubber compounded with exfoliated vermiculite used as a 20 mm coating that can be applied in tire inner liners. Thinner coating allows recycling easier and more environmentally friendly.

Several patents (EP 1,193,085, EP 1,273,616, US 2003/0004250 A1) from well-known tire company of Goodyear deals with the improvement of the dynamic and static modulus of tire compound by using Nanofillers loaded between 5 and 20 phr. Pirelli was announced first application of nanocomposite technology in 2004 for their winter segment tires. Scientist of Yokohama Research Center has developed a tire compound by bonding nanosilica (extra-fine structure), polymer and carbon-fillers to improve the grip properties of Advan Sport Tire. Kumho has a patent on nanocomposites for thermal stability improvement (KR 2002/047892).

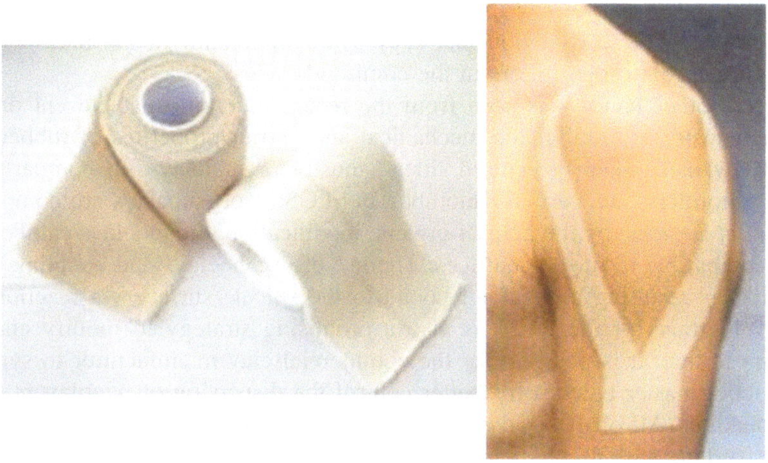

**Fig. 39** The pictures of Elastoguard [127]

Another interesting non-tire application of Nanofillers is nanosized $CaCO_3$ [125]. The reinforcing and barrier effect of nano-$CaCO_3$ is combined with NR and NR/NBR blends to produce laminated sheets for inflated balls and NR based cycle tubes.

Graphite nano-sheets found also an application in sports. A golf ball contains an intermediate nanocomposite layer of exfoliated graphite with a thickness $<5$ nm dispersed in elastomeric matrix of NR, polyisoprene, polybutadiene, SBR, styrene-propylene or ethylene-diene block copolymer rubbers [126].

Nano-antimicrobial additives are widely used in polymer industry to produce antimicrobial medical goods. Milliken Chemical Speciality Elastomers came up with a pro-active antimicrobial rubber (HNBR- or EPDM-based, Trade name: Elastoguard) that provides residual protection against microbial contamination due to silver nanoparticles. The advantage of Elastoguard is to reduce the necessity for a traditional routine decontamination service agenda [127]. This product includes a zirconium phosphate-based ceramic ion-exchange resin with silver, which can be used in food contact applications and medical purposes (Fig. 39).

## 6 Conclusions

Today, the most commonly used filler in rubber systems is carbon black with many different grades and amorphous silica. The one of the most important reasons in utilization of carbon black and silica in rubber systems, especially in tire industry, is to achieve longer-wearing products, increased tire strength and longevity and, especially for silica, decreased greenhouse gas emissions of the vehicles using these tires. In order to achieve these properties, the dispersion state and the filler/polymer adhesion have crucial importance. These two filler types are still in the same physical form as since the start of their use by the rubber industry and for the coming decade. It is obvious that they will dominate the rubber industry as the conventional reinforcing fillers in the coming decades.

On the other hand, it is seen from the recent literature that current drive in rubber industry is to tailor the mechanical and barrier properties of rubber compounds by using nano-structured fillers, mostly organoclays. In comparison to other Nanofillers, i.e., carbon nanotubes and POSSs, organoclays are comparably economic. Besides, by the help of organic modification, organoclays can be made more compatible to rubber matrices. The use of organoclays and coupling agents like silanes, resulting a covalently bond to the silicate surface via reaction with hydroxyl groups of the latter, is also a promising strategy to modify clays. In addition to that, it is possible for the commercial clay manufacturer to synthesis clay/rubber master batches for better control the dispersion of nanolayers during compounding. All of these advantageous indicate that the usage of organoclay in rubber system will be challenging in the coming years for both academy and industry.

## References

1. Savran, H.Ö.: Elastomer Teknolojisi I. Kauçuk Derneği Yayınları, İstanbul (2001)
2. Xanthos, M.: Functional Fillers for Plastics. Wiley-VCH, Germany (2005)
3. Simpson, R.B.: Rubber Basics. Rapra Technology Ltd., London (2002)
4. Rodgers, B.: Rubber Compounding Chemistry and Applications. Marcel Dekker Inc., New York (2004)
5. Ghosh, P.: Polymer Science and Technology, Plastics, Rubbers, Blends and Composites. Tata McGraw-Hill Publishing, New Delhi (2002)
6. White, J.R., De, S.K.: Rubber Technologist's Handbook. Rapra Technology Ltd., Shawbury (2001)
7. ASTM D 1765-00a.: Standard Classification System for Carbon Black Used in Rubber Products. American Society for Testing and Materials, West Conshohocken (2000)
8. Gachter, R., Müller, H.: Plastics Additives Handbook, 4th edn, pp. 525–561. Hanser/Gardner Publications, Munich (1993)
9. Vansant, E F., Van Der Voort, P., Vrancken, K.C.: Characterization and Chemical Modification of the Silica Surface. Elsevier Science, Amsterdam (1995)
10. Ciullo, P.A., Hewitt, N.: The Rubber Formulary. Noyes Publication/William Andrew Publishing, Norwich (1999)
11. Yan, H., Sun, K., Zhang, Y., Zhang, Y.: Effect of nitrile rubber on properties of silica-filled natural rubber compounds. Polym. Testing **24**, 32–38 (2005)
12. Choi, S.S.: Improvement of properties of silica-filled natural rubber compounds using polychloroprene. J. Appl. Polym. Sci. **83**, 2609–2616 (2002)
13. Mareri, P., Bastide, S., Binda, N., Crespy, A.: Mechanical behaviour of polypropylene composites containing fine mineral filler: Effect of filler surface treatment. Compos. Sci. Technol. **58**, 747–752 (1998)
14. Gonzales, J., Albano, C., Ichazo, M., Diaz, B.: Effects of coupling agents on mechanical and morphological behavior of the PP/HDPE blend with two different $CaCO_3$. Eur. Polymer J. **38**, 2465–2475 (2004)

15. Ismail, H., Schuhelmy, S., Edyham, M.R.: The effect of silane couple agent on curing characteristics and mechanical properties of bamboo-filled rubber composites. Eur. Polymer J. **38**, 39–47 (2002)
16. Metin, D., Tihminlioğlu, F., Balköse, D., Ülkü, S.: The effect of interfacial interactions on the mechanical properties of polypropylene/natural zeolite composites. Compos. A Appl. Sci. Manuf. **35**, 23–32 (2004)
17. Hewitt, N.: Compounding Precipitated Silica in Elastomers. William Andrew Publishing, Norwish (2007)
18. Ozkoc, G., Bayram, G., Bayramlı, E.: Effects of polyamide 6 incorporation to the short glass fiber reinforced ABS composites: an interfacial approach. Polymer **45**, 8957–8966 (2004)
19. Plueddemann, E.P.: Silane Coupling Agents, 2nd Edn. Plenum Press, New York (1991)
20. Whelan, A., Lee, K.S.: Developments in Rubber Technology 1. Applied Science Publishers Ltd, London (1979)
21. Susmita, S., Bhowmick, A.K.: Effect of chain length of amine and nature and loading of clay on Styrene-Butadiene rubber-clay nanocomposites. Rubber Chem. Technol. **76**, 860 (2003)
22. Wypych, G.: Fillers. ChemTec Publishing, Toronto (1993)
23. Paul, D.R., Robeson, L.M.: Polymer nanotechnology: Nanocomposites. Polymer **49**, 3187–3204 (2008)
24. Das, A., Stöckelhuber, K.W., Jurk, R., Jehnichen, D., Heinrich, G.: A general approach to rubber-montmorillonite nanocomposites: Intercalation of stearic acid. Appl. Clay Sci. **51**, 117–125 (2011)
25. Dennis, H.R., Hunter, D.L., Chang, D., Kim, S., White, J.L., Cho, J.W., Paul, D.R.: Effect of melt processing conditions on the extent of exfoliation in organoclay-based nanocomposites. Polymer **42**, 9513–9522 (2001)
26. Whelan, T.: Polymer Technology Dictionary, 1st edn. Chapman and Hall Publishing, London (1994)
27. Wang, L.L., Tong, Y.P., Zhang, L.Q., Tian, M.: Comparison of Ethylene-Propylene Diene terpolymer composites filled with natural and synthesized micas. J. Appl. Polym. Sci. **116**, 3184–3192 (2010)
28. Miles, I.S., Rostami, S.: Multicomponent Polymer Systems. Wiley, New York (1992)
29. Oscar, F., III Noel.: Talc, the solution to challenges in automotive. Rubber World **28**(7), 244 (2011)
30. Noel, O., Meli, G.: Synergism of talc with carbon black, Technical paper No: 13 presented at the 174th ACS Rubber Division Meeting in Louisville, KY, (October, 2008)
31. Noel, O., Meli, G., Thakkar, H.: Talc as a dispersion aid for reinforcing fillers in rubbers. Rubber World, Technical meeting of the rubber division, American Chemical Society (2008)
32. Harper, C.: Handbook of Plastics Technologies. McGraw Hill Handbooks, New York (2006)
33. Dick, J.S.: Rubber Technology, Compounding and Testing for Performance. Hanser Gardner Publication, Cincinnati (2001)
34. Morton, M.: Rubber Technology, 3rd edn. Kluwer Academic Publishers, Boston (1987)
35. Fina, A., Tabuani, D., Frache, A., Camino, G.: Polypropylene-polyhedral oligomeric silsesquioxanes (POSS) nanocomposites. Polymer **46**, 7855–7866 (2005)
36. Li, G., Wang, L., Ni, H., Pittman, C.U.: Polyhedral Oligomeric Silsesquioxanes (POSS) polymers and copolymers: A review. J. Inorg. Organomet. Polym. **11**, 123 (2001)
37. Luecke, S., Stoppek-Langner, K.: Polyhedral oligosilsesquioxanes (POSS)-building blocks for the development of nano-structured materials. Appl. Surf. Sci. **713**, 144–145 (1999)
38. Chen, D., Yi, S., Fang, P., Zhong, Y., Huang, C., Wu, X.: Synthesis and characterization of novel room temperature vulcanized (RTV) silicone rubbers using octa[(trimethoxysilyl)ethyl]-POSS as cross-linker. React. Funct. Polym. **71**, 502–511 (2011)
39. www.hybridplastics.com, "POSS User's Guide"
40. Rothon, R.N.: Particulate Fillers for Polymers. Rapra Review Reports (2001)

41. Alger, M.: Polymer Science Dictionary, 2nd edn. Chapman and Hall, London (1997)
42. Dick, J.S.: Basic Rubber Testing: Selecting Methods for a Rubber Test Program. ASTM International, Pennsylvania (2003)
43. Reich, S., Thomsen, C., Maultzsch, J.: Carbon Nanotubes, Basic Concepts and Physical Properties. Wiley-VCH, Germany (2004)
44. Mahmoud, W.E., El-Mossalamy, E.H., Arafa, H.M.: Improvement of the mechanical and electrical properties of waste rubber with carbon nanotubes. J. Appl. Polym. Sci. **121**, 502–507 (2011)
45. Zhou, X., Zhu, Y., Liang, J., Yu, S.: New fabrication and mechanical properties of Styrene-Butadiene Rubber/Carbon. J. Mater. Sci. Technol. **26**(12), 1127–1132 (2010)
46. Kane, C.L., Mele, E.J.: Size, shape and low energy electronic structure of carbon nanotubes. Phys. Rev. Lett. **78**(10), 1932 (1997)
47. Raymond, M.R.: Carbon nanotubes: Potential benefits and risks of nanotechnology in nuclear medicine. J. Nucl. Med. **48**, 1039–1042 (2007)
48. Perez, L.D., Zuluaga, M.A., Kyu, T., Mark, J.E., Lopez, B.L.: Preparation, characterization, and physical properties of multiwall carbon nanotube/elastomer composites. Polym. Eng. Sci. **49**, 866–874 (2009)
49. Kundu, S.: Surface Modification of Carbon Nanotubes and their Application in Electro-Catalysis, Degree of doctor of natural science. Ruhr University Department of Chemistry, Bochum (2009)
50. Gong, K., Zhu, X., Zhao, R., Xiong, S., Mao, L., Chen, C.: Rational attachment of synthetic triptycene orthoquinone onto carbon nanotubes for electrocatalysis and sensitive detection of thiols. Anal. Chem. **77**, 8158–8165 (2005)
51. Fu, S., Y., Feng, X.Q., Lauke, B., Mai, Y.W.: Effects of particle size, particle/matrix interface adhesion and particle loading on mechanical properties of particulate-polymer composites. Composites. B **39**, 933–961 (2008)
52. Li, Z.H., Zhang, J., Chen, S.J.: Effects of carbon blacks with various structures on vulcanization and reinforcement of filled ethylene-propylene-diene rubber. Express Polym. Lett. **2**(10), 695–704 (2008)
53. Soltani, S., Sourki, F.A.: Effect of carbon black type on viscous heating, heat build-up and relaxation behaviour of SBR compounds. Iran. Polym. J. **14**(8), 745–751 (2005)
54. Saatchi, M.M., Shojaei, A.: Mechanical performance of styrene-butadiene-rubber filled with carbon nanoparticles prepared by mechanical mixing. Mater. Sci. Eng. A **528**,7161–7172 (2011)
55. Fröhlich, J., Niedermeier, W., Luginsland, H.: The effect of filler-filler and filler-elastomer interaction on rubber reinforcement. Compos. A **36**, 449–460 (2005)
56. Nah, C., Lim, J., Y., Cho, B.H., Hong, C.K., Gent, A.N: Reinforcing rubber with carbon nanotubes. J. Appl. Polym. Sci. **118**, 1574–1581 (2010)
57. Nah, C., Lim, J.Y., Sengupta, R., Cho, B.H., Gent, A.N.: Slipping of carbon nanotubes in a rubber matrix. Poly. Int. **60**, 42–44 (2011)
58. Vinod, V.S., Varghese, S., Alex, R., Kuriakose, B.: Effect of aluminum powder on filled natural rubber composites. Rubber Chem. Technol. **74**(2), 236–248 (2001)
59. Zhou, W., Yu, D., Wang, C., An, Q., Qi, S.: Effect of filler size distribution on the mechanical and physical properties of alumina-filled silicone rubber. Polym. Eng. Sci. **48**, 1381–1388 (2008)
60. Zhou, W.Y., Qi, S.H., Tu, C.C.: Novel heat-conductive composite silicone rubber. J. Appl. Polym. Sci. **104**, 2478 (2007)
61. Bae, J.W., Kim, W., Cho, S.H.: The properties of AlN-filled epoxy molding compounds by the effects of filler size distribution. J. Mater. Sci. **35**, 5907 (2000)
62. Zhou, W., Qi, S., Tu, C., Zhao, H., Wang, C., Kou, J.: Effect of the particle size of $Al_2O_3$ on the properties of filled heat-conductive silicone rubber. J. Appl. Polym. Sci. **104**, 1312–1318 (2007)
63. Zhang, Q., Tian, M., Wu, Y., Lin, G., Zhang, L.: Effect of particle size on the properties of $Mg(OH)_2$-filled rubber composites. J. Appl. Polym. Sci. **94**, 2341–2346 (2004)

64. Ismail, H., Rozman, R.H., Jaffri, R.M., Mohd Ishak, Z.A.: Oil palm wood flour reinforced epoxidized natural rubber composites: the effect of filler content and size. Eur. Polym. J. **33**, 1627–1632 (1997)
65. Mostafa, A., Abouel-Kasem, A., Bayoumi, M.R., El-Sebaie, M.G.: The influence of CB loading on thermal aging resistance of SBR and NBR rubber compounds under different aging temperature. Mater. Des. **30**, 791–795 (2009)
66. Mostafa, A., Abouel-Kasem, A., Bayoumi, M.R., El-Sebaie, M.G.: Effect of carbon black loading on the swelling and compression set behavior of SBR and NBR rubber compounds. Mater. Des. **30**, 1561–1568 (2009)
67. Ain, Z.N., Azura, A.R.: Effect of different types of filler and filler loadings on the properties of Carboxylated Acrylonitrile-Butadiene rubber latex films. J. Appl. Polym. Sci. **119**, 2815–2823 (2011)
68. Daniele, F.C., Joao, C.M.S., Regina, C.R.N., Leila, L.Y.V.: Effect of mica addition on the properties of natural rubber and polybutadiene rubber vulcanizates. J. Appl. Polym. Sci. **90**, 2156 (2003)
69. Mostafa, A., Abouel-Kasem, A., Bayoumi, M.R., El-Sebaie, M.G.: Insight into the effect of CB loading on tension, compression, hardness and abrasion properties of SBR and NBR filled compounds. Mater. Des. **30**, 1785–1791 (2009)
70. Nakamura, Y., Honda, H., Harada, A., Fujii, S., Nagata, K.: Mechanical properties of silane-treated, silica-particle-filled polyisoprene rubber composites: Effects of the loading amount and alkoxy group numbers of a silane coupling agent containing mercapto groups. Inc.J Appl. Polym. Sci. **113**, 1507–1514 (2009)
71. Suzuki, N., Ito, M., Yatsuyanagi, F.: Effects of rubber/filler interactions on deformation behavior of silica filled SBR systems. Polymer 193–201 (2005)
72. Wang, J., Wu, W., Wang, W., Zhang, J.: Effect of a coupling agent on the properties of Hemp-Hurd-Powder-Filled styrene-butadiene rubber. Inc. J. Appl. Polym. Sci. **121**, 681–689 (2011)
73. Park, S.J., Cho, K.S.: Filler-elastomer interactions: influence of silane coupling agent on crosslink density and thermal stability of silica/rubber composites. J. Colloid Interface Sci. **267**, 86–91 (2003)
74. Idrus, S.S., Ismail, H., Palaniandy, S.: Study of the effect of different shapes of ultrafine silica as fillers in natural rubber compounds. Polym. Testing **30**, 251–259 (2011)
75. Leblanc, J.L.: Rubber-filler interactions and rheological properties in filled compounds. Prog. Polym. Sci. **27**, 627–687 (2002)
76. Montes, S., White, J.L., Nakajima, N.: Rheological behavior of rubber carbon black compounds in various shear histories. J. Non-Newtonian Fluid Mech. **28**, 183–212 (1988)
77. Medalia, A.I., Kraus, G.: Reinforcement of elastomers by particulate fillers. In: Mark, J.E., Erman, B., Eirich, F.R. (eds.) Science and Technology of Rubber, 2nd edn, pp. 387–418 (Chap. 8). Academic, New York (1994)
78. Saad, I.S., Fayed, M.Sh., Abdel-Bary, E.M.: Effects of carbon black content on cure characteristics, mechanical properties and swelling behaviour of 80/20 NBR/CIIR blend. In: 13th International Conference on Aerospace Sciences and Aviation Technology, Paper: ASAT-13-CA-06 (2009)
79. Kemaloglu, S., Ozkoc, G., Aytac, A.: Properties of thermally conductive micro and nano size boron nitride reinforced silicon rubber composites. Thermochim. Acta **499**, 40–47 (2010)
80. Medalia, A.I.: Electrical conduction in carbon black composites. Rubber Chem. Technol. **59**, 432–454 (1986)
81. El-Wakil, A.A., Abd El-Megeed, A.A.: Effect of calcium carbonate, silitin N85 and carbon black fillers on the mechanical and electrical properties of the EPDM. ARPN J. Eng. Appl. Sci. **6**, 24–29 (2011)
82. Wagner, M.P: Reinforcing silicas and silicates. Rubber Chem. Technol. **49**, 704 (1976)
83. Butler, J., Freakley, P.K.: Effect of humidity and water content on the cure behavior of a natural-rubber accelerated sulfur compound. Rubber Chem. Technol. **65**, 374 (1991)

84. Ismail, H., Shaari, S.M., Othman, N.: The effect of chitosan loading on the curing characteristics, mechanical and morphological properties of chitosan-filled natural rubber (NR), epoxidised natural rubber (ENR) and styrene-butadiene rubber (SBR) compounds. Polym. Testing **30**, 784–790 (2011)
85. Zhang, Y., Liu, Q., Zhang, Q., Lu, Y.: Gas barrier properties of natural rubber/kaolin composites prepared by melt blending. Appl. Clay Sci. **50**, 255–259 (2010)
86. Sadhu, S., Bhowmick, A.K.: Effects of chain length of amine and nature loading of clay on Styrene-Butadiene Rubber-clay nanocomposites. Rubber Chem. Technol. **76**, 860 (2003)
87. Sadhu, S., Bhowmick, A.K.: Preparation and properties of styrene-butadiene rubber based nanocomposites: The influence of the structural and processing parameters. J. Appl. Polym. Sci. **92**, 698 (2004)
88. Kojima, Y., Fukumori, K., Usuki, A., Okada, A., Kurauchi, T.: Gas permeabilities in rubber-clay hybrid. J. Mater. Sci. Lett. **12**, 889 (1993)
89. Vu, Y.T., Mark, J.E., Pham, L.Y.H., Engelhardt, M.: Clay nanolayer reinforcement of cis-1,4-polyisoprene and epoxidized natural rubber. J. Appl. Polym. Sci. **82**, 1391–1403 (2001)
90. Sengupta, R., Chakraborty, S., Bandyopadhyay, S., Dasgupta, S., Mukhopadhyay, R., Auddy, K., Deuri, A.S.: A short review on rubber/clay nanocomposites with emphasis on mechanical properties. Polym. Eng. Sci. **47**, 1956–1974 (2007)
91. Joly, S., Garnaud, G., Ollitrault, R., Bokobza, L.: Organically modified layered silicates as reinforcing fillers for natural rubbers. Chem. Mater.,**14**, 4202 (2002)
92. Magaraphan, R., Thaijaroen, W., Lim-Ochakun, R,: Structure and properties of natural rubber and modified montmorillonite nanocomposites. Rubber Chem. Technol. **76**, 406 (2003)
93. Lim, S.K., Kim, J.W., Chin, I.J., Choi, H.J.: Rheological properties of a new rubbery nanocomposite: Polyepichlorohydrin/organoclay nanocomposites. J. Appl. Polym. Sci. **86**, 3735 (2002)
94. Wang, Y., Zhang, L., Tang, C., Yu, D.: Preparation and characterization of rubber-clay nanocomposits. J. Appl. Polym. Sci. **78**, 1879 (2000)
95. Wang, Y., Zhang, H., Wu, Y., Yang, J., Zhang, L.: Preparation, structure, and properties of a novel rectorite/styrene-butadiene copolymer nanocomposite. J. Appl. Polym. Sci. **96**, 324 (2005)
96. Vaia, R.A., Ishii, H., Giannelis, E.P.: Synthesis and properties of two-dimensional nanostructures by direct intercalation of polymer melts in layered silicates. Chem. Mater. **5**, 1694 (1993)
97. Arroyo, M., Lopez-Manchado, M.A., Herrero, B.: Organo-montmorillonite as substitute of carbon black in natural rubber compounds. Polymer **44**, 2447 (2003)
98. Teh, P.L., Mohd-Isak, Z.A., Hashim, A.S., Karger-Kocsis, J., Ishiaku, U.S.: On the potential of organoclay with respect to conventional fillers (carbon black, silica) for epoxidized natural rubber compatibilized natural rubber vulcanizates. J. Appl. Polym. Sci. **94**, 2438 (2004)
99. Tian, M., Qu, C., Feng, Y., Zhang, L.: Structure and properties of fibrillar silicate/SBR composites by direct blend process. J. Mater. Sci. **38**, 4917 (2003)
100. Usuki, A., Tukigase, A., Kato, M.: Preparation and properties of EPDM-clay hybrids. Polymer **43**, 2185 (2002)
101. Gatos, K.G., Thomann, R., Karger-Kocsis, J.: Characteristics of ethylene propylene diene monomer rubber/organoclay nanocomposites resulting from different processing conditions and formulations. Polym. Int. **53**, 1191 (2004)
102. Treacy, M., Ebbesen, T.W., Gibson, J.M.: Exceptionally high Young's modulus observed for individual carbon nanotubes. Nature **381**, 678, 1, (1996)
103. Ijima, S.: Helical microtubules of graphitic carbon. Nature **354**, 56 (1991)
104. Saito, R., Dresselhaus, G., Dresselhaus, M.S.: Physical Properties of Carbon Nanotubes, pp. 1–4. Imperial College Press, London (1999)

105. Nazlia, G., Fakhru'l-Razi, A., Suraya, A.R., Muataz, A.A.: Multi-walled carbonnanotubes/ styrene butadiene rubber (SBR) nanocomposite. J. Fullerenes, Nanotubes, Carbon Nanostruct. **15**(3), 207–214 (2007)
106. Muataz, A.A., Nazif, N., Faridah, Y., Mohammed, F., Chantara, T.R., Mamdouh, A., Faraj, A.A., Khalid, M., Adnan, M.AL-A.: Radiation vulcanization of natural rubber latex loaded with carbon nanotubes. J. Fullerenes, Nanotubes, Carbon Nanostruct. **18**, 61–75 (2010)
107. Kannan, B., Burghard, M.: Chemically functionalized carbon nanotubes. Small **1**, 180–192 (2005)
108. Atieh, M.A.: Effects of functionalized carbon nanotubes with carboxylic functional group on the mechanical and thermal properties of styrene butadiene rubber. J. Fullerenes, Nanotubes, Carbon Nanostruct. **19**, 617–627 (2011)
109. Subramaniam, K., Das, A., Heinrich, G.: Development of conducting polychloroprene rubber using imidazolium based ionic liquid modified multi-walled carbon nanotubes. Compos. Sci. Technol. **71**, 1441–4149 (2011)
110. Aso, O., Eguiazabal, J.I., Nazabal, J.: The influence of surface modification on the structure and properties of a nanosilica filled thermoplastic elastomer. Compos. Sci. Technol. **67**(13), 2854–2863 (2007)
111. Peng, Z., Kong, L.X., Li, S.D., Chen, Y., Huang, M.F.: Compos. Sci. Technol. **67**, 3130–3139 (2007)
112. Marosi, G., Marton, A., Szep, A., Csontos, I., Keszei, S., Zimonyi, E., et al.: Fire retardancy effect of migration in polypropylene nanocomposites induced by modified interlayer. Polym. Degrad. Stabil. **82**, 379–385 (2003)
113. Likozar, B., Major, Z.: Morphology, mechanical, cross-linking, thermal and tribological properties of nitrile and hydrogenated nitrile rubber/multi-walled carbon nanotubes composites prepared by melt compounding: The effect of acrylonitrile content and hydrogenation. Appl. Surf. Sci. **257**, 565–573 (2010)
114. Chen, S., Yu, H., Ren, W., Zhang, Y.,: Thermal degradation behavior of hydrogenated nitrile-butadiene rubber (HNBR)/clay nanocomposite and HNBR/clay/carbon nanotubes nanocomposites. Thermochim. Acta **491**, 103–108 (2009)
115. Brasa, J., Hassan, M.L., Bruzessea, C., Hassan, E.A., El-Wakil, N.A., Dufresne, A.: Mechanical,barrier, and biodegradability properties of bagasse cellulose whiskers reinforced natural rubber nanocomposites. Ind. Crops Prod. **32**, 627–633 (2010)
116. Arroyo, M., Lopez-Manchado, M.A., Valentin, J.L., Carretero, J.: Morphology/behavior relationship of nanocomposites based on natural rubber/epoxidized natural rubber blends. Compos. Sci. Technol. **67**, 1330–1339 (2007)
117. Gibala, D., Laohapisitpanich, K., Thomas, D., Hamed, G.R.: Cure and mechanical behaviour of rubber compounds containing ground vulcanizates. 2. Mooney viscosity. Rubber Chem. Technol. **69**(1), 115–119 (1996)
118. Choi, S.S., Nah, C., Lee, S.G., Joo, C.W.: Effect of filler–filler interaction on rheological behaviour of natural rubber compounds filled with both carbon black and silica. Polym. Int. **52**(1), 23–28 (2003)
119. Sadhu, U., Bhowmick, A.K.: Unique rheological behavior of rubber based nanocomposites. J. Polym. Sci. B: Polym. Phys. **43**, 1854–1864 (2005)
120. Takahashi, S., Goldberg, H.A., Feeney, C.A., Karim, D.P., Farrell, M., O'Leary, K., Paul, D.R.: Gas barrier properties of butyl rubber/vermiculite nanocomposite coatings. Polymer **47**, 3083–3093 (2006)
121. Wang, M., Kutsovsky, Y., Reznek, S.R., Mahmud, K.: Elastomeric compounds with improved wet skid resistance and methods to improve wet skid resistance. US Patent 6469089, Cabot Corp
122. http://www.azonano.com/article.aspx?ArticleID=1351
123. Rogers, B., Webb, R., Wang, W.: Nanocomposite technology in tire inner liners. ACS Central Region al Meeting, Frankenmuth, Mich., 16–20 May, paper no. 140

124. Feeney, C.A., Goldberg, H.A., Farrell, M., et al.: Barrier coating of a non-butyl elastomer and a dispersed layered filler in a liquid carrier and coated articles. US Patent 10741251, to InMat Inc
125. Chakravarty, S.N., Chakravarty, A.: Reinforcement of rubber compounds with Nanofiller. KGK-Kautschuk Gummi Kunststoffe **11**, 619–622 (2007)
126. Sullivan, M.J., Ladd, D.A.: Golf ball containing graphite nanosheets in a polymeric network. US Patent 715756, issued on 11 April, to Acushnet Co
127. http://www.guardee.com/en/

# Magnetorheological Elastomers and Their Applications

**W. H. Li, X. Z. Zhang and H. Du**

**Abstract** Magnetorheological elastomers (MRE) are smart materials whose modulus or mechanical performances can be controlled by an external magnetic field. In this chapter, the current research on the MRE materials fabrication, performance characterisation, modelling and applications is reviewed and discussed. Either anistropic or isotropic or MRE materials are fabricated by different curing conditions where magnetic field is applied or not. Anistropic MREs exhibit higher MR effects than isotropic MREs. Both steady-state and dynamic performances were studied through both experimental and theoretical approaches. The modelling approaches were developed to predict mechanical performances of MREs with both simple and complex structures. The sensing capabilities of MREs under different loading conditions were also investigated. The review also includes recent representative MRE applications such as adaptive tuned vibration absorbers and novel force sensors.

## 1 Introduction to Magnetorheological Materials

Magnetorheological (MR) material is a class of smart materials whose rheological properties can be controlled rapidly and reversibly by the application of an external magnetic field. Traditionally, it is composed of MR fluids and MR foams, while Magnetorheological elastomers (MREs) became a new branch of this kind of smart

W. H. Li (✉) · X. Z. Zhang
School of Mechanical, Materials and Mechatronic Engineering,
University of Wollongong, Wollongong NSW 2522, Australia
e-mail: weihuali@uow.edu.au

H. Du
School of Electrical, Computer and Telecommunications Engineering,
University of Wollongong, Wollongong NSW 2522, Australia

P. M. Visakh et al. (eds.), *Advances in Elastomers I*,
Advanced Structured Materials 11, DOI: 10.1007/978-3-642-20925-3_12,

material. MR materials typically consist of micron-sized magnetic particles suspended in a non-magnetic matrix [1]. In MR fluids, magnetic particles, such as iron or carbonyl iron particles, are suspended in a liquid carrier fluid. MR foams, in which the controllable fluid is contained in an absorptive matrix or magnetic particles are dispersed in a foam-like matrix, are solid-state materials with very low intrinsic modulus [1]. MREs are composites where magnetic particles are suspended in a non-magnetic solid or gel-like matrix. The particles inside the elastomer can be homogeneously distributed or they can be grouped (e.g. into chain-like columnar structures). To produce an aligned particle structure, the magnetic field is applied to the polymer composite during crosslinking so that the columnar structures can form and become locked in place upon the final cure [2–10].

The magnetic interactions between particles in these composites depend on the magnetization orientation of each particle and on their spatial relationship, coupling the magnetic and strain fields in these materials and giving rise to a number of interesting magneto-mechanical phenomena. For example, Shiga et al. [2] prepared a kind of composite gel with magnetic properties. Then, Jolly et al. [3] tested and analysis the mechanical properties of silicon rubber based magneto-rheological elastomer. When the magnetic field was 0.8 T, the shear modulus ratio increased about 40 % of the initial value. A number of groups have studied the effect of volume fraction on the MR effect and concluded the optimal volume fraction is around 27 % [4–7]. Bossis et al. [8] investigated the electromagnetic, optical and other physical properties of MR elastomers. Lokander and Stenberg [9, 10] prepared isotropic MR elastomres and investigated their magnetic, mechanical properties and their oxidation resistance, chemical and physical stability.

Although MR materials have many analogical mechanical behaviors, MREs have a unique mechanical performance different from others material: MREs have a controllable, field-dependent modulus while MR fluids and MR foam have a field-dependent yield stress. This makes the two groups of materials complementary rather than competitive to each other. In other words, the strength of MR fluids is characterized by their field dependent yield stress while the strength of MREs is typically characterized by their field dependent modulus. The unique mechanism difference of MR fluids and MREs have made them to find different applications. MR fluids are widely used to develop semi-active damping devices, such as dampers, clutches and brakes [1]. MREs have found applications in developing adaptive tuned vibration absorbers [11–15].

In addition, MRE exhibits other obvious advantages: the particles in MREs are not able to settle with time and thus that there is no need for containers to keep the MR material in place; the response time of MREs is very short (several milliseconds) because the particles, locked in the matrix, have no time to arrange again while MREs are applied an external magnetic field. All these characteristics have made MREs have a huge market potential.

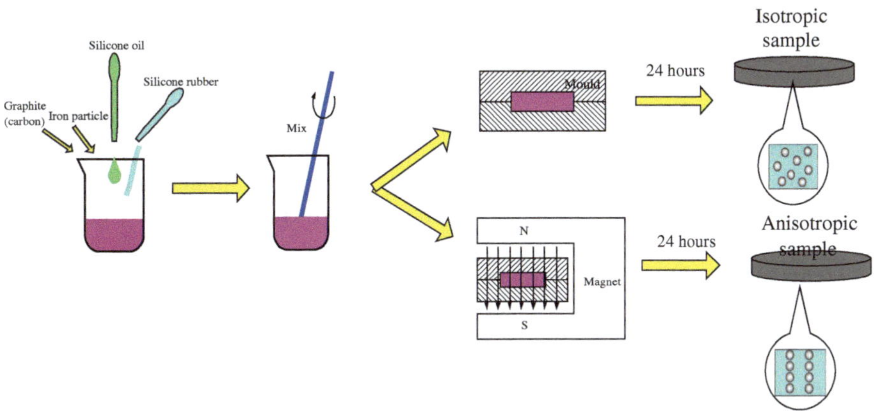


**Fig. 1** Schematic of fabrication of both isotropic and anistropic MREs

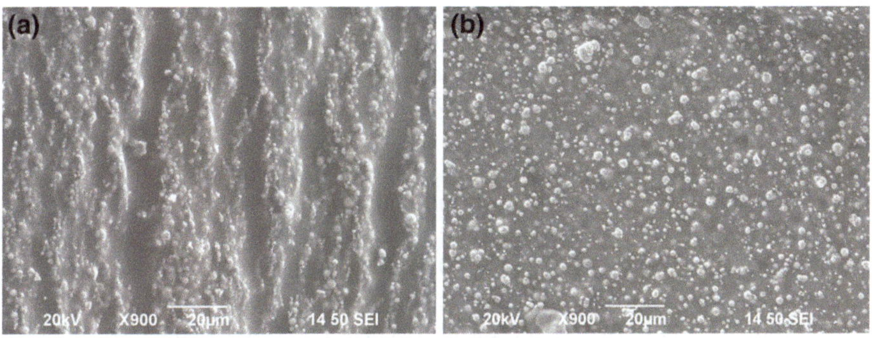


**Fig. 2** SEM images of MREs (**a**) anistropic MRE, and (**b**) isotropic MRE [16]

## 2 Fabrication of Isotropic and Anistropic MR Elastomers

There are two main types of MREs: isotropic and anistropic. The fabrication of these two kinds of MREs depends on whether an external magnetic field is applied or not. Figure 1 is a schematic of the fabrication methods.

The anistropic MRE is a kind of pre-structured magnetic elastomer. It is the main investigated MREs in the literature. During the curing process, an external magnetic field is applied to the mixture of elastomer matrix and magnetic particles. Before the matrix is not cured yet, the magnetic particles in the liquid still can move. They are able to align in the direction of magnetic field to form a chain-like or column structure (see Fig. 2a). After curing, these structures were locked in the matrix. Therefore, pre-structured MRE (also known as prepared under a magnetic field MR elastomer) is an anisotropic MR elastomer.

The isotropic MR elastomer is a kind of unstructured magnetic elastomers. During the curing, No external magnetic field was applied on the mixture, and particles do not form chains or columnar structure (see Fig. 2b). Therefore, unstructured MRE (also known as prepared without magnetic field MR elastomer) is an isotropic MRE.

Curing without magnetic field can greatly simplify the manufacture process, which is a significant advantage for manufacture in large quantities in industry [17–19]. However, the isotropic MREs only have about half field-dependant modulus compared with anisotropic MREs [19–21].

# 3 Steady-State and Dynamic Properties of MREs

## 3.1 Steady-State Properties

The MR effect can be evaluated by measuring the shear strain-stress curve of sample with and without an applied magnetic field. The quasi-static shear method is generally employed to evaluate the shear modulus of MREs. Figure 3 shows the strain-stress curve of a MRE sample at 7 different magnetic field intensity ranging from 0 to 750 mT with a magnetic field step increase of 125 mT. The slope of the strain-stress curve is the shear modulus of the material. As the figure shows, the shear modulus of the increased with the magnetic fields, proves that the MREs exhibited obvious MR effects. The shear stress shows a linear relationship with the shear strain when the strain is within 10 %. This means the MRE acts with linear viscoelastic properties when the strain is below 10 %. This finding is totally different from MR fluids where the linear viscoelastic region is only below 0.1 % [22]. This also demonstrated that MREs generally operate at the pre-yield region

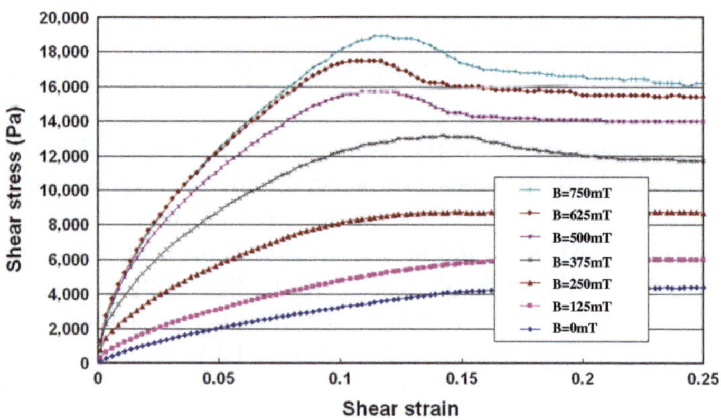


**Fig. 3** Shear stress–strain curves of MRE sample under different magnetic fields

while MRF operates at the post-yield region. When the strain is above 10 % the modulus reaches a maximum value and then steadily decreases, which implies that MRE reaches nonlinear-viscoleastic regime.

## 3.2 Dynamic Properties

Under dynamic loading, MREs exhibit viscoelastic properties. In our experimental study [23], harmonic loadings with various strain amplitudes ranging from 1 to 50 %, and frequencies from 1 to 10 Hz, were used to study dynamic properties of MRE samples. Figure 4 shows the stress–strain relationships of the MRE sample at a constant strain amplitude of 10 % but at various magnetic fields from 0 to 440 mT. It can be seen from this figure that all stresses and strains from nice elliptical shapes, the areas of which increase steadily with the increment of the magnetic fields. These results demonstrate that MRE materials have controllable mechanical properties. An increase in the stress–strain loop area with the magnetic field demonstrates that the damping capacity of the MR elastomers is a function of the applied magnetic field. These experimental results demonstrate that not only are the areas dependent on the magnetic fields, the shape of the ellipse loops are also different. In particular, the slope of the main axis of the elliptical loops varies with the magnetic field, which means that the modulus of MREs varies with magnetic field. Therefore, MREs exhibit variable stiffness and damping properties. This feature is totally different from ER and MR fluids, which mainly exhibit damping controllable properties.

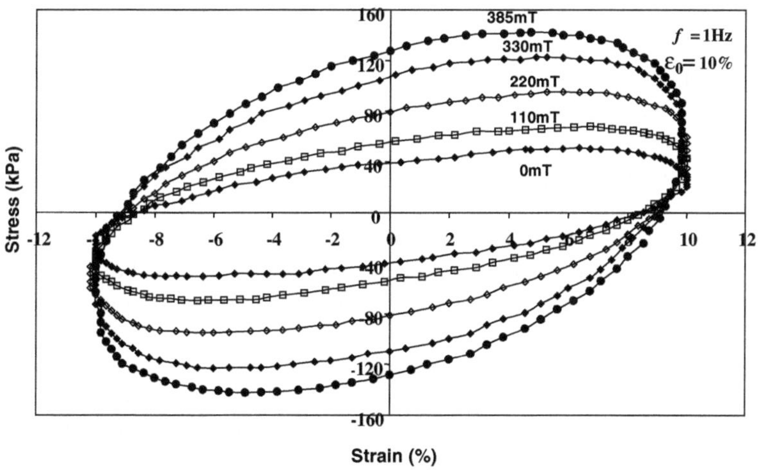


**Fig. 4** Stress–strain relationships with various magnetic fields at a constant strain amplitude of 10 % [23]

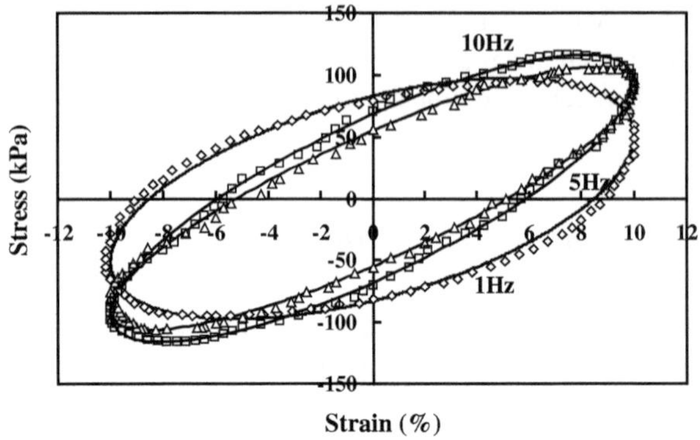


**Fig. 5** Stress–strain relationships for different frequency inputs in 220 mT [23]

Figure 5 shows the effects of shear strain frequency inputs on MRE performance. The slope of the main axis of the elliptical loops increased with an increasing shear strain frequency inputs while the maximum stress amplitudes of different strain frequency inputs changed steadily. The shear strain frequency inputs mainly influences the slope of main axis of the ellipses. When the strain frequency inputs change from 1 to 5 Hz, the slope of the main axis increases steadily. When the strain frequency inputs change from 5 to 10 Hz, the slope of the main axis increases slightly. This implies that the stiffness of the system shows a slowly increasing trend with the frequency.

### *3.3 MRE Modeling*

A number of non-parametric and parametric models have been developed to describe MREs performances. Jolly et al. [24] proposed a quasi-static model to explain the modulus increase by calculating the magnetic interaction between the adjacent particles. Davis [25] used finite element methods to analyze the modulus increase under varied magnetic fields. Shen et al. [26] developed a model that takes into account the magnetic interactions of dipoles in the same chain and the nonlinear properties of the host composites. Most of models are based on the basis of the dipole model for particle energy interaction, and they have the assumption that the particles are the same size and shape. These models are based on the previous studies on MR fluids and particle chains structure. Their results showed that the field-dependent modulus of MREs varies with the square of the saturation magnetization of the particles. Zhang et al. [27] analyzed the complex structure with different size particles or coating particles. The saturation of MREs has also been taken into account. Their theoretical and experimental results indicated that the MREs fabricated with particles have different sizes can provide larger field-dependent modulus.

Recently, we developed a parametric model to describe MRE dynamical performances. The experimental results, shown in Figs. 4 and 5, indicate that the response stresses and input strains formed perfect elliptical loops when the strain

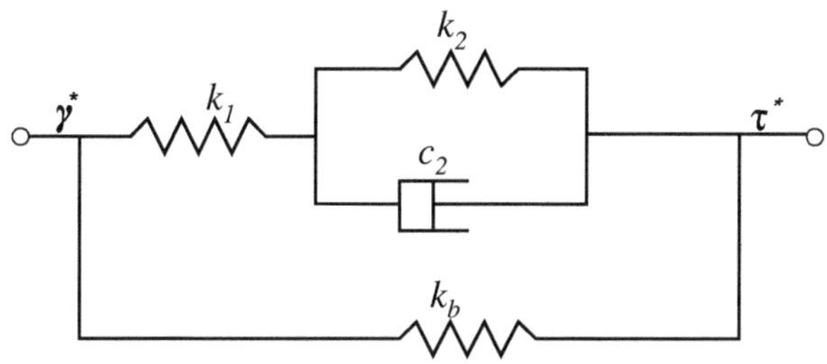


**Fig. 6** Four-parameter viscoelastic model for MR elastomers

amplitude is not above 10 % and the excitation frequency is lower than 10 Hz. These results demonstrate that the MREs behave as linear viscoelastic properties. Viscoelastic properties of MR fluids have obtained intensive studies through experimental and modeling approaches [28, 29]. In particular, a classical Kelvin-Voigt model and three-parameter standard solid model were used to model MR fluids at pre-yield region. Such linear viscoelastic model was mainly developed to process damping capabilities of ER and MR fluids. Compared with MR fluids, the MRE material exhibits a feature that its modulus and damping capability are both field dependent. To predict these characteristics, a four-parameter viscoelastic model, as shown in Fig. 6 was proposed by extending the classical three-parameter standard solid model. In this model, a spring element is introduced, which is in parallel with the standard model, for representing the field dependence of modulus.

In this model, $k_1$, $k_2$, and $C_2$ form a standard viscoelastic solid model, which mainly deals with the damping capabilities of the model, while $k_b$ represents the field dependence of modulus. Suppose the input complex strain is $\gamma^*$ while the output complex stress is $\tau^*$, and the complex modulus is $G^*$, the stress–strain relationship is given by

$$\tau^* = G^* \gamma^* = (G_1 + iG_2)\gamma^* \tag{1}$$

where $G_1$ and $G_2$ are real and imaginary parts of the complex modulus, respectively. They can be derived by using the theory of linear viscoelasticity.

$$G_1 = \frac{(k_1 k_b + k_2 k_b + k_1 k_2)[(k_1 + k_2)^2 + c_2^2 \omega^2] + c_2^2 \omega^2 k_1^2}{(k_1 + k_2)[(k_1 + k_2)^2 + c_2^2 \omega^2]} \tag{2}$$

$$G_2 = \frac{c_2 \omega k_1^2}{[(k_1 + k_2)^2 + c_2^2 \omega^2]} \tag{3}$$

where $\omega$ is the driving frequency.

Suppose that the shear strain input $\gamma$ is a harmonic input

$$\gamma(t) = \gamma_0 \sin(\omega t) \tag{4}$$

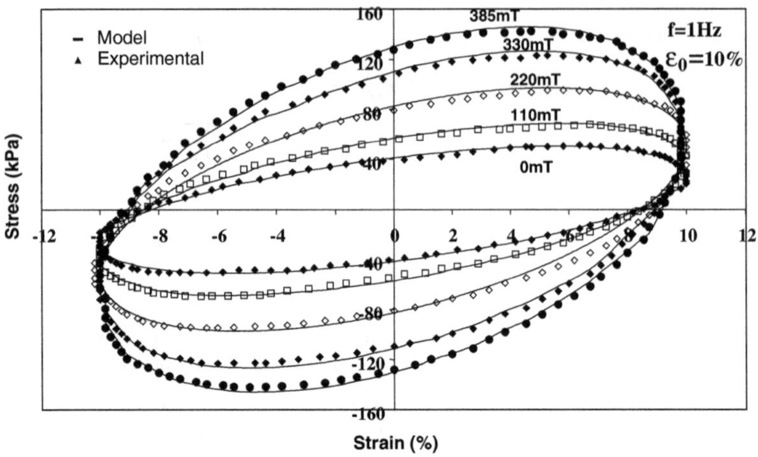


**Fig. 7** A comparison between experimental data with model-predicted results with the amplitude input of 10 %

The steady-state response of the stress can be obtained as

$$\tau(t) = \gamma_0 \sqrt{G_1^2 + G_2^2}\, \sin(\omega t + \phi) \tag{5}$$

where $\phi$ is the phase angle difference between the input and output, which can be calculated as $\phi = \tan^{-1}(G_2/G_1)$.

The developed viscoelastic model includes four parameters, i.e. $k_b$, $k_1$, $k_2$, $c_2$. The model uses the shear strain as an input, calculates the modulus $G_1$ and $G_2$ needed for the model, and then gives the shear stress, given by the Eq. (5). The four parameters are estimated on the base of the least squares method to minimise the error between the model-predicted stress and the experimental result. Using the parameters estimated from the system identification process, the stress versus strain was reconstructed and compared with the experimental data curve. Figure 7 shows the reconstructed stress versus strain loops compared with the practical experimental curves. It can be seen from the plots that the model can simulate the experimental data very well.

# 4 Sensing Properties of MREs

## 4.1 MRE Resistances

Unlike the study of MR effect of MREs, the research on MRE's sensing capabilities is quite few. MREs are generally insulating materials, the resistances of which are quite high. However, the MRE resistance value is dependent on the deformation strain and applied magnetic field. Bossis et al. [8] found that the MRE resistance can have a huge resistance drop by five orders of magnitude when the magnetic field had a decrease of 6,000 Oersted. Following this research, a number of groups have reported the sensing capabilities of different MRE materials. Kchit and Bossis [30] found that the initial resistivity of metal powder at zero pressure is about 108 Ωcm for pure nickel powder and 106 Ωcm for silver coated nickel particles. The change in resistance with pressure was found to be an order of magnitude larger for a MRE composite than for the same volume fraction of fillers dispersed randomly in the polymer. Wang et al. [31] proposed a phenomenological model to understand the impedance response of MREs under mechanical loads and magnetic fields. Their results showed that MRE samples exhibit significant changes in measured values of impedance and resistance in response to compressive deformation, as well as applied magnetic field. Bica [32] found that MRE with graphite micro particles ($\sim$14 %) is electroconductive. The magnetoresistance has an electric resistance whose value diminishes with both the increase of the intensity of the magnetic field and with the compression force. The variation of resistance with magnetic field intensity is due to the compression of MRE with graphite microparticles. In the approximation of the perfect elastic body, the sums of the main deformations and the compressibility module of MRE with graphite microparticles, depend on the magnetic field intensity. In our study [33], graphite was introduced into conventional MREs to change the overall resistances so that it can be detected by a multimeter. We found a MRE sample with 55 % carbonyl iron, 20 % silicon rubber and 25 % graphite powder has the best performance. The test result showed that at a normal force of 5 N, the resistance decreases from 4.62 kΩ without a magnetic field to 1.78 kΩ at a magnetic field of 600 mT. The decreasing rate is more than 60 %. This result also demonstrated the possibility of using MREs to develop a sensor for measuring magnetic fields. This result indicates that the detection is very sensitive to the normal force. When the normal force is 15 N, the field-induced resistance only has less than 28 % change from 0.65 kΩ at 0 mT–0.47 kΩ at 600 mT.

## 4.2 Modeling Approach

Depending on an elastic-plastic asperity microcontact model for contact between two nominally flat surfaces, Kchit and Bossis developed a model to analyse the contact of two rough surfaces. They used two kinds of magnetic particles: nickel and nickel coated with silver, dispersed in a silicone polymer as the polarised particles [30]. To understand the complex conductivity of particle embedded composites, quantitative or semi-quantitative models were developed by Zhao et al. [34]. The Maxwell–Wagner and the Bruggeman symmetric and asymmetric media equations were introduced by McLachlan [35] to model the electrical behaviour of conductor-insulator composites. The microstructures for which these effective media equations apply are considered in simulating the measured impedance and modular spectra of these composites. Woo et al. [36] developed a universal equivalent circuit model in modelling the impedance response of

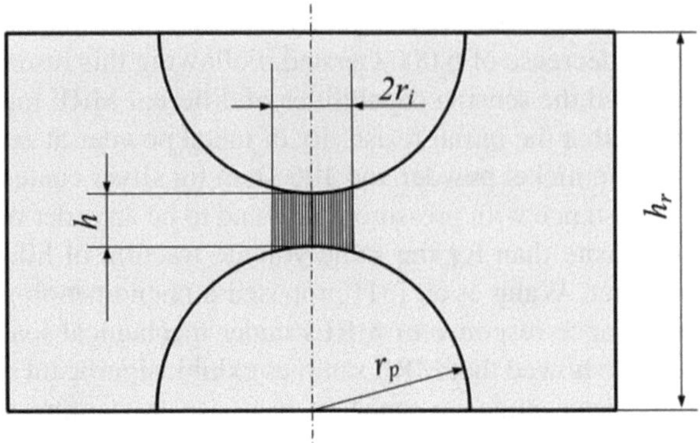


**Fig. 8** The longitudinal section of RVU

composites with insulating or conductive particles or fibres. Based on the micro-structure of MREs, Wang et al. [31] proposed an equivalent circuit model to interpret the impedance measurement results.

By supposing the anisotropic MRE has a single chain structure, a representative volume unit (RVU) as shown in Fig. 8, which consists of two neighboring hemispheres and the surrounding polymer matrix, is used to derive currents passing through the particles. Most of the current flowing through the RVU concentrates on the small area between the two adjacent hemispheres. Meanwhile the conductivity of the iron particles is much higher than that of the polymer therefore the electric potential drops on the particles could also be negligible. The insulating polymer film between two neighbouring iron particles is very thin because of the magnetic attraction during preparation and it is across this film that the electrical field assisted tunnel current can occur. The Fowler-Nordheim equation [37, 38] can be used to express the tunnel current. Meanwhile the iron particles dispersing in the polymer matrix contribute to the conductivity of the polymer and then the total current density is the sum of the tunnel density $j_t$ and the conduction density $j_c$, or

$$j = j_t + j_c = \alpha E_{\mathrm{loc}}^2 \exp\left(-\frac{\beta}{E_{\mathrm{loc}}}\right) + \sigma_f E_{\mathrm{loc}} \tag{6}$$

in which $\sigma_f$ is the conductivity of the polymer film.

The total current density $j$ is for the current flowing through the small area between the tips of two adjacent iron particles, however, the density of RVU $j_r$ should be derived from its cross section. So from the total density of RVU $j_r$ and the electric field intensity $E$, the conductivity of typical MRE $\sigma_r$ can be represented as

$$\sigma_r = \frac{j_r}{E} = \frac{\pi \cdot r_i^2 \cdot j}{S_r \cdot E} = 3\phi \cdot r_i^2 \left[\frac{2\alpha}{h^2} E \exp\left(-\frac{h\beta}{2r_p E}\right) + \frac{\sigma_f}{r_p h}\right] \tag{7}$$

in which the $r_i$ is the radius of the area between the two adjacent iron particles.

When a MRE sample is compressed its conductivity increases. This phenomenon is explained by two factors, one of which is the increments of the conductive area induced by the deformation of MREs and the other is the reduction of the thickness of the polymer membrane between the two adjacent iron particles.

From the Hertz Theory [39, 40], when the initial compressed stress applied on MRE is $\sigma_0$, corresponding to which there is an initial contact area radius $r_{i0}$

$$r_{i0} = \left[\frac{3\pi\sigma_0(1-v^2)}{2E_p}\right]^{1/3} \cdot r_p \tag{8}$$

So the radius $r_i$ increases along with the increment of compressed stress

$$r_i = r_{i0} + r_p\left((\sigma_0+\sigma)^{1/3} - \sigma_0^{1/3}\right) \cdot \left(\frac{3\pi(1-v^2)}{2E_p}\right)^{1/3} \tag{9}$$

Meanwhile the magnetic field also contributes to the resistance of MREs. When the external magnetic field is applied to MREs, the carbonyl iron particles are attracted by the poles of magnetic field. The nearer magnetic pole contributes much more powerful magnetic force than the other pole. So the attraction from the farther magnetic pole can be negligible.

For the two iron particles in each RVU, the magnetic attraction from the pole applied to the father particle compresses the thin film between the two adjacent iron particles. Similar to the piezoresistivity, the increment of the conductive area is the main cause for the conductivity increasing.

Thus, the radius $r_i$ can be updated as

$$r_i = r_{i0} + r_p\left((\sigma_0+\sigma_1+\sigma_2)^{1/3} - \sigma_0^{1/3}\right) \cdot \left(\frac{3\pi(1-v^2)}{2E_p}\right)^{1/3} \tag{10}$$

in which, $\sigma_1$ is the compressive stress, $\sigma_2$ is the stress from the magnetic attraction.

So the dependence of the conductivity of MREs on electric field intensity and compressive stress is

$$\begin{aligned}\sigma_m = 3\phi \cdot &\left[\frac{2\alpha}{h^2}E\exp\left(-\frac{h\beta}{2r_pE}\right) + \frac{\sigma_f}{r_ph}\right] \\ &\cdot \left[r_{i0} + r_p\left((\sigma_0+\sigma_1+\sigma_2)^{1/3} - \sigma_0^{1/3}\right) \cdot \left(\frac{3\pi(1-v^2)}{2E_p}\right)\right]^2\end{aligned} \tag{11}$$

When the initial condition $\sigma_0$ and $r_{i0}$ are set, except E and $\sigma$ the other parameters in this equation are all constants. So the conductivity of MREs $\sigma_m$ is dependant on the intensity of the electric field $E$ and the compressed stress $\sigma$.

For our fabricated graphite MRE based samples, the final resistance is given by

$$\begin{aligned}R_g &= \lambda_g\lambda_i R = \frac{\lambda_g\lambda_i\rho l}{A} = \frac{\lambda_g\lambda_i l}{A\sigma_m} \\ &= \frac{\lambda_g\lambda_i l}{3A\phi \cdot \left[\frac{2\alpha}{h^2}E\exp\left(-\frac{h\beta}{2r_pE}\right) + \frac{\sigma_f}{r_ph}\right] \cdot \left[r_{i0} + r_p\left((\sigma_0+\sigma_1+\sigma_2)^{1/3} - \sigma_0^{1/3}\right) \cdot \left(\frac{3\pi(1-v^2)}{2E_p}\right)\right]}\end{aligned} \tag{12}$$

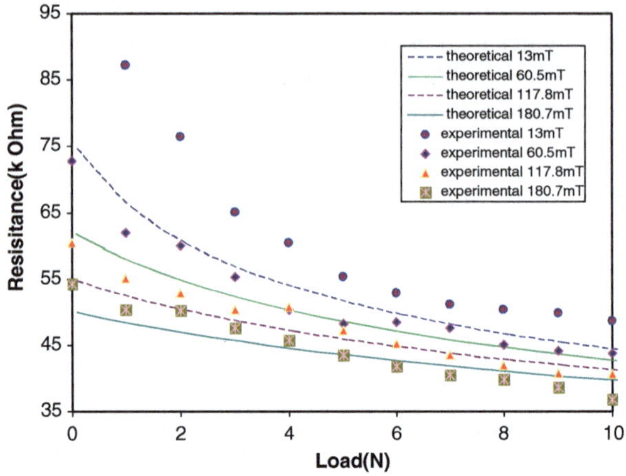


**Fig. 9** Comparison between experimental result and theoretical result (anisotropic MRE Gr 21.95 %)

The comparison between experimental results and modelling predictions is shown in Fig. 9. Though they couldn't match each other perfectly, still the trends of both the experimental result and the theoretical result are the same.

# 5 MRE Applications

## *5.1 Tunable Dynamic Vibration Absorbers*

MREs hold promise in enabling simple variable stiffness devices. The pioneering application of MREs was presented in vehicle industry. In US Patent 5816587 [41], as shown in Fig. 10, a novel method and apparatus for varying the stiffness of a suspension bushing having a MRE disposed therein was presented. This allows for improved customer satisfaction by reducing dissatisfaction associated with braking events, such as shudder, which can result in increased noise, vibration and harshness. It solves these problems by controlling relative displacements of a longitudinal suspension member relative to a chassis member in a motor vehicle. The control method is: communicating a brake actuation signal from a brake actuator to a suspension control module; determining from the brake actuation signal a desired suspension bushing stiffness and generating an electrical current in response thereto; and communicating the electrical current to an electrical coil operatively associated with the MRE, thereby generating a magnetic field so as to vary the stiffness characteristics of the MRE in a way to reduce an operator's perception of brake induced shudder. As shown in Fig. 10, the apparatus comprises a suspension bushing shaft, an inner steel cylinder annularly surrounding the suspension bushing shaft, a MRE annularly surrounding the annular inner steel cylinder, and an outer

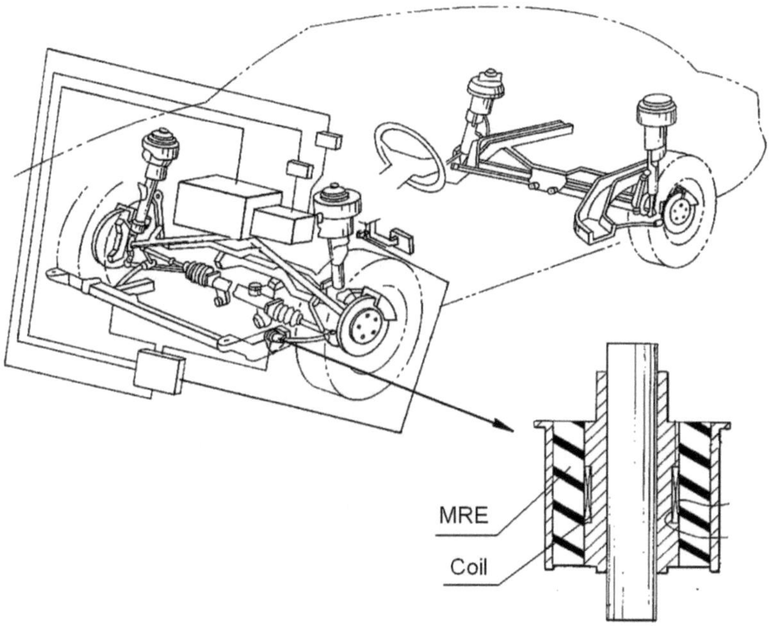


**Fig. 10** MREs suspension bushing [41]

steel cylinder annularly surrounding the annular MRE. An annular coil is disposed about an outer peripheral surface portion of the annular inner steel cylinder so as to be interposed between the annular inner steel cylinder and the annular MRE, and the coil is adapted to be electrically connected to the automotive vehicle electrical system by means of suitable electrical leads. By applying suitable predetermined amounts of electrical current to the suspension bushing coil, a variable magnetic field is generated which variably controls the stiffness values of the suspension bushing. The suspension bushings stiffness can therefore be adjusted in response to actuation of the brake system to reduce brake shudder.

In US Patent 7086507 and 20050011710 [42, 43], the inventors developed MRE devices and methods for their use for vibration isolation by changing the storage and loss moduli of one or more MRE layers. As shown in Fig. 11, the MREs are sandwiched between magnetic activation layers. The magnetic activation layers contain embedded magnetic nodes that control the magnetic field, and thus the stiffness, of the MRE. The current and voltage supplied to the electromagnets will affect the magnetic field strength within the MRE and, hence, the stiffness of the MRE. The current and voltage are determined by the external vibration and associated forces imparted to the MRE device. This system can detect, measure and signal one or more physical manifestations of these external forces, and transforming this signal into the appropriate current and voltage to send to the magnets to obtain optimal vibration isolation energy dissipation for a given shock event. These MRE devices are used for vibration isolation of mechanical systems for random shock events.

Fig. 11 MRE devices for shock isolation [42, 43]

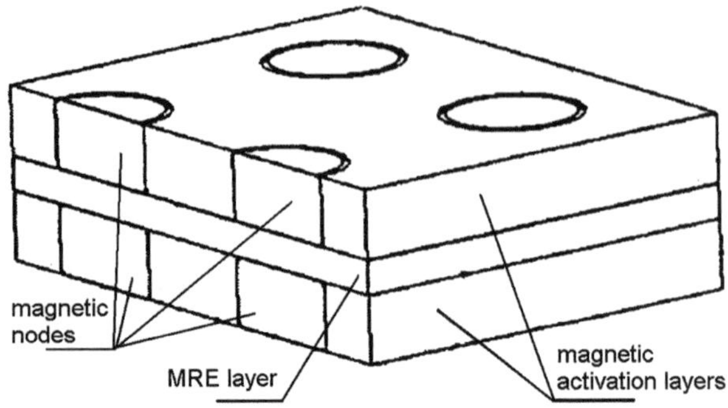


Fig. 12 AVA having MRE elements [44, 45]

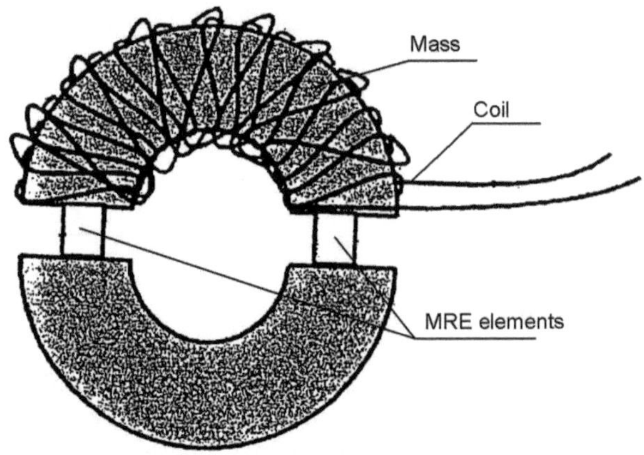


MREs can also be used as the smart spring in dynamic vibration absorber. In US Patent 7102474 and 20050040922 [44, 45], the inventors presented an adaptive vibration absorbers (AVAs) working with MREs. As shown in Fig. 12, the AVA is designed to operate selectively over a range of frequencies rather than a single frequency. It is configured of the elements including a base mass and an absorber mass connected by a pair of MREs, which function effectively as tunable springs and are also held responsible for the advantageous bandwidth increase in vibration suppression by the AVA. The configuration and composition of the elements of the exemplary AVA provide a path (also referred to as magnetic circuit) for magnetic flux that may be induced by a magnetic field source. Specifically, the magnetic circuit passes through MREs. When the source provides the magnetic field and flux travels through the described magnetic circuit, the change of MREs' properties will make the system's natural frequency shift. This change may be controlled as necessary or desired via a control algorithm applied through a processor. A number of research groups were focused on such applications, for example, Albanese [46] and Holdhusen [47] presented MRE absorbers working in a compression mode. Deng et al. [11, 48], Zhang and Li [14] presented MRE absorbers worked in a shear mode. Their results indicated that the frequency of vibration absorber can be tuned in wide frequency range, and the controlled frequency band was expanded too.

### *5.2 MRE Force Sensors*

Because of their magnetic field sensitive behavior, MREs were also considered as one of idea materials for sensor and actuator. In US Patent 5814999 [49], a method and apparatus for measuring displacement and force applied to an elastomeric body having a MRE disposed therein has been discovered. A transducer apparatus comprises two structural members and a MRE interposed between them. A module is provided for acquiring measurement data by applying a drive signal to an electrode disposed within the MRE and monitoring a preselected electrical state of the MRE. The module generates an output signal corresponding to variations the preselected electrical state caused by deflections of the MRE. It increases the usefulness of conventional elastomeric members by providing within them additional materials so that they may serve as transducers in addition to serving as isolators and pivotable joints, etc. This combination of functionality reduces part and assembly complexity as well as improving overall package efficiency for the assembled product.

## 6 Conclusions

This chapter presented a review of the state of the art in MREs technology. The basic materials, fabrication methods and many models to describe MREs' behavior were discussed. There is a growing need to understand and model their behavior and improve them by using optimal components and fabrication methods. The potential applications of this smart material were introduced. The patent review indicated that researchers continue to develop new MREs with new materials to meet new requirements. There were many applications prospect, which provide an impetus for continued research in this area. Numerous applications, which make use of controllable stiffness and the unique anisotropic characteristics of these elastomers, will be developed and the research efforts of the past decade in MREs will be paid off in the near future.

## References

1. Carlson, J.D., Jolly, M.R.: MR fluid, foam and elastomer devices. Mechatronics. **10**, 555–569 (2000)
2. Shiga, T., Okada, A., Kurauchi, T.: Magnetroviscoelastic behavior of composite gels. J. Appl. Polym. Sci. **58**, 787–792 (1995)

3. Jolly, M.R., Carlson, J.D., Munoz, B.C., Bullions, T.A.: The magnetoviscoelastic response of elastomers composites consisting of ferrous particles embedded in a polymer matrix. J. Intell. Mater. Syst. Struct. **7**, 613–622 (1996)
4. Ginder, J.M., Clark, S.M., Schlotter, W.F., Nichols, M.E.: Magnetostrictive phenomena in magnetorheological elastomers. Int. J. Mod. Phys. B. **16**(17–18), 2412–2418 (2002)
5. Zhou, G.Y.: Shear properties of a magnetorheological elastomer. Smart Mater. Struct. **12**, 139–146 (2003)
6. Chen, L., Gong, X.L., Jiang, W.Q., et al.: Investigation on magnetorheological elastomers based on natural rubber. J. Mater. Sci. **42**, 5483–5489 (2007)
7. Hu, Y., Wang, Y.L., Gong, X.L., et al.: New magnetorheological elastomers based on polyurethane/Si-rubber hybrid. Polym. Test. **24**, 324–329 (2005)
8. Bossis, G., Abbo, C., Cutillas, S., et al.: Electroactive and electrostructured elastomers. Int. J. Mod. Phys. B. **15**(6–7), 564–573 (2001)
9. Lokander, M., Stenberg, B.: Performance of isotropic magnetorheological rubber materials. Polym. Test. **22**, 245–251 (2003)
10. Lokander, M., Stenberg, B.: Improving the magnetorheological effect in isotropic magnetorheological rubber materials. Polym. Test. **22**, 677–680 (2003)
11. Deng, H.X., Gong, X.L., Wang, L.H.: Development of an adaptive tuned vibration absorber with magnetorheological elastomer. Smart Mater. Struct. **15**, N111–N116 (2006)
12. Ni, Z.C., Gong, X.L., Li, J.F., et al.: Study on a dynamic stiffness-tuning absorber with squeeze-strain enhanced magnetorheological elastómer. J. Intel. Mater. Syst. Struct. **20**, 1195–1202 (2009)
13. Xu, Z.B., Gong, X.L., Liao, G.J., et al.: An active-damping-compensated magnetorheological elastomer adaptive tuned vibration absorber. J. Intel. Mater. Syst. Struct. **21**, 1039–1047 (2010)
14. Zhang, X.Z., Li, W.H.: Adaptive tuned dynamic vibration absorbers working with MR elastomers. Smart Struct. Syst. **5**(5), 517–529 (2009)
15. Hoang, N., Zhang, N., Du, H.: An adaptive tunable vibration absorber using a new magnetorheological elastomer for vehicular powertrain transient vibration reduction. Smart Mater. Struct. **20**, 015019 (2011)
16. Tian, T.F., Li, W.H., Alici, G., et al.: Microstructure and magnetorheology of graphite based MR elastomers. Rheologica Acta, print online. doi:10.1007/s00397-011-0567-9
17. Lokander, M., Stenberg, B.: Performance of isotropic magnetorheological rubber materials. Polym. Test. **22**, 245–251 (2003)
18. Lokander, M., Stenberg, B.: Improving the magnetorheological effect in isotropic magnetorheological rubber materials. Polym. Test. **22**, 677–680 (2003)
19. Gong, X.L., Zhang, X.Z., Zhang, P.Q.: Fabrication and characterization of isotropic magnetorheological elastomers. Polym. Test. **24**, 669–676 (2005)
20. Davis, L.C.: Model of magnetorheological elastomers. J. Appl. Phys. **85**(6), 3348–3351 (1999)
21. Ginder, J.M., Nichols, M.E., Elie, L.D., Clark, S.M.: Controllable stiffness components based on magnetorheological elastomers. In: Wereley, N.M. (ed.) Smart Structures and Materials 2000: Smart Structures and Integrated Systems, Proceedings of SPIE 3985, pp. 418–425. (2000)
22. Li, W.H., Du, H., Chen, G., et al.: Nonlinear viscoelastic properties of MR fluids under large-amplitude oscillatory shear. Rheol. Acta. **42**(3), 280–286 (2003)
23. Li, W.H., Zhou, Y., Tian, T.F.: Viscoelastic properties of MR elastomers under harmonic loading. Rheol. Acta. **49**, 733–740 (2010)
24. Jolly, M.R., Carlson, J.D., Munoz, B.C.: A model of the behaviour of magnetorheological materials. Smart Mater. Struct. **5**, 607–614 (1996)
25. Davis, L.C.: Model of magnetorheological elastomers. J. Appl. Phys. **85**(6), 3348–3351 (1999)
26. Shen, Y., Golnaraghi, M.F., Heppler, G.R.: Experimental research and modeling of magnetorheological elastomers. J. Intel. Mater. Syst. Struct. **15**, 27–35 (2004)

27. Zhang, X.Z., Li, W.H., Gong, X.L.: An effective permeability model to predict field-dependent modulus of magnetorheological elastomers. Commun. Nonlinear Sci. Numer. Simul. **13**(9), 1910–1916 (2008)
28. Li, W.H., Du, H., Chen, G., et al.: Nonlinear viscoelastic properties of MR fluids under large-amplitude oscillatory shear. Rheol. Acta. **42**, 280–286 (2003)
29. Li, W.H., Du, H., Chen, G., et al.: Nonlinear rheological behavior of MR fluids: step strain experiments. Smart Mater. Struct. **11**, 209–217 (2002)
30. Kchit, N., Bossis, G.: Electrical resistivity mechanism in magnetorheological elastomer. J. Phys. D-Appl. Phys. **42**(10), 5505 (2009)
31. Wang, X.J., Gordaninejad, F., Calgar, M., et al.: Sensing behavior of magnetorheological elastomers. J. Mech. Des. **131**(9), 6 (2009)
32. Bica, I.: Influence of the transverse magnetic field intensity upon the electric resistance of the magnetorheological elastomer containing graphite microparticles. Mater. Lett. **63**(26), 2230–2232 (2009)
33. Li, W.H., Kostidis, K., Zhang, X.Z., et al.: Development of a force sensor working with MR elastomers. IEEE/ASME Int. Conf. Adv. Intel. Mechatron. **1–3**, 233–238 (2009)
34. Zhao, Y., Maietta, D.M., Chang, L.: An asperity microcontact model incorporating the transition from elastic deformation to fully plastic flow. J. Tribol. **122**(1), 86–93 (2000)
35. McLachlan, D.S.: Analytical functions for the dc and ac conductivity of conductor-insulator composites. J. Electroceram. **5**(2), 93–110 (2000)
36. Woo, L.Y., Wansom, S., Hixson, A.D., Campo, M.A., Mason, T.O.: A universal equivalent circuit model for the impedance response of composites. J. Mater. Sci. **38**(10), 2265–2270 (2003)
37. Weinberg, Z.A.: On tunneling in metal-oxide-silicon structures. J. Appl. Phys. **53**(7), 5052–5056 (1982)
38. Serdouk, S., Hayn, R., Autran, J.L.: Theory of spin-dependent tunneling current in ferromagnetic metal-oxide-silicon structures. Journal of Applied Physics, vol. 102(11), p. 113707-1-113707-5 (2007)
39. Zhupanska, O.I., Ulitko, A.F.: Contact with friction of a rigid cylinder with an elastic half-space. J. Mech. Phys. Solids. **53**(5), 975–999 (2005)
40. Etsion, I., Levinson, O., Halperin, G., Varenberg, M.: Experimental investigation of the elastic-plastic contact area and static friction of a sphere on flat. J. Tribol. **127**(1), 47–50 (2005)
41. Stewart, W.M., Ginder, J.M., Elie, L.D.: Method and apparatus for reducing brake shudder. US Patent 5816587, 1998
42. Hitchcock, G.H., Gordaninejad, F., Fuchs, A.: Controllable magneto-rheological elastomer vibration isolator. US Patent 7086507, 2006
43. Hitchcock, G.H., Gordaninejad, F., Fuchs, A.: Controllable magneto-rheological elastomer vibration isolator. US Patent 20050011710, 2005
44. Lerner, A.A., Cunefare, K.A.: Adaptable vibration absorber employing a magnetorheological elastomer with variable gap length and methods and systems therefore. US Patent 7102474, 2006
45. Lerner, A.A., Cunefare, K.A.: Adaptable vibration absorber employing a magnetorheological elastomer with variable gap length and methods and systems therefore. US Patent 20050040922, 2005
46. Albanese, A.M.: The design and implementation of a magnetorheological silicone composite state-switched absorber. A Thesis for the Degree Master of Science, Georgia Institute of Technology (2005)
47. Holdhusen, M.H.: The state-switched absorber used for vibration control of continuous systems. A Dissertation for the Degree Doctor of Philosophy, Georgia Institute of Technology (2005)

48. Deng, H.X., Gong, X.L.: Adaptive tuned vibration absorber based on magnetorheological elastomer. J. Intel. Mater. Syst. Struct. **18**(12), 1205–1210 (2007)
49. Elie, L.D., Ginder, J.M., Mark, J.S., Nichols, M.E.: Method and apparatus for measuring displacement and force. US Patent 5814999, 1998

# Radiation Processing of Elastomers

**Z. P. Zagórski and E. M. Kornacka**

**Abstract** The chapter provides introduction to radiation processing of solid state materials, using commercially available sources of ionizing radiation, i.e., radioisotopic and/or accelerated electron beam installations. Dosimetry is described as the method of controlling progress of changes in irradiated material. Distribution of doses in irradiated material is described, allowing proper processing of polymers. Basics of radiation chemistry of polymers is explained, in particular of elastomers. Radiation-induced crosslinking is most interesting reaction, but it can be accompanied by undesired phenomena like chain scission. Specific phenomena like energy transfer occur in radiation processing; therefore, composites of elastomers with components of different radiation characteristics may show unexpected results. Examples of selected cases are described in details. Comparisons between traditional methods of crosslinking with these using ionizing radiation allow consideration of introduction of the latter into industrial praxis.

## 1 Introduction

Application of high energy chemistry in processing of elastomers consists in the supply of specific energy, which, in contrary to thermal energy effects, changes the position of electrons in the outer shell of atoms in the molecule. Particularly important is the abstraction of the electron from the outer orbital, leading to ionization and subsequent chemical reactions. High energy chemistry is that branch of chemistry which applies a physical source of energy to achieve reactions of chemical compounds with the medium and between compounds. There are different high energy chemical effects, but in processing of elastomers, only the

Z. P. Zagórski (✉) · E. M. Kornacka
Centre for Radiation Research Institute of Nuclear Chemistry and Technology, PL-03-195 Warsaw, Poland
e-mail: z.zagorski@ichtj.waw.pl

P. M. Visakh et al. (eds.), *Advances in Elastomers I*,
Advanced Structured Materials 11, DOI: 10.1007/978-3-642-20925-3_13,

radiation chemistry approach, discussed in the chapter, can be quantified perfectly by precise dosimetry of applied and really absorbed ionizing radiation. Confrontation of deposited energy with products of irradiation allows exact definition of the radiation yield, i.e. relation between delivered energy and resulting change of chemical energy and achieved effects.

Basics of radiation chemistry and radiation processing by ionizing radiation are rather far from chemical technology and conventional processing; therefore a detailed approach is presented. It is needed for understanding of mechanisms and/or to find a common language with specialist of irradiation devices and technologies collaborating with rubber industry. The knowledge of basics is useful both in research and development and in the application of acquired procedures or patents.

Radiation chemistry is a multidisciplinary field, but is most properly considered as a part of physical chemistry. Certainly it is not a part or clone of radiochemistry, which describes rather the chemistry of radioactive elements. It is therefore improper to call chemical changes due to the absorption of ionizing radiation "radiochemical changes" because reactions induced by ionizing radiation are not necessarily caused by radiation emitted by radioactive isotopes. They can be induced by ionizing radiation produced via electrical means, e.g. in natural electrical discharges and man made in electron and positive ions accelerators. Let us define radiation chemistry as the branch of chemistry dealing with chemical effects of the absorption of ionizing radiation, whatever its origin.

Contrary to photochemistry, a fast particle or electromagnetic quantum of energy of several orders of magnitude higher than needed for ionization of the most difficult to be ionized atom, helium (24 eV), is not absorbed by a molecule as it is the case in photochemistry. The collision of an ionizing particle with the atom in the molecule is an inelastic one and the atom is ionized, loosing one outer electron and a positive ion, sometimes called a hole is left in the chain. According to the definition above, one already can call it a chemical reaction. Indeed, such positive holes can enter further reactions, e.g. in polymers, after traveling along the chain. The detached electron usually has enough energy to move far and to start new ionizations. There is neither the place, nor the need to discuss mechanisms in which subsequent ionizations proceed. For the purpose of explanation of developing chemistry, one should only mention that at the final generation of the cascade of electrons, they do not have enough energy to travel far, and they form a so called spur containing several ionizations, one not very far from the other. The chemistry here is therefore different in comparison to single ionization spurs formed with higher energy electrons. The phenomenon is most important in the radiolysis of water, being responsible for the basic formation of hydrogen peroxide. In our case of elastomers the multi-ionization spur is also important, because causes irreparable break of the chain.

Multi-ionization spurs are common to both, low molecular weight molecules and high molecular weight compounds—their chemical products stay where they have been formed. Single ionization spurs formed in a small molecule release an electron which usually leaves the molecule rapidly, but the positive ion remains. In polymers, both products of single ionization—positive hole and the electron can travel from the site of origin to another, even distant places of the macromolecule, not leaving it.

Radiation chemistry deals with the interaction of ionizing radiations with chemical compounds. A quant of electromagnetic radiation of energy higher than 24 eV is ionizing any atom or molecule. An electron is abstracted from the outer shell, starting sequences of chemical phenomena. That is, of course, the field of chemistry and the background of materials chemistry in which practically everything happens on the outer shell of molecules. Electromagnetic quanta of lower energy, e.g. infrared radiation, and microwave radiation cannot ionize constituents of the matter, but, if absorbed, can cause indirectly chemical effects, e.g. due to the increase of temperature, like in the microwave field.

On the other side of the UV-Vis spectrum, i.e. at shorter wave length and higher energy of quanta, a possibility of ionization can occur. Going towards shorter waves from deep UV, sometimes called UVc, one reaches so called vacuum UV, under terrestrial conditions absorbed mostly by air of the atmosphere. Chemical effects of UV on elastomers are connected with the resistance of rubber to the influence of sunlight. In spite of some relations to the ionizing radiation influence, that topic will be discussed in the present chapter only shortly.

With increasing energy, the first ionization starts the sequence of secondary and next generations of ionizations by degraded quanta and secondary electrons.

The ionization capability described by the energy of quanta applies also to charged particles, e.g. electrons and protons and also to neutrons, which are classified as ionizing indirectly.

Phenomena of ionization, described above, were related to single photons or high energy particle. In radiation chemistry, we are not dealing with single photon but with enormous flux of them or with a beam of electrons or positive ions. One cannot estimate in what place of the medium a single photon will ionize; it may do it close to the source or far from it. However, when in great number, the distribution of energy deposited by ionizing radiation is precisely defined, and shown on Fig. 1.

The described chain of ionization events proceeds statistically on the full length of the range diminishing, in some cases with complications, as shown on the Fig. 1 for the energy of gamma quanta, e.g. 1.17 and 1.33 MeV (from cobalt-60). The range of gamma radiation is infinite and even most thick objects are penetrated with chemical effects, time of irradiation permitting. Absorbing medium in the case of Fig. 1 is of density equivalent to water and it is obvious that the range of radiation is substantial and is deep penetrating. Even in aluminum, if we recalculate the curves by the increased density of the medium.

Electrons of 10 MeV energy, typical in radiation processing of elastomers, give a more complicated depth dose curve, in this case describing the broad geometry irradiation, as it is the case with scanned beam of electrons. Straight beam of electrons, as it is used in basic research, yields simpler depth dose curve.

Density of ionization for particular kind of radiations is described by LET (linear energy transfer) showing well differences between different radiations, gamma radiation (for $\gamma$ radiation from cobalt 60 the value is 0.22 keV/$\mu$) and high

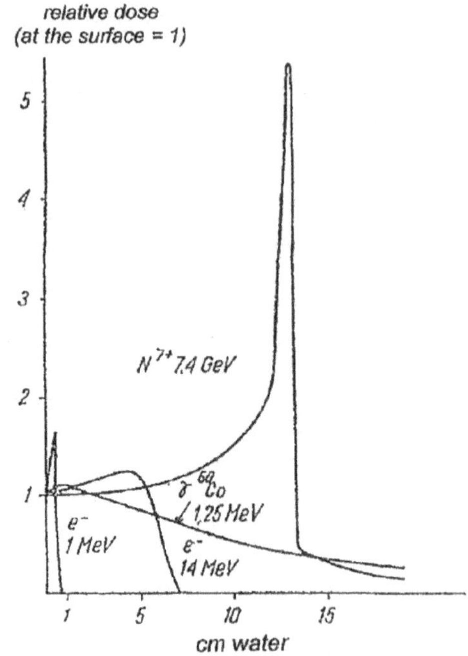


**Fig. 1** Depth-dose curves of different ionizing radiations. Quantity of energy (any unit of energy per mass of water, or equivalent material after recalculation of electron density) deposited on particular depth from the irradiated surface. The dose close to the surface is taken as unity (1). Comparatively deep range of gamma rays is visible. Curves for electrons refer to a broad beam—electron beam scanned over the material

energy electron ($e^-$ of energy 10 MeV the value is 0.20 keV/$\mu$). When a high-energy electron passes through material, it loses its energy gradually by interaction with the electrons and atoms of the absorbing material. Sometimes used in processing, X-rays show higher LET values, e.g. 200 keV rays have the value of 0.7 keV/$\mu$. Alpha particles emitted by polonium-210 have LET value of 90 keV/$\mu$ what indicates no meaning for radiation processing. However, $\alpha$ particles of similar energy, emitted by radon are penetrating the surface of polymers. In the case of polycarbonates, shallow penetration is sufficient to degrade the polymer allowing to etch the hole in it. The phenomenon is used for the determination of concentration of radon in the air, using a polymer detector.

Interactions of ionizing radiations with polymers are seen as ionizations, as electrons in the material are ejected by the field of the passing primary electron. The excited molecules eject secondary electrons, which are normally energetic enough to cause further excitations and ionizations, these processes eventually leading to the chemical reactions.

While the rate at which the linear density of events in which radiation energy is imparted to a material may vary substantially from one type of radiation to another (e.g., from $\alpha$ rays to $\gamma$ rays), it is almost constant for electrons and $\gamma$ rays as obtained from high-energy electrons and $^{60}$Co or $^{137}$Cs sources. The dose absorbed, however, is a function also of the atomic composition of the absorbing material. Therefore, different materials similarly exposed will absorb different doses. When evaluating the absorbed dose in a product or sample it is therefore necessary either to use a dosimeter of similar atomic composition or to correct for the different dose absorbed in the dosimeter and in the product itself. For these reasons aqueous chemical dosimeters are often preferred for dosimetry in radiation chemistry experiments, as they can be made to match the absorption characteristics of the object, e.g. piece of rubber. For use in cases where such a "match" cannot be conveniently established, a simple method for derivation of the necessary corrections for $^{60}$Co irradiation is: The ratio between the doses for any two media equals the ratio of their mass-energy absorbtion coefficients.

The initial interactions between radiation and matter, i.e., the ionizations and excitations referred to above take place over a very short time period and lead to formation of thermalized chemically active species. In condensed systems these active species are initially distributed nonhomogeneously in *spurs* in the irradiated medium along the path of the ionizing particle track.

For low LET radiations, such as high energy electrons and gamma rays, where the spurs are widely separated, high radical yields and low molecular yields are experienced since the reactions leading to molecular products are taking place in the spurs at a relatively low radical concentration What is said about high LET radiations also resembles what happens with low LET radiations at the extremely high dose rates obtainable with electron accelerators used commercially. At low dose rates, we had relatively low molecular yields. By increasing the dose rate, we increase the instantaneous particle density, thereby decreasing the distance between the spurs until some come close enough to each other to result in spur overlap, thus increasing the molecular yields. In the case of water they are hydrogen peroxide and dihydrogen. In polymers these are debris of low molecular weight, including dihydrogen [1].

Dealing with radiation chemistry we are asking first for sources of radiation—next for the nature of the irradiated object, and eventually for the geometry of distribution of radiation energy in it. Recognition of mechanisms of chemical reactions helps to find general rules and make predictions concerning the final results, sometimes not feasibly achieved via experiments. At the end of our fundamental for processing discussions, the basics of phenomena will be outlined starting with radiation induced chemistry.

Research in radiation chemistry related to elastomers involves answering three questions: what is the composition and structure of the irradiated object, what are stable, final products of irradiation and, if possible, what are the intermediate products of irradiation and the mechanisms that produce particular results. Precise answers to all these questions justify the veracity of the subsequent conclusions.

## 2 Units in Radiation Chemistry and Processing

The absorption of ionizing radiation leaves reactive intermediates and semistable and stable products. The fate of intermediates depends on the nature of the chemical system—main and minor constituents. The yield of primary products depends on the amount of deposited energy and not on the composition of the system, therefore from the point of view of the order of chemical reaction (elementary definition in chemical kinetics), radiation induced reactions are of 0 (zero) order.

The absorbed energy can be given in any units of energy, even in classical, old calories, but according to SI system it should be given in joules, and if needed in milijoules (mJ), kilojoules (kJ) and megajoules (MJ). Therefore the unit of absorbed dose is 1 joule per kilogram. That unit is called 1 gray (Gy), after Louis Harold Gray (1905–1965) and accordingly kilogray (kGy) and megagray (MGy).

Still another unit can be found, even in the present day literature, namely rad (**r**adiation **a**bsorbed **d**ose) and accordingly kilorad (krad), megarad (Mrad). One rad has been defined in early radiobiological and chemical experiments as 100 erg/g. The not very convenient recalculation is 100 rad equals 1 gray, and respectively 1 Mrad is 10 kGy. In the present chapter the dose is no longer expressed in rads, even if cited paper uses rad. The present Authors appeal to collegues for translation of rads into grays for the sake of universal understanding in the vast field of radiation-connected processing.

The last, but not least, question of units in the multidisciplinary field of radiation chemistry is connected with basic research. In the beginnings of radiation chemistry, at the start of the atomic age, the absorbed energy was expressed in electronvolts, eV. Relation of electronvolt to joule is $1\ \text{eV} = 1.6020 \times 10^{-19}$ J. The main argument for that unit of energy was the close relation to energies of bonds discussed in chemistry on atomic and molecular levels. Chemical effects were related to number of chemical individuals changed (formed or decayed) per 100 eV of absorbed energy. That value was called radiation yield, symbol capital G (one of few letters left free in science at that time) with the index denoting the chemical entity. Such designation allowed chemists to easily relate the effect to energy; e.g. $G_H = 2$ means that two atoms of hydrogen were formed after absorption of 100 eV of energy. These precise definitions helped to interpret quickly the results and, what is very important, to recognize immediately the presence of chain reaction. It was noticed very early, that the chemical yield appears in single G units. Taking into account energetics of chemical reactions, it became clear that a substantial part of absorbed energy is transformed into thermal energy of low quality, rendering it chemically useless.

Already values of G exceeding 10 are drawing attention to possible deviations from simple mechanisms deriving from energy of radiation-induced intermediates and are supported by internal energy of reagents present in the system.

Yields exceeding 100 are certainly chemical chain reactions, in which ionizing radiation plays the role of an initiator. They are not expected to happen very often in radiation chemistry but they can occur, in radiation induced polymerization of monomers and in radiation induced grafting of a monomer on the chain of a polymer or other, even inorganic surface.

Calculations in old units of G probably will survive, due to their practical meaning. Radiation yields expressed in that way suggest immediately the relation of phenomena to the energy of bonds measured in electronvolts. On the other hand, yields given in SI units show an easy link to doses expressed in joules.

In any case, yields of “x” expressed in old units of the effect per 100 eV are easy to recalculate into μmol per J:

$$G_x(1/100\ \text{eV}) \times 0.10364 = G_x(\mu\text{mol/J})$$

$$\text{or } G_x(\mu\text{mol/J}) \times 9.648784 = G_x(1/100\text{eV})$$

All previous information above, on expression of dose and radiation yields, is a common textbook description. However, some authors describe their radiation chemistry experiments in a way connecting the dose with deposition of energy on defined chemical entity. For instance, one can meet characterization of the experiment of irradiation with 75 keV protons, as the deposition of 180 eV/C-atom.

The present section on units and recalculations is limited to radiation chemistry and radiation processing. It does not contain units of radioactivity, because the activity cannot be easily translated into the dose absorbed by a chemical compound on the object in a radiation source. For the sake of completeness: the Standard International unit of radioactivity is 1 becquerel (Bq), after Henri Becquerel. One becquerel means one decay per second, the older unit, one curie (Ci, after Maria and Pierre Curie) equals to $3.7 \times 10^{10}$ becquerels. The legal unit is very small and is useful in the description of low activities, frequent in radiological protection, when expressing, e.g. the dose rate outside an industrial gamma source. The curie (Ci) is more practical in description of large sources, natural and artificial, used in radiation processing of polymeric products. The typical industrial source is loaded up to the order of megacuries (MCi) of cobalt-60.

## 3 Sources of Ionizing Radiation

Large scale, man made sources of ionizing radiation are most important in radiation processing of elastomers. The second half of the 20th century witnessed production of man-made radioactive isotopes in activities equivalent to thousands of tons of natural radioactive isotopes. The latters were at the beginning of 20th century produced in gram quantities only, and the first unit of activity was called not very precisely “gram equivalent of radium”. One gram of radium valued a fortune and was hardly sufficient to make basic research in radiochemistry and radiation chemistry and provided only the base for therapy in oncology. In the 50s of the 20th century nuclear reactors dedicated to production of cobalt 60 were build and sources of activity measured in thousand of Ci (curie), i.e. equivalent roughly to thousands of gram of radium. Artificial radioactive isotopes, mainly cobalt 60, produced in dedicated nuclear reactors, are used in industry, mainly for sterilization of medical products and other kinds of radiation processing like crosslinking of polymers. There is some use of radioactive isotopes recovered from spent nuclear reactor fuel, e.g. caesium 137 and strontium 90. Demand for ionizing radiation is so big, that nowadays the industrial ionizing radiation is obtained from electrical sources, mainly electron accelerators without any connection to radioactive isotopes whatsoever.

There is important connection of large industrial sources of ionizing radiation to topics of the present chapter. Before the technological version of radiation processing is developed, the research using laboratory sources of ionizing radiation is needed. Common laboratory sources of ionizing radiation can be considered as the small versions of industrial sources described above. They are more convenient, in comparison to large, industrial sources, as sources of ionizing radiation for simulation of chemical phenomena. These are cobalt 60 sources of activity up to 20 kCi, selfshielded, located in a laboratory without any special shielding, or sources of lower activity, loaded with cesium 137, used for special purposes like irradiation of blood for transfusion.

From the perspective of basic research more important are complex installations for realization of pulse radiolysis experiments. That specific technique of irradiations consists in application of a comparatively large dose of ionizing radiation in microseconds, or even nanoseconds, achieves the formation of measurable concentrations of short lived intermediates of chemical reactions. They can be studied spectrophotometrically for identification and further reactions, including kinetics. Such conditions cannot be realized with isotopic sources of ionizing radiation, because of their limited intensity. Realization of pulse radiolysis is rather simple with electron accelerator techniques, which already in the 50s allowed administration of microsecond pulses of electrons of up to 10 MeV energy, from linear accelerators. Pulses of lower energy electrons (single MeV's) were available from electrostatic machines, Febetron type, using the discharge of the bank of condensers, but the shape of pulses is not a perfect one, preventing proper, quantitative investigation of kinetics of intermediates generated during the irradiation period. Already 50 years ago the technique become a well established field around growing number of installations serving basic research in radiation chemistry.

The purpose of pulse radiolysis experiments is the recognition of intermediates and their kinetics features, of chemical reactions, known only from their starting and final compounds. Those which are absorbing are excited and decay into fragments and/or are entering reactions with other components of the system, not necessarily absorbing light of the main pulse. At the same time the pulse irradiated polymer is crossed by the analytical beam of continuous UV/VIS light, spectrophotometrically analyzing the solution. As the result, kinetic decay or build up curves of the spectrum are obtained. The range of the spectrum is determined by the applied optoelectronic hardware. In the present version of the technique, all parts are digitalized and computerized.

Radiation research and radiation processing laboratories and industries are usually equipped also with electron accelerators working parallel with gamma sources. As described before they usually operate linear electron accelerators of energy not exceeding 10 MeV to avoid nuclear activation of irradiated material. The typical power of such accelerators is 10 kW, which is equivalent almost to several megacuries of cobalt 60. The application of the accelerator is similar to gamma irradiation, but procedures of irradiation are different in both cases, due to different dose rates and adiabatic character of accelerator irradiation. Depending on particular technical solution of the irradiation technology, the sample of tested material is put on the conveyor and passed under the scanned beam of electrons. According to standard dosimetry, the sample is irradiated at predetermined speed of the conveyor at constant intensity of the beam. Usually the sample is accompanied by solid dosimeter—many types are available. The dose rate is usually much higher than is the case with gamma radiation—the dose of ca 30 kGy, typical for sterilization of medical devices made of polymers, is delivered in few minutes. Depending on specific heat of particular elastomer it is usually the highest dose permitted to avoid excessive heating of the material. The same doses delivered in gamma sources demands hours, and during that time the material is cooled efficiently. Doses needed for crosslinking are usually many times higher than in the case of radiation sterilization and the irradiation of one batch with electron beam has to be repeated several times, with cooling of the material in the mean time, inbetween repeating the cycle of irradiation.

The higher dose rate of accelerated electrons than gammas from isotopic source can demand more care in the application of doses higher than for sterilization. The application of electron dose, in contrast to the gamma radiation, is closer to adiabatic condition that means the heat produced cannot be removed quickly enough. The temperature of the irradiated object is increasing to the level determined by the specific heat (heat capacity) of the object. It may complicate the interpretation of effects and it is better to remain at the ambient temperature. That is possible by application of interrupted irradiations with cooling periods. The technique is called split irradiation and it does not complicate the interpretation, because the effects are additive and run linearly with the dose.

There are additional questions asked by polymer chemists. One is concerning the atmosphere in which the sample is irradiated, i.e. the gaseous atmosphere surrounding the sample. There is no problem in basic research to keep a small sample in vacuum during irradiation, as well as in any gas. Large scale irradiations have to be performed in simple air surrounding, because already simplest keeping off of oxygen with nitrogen atmosphere would be too expensive on industrial scale. Therefore introductory experiments leading to formulation of procedures of industrial procedures must answer the question of possible oxidation effects during irradiation and in the posteffect.

Sources producing low LET ionizing radiation, i.e. gamma from isotopes, mainly cobalt 60 and caesium 137 as well as electrically produced accelerated electrons play a key role in radiation processing of elastomers. Radiation produced by these sources has deep penetration, allowing to irradiate technologically reasonable large objects. Although van de Graaff, or transformer type electron accelerators can also be used, linacs are preferred, securing higher energy of electrons, what gives more experimental freedom in the construction of radiation location—conveyor with the material. The most advanced version of electron accelerators, the Rhodotron, represents the excellent accelerator for radiation processing.

High LET sources—isotopic alpha emitters and of medium LET beta emitters, as well as high LET radiations produced electrically—positive ions accelerated to high energies of particles are sometimes used in radiation chemistry of elastomers. They are of modest importance in basic research and of none in radiation processing, because they cause only degradation of the surface of the elastomer.

Practical conclusions. Rather small but important chemical changes in irradiated elastomers justify buying expensive sources. There are mainly electron accelerators producing beams of electron energy from hundred thousands of electronvolts till 13 MeV. Higher energies are not applied because they can induce artificial radioactivity which complicates the application of the product. New generation of electron accelerators, which are until now mainly linacs, is Rhodotron mentioned above. There is still another option of sources—those with cobalt 60, produced in dedicated nuclear reactors, e.g. by Nordion in Canada. Caesium 137, an abundant isotope from reprosessing spent nuclear fuel is seldom used, because of dangerous form of the isotope—CsCl, light soluble in water, demanding dry storage in contrast to convenient cobalt 60 sources, stored under water. Radiation energy of ceasium 137 is lower than gammas from cobalt 60, what also speaks against it.

Electromagnetic radiation like gamma has its own advantages in comparison to electron beam, therefore the latter can be conveniently transformed to electromagnetic Brehmstrahlung of deeper penetration, similar to gamma. New types of accelerators, like Rhodotron are excellent for that purpose, because they have exceptionally high power. Conversion of electron beam into electromagnetic radiation by hitting a target is connected with loss of energy. It is justified if the penetrating radiation is really more convenient and acceptable than electron beam. Comparisons between isotopic and electric sources can be continued. The background of discussion is sometimes very far from the pure physical, and reaches such circumstances like world ecology and climate, danger of terrorism etc. Sometimes the arguments are very close one to another and result is, that a big irradiation company buys an accelerator and a big cobalt 60 source. They are operated parallely.

How expensive is radiation processing? Introduction of the dose expressed in joule per kilogram (unit called gray, 1 Gy = 1 J/kG) tends to recalculate radiation energy into electric energy. The crosslinking dose of 250 kGy would mean delivery of only 69.4 watt-hours. It would be true if the electric energy from the net would be directly converted into electron beam. That is not the case; on the contrary, conversion of electric energy into the high energy electrons beam proceeds, depending on the type of the accelerator, with the yield around 0.1 % of the input. One has to realize in plans, that the radiation processing is not cheap. Exact costs are delivered by the producer of sources or by the firm irradiating the material on the basis of a price list. Joules taken "from the wall" have nothing to do with joules delivered by the beam. Deep reaching comparisons have even less sens, e.g. in one paper one can find a statement that "the energy needed for radiation crosslinking of rubber is 80 kJ/kg, what means that it is 3.5 times lower than classic crosslinking with peroxides, which is 280 kJ/kg". The cost of producing electron beam is thousand times higher and therefore energetically speaking the expence of radiation crosslinking is by that proportion higher (exact factor depends on type of accelerator). From the known input of energy needed to produce the beam and the not full absorption of the electron beam by objects one can calculate that the energetical yield of radiation processing is expressed in single percent points. Radiation processing is expensive and should be applied after consideration of the increase of value of the product. There are cases in which the electrical energy for production of the beam could be better used for the production of similar product. Such cases are rare.

## 4 Dosimetry of Ionizing Radiation

A quantitative work in simulations of phenomena in radiation processing of polymers is not possible without dosimetry [2]. It is obvious that during the routine processing the dosimetric control is vital part of procedures. The deviation downwards from prescribed dose can cause insufficient effect, e.g. too low a degree of crosslinking and deviation upwards causes the destruction of the material. Dosimetry is carried out with the objective of determining the absorbed dose in an irradiated material. The absolute reference dosimeter is calorimeter, in which the physical principle is change of temperature, of its medium, e.g. water, due to the change of absorbed ionizing radiation energy into heat. Figure 2 shows a most popular version of the device. Unfortunately its use is limited to high dose rate sources i.e. electron accelerators, where the energy is deposited adiabatically and the rapid exchange of heat with the surrounding is limited by thermal insulation. In gamma sources the dose rate is too low to keep the process out the thermal exchange. Part of absorbed radiation energy is changed into chemical

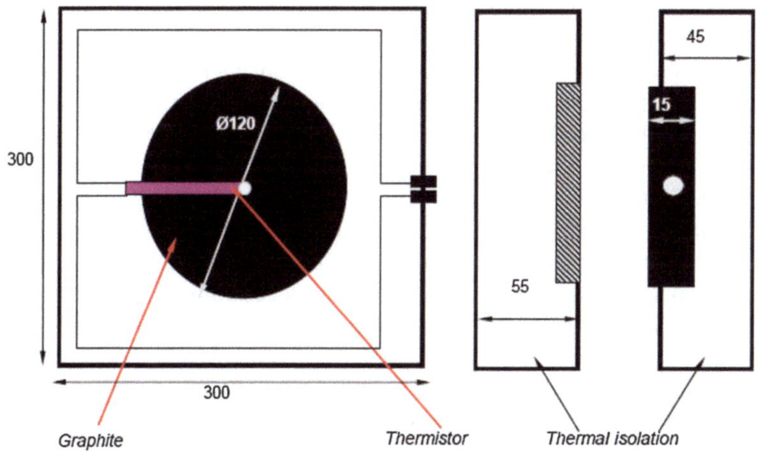


**Fig. 2** Calorimeter for determination of the dose of electron beam

energy of formed or destroyed compounds in the calorimeter. If the body of calorimeter is pure water, the heat of chemical change is negligible. If the body of calorimeter is graphite, the change of chemical energy is nil.

Chemical dosimetry is free from limitation of use, because chemical changes in irradiated systems proceed independent from the temperature in the irradiation chamber and they are additive, with no dependence on the dose rate of irradiation. Any system in which a measureable change in a chemical property takes place upon ionizing irradiation, may in principle be termed a chemical radiation dosimeter and used as such. Solid as well as liquid and gaseous systems have been applied for radiation dosimetry. Examples of solid dosimeters are inorganic crystals, like lithium fluoride, plastic-film dosimeters as plain polymers like poly(vinylchoride), or mixed systems incorporating organic dyes, and photographic films; these are mentioned elsewhere in this chapter. Most liquid dosimeters are based on dilute aqueous solutions of various compounds. Best known are the Fricke dosimeter ($FeSO_4$ in 0.8 *N* sulphuric acid), the ferrous-cupric dosimeter and the ceric sulfate dosimeter and others. The use of gaseous systems is generally limited to research applications, of indirect connection to polymers exposed to atmospheric conditions.

The systems referred to above, plus a multitude of other systems, can in principle be characterized as chemical dosimeters as the reactions taking place in them result only in changes in the outer electron shells of the atoms involved. In this section, however, we shall limit ourselves to discussion of practical dosimeters, starting with aqueous chemical dosimeters. Dosimetry based on films, dyes, and photographic systems is dealt with later. Solid systems utilizing radioluminescence or thermoluminescence phenomena, are discussed rather in dosimetric manuals. Nonaqueous liquid systems have not quite sustained their early reputation, as they have in general been found to be too sensitive to external factors such as temperature, dose rate, etc.

Procedures ("recipes") for eight of the most widely used aqueous chemical dosimeters are given in handbooks and standards (the *Fricke* dosimeter, the ferrous—cupric dosimeter, the ceric sulfate dosimeter, the *water* dosimeter, the *hydrated electron* dosimeter, the *oxalic acid* dosimeter, the benzene—water dosimeter, and the ethanol—chlorobenzene dosimeter). Three of these systems, namely the Fricke, the ferrous—cupric, and the oxalic acid dosimeters have previously been written into tentative standard procedures, e.g. at ASTM (American Society for Testing and Materials). The availability of such a step-by-step procedure certainly facilitates the introduction of a particular dosimeter in the laboratory. It is, however, the experience of the authors that the beginner in the field needs additional information in order to obtain successful results, in the following, emphasis is placed on (1) establishing proper criteria for selection of a dosimeter for a given application, (2) preparation of the dosimeter, (3) the relevant factors influencing the absorbed dose in the dosimeter and influencing the radiolytic reaction in the irradiation procedure, (4) the analytical measurements, and (5) the calculation of dose. Accordingly only few references to the literature are given, and the reader who wants to dig deeper is referred to the references given in the recipe section and to the monographs on chemical analysis.

Proper criteria for selection of dosimeter system in particular radiation processing problem to be solved must be chosen. General concepts and reaction mechanisms help in proper choice. All radiation effects produced in a chemical dosimeter exposed to ionizing radiation are caused by fast electrons; formed either directly from an electron beam from the accelerator, or indirectly as photoelectrons or Compton electrons by X rays or by gamma rays from an isotope source.

In dosimeter systems where the solute reacts differently with the radicals and the molecular products, large changes in LET or dose rate will therefore cause a change in the yield of the chemical dosimeter and investigated system. Similarly the solute concentration may influence the yield of the primary products since, e.g., reactions are suppressed by solutes which compete for free radicals.

It is beyond the scope of this chapter to proceed into a detailed discussion of the reaction mechanism of individual aqueous chemical dosimeters, but some general comments relating to these characteristics may be of help when selecting a suitable system.

The dose absorbed in a chemical dosimeter is measured by analyzing the quantitative change in a given parameter in the system. In the case of the Fricke dosimeter, one measures spectrophotometrically the number of ferric ions formed by radiation induced oxidation in the ferrous sulfate solution; in the ceric sulfate dosimeter one measures the decrease in number of ceric ions; in the oxalic acid system the number of oxalic acid molecules decomposed by the irradiation is followed; and in a solid polyethylene dosimeter the parameter followed may be the number of double bonds introduced by the irradiation or the dose-dependent evolution of hydrogen gas. In the aqueous systems oxidation or reduction is caused by the primary species produced during the water radiolysis described above. In the case of the Fricke dosimeter and the ceric sulfate dosimeter the chemical reactions proceed linearily with dose, i.e., the primary species from the radiolysis keep reacting with the solutes in the same way as long as the supply of the reacting constituents are not exhausted (in the case of the Fricke dosimeter, ferrous ions and oxygen, and in the ceric sulfate dosimeter, ceric ions).

In the case of the oxalic acid dosimeter the situation is somewhat different as decomposition products of oxalic acid such as glyoxal and glyoxylic acid compete with the remaining oxalic acid in reacting with the primary species. As radiolytic by-products build up in the system during irradiation, the oxalic acid decomposition yield decreases continuously, and the dose versus response function of the dosimeter can be approximated by first-order kinetics.

These examples demonstrate, that depending on the reaction mechanism of the particular dosimeter system, one may in some cases obtain a linear dose versus response relationship and in some cases not. While it is conceded that systems with a linear relationship are more convenient in practice, it is felt that this criterion has been somewhat overemphasized in the past.

The liquid chemical dosimeters described in the practical recipes combined cover a dose range from approximately 10 Gy and up to more than 1 MGy, and it should therefore be easy to select a dosimeter for the dose range to be applied. Some of the systems, of course, are more problem-free than others and this is particularly the case with the Fricke dosimeter, which for many users is the standard dosimeter. Unfortunately, this system operates only in the range 10 Gy–4 kGy. It may, however, be used for higher doses in the form of the "super-Fricke" dosimeter (described together with the Fricke dosimeter in standards).

In cases where a dosage within the applicable range of the Fricke dosimeter can be precisely timed, it is recommended to do the primary dosimetry with this system and to scale to higher doses by appropriate extrapolation on a dose versus irradiation-time plot. In cases where such small doses cannot be given precisely, e.g., at very high intensity irradiation sources, or in production facilities where the combination of radiation field and the range of conveyor speeds available makes it impossible, other systems must be applied. The dose range from approximately 0.5–50 kGy can be handled by the ferrous-cupric dosimeter, which may be operated with an accuracy corresponding to that of the Fricke dosimeter provided that a calibration curve is made, and that the dosimeter is prepared under well standardized conditions. The response of this dosimeter is not completely linear with dose as some competition reactions do take place in the system during irradiation. The mechanism has not yet been fully explained, but it has been shown that the dose vs response curve deviates from linearity by only a few percent. Also the ethanol—chlorobenzene dosimeter may be recommended for use in this range.

At higher doses several systems may come into consideration. Among those often used are the ceric sulfate dosimeter and the oxalic acid dosimeter. The strong influence of the photon spectrum on the dose absorbed in the ceric sulfate system under cobalt irradiation and pronounced sensitivity to impurities has given rise to increasing interest in the oxalic acid dosimeter, the response of which is independent of the energy spectrum. Dose-rate requirements can influence the choice of dosimeter.

The radiolytic yield (G value) of a chemical dosimeter may vary with dose rate. All the aqueous systems described here are subject to changes in yield at excessively high dose rates, i.e., rates of the order of 100 kGy/s and higher. In essence, it is caused by the very dense distribution of primary species from the radiolysis of the water, which in turn gives rise to radical–radical reactions in competition with the conventional reactions taking place between the radicals and the solute. A somewhat similar type of dose-rate dependence may be seen in systems based on, e.g., organic chain reactions. Such systems may exhibit dose-rate dependence at much lower dose-rate levels even such as those obtained in normal $^{60}$Co irradiation units.

Dose-rate problems of the first category are experienced only with high power pulsed electron accelerators and single-pulse electron and x-ray generators of the field-emission type. The effect can be suppressed somewhat by using a high solute concentration, and this makes the oxalic acid dosimeter, and even more so the "super-Fricke" dosimeter, suitable systems for high dose-rate work. The dose response of a 100 mM oxalic acid solution, for example, is independent of dose rate up to at least $2 \times 10^7$ Gy/s corresponding to a dose of 20 Gy delivered over a time period of 1 μs. Once calibrated at an exposure time of the same order of magnitude as those used in the practical applications, the dosimeter may be applied over a fairly broad dose-rate range. The "super Fricke" dosimeter is independent of dose rate up to 100 Gy/μs.

Other systems have more pronounced changes in yields at these high dose rates. In the case of the *Fricke* dosimeter, the yield decreases by approximately 15 % at dose rates around 100 Gy/μs. In the case of the ceric sulfate dosimeter there is a 50 % increase in yield at 100 Gy/μs, and in the case of the ferric—cupric dosimeter, there is a 20–60 % increase in yield depending on the composition of the system.

At moderate dose rates, such as those obtained in a $^{60}Co$ facility, most of the systems referred to in this chapter are independent of dose rate. An exemption is the oxalic acid dosimeter, where the formation of semistable compounds, such as glyoxal and glyoxylic acid seems to influence the reaction scheme over extended irradiation periods. In a case like this a calibration of the system against a rate-independent dosimeter is recommended.

It is quite evidently an advantage to select a dosimeter which has absorption characteristics similar to that of the sample of elastomeric object to be irradiated. When this cannot be obtained, corrections can be made as described below. It is necessary, however, to exercise much caution, when using such corrections.

In $^{60}Co$ irradiation, softening of the $\gamma$ rays if they are scattered by passage through substantial layers of material (e.g. as in a tire) may cause additional differences in the radiation doses absorbed in a sample and in a dosimeter. This phenomenon is due to the fact that (particularly in high $Z$ materials) the absorption cross sections increase significantly with decrease in photon energy. If one compares the dose absorbed in a given dosimeter to that of the dose absorbed in water irradiated for the same time in the same geometry, the absorbed doses will of course be identical only if the dosimeter has the same average atomic number as water.

In the case of the oxalic acid dosimeter the difference in absorption characteristics is so small that the difference in absorbed dose is negligible even with heavily degraded spectra. The concentrated ceric sulfate dosimeter (0.4 M) represents the other extreme. In this case the high concentration of heavy cerium ions gives rise to relatively much higher absorbed doses in locations where the spectrum is degraded. If, for example, there is approximately 15 cm of water between the source and the sample, the dose absorbed in the concentrated ceric sulfate dosimeter will be a factor of 1.7 times higher than the dose absorbed in water irradiated under the same conditions. Relevant information for the systems is covered in the standard recipe. While many irradiations in practice may take place under unperturbed spectral conditions, one should always be cautious of this factor which may play a particularly important role in radiation facilities where a high radiation economy is attempted, such as industrial processing plants. The troubles connected with the PVC-film dosimeter may be caused by the fact as the high chlorine content in this system makes its response relative to dose in water quite dependent on the composition of the energy spectrum.

In electron irradiation dosimetry, the spectral degradation problems described above do not occur. Corrections for differences in dose absorbed by dosimeter and sample can be based directly on stopping power data for the initial electron energy.

A number of other factors may be taken into account when selecting a dosimeter for a given application. Some of these factors are more or less self-explanatory: If the product irradiation is to take place at elevated temperatures, or at extremely low temperatures, it should of course be ascertained that the dosimeter functions properly under those conditions. Simulation of cryogenic irradiations under liquid nitrogen (needed in basic research on radiation chemistry of elastomers, by the EPR method) can be done at room temperature, placing the dosimeter in proper vessel in methanol, which has similar density, close to that of liquid nitrogen. Similarly the radiation-induced chemical yield of the dosimeter should be constant and known under possible variations in light, humidity, or gaseous environment, etc., as they may occur in the course of the irradiation process. Important factors still to be considered are, however, the requirements for accuracy and ease of operation. Industrial irradiations tolerate a narrow set of limits for maximum and minimum doses to be applied the dosimeter has to be very accurate and very precise. Assumptions of simulations involve not very accurate starting parameters. Ease of operation of dosimetric procedure is not always combined with high accuracy, and in most situations the optimum system must be chosen by compromise between these two characteristics. While it has been stressed repeatedly that the Fricke dosimeter is the most accurate and most reliable system available today, it remains the opinion of the authors that any of the systems described here may render high accuracy once the system has been well adapted by the laboratory and a good and clean procedure has been worked into routine operation. General requirements in preparation of a dosimetric system supply also orientation in polymer chemistry what changes in composition of the object can occur in the result of absorption of that form of energy.

If the G value for a given reaction is 1 per 100 eV of absorbed energy, an absorbed dose of 10 Gy will lead to the formation of approximately 1 μM of product. This demonstrates that much of radiation chemistry, and by that also chemical dosimetry, is really a trace chemistry. It also demonstrates why most radiolytic reactions can be so highly sensitive to inorganic and organic impurities and accidental individual additives, as these in general can have high reaction-rate constants with the primary species formed by the radiolysis of the main constituent of irradiated system. An example is the ceric sulfate system, where as little as 1 μM of organic impurity will cause a significant change in the observed G value.

In all chemical dosimeters referred to in this section, analytical grade chemicals can be used without further purification. Purification of lower grade chemicals in the laboratory is not recommended; it is generally experienced that the conventional purification procedures may well remove some inorganic impurities (which by the way are generally harmless), but they often introduce new organic impurities. In the case of the ceric sulfate dosimeter, not even quality brand chemicals may suffice, and an adequate brand may have to be selected by "trial and error".

It has to be ascertained that the combination of chemicals used in preparing a chemical dosimeter leads to a correct G value for the system, and the test should include a dose versus response curve with several points on the curve. Deviations from linearity or from the generally accepted dose versus response relationship indicates the presence of harmful impurities either in the chemicals or in the water.

Dosimetric work starts with water purification when one of aqueous dosimeters will be applied. This part of the preparation procedure may be the one which has given rise to most practical problems. Water originating from an organic ion exchanger may be a pure poison in chemical dosimetry. Single distilled water from metal equipment is satisfactory in some cases, but cannot be recommended in general. Not even double distilled water from Pyrex or quartz apparatus suffices in all cases as some organic impurities (like, e.g., humic acids), inherently present in most waters, are indestructible by heat. Two methods of water purification may be recommended: a triple distillation in Pyrex or quartz equipment, where strong oxidizing agents are added in the first two distillation stages and where a stream of oxygen through the system may help in oxidizing the volatile organic compounds. Another method for purification is to irradiate the water in Pyrex or quartz containers with doses of approximately 10–20 kGy. By this procedure, $H_2O_2$ is formed in quite substantial quantities, but it may be destroyed by subsequent exposure of the solution to ultraviolet light. The water purity should be tested, e.g., by ascertaining that the dosimeter prepared from the water yields the correct G value and dose versus response curve, or by measuring the steady state concentration of hydrogen, hydrogen peroxide, and carbon dioxide developing after prolonged $\gamma$ irradiation of the deaerated water.

In cases where sufficient knowledge of the radiation yield under the given dose-rate conditions is not established for a dosimeter to be applied routinely, it is recommended that a calibration of the system be carried out under the conditions in which it is to be used. The accurate calibration of the yield of the Fricke dosimeter by calorimetric and other physical methods has led to the evaluation of a G value, which is believed to be highly accurate. This implies that the Fricke dosimeter is well suited for calibration of other systems in situations where precise dosage within the ranges of applicability for the systems can be obtained. In other situations it is recommended that the routine system be calibrated by an absolute method. Much work has been done over the years with the objective of designing calorimeters suited for calibration of aqueous systems, and convenient methods are available today.

The dosimetry is closely connected with accuracy and precision in chemical analysis. An analytical method used for quantitative measurement of the chemical change in a dosimeter should be accurate as well as precise. That the measurement is accurate means that the sum of systematic errors introduced through the analysis is low. The best analytical methods in this respect are the preparative methods in which the species to be determined are isolated, from all accompanying compounds. In practice, only gravimetric methods meet such requirements, and unfortunately they can only be applied for rather high compound concentrations (more than 0.1 N for a 10 ml sample). A titrimetric method in which the titrating agent is well standardized may also yield good accuracy provided that it is specific for the product to be measured. Titration is often used for determination of the remaining oxalic acid in the oxalic acid dosimeter, and this is possible because only negligible amounts of other organic acids are formed during the radiolysis. Spectrophotometry, which is the preferred analytical method in chemical dosimetry, may also lead to systematic errors, if a foreign constituent absorbing at the wavelength chosen for the product determination is present in the irradiated solution. Fortunately, this is not the case in the systems usually applied, but it has caused problems when spectrophotometry is used with the oxalic acid dosimeter.

The precision of a given procedure is the same as the reproducibility of the results. Good precision is of course necessary but it is not by any means sufficient and it may therefore be recommended, whenever doubt arises, to check the accuracy by using more than one analytical method. Methods such as gas chromatography, polarography, conductivity measurements, etc., have also been used in chemical dosimetry, but not to an extent that would justify their inclusion in this section. Spectrophotometry is the most important method of investigation in chemistry related to problems of radiation processing. In the following it will be assumed that the absorbance of the sample, as read in the spectrophotometer at a given wavelength, is specific for the product to be measured.

As it has been experienced that quite significant variations may occur in the results obtained with various spectrophotometers, it is recommended that the experimenter checks his spectrophotometer carefully. Let us demonstrate what such a testing procedure might include in case of the Fricke dosimeter. The peak wavelength applied for the measurement of the ferric ion is usually given in the literature as 305 nm. Instead of just setting the wavelength scale at 305 nm and reading the O.D. (optical density), it is recommended that an absorption spectrum be made for the system from, e.g., 280–320 nm. It may be found that the peak is at 303 rather than at 305 nm, and this result only proves that the wavelength calibration may be wrong in one of the instruments. The measurement should of course take place exactly at the peak on the instrument used in practice. Furthermore the linearity of the spectrophotometer should be tested. This may be done either by using standard absorbing filters supplied by the manufacturer of the instrument or by preparing a number of solutions with known concentrations of ferric sulfate. The optical density versus concentration curve should be linear within the concentration range applied in the actual measurements; thus demonstrating that Lambert–Beer's law is valid. The drift of the instrument should be followed over a time period of, e.g., 10 min in order to check how often it is necessary to control the 0- and 100-point adjustments. Furthermore the noise level should be checked so that the instrument can be adjusted to give the highest sensitivity at the lowest noise level. Finally the extinction coefficient should be checked by preparing a solution with an accurately known concentration of the compound to be followed. In the case of the Fricke dosimeter highly accurate extinction-coefficient determinations are reported in the literature (the best average is 2,197 at 25 °C).

Measurements should take place in the optical density range where the instrumental accuracy is best, i.e., from 0.4 to 0.6. It should be checked that the spectrophotometric cells are in good condition, i.e., that all optical surfaces are clean. A good way to clean cells is with dilute nitric acid. Most cells will tolerate boiling with diluted nitric acid, but this should, of course, be done only when necessary. The cleanliness of the cells may be maintained by storing them filled with dosimeter solution. For most systems, the optical density is a function of the temperature and this parameter should therefore be under control. It is also important to check that the sample is stable to UV light in the form it is exposed to it during the spectrophotometric measurement. For example, it has been found that the cupric-benzidine complex of oxalic acid, which is used in the spectrophotometric determination of oxalic acid, is sensitive to intense UV light.

The titration method is used by many experimenters for oxalic acid dosimetry, and it has been applied previously also in Fricke dosimetry and in ceric sulfate dosimetry. There is no need to give detailed information on titrimetric procedures in this section, as such information may be found in monographs on chemical analysis. In the case of titration, A molecules/ml is deducted similarly from molarity of titrant, milliliter of titrant consumed per milliliter sample, etc.

It should be noticed, however, that by careful calibration of volumetric glassware (or by weighing all samples), by clean operation, and by good working standards, the accuracy obtainable with this method can be much better than that normally required in conventional titration analysis. In the experience of the author, a reproducibility of better than 0.2 % can quite easily be obtained.

Eventually the calculation of dose follows. Knowing the G value of the dosimeter, and knowing the number of molecules changed per 1 ml of solution, the dose can be readily calculated. In cases where the G value is a function of dose, a dose versus response calibration curve has to be applied.

All dosimeters working in vessels or other containers, like calorimeters belong to the group of "unintentional composite" (See later the section on "radiation chemistry of composites"). The classical radiation dosimeter used in EB radiation processing consists of flat container with water, with inserted thermocouples or other temperature measuring, thin electronic device. It is used just before putting the dosimeter on the conveyor, and immediately when it appears at the loading–unloading station. It can be called unintentional composite, because the aqueous body of the calorimeter is placed in polymer, usually polystyrene container, which absorbs also the energy of ionizing radiation and reaches temperature higher than water. The veracity of dose measurements is saved by keeping the walls of the calorimeter as thin as possible and introducing a correction. Nevertheless, other solutions are tried, i.e. using the solid body of the calorimeter, which does not need any walls. Such a body is made of polystyrene or carbon (graphite). It can be kept thinner than water calorimeter, but the temperature equalization is slower and cannot be make faster by shaking before the measurement, what is the usual praxis in the case of water calorimeter.

## 5 Radiation Chemistry of Solid and Rigid Matter

What is "solid"? At the beginning of the section one can meet terminological problems what is, and how is it called, the solid matter, opposite to liquid. Physicists call "solid" the crystalline material only. The remaining rigid matter has to be categorized in a way, what is a difficult procedure, if even not possible. One proposal is to call the amorphous, but visibly rigid group of "noncrystalline solids" which is the title of one of scientific journals devoted to this field. In the field of polymers one cannot accept that category immediately, because strictly speaking, only elastomers are almost free from the crystalline phase, but the rest of polymers can contain up to 50 % of crystallites in the mass of amorphous material. Crystallites of well determined structures determine the properties of the material. Physicists have introduced another term describing non-crystalline solids, especially polymers: "soft matter" (it is even a title of a general journal and the subtitle of a physical journal). Chemist avoids that term, because dealing with polymers as materials, softness means something defined, which can be measured, like softness of the upholstery in the car sittings. As concerns the term "soft" one can have more objections like association with "soft sciences" as distinguished from "hard science". Chemists are specially sensitive in this point about their field. One can agree that there is no precise term, how the vast kind of matter outside real solid state should be called, and it is better to name it according to particular case.

Nevertheless, we cannot escape from looking separately at radiolysis in crystalline and amorphous state, as the differences in radiolytic answer of reactions is enormous. In the rigid state, diffusion controlled reactions have no meaning, but easy movements and travel distances of small reactant like atomic hydrogen gain on importance in comparison to larger reactive radicals. Only electrons move rather freely, usually reaching the trap, sometimes at quite a distance from the site of ionization.

Also the basic primary acts which start radiolysis, i.e. the ionization of a molecule as described in the Introduction, are the same, in liquids and in solids. Identical is the cascade of ionizations, formation of single and multiple spurs are the same, because the primary and secondary electrons are moving too fast to "see" the movement of molecules in the liquid, versus the practically frozen state in solids [3].

As stressed already, primary ionizations in irradiated medium are divided according to percentage of electrons belonging to particular compounds. It means, as concerns polymeric systems, that all components of the material will obtain the dose in proportion to their content and electron density.

Radiolysis of elastomers has many features of mechanisms involved in radiolysis of organics. Typical is here the phenomenon of energy transfer. Before starting the discussion of it in elastomers, let us describe the simple case of the phenomenon, discovered in the laboratory of radiation chemistry.

One of energy transfer systems, which works in liquids and solids, consists in the increased resistance to ionizing radiation shown by aromatic compounds. They undergo radiolysis with low radiation yield value's. Classical case of benzene ($C_6H_6$) shows only G = 0.005 molecules of hydrogen per 100 eV of absorbed radiation energy. The apparently similar compound, cyclohexane ($C_6H_{12}$) produces hydrogen with the yield of $G_{H2} = 6$. What is more striking, benzene is lowering the yield of radiolysis of cyclohexane, if present together, in higher proportion than it could be concluded from the relation of particular electrons. Already 2 % of benzene added to cyclohexane lowers the yield of hydrogen to $G_{H2} = 3$, but 2 % of cyclohexane added to benzene is not increasing the yield of hydrogen from benzene in measurable degree. The mechanism of the protection effect consists in the ability to accept the excitation energy from the affected molecule or its part and change it into thermal vibrations and rotations, which result in transformation of energy into heat of low value, without chemical effects.

Phenomenon of protection among polymers has been also noticed. In wood, a biopolymer, there is a combination of cellulose and aromatic lignin present. Wood is comparatively resistant to ionizing radiation. Cellulose deprived of lignin in the industrial process, is easily radiolysed, being not protected by aromatic compound. The same applies to textiles applied sometimes in elastomeric composites. Phenomenon of specific property of aromacity, responsible for transformation of energy explains some facts known in radiation processing of synthetic polymers. It was shown [4] that protection action is also exhibited by aromatic polymers, if closely mixed with aliphatic, on the example of polystyrene composite with polypropylene. The protective action is due to energy transfer, in this case working on the distance of 8–12 m.

Radiation induced dehydrogenation of organics is most important in discussing radiolysis of elastomers. It has been shown, that determination of radiation yield of $H_2$ in many irradiated systems helps to elucidate the mechanisms of reactions involved. In particular, examples are presented from the radiation chemistry of elastomers. Conclusions are as far reaching, as concerning problems mechanism of crosslinking.

Abstraction of atoms of hydrogen (which appears eventually after dimerization as gaseous $H_2$), due to absorption of ionizing radiation is specific to this kind of high energy supply to the system. It is intriguing, if considered from the point of view of classic chemistry, because it occurs already at room temperature and even at cryogenic temperatures, due to low activation energy of basic reactions. First observations of hydrogen as the product of radiolysis in aqueous solutions of radioactive substances have been made already by Maria Curie, but systematic measurements were possible after convenient, powerful laboratory sources of $\gamma$ and $\beta$ and fast electron radiations became available. Determinations of radiation yield of hydrogen helped to formulate basic mechanisms, because $H_2$ could be called a molecular products of radiolysis, in contrast to radical products. Molecular hydrogen are products of both, single and multi-ionization spurs, the latter being the only site of high concentration of radical species. Radiation yield of directly formed $H_2$ depends on the LET value of used radiation, what is treated in textbooks as elementary phenomenon.

As concerns organics, the most important case of hydrogen formation is the model system of benzene/cyclohexane, in which already a small participation of benzene is diminishing the $H_2$ production from the mixture, in nonlinear way. Already a low concentration of benzene reduces the hydrogen yield, expected from supposed linear function. The protection effect, explained by energy transfer, is observed also in frozen system, if a perfect homogeneity of the system $C_6H_6$–$C_6H_{12}$ is achieved on the molecular level. All these observations were made using primitive gamma-radiation sources. Frozen benzene/cyclohexane mixture, if really homogenous one, simulates well absorption characteristics of a polymer.

In next decades, the interest in hydrogen yields dropped down, in spite of the important results which could have been gained. We have recognized them using modern electron beam irradiation and advanced analytics. Only in some cases $\gamma$ radiation was used.

All irradiations in Authors laboratory have been made with electron beam of 10 MeV energy, supplied as straight beam or scanned beam of more narrow energy spectrum. Low temperature irradiations (down to −20 °C) were made in a special box placed on the conveyor under liquid nitrogen in special vials. Results helped to understand that low temperatures do not act as protective factor preventing the radiation damage. Irradiated samples were placed on the bottom of 3 mL vials, closed with septum, protected by a hood made of lead. Because the hydrogen formed during irradiation is rapidly equilibrated with the argon space in the vial, it is ready to immediate determination by gas chromatography, in the procedure described elsewhere. Proper choice of gas chromatograph and the column is important to obtain maximum possible sensitivity; we have started investigations with Shimadzu gas chromatograph type GC-14B but later we have switched to the type GC-2014 with another column, thus improving the sensitivity by one order of magnitude. That means the possibility to lower ten times the dose. The time elapsed between irradiation and measurement does not matter, as the release of hydrogen is fast and irreversible.

Every polymeric material has been tried for hydrogen production at different doses, from low doses determined by limits of sensitivity, to high doses. In the case of simple systems, the abstraction of hydrogen runs linearly with dose and there is only one value of radiation yield rather easy to understand. Nonlinear cases are to be treated specially and for the case of interpretation confronted with other investigations, like DRS.

All yields are given in old units of indexed compound (hydrogen) formed per 100 eV of absorbed energy, because it is better understandable in the interpretation than in SI units (μmol/J).

Presented data are collected from several hundred systems investigated until now.

The cases of simple short chain hydrocarbons and polyolefins are rather exotic in context of elastomers, but for the sake of completeness should be mentioned. These hydrocarbons contain maximum possible number of hydrogen atoms. Due to determinations of radiation induced dehydrogenation, we know why. Our determinations of radiation yield show high values: The most extensively investigated, all kinds of polyethylenes, release hydrogen with $G_{H2}$ in the range from 4 to 5, per 100 eV. In that case the release of hydrogen does not mean the degradation of oligoethylens. Vice versa, it is connected with the crosslinking and increase of molecular weight. Basic radiation induced crosslinking reaction consists of formation of positive hole remaining after ionization, which travels along the chain to the site, where the chain with the ionization meets another chain, not participating in ionizations. At the meeting site of chains, the crosslink bond is formed, with the release of $H_2$.

The case of energy transfer in mixtures of polymers should be mentioned again. The classical case ($C_6H_6/C_6H_{12}$ i.e. aromatic/aliphatic system) is well adapted for study by changes in radiation yield of hydrogen abstraction. In our experiments, the radiation yield of hydrogen in pure polystyrene (0.04 $H_2$/100 eV) changed nonlinearly to (3.20 $H_2$/100 eV) in the case of pure polypropylene. Thus the energy transfer, also in polymers has been demonstrated, along with the role of homogeneity on the molecular level. Therefore a well pronounced protection effect, due to the presence of aromatic rings, can be considered as a fact of general importance. However, again the role of homogeneity of the polymer blend on the molecular level has to be considered. Interesting natural combination of two polymers, one aromatic (lignine) the second non-aromatic (cellulose) is intimate enough to exhibit the energy transfer and protection. Ordinary timber and cheap paper containing the mixture of cellulose and lignine, are well resistant to ionizing radiation, whereas cellulose, separated from lignine and not protected, is radiolysed with high radiation yield.

## 6 Comparison of Radiation Chemistry with Photochemistry

Photochemistry is mentioned in connection with effects occurring during exposure of elastomers to sunlight. Photochemistry is usually not classified in the high energy chemistry, as the energies involved are close to energies of bonds and excitations are usually not the cause of ionizations. However, photochemistry is close to radiation chemistry due to precise definition and measurement of energies involved, in the contrary to poor quantitative approach of majority of high energy chemistry, except radiation chemistry, to these questions. Photochemical effects in elastomers will not be discussed here in deep, because they do not belong to radiation processing. Basic differences consist in low energies and a short range of penetration into the material, especially when the latter contains light absorbing additives like carbon black.

Photochemistry: The main source, sunlight, is photochemical active, if illuminated object contains chromophoric groups. Range of penetration at absorbing wavelength is very short; with typical $\varepsilon$'s, molecular extinction coefficients—half-depth is of microns ($\mu$'s) to millimeters (mm's). Otherwise there are negligible chemical effects. e.g. during daylight illumination of surfaces of elastomeric products. Illumination produces in that cases mainly the low grade heat only and chemical products at the very shallow depth. The Sun-energy received would be sufficient to support the decay, but the energy is lost due to the lack of properly absorbing chemical compounds.

Selective absorption of photochemical energy causes different effects on. Applications of elastomers in space research show sad fate of polymers. Anoxic atmosphere or vacuum in the outer space is passing more chemically active UV light to the surface than oxygen-containing atmosphere with additional ozone filter. The spectrum of solar UV emission is generally divided into the UVA (315–400 nm), the UVB (280–315 nm), the UVC (200–280 nm) and the VUV (<200 nm) so called vacuum UV. Sometimes other names for particular ranges for ultraviolet are used: One can meet following subdivisions of the UV spectral range: NUV—near UV 300–400 nm, MUV—middle ultraviolet 200–300 nm, already mentioned FUV—far ultraviolet—120–200 nm. Among other proposals: Some researchers introduce EUV—50–150 nm (E for extreme) and FUV, this time defined as 145–330 nm (F for far). In addition, to be closer to radiation chemistry, one can meet definition of X-ray ultraviolet as 0.1–30 nm, contrasting with hard X-rays at wavelengths less than 0.1 nm. These radiations belong already to the category of ionizing radiations, however, penetrating vacuum only and being fully absorbed by the subsurface of irradiated object.

More precise definition of the UV wavelengths is connected with the kind of the source applied in experimental simulations, most close to ionizing radiations; e.g. $H_2$-flow discharge lamp providing mainly Lyman-$\alpha$ (121.6 nm or 10.25 eV) and 160 nm (7.75 eV) photons, as well as UV/EUV light provided by a synchrotron beamline in the 4–20 eV range with an average energy of 6.07 eV (204 nm), etc.

Photochemistry: sunlight as the main source interacts only with solutes absorbing UV/Vis (e.g. cations with proper absorption spectra, inorganic products of radiolysis, organic products of “other sources” (above) if absorbing UV/Vis). Otherwise there is production of low grade thermal energy,

Photochemical energy dominates over other active sources at the surface of Earth during daylight.

The average dose rate in space is not sufficient to induce changes in properties of majority of materials of construction immediately. Proper functioning of vehicles (containing elastomers) on Mars show, that materials of construction were properly chosen from the point of view of radiation chemistry. However, some organic materials are not resistant enough, as the example of fluorine containing polymeric material shows. Parts of Hubble telescope were made of polymer Teflon, which become brittle after comparatively short stay in outer space. In this case the UV was playing additional destructive role on parts exposed to sunlight.

Simulation of behavior of polymers in ionizing radiation fields is hardly necessary in view of well known radiation chemistry and processing of these macromolecules. Teflon belongs to polymers degrading after absorption of ionizing radiation. Doses are known and the stage of easy degradation of Teflon into powder can be easily reached. Precise value of dose after which a defined failure of mechanical properties will occur, can be determined by application of the procedure developed for different purposes in present Authors Laboratory.

## 7 An Example of Radiation Chemistry of Single Component Elastomer

The example is a thoroughly investigated case of electron beam crosslinking of hydrogenated acrylonitrile-butadiene rubber [5, 6]. Radiation induced crosslinking of hydrogenated acrylonitrile-butadiene rubber (HNBR) is proposed as an alternative to chemical curing methods. Radiation-induced crosslinking of polymers is a common method, applied also commercially, e.g. for the production of polyethylene shrinkable tubes and tapes, as well as hot water pipes. This time it has been applied also to a synthetic elastomer. The radiation crosslinking technique was proposed as an alternative curing method for hydrogenated acrylonitrile-butadiene rubber (HNBR). HNBR is a high-performance elastomer requiring unconventional curing methods. Partly hydrogenated HNBR (hydrogenation degree <96.5 mol %) is suitable for crosslinking with sulphur in the presence of rubber accelerators [7–9] or with selected tetraalkylthiuram disulfides. Fully hydrogenated products (hydrogenation degree >99 mol %) can be crosslinked only with radical donors such as organic peroxides. Properties of crosslinked HNBR depend on the curing method. Peroxide-cured HNBR exhibits better hot air and ozone resistance but it shows lower tensile strength and poorer dynamic behaviour than the sulphur-cured rubber. The new approach was to investigate the radiation induced crosslinking and degradation processes of HNBR (bound acrylonitrile content 43 wt %, hydrogenation degree >99.0 and 94.5 mol %) caused by electron beam radiation. The influence of residual carbon–carbon double bonds content, electron beam dose value and radiation environment on the radiation effects was also determined.

HNBR of trade names Therban A4307 and C4367 (bound acrylonitrile content 43 wt %, hydrogenation degree: >99.0 and 94.5 mol %, denoted as H43 and S43 respectively) from BAYER were used for the investigations. Rubber plates of 1 mm thickness were prepared from cold masticated HNBR by press moulding under pressure in heated stainless steel forms. Samples were evacuated, saturated with oxygen or argon, stored in oxygen or argon atmosphere and finely irradiated in oxygen or argon atmosphere and hence they are denoted by Ox and Ar symbols respectively. Irradiation was performed by 10 MeV electrons, monoenergetical, due to selection during the 270° bending [10]. The resulting 6 kW power electron beam of 6 kW power was scanned over the conveyor, securing homogenous distribution of the dose. The applied doses varied from 20 up to 300 kGy. A split dose technique (20 kGy increments) was applied to avoid excessive heating of samples caused by adiabatic nature of the process [11]. Dosimetry was typical for radiation processing of polymers used in Authors Centre for Radiation Research,

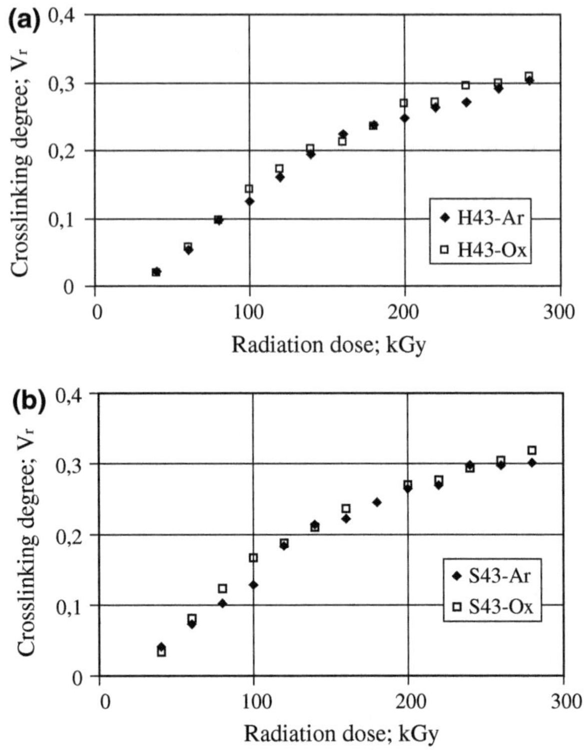


**Fig. 3** Crosslinking degree $V_r$ for irradiated: **a** *Therban A4307* (H43) and **b** *Therban C4367* (S43) as function of radiation dose *D;* swelling in MEK; samples were irradiated after conditioning in argon (*Ar*) or in oxygen (*Ox*) atmosphere; hydrogenation degree >99.0 (H43) and 94.5 mol % (S43) respectively

Institute of Nuclear Chemistry and Technology, Warsaw (INCT); it is traced to absolute calorimetric dosimetry, according to the ASTM standards.

The equilibrium volume swelling $Q_v$, Mooney-Rivlin elasticity constant $2C_1$ and sol content $S$ were determined according to the standard methods used in previous studies on HNBR. The determined $2C_1$ values were corrected for the presence of non-rubber constituents and sol-fraction in the cured samples.

The irradiated HNBR samples were exposed to solvents by swelling to elastomer with a thermodynamical affinity similar to that of an elastomer ($\delta_{\mathrm{HNBR}} = 21.9\ \mathrm{MPa}^{1/2}$). It was noticed that irradiated samples lost their solubility and swell in tetrahydrofurane (THF), methyl ethyl ketone (MEK) and in other solvents to a limited extent only. Figure 3 shows the crosslinking degree, expressed as a volume fraction of rubber ($V_r$) in a swollen gel as a function of radiation dose ($D$). The increase in the degree of crosslinking with rising of the radiation dose is evident from these measurements.

The measurements of Mooney-Rivlin elasticity constant $2C_1 = nRT$ which at temperature $T$ is a linear function of active network chains $n$ and network crosslink density $N = 2n/f$ (where: $f$–crosslink functionality; $R$–universal gas constant), enabled a more precise analysis of crosslinking density. It was found that the crosslinking density for all investigated rubbers is directly proportional to the radiation dose (Fig. 4). These results correspond to theses from the preliminary

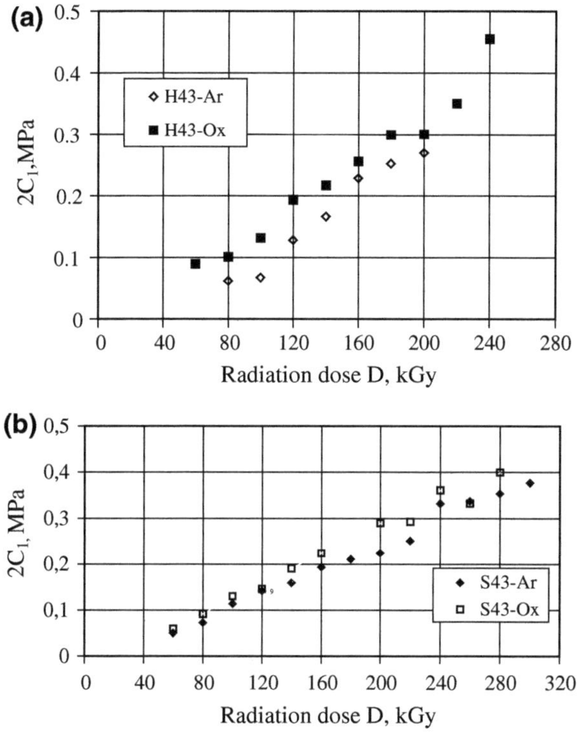


**Fig. 4** Mooney-Rivlin elasticity constant $2C_1$ for irradiated: **a** *Therban A4307* (H43) and **b** *Therban C4367* (S43); samples were irradiated after conditioning in argon (*Ar*) or in oxygen (*Ox*) atmosphere

studies of the influence of residual carbon–carbon double bonds content in HNBR with 34 wt % acrylonitrile content. The similar results for HNBR with hydrogenation degree: >99.0 and 94.5 % and irradiated in oxygen or oxygenfree atmosphere, indicate that both the residual carbon–carbon double bonds content and the irradiation environment do not influence the crosslinking process in HNBR under study.

From the similar slope values of a linear function $2C_1 = f(D)$ it follows that the crosslinking efficiency ($E$), expressed as a number of crosslinks formed by 100 eV dose, is the same in all investigated rubbers. For the quantitative determination of crosslinking yield (Table 1) Eq. (1) was applied.

$$E = C_1 N_a / RTdD_e[\text{network crosslinks}/100\ \text{eV}] \quad (1)$$

where: $C_1$—Mooney-Rivlin elasticity constant; $d$—density of cured rubber; $R$ = universal gas constant; $D_e$—radiation dose expressed as multiple of 100 eV.

The extrapolation of relationship of $V_r$—$D$ and $2C_1$—$D$ to $D = 0$ leads to negative values both of $V_r$ and $2C_1$. It means that a certain part of radiation dose,

**Table 1** Crosslinking efficiency (*E*), gel dose ($D_g$) and ratio of chain scission acts to crosslinking ones (*p/q*) for radiation cured HNBR: Therban A4307 (H43) and Therban C4367 (S43); hydrogenation degree: >99.0 % and 94.5 mol % respectively

| Elastomer | Irradiation environment | E (crosslinks/ 100 eV) | $D_g$ (kGy) | p/q |
|---|---|---|---|---|
| Therban A4307 (H43) 0.062 mol >C=C</ kg | Oxygen (Ox) | 2.49 | 21.6 ± 2.7 | 0.086 ± 0.013 |
| | Argon (Ar) | 2.69 | 29.1 ± 3.1 | 0.060 ± 0.005 |
| Therban C4367 (S43) 0.682 mol >C=C</ kg | Oxygen (Ox) | 2.68 | 25.4 ± 2.5 | 0.059 ± 0.012 |
| | Argon (Ar) | 2.77 | 24.8 ± 2.6 | 0.088 ± 0.019 |

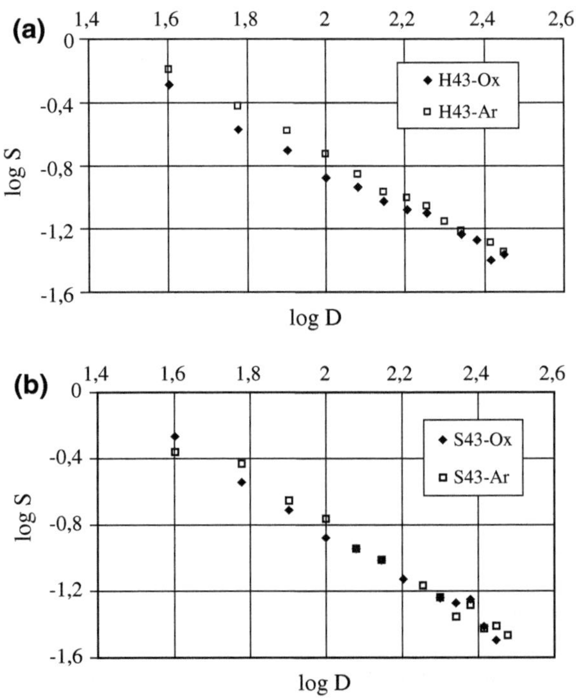


**Fig. 5** The relationship between sol fraction content *S* and radiation dose *D* for irradiated: **a** *Therban A4307* (H43) and **b** *Therban C4367* (S43); samples were irradiated after conditioning in argon (*Ar*) or in oxygen (*Ox*) atmosphere

called gel dose ($D_g$) has to be applied before crosslinking process can start. The $D_g$ values (Table 1) were determined from the linear dependence of the sol content *S* from the dose *D* in double logarithmic coordinates (Fig. 5). These values are similar for all polymers studied and it is evident that $D_g$ is only slightly influenced by the presence of oxygen during irradiation and residual carbon–carbon double bonds content in HNBR. This gel dose is probably used up in reactions with antioxidants and other non-rubber constituents present in commercial products, acting as scavengers of free radicals.

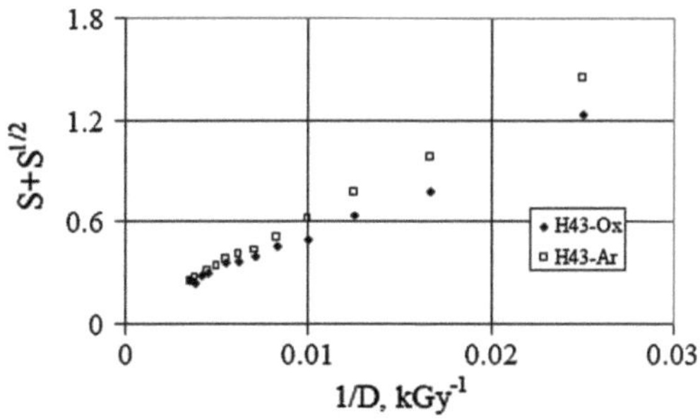


**Fig. 6** The sol fraction content *S* in radiation cured *Therban A4307* in the function of radiation dose *D*; coordinates according to Charlesby-Pinner equation; samples were irradiated after conditioning in argon (*Ar*) or in oxygen (*Ox*) atmosphere

Both crosslinking and degradation occur simultaneously in the polymer. These acts relate to the partition of deposited ionizing energy between single- and multi-ionization spurs. As in all polymers about 80 % of deposited radiation energy appears in single-ionization spurs, located far from each another. The energy accumulated in those spurs (<100 eV) is sufficient for excitation or ionization only, without the chain scission. Transfer of those primary radiolysis effects along the chain leads to the forming of tetrafunctional (X-types) crosslinks. The remaining energy (about 20 %) is deposited in multi-ionization spurs. They are formed in random places by electrons of final generations, of energies next to subexcitation, being not able to travel far from this site. The accumulation of energy >100 eV in centres of a small volume, causes the instant chain scission. Loose ends of broken chains can form trifunctional crosslinks (Y-type) with neighbouring chains. Some of them cannot find a partner for reaction and result in a reduction in the average crosslinking degree [12, 13].

For the calculation of ratio of chain scissions acts *p* to the crosslinking acts *q* the Charlesby-Pinner Eq. (2) for polymers with the most probable distribution of molecular weight ($M_w/M_n \approx 2$, as in the cased of used HNBR [19]) was applied (Fig. 6) [14]:

$$S + S^{1/2} = p/q + A/D \tag{2}$$

where: *S*—sol content, *A*—constant connecting crosslinking efficiency and degree of polymerization, *D*—radiation dose.

The extrapolation of linear function $S + S^{1/2} = f(1/D)$ to $D = \infty$ gives *p/q* values. From the estimated values it follows that every 100 crosslinking acts are accompanied by 6–9 scission acts ($p/q = 0.06 - 0.09$). These values are very similar, hence the conclusion that the hydrogenation degree and presence of oxygen during the irradiation do not influence the degradation processes of the HNBR under study. The results of degradation are lower than the ones reported by Zhao et al. [15] ($p/q = 0.41$). However, HNBR studied by Zhao (34 % bound acrylonitrile content) was filled with carbon black and cured with gamma radiation.

In order to investigate post-radiation effects in HNBR studied, samples after 2 weeks and 6 months following the irradiation were exposed to equilibrium

**Table 2** Crosslinking degree ($V_r$ in MEK) for irradiated HNBR (43 wt % bound acrylonitrile content; hydrogenation degree >99.0 and 94.5 mol %, denoted H43 and S43 respectively) in 1 weeks or in 6 months after irradiation; samples were irradiated after conditioning in argon (Ar) or in oxygen (Ox) atmosphere

| Elastomer | Swelling (solvent: MEK) | Crosslinking degree ($V_r$ in MEK) | | |
|---|---|---|---|---|
| | | Radiation dose, kGy | | |
| | | 80 | 180 | 280 |
| H43-Ar | in 2 weeks after irradiation | 0.0895 | 0.2175 | 0.2848 |
| | in 6 months after irradiation | 0.0899 | 0.2186 | 0.2723 |
| H43-Ox | in 2 weeks after irradiation | 0.0838 | 0.2096 | 0.2848 |
| | in 6 months after irradiation | 0.1057 | 0.2143 | 0.2825 |
| S43-Ar | in 2 weeks after irradiation | 0.0821 | 0.2160 | 0.2616 |
| | in 6 months after irradiation | 0.0847 | 0.2172 | 0.2728 |
| S43-Ox | in 2 weeks after irradiation | 0.1067 | 0.2185 | 0.2870 |
| | in 6 months after irradiation | 0.1046 | 0.2155 | 0.2856 |

swelling in MEK. The estimated values of crosslinking degree (Table 2) are similar, therefore it could be assumed that no post-radiation effects occurred in investigated samples. However, it seems to be reasonable to expect that some kind of post-radiation effects can occur immediately after irradiation.

The mechanical properties are very important from the point of view of applications. The investigations were focused on the dependence: tensile strength $TS_b$ and elongation at break $E_b$ versus crosslinking density or radiation dose $D$. Elongation at break ($E_b$) for peroxide or sulphur-cured elastomers is a decreasing function of crosslinking density [16]. Similar dependence for irradiated HNBR was found (Fig. 7).

Figure 8 shows the tensile strength in the function of radiation dose. The estimated values of $TS_b$ are varied from 15 up to 20 MPa and it could be assumed that they only slightly depend on the crosslinking density. It should be mentioned that tensile strength for conventionally cured elastomers is an extreme function of network density [17]. There are several reasons why such dependence is not observed. It could be caused by different susceptibility of peroxide or sulfur and radiation cured HNBR to crystallization during the axial elongation.

From the results it also follows, that both elongation at break and tensile strength are only slightly influenced by the presence of oxygen during irradiation and residual carbon–carbon double bonds content in HNBR studied.

Both fully and partly hydrogenated HNBR (43 wt % bound acrylonitrile content) are suitable for radiation induced crosslinking. The crosslinking density of hydrogenated acrylonitrile-butadiene rubber is a linear, increasing function of electron radiation dose. It was found that both the crosslinking density and radiation yield are not influenced by the radiation environment (oxygen or argon atmosphere) and by residual carbon–carbon double bonds content in HNBR studied. Every 100 crosslinking acts is accompanied by the 6–9 chain scissions acts. The values of gel dose determined for *Therban* A4307 and *C*4367

**Fig. 7** Elongation at break $E_b$ for: **a** *Therban A4307* (H43) and **b** *Therban C4367* (S43) irradiated with different radiation dose *D* (hydrogenation degree >99.0 and 94.5 mol % respectively); samples were irradiated after conditioning in argon (*Ar*) or in oxygen (*Ox*) atmosphere

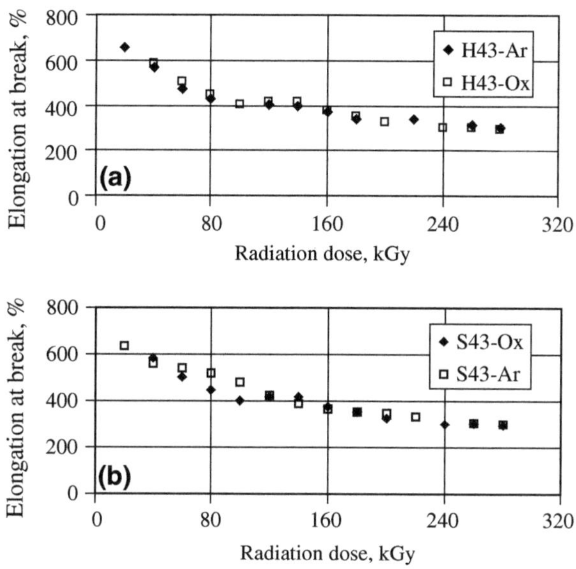


(hydrogenation degree: >99.0 and 94.5 mol % respectively) are varied from 22 up to 29 kGy. From the comparison of crosslinking degree for HNBR, measured in 2 weeks or in 6 months after irradiation no post-irradiation effects were found.

This most important investigation dealing with radiation chemistry of HNBR is continued in another laboratory by Perraud et al. [18] in 2010. Results of both papers quoted above have been confirmed. Interest in the HNBR is justified by excellent resitance of that rubber to solvents, oils, hydrocarbons, ozone and other oxidants, all in a wide range of temperatures.

For NBR the relation of radiation yield of chain scission to the yield of crosslinking was [G(S)/G(X)] from 0.12 to 0.42, depending on molecular weight, presence of polyfunctional monomers. Differences between gamma and electron irradiation are small. Authors of new publication confirm the mechanism proposed by Zagórski [12, 13] as similar to classic crosslinking with a peroxide: The peroxide adsorbed at the energetically proper place at the chain, during preparation of the blend, removes hydrogen during heating. The path to the formation of crosslinking bond is opened. In the case of irradiation and in the absence of peroxide, positive hole travels along the chain and reaches the point where two chains are close one to another and crosslinking bond is formed with the release of hydrogen. Extrapolation of the Charlesby-Pinner diagram to the infinite dose shows wide range of acts of chain scission 12–48 ($p_0/q_0 = 0.06 - 0.238$ for 100 acts of crosslinking). Radiation yield of crosslinking increases with the content of acrylonitryl due to the limitation of cyclization of butadiene units. Such reaction is impossible in the HNBR but absence of double bonds can reduce the yield of crosslinking, as it has been shown in the case ethylene-propylene—diene copolymer. Thus observations of the influence of hydrogenation range on the yield of crosslinking. FTIR investigations of the irradiated systems have shown

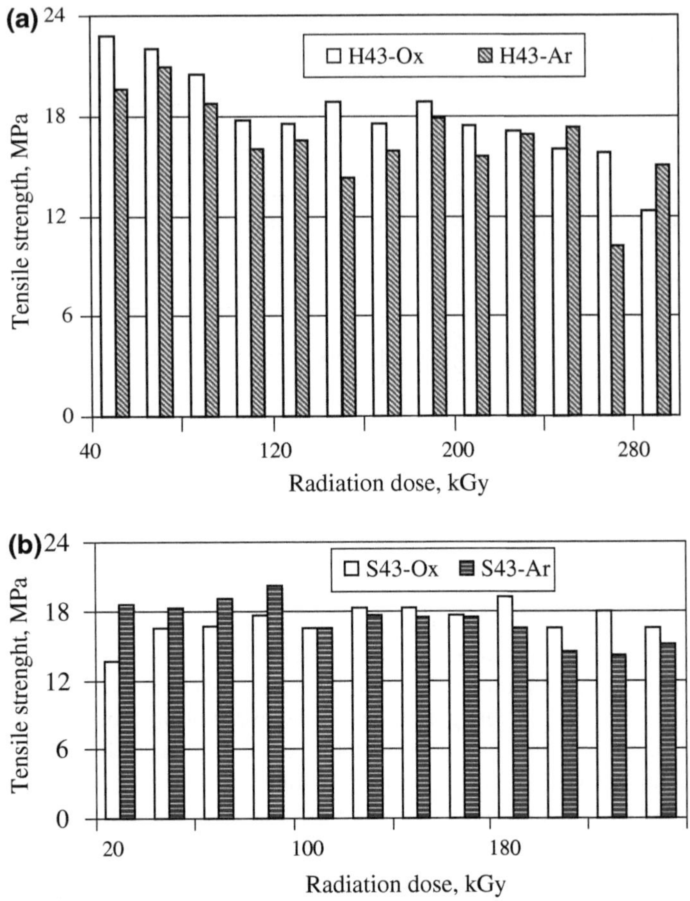


**Fig. 8** Tensile strength $TS_b$ for: **a** *Therban A4307* (H43) and **b** *Therban C4367* (S43) irradiated with different radiation dose *D* (hydrogenation degree >99.0 and 94.5 mol % respectively); samples were irradiated after conditioning in argon (*Ar*) or in oxygen (*Ox*) atmosphere

formation of trans-vinylene double bonds. Their concentration proportional to the irradiation dose. Introduction of new analytical method has shown the appearence of a new group—first order amine group and alkene-imine, which disppear in time.

## 8 Basic Radiation Physics and Chemistry of Elastomer Composites, Subjected to Radiation

The section starts with definitions of a composite, encountered in scientific and engineering literature and proposed definition from the point of view of radiation chemistry and processing is formulated. Composites are more and more important in applied and fundamental polymer science, the participation of radiation processing will increase. In the present chapter, specific features of composites are discussed from the point of view of radiation processing of elastomers. The discussion is in terms of: (1) dose distribution, which due to basics of radiation physics cannot be like in homogenous medium, (2) of thermal effects, because both phases have different specific heat capacity, (3) energy and reactive species exchange between the matrix and the dispersed phase. The first phenomenon results in complicated DUR (dose uniformity ratio) in comparison to homogenous systems, the second in unfavorable phenomena if the dispersed phase is a high Z compound, or even a metal like steel in the tire, the third results in effects which can be different from the sum of radiation chemistries of both materials (matrix and the second phase).

Elastomers in pure state are used only in basic research of their radiation chemistry features. In technological approach they occur only as composites of specific response to ionizing radiation. What is a composite? We have to answer the question before one can start the discussion of the radiation chemistry of the system. That is a routine question, because the energy of ionizing radiation supplied to the specific system is not like putting the material into a traditional laboratory oven or drying box. Results of deposition of ionizing radiation energy depend on the composition, structure and complexity of the system. Already effects of non-ionizing radiation, like those in the microwave oven are more complicated and depending on the material, than ordinary thermal treatment.

Looking into general encyclopedia's of technology like McGraw [19] and Ullmans Industrial Chemistry [20] one sees that the philosophy of composites is not limited to polymeric compositions, but is also applied to practically all complex materials, starting with inorganics, like strong, creep resistant fibers and metals used also in rubber industry. Far from the case of reinforced polymers are such components like plasticizer or low concentrations of pigments or processing aids. Typically, the main goal of composites is to improve strength, stiffness or toughness, or dimensional stability by embedding particles or fibers in a matrix or binding phase. A second goal is to use inexpensive, readily available fillers to extend a more expensive or scarce resin. In rubber industry, components secure proper mechanical properties of the composite. From the point view of radiation chemistry which deals with the distribution of specific energy, definition of a composite involves also elastomeric foams, in which empty spaces change the geometry of absorption of energy.

There can be a difference between composite in which added phase serves to reinforce the material and a filler which serves other purposes. The reason can be also quite nontechnical, like lowering the price of the product. In that sense the filler is categorized as a neutral additive to the elastomer. However, in spite of to be composed for trivial reason, such material, when irradiated, it will behave like a composite.

Composites are drawing increased general attention, not limited to radiation chemists. It resembles the interest paid about one 100 years ago to colloids, with their specific behavior, nowadays explained in categories of nano-dimensions.

There were proposals to treat composites as a hybrid material, which is a creation something different than the sum of constituents. Size, shape and chemical identity of the nanoparticle and interaction of the nanoparticles with the polymer matrix can affect significantly the final properties of a hybrid material.

The present attitude, especially nano-size oriented, treats composites as a hybrid material, which is a creation something different than the sum of constituents. The size, shape and chemical identity of the nanoparticle and interaction of the nanoparticles with the polymer matrix can affect significantly the final properties of a hybrid material. Radiation chemistry helps in fundamental understanding of hybrid materials containing inorganic nanoparticles embedded in an organic macromolecular matrix, in terms of the formation and intrinsic properties of the nanoparticles and the structure and properties inherent to the polymer.

We can now suggest a definition, what is a composite from the point of view of radiation processing and radiation chemistry: A composite is any heterogenic material in which the shortest size of a 3D, dispersed phase in the matrix is of the order of few nanometers. Shorter sizes do not qualify to the category of composites, because from the point view of radiation chemistry they form homogenous material. Such small sizes are comparable with sizes of spurs, in looking into radiation chemistry on the molecular level. Upper limit of size of the dispersed phase is not defined. Due to the present fashion of looking for small dimensions and calling them nanotechnology, the philosophy of composites fits into presented approach. However, in the most cases, composites show the dispersed objects, particles, fibers, platelets, films the smallest one size much bigger then nanometers. They are usually well over micrometer size, but some authors underestimate them in unknown purposes. Proposed classification from the point of view of radiation processing does not distinguish between a composite (e.g. elastomer with reinforcing fiber) and polymer with chemically and mechanically neutral filler, even if in other classifications such differences occur.

Proposed classification is justified, because the response of two phases present in a composite to radiation is always different what will be shown below on many examples. The behavior of dispersed, small size phase, without radiation, drew attention already 100 years ago, when colloid chemistry started its existence. At that time the concept of nanotechnology did not exist yet, but the same sizes were functioning in colloid chemistry. For instance the diameters of metallic particles created in a material were of nanometer orders. No connection to ionizing radiation was looked for, but visible light interacted, observed in beautiful colors. Their behavior was different in comparison to larger sizes: due to their small size, they were stable, like metallic colloids in virtual solution.

The proposed definition of composites from the point of view of radiation physics and chemistry does not define the state of aggregation of both components of composites. Although the most common combination is solid–solid, e.g. fiber reinforced elastomer, possible are combinations solid–gas, e.g. porous foams. During radiation processing, the air gaps are also absorbing ionizing radiation, however, with intensity by three orders of magnitude lower effect. Finally, the combination liquid–liquid is possible, e.g. in rubber latex, as well as solid–liquid, in which polymer is soaked with monomer with the intention of radiation induced curing, grafting or copolymerization.

Systematics of composites used in the present paper can be incompatible with other proposals. For instance, the paper by Marwanta et al. [21], already in the title speaks about "liquid composites". However, the elastomer poly(acrylonitrile-co-butadiene) rubber, NBR, has been mixed with ionic liquid (N-ethylimidazolium bis[trifluoromethanesulfonyl]imide) up to 60 %, forming homogenous and transparent film. Thermally stable elastomeric ion conductors were obtained. As the material is homogenous and will respond as such to ionizing radiation, there is no need to treat it as a composite. If the term "liquid composite" can be used, it is rubber latex in which both matrix and dispersed phase is liquid.

Closing the discussion on definitions, we would like to draw the attention to relations of words "processing" and "curing". The term "curing" originated in clothing manufacture and means baking a garment and similar operations. It is used also in food preservation technology—e.g. curing of ham. The term curing is popular in photochemistry of polymers, because it means the initiation of polymerization of monomer composition, if it absorbs light. There are even special devices for that purpose called UV-curer, not to mention mass produced small devices used in dentistry to cure fillings in teeth. In the present chapter the term curing will be used in the sense of radiation induced polymerization, e.g. radiation curing of ink, not to be confused with radiation induced crosslinking and grafting. The term "radiation processing" is a general one, involving many different chemical consequences of absorbtion of ionizing energy; even of biological ones, like in radiation induced sterilization of medical and food supplies. From the point of view of radiation physics and chemistry the rules are the same.

Specific radiation chemistry of composites consists in:

1. Different electron density of both substances results in different density of ionizations. This effect is not high, if both substances are chemical compounds of low Z number. However, steel parts in a tire introduce serious complication in the distribution of energy, and the final temperature after irradiation, in the material.
2. Different specific heat capacities of both phases, the dispersed and the main one, result in different temperatures reached in adiabatic irradiation, i.e. by electron beam. The effect at low dose rates, as it is the case in gamma-irradiation, is not always visible, due to the heat transfer which can be faster then the supply of energy.
3. Large surface area of the interphase between two phases. It can reach enormous values with diminishing of the size of dispersed phase. For instance; let us assume one liter of latex emulsion of 30 % concentration; the reduction of the diameter of the latex emulsion spheres from 1 μm, 100 to 10 nm means the increase of the surface area resp. from 1,800, 18,000, to 180,000 $m^2$ (equivalent to 900 × 200 m "lot"). These calculations are made for spheres, which show the smallest surface area for a given volume. Spheres occur in rather few composites, like raw latexes, and in most cases the shape of additional phase is

far from the spherical. Therefore any other shape means even higher surface area of the interface, with many consequences. In both phases independent radiation chemistries are running, with a variety of energy and material transfer in both directions. An important role is played by the quality of the interface. It is usually modified at the stage of preparation of a composite. Usually the goal is to have the surface of dispersed phase as friendly to the host as possible, e.g. to make radiation induced grafting possible, i.e. compatible. The consequences of enormous increase of the interphase, occurring with reduction of the size of dispersed phase, are well know in catalysis and adsorption science.

Composites are sometimes (e.g. Nastase [22]) processed by specific technique of high energy chemistry, i.e. plasma treatment, glow discharge reactor or similar device. In spite of apparent similarities to radiation processing, that technique has nothing to do with penetrating ionization and consists in superficial treatment of material with gases, in which reactive species has been formed. Chemical action is limited to the surface, whereas the ionizing radiation, even of not mega-electronvolt energy of quanta or particles, is penetrating well below the surface of the material.

Points 1 and 2 are comparatively easy to realize, but the last mentioned point 3 is still open for many experiments and successful applications.

Dose distribution in composites is determined by heterogeneity of the system. Proper explanation of dose distribution in composites demands description of basic experimental approach allowing the optimal condition for applications. Leaving aside gamma irradiations as not frequently used in processing of composites, let us concentrate on electron accelerators. In gaining basic facts, accelerators with controlled energy distribution spectrum of the beam are preferred. Such linear accelerator "LAE 13/9", was used in Authors research, and its beam exits are shown on Fig. 9. Veracity of depth-dose curves is secured by using monoenergetic electrons. The straight beam of initially not satisfactory spectrum of energy is bent

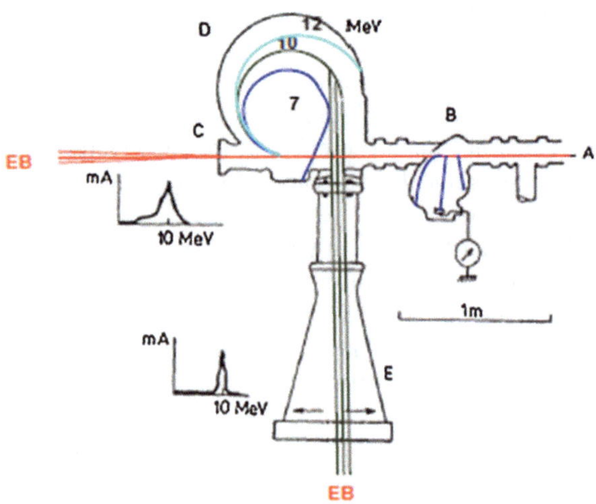


**Fig. 9** Exits of electrons in double function, the first linear electron accelerator in the Institute of Nuclear Chemistry and Technology in Warsaw, Poland. *A* the primary beam of electrons, *B* Scanning by 90° magnet for determination energy spectrum of electrons, *C* exit window for the straight beam of electrons, *D* scanning by 270° magnet for processing beam, *E* horn for scanned electrons

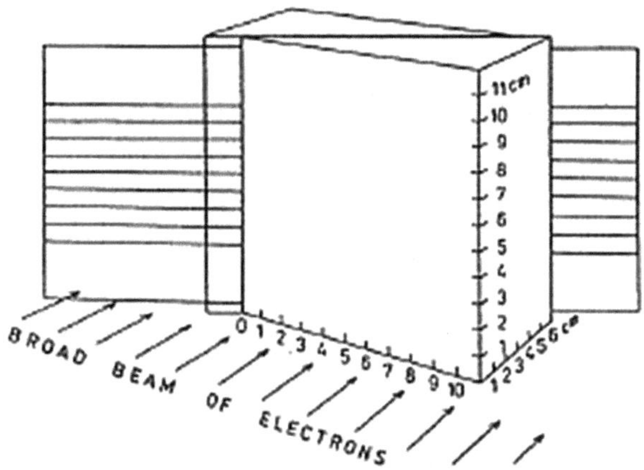


**Fig. 10** PVC wedge-blocks for determination of depth dose curves by PVC dosimetric foil, inserted between blocks

by $270^0$, separating electrons of lower and higher energy from desired. One-third of power is lost, but the bent and scanned beam can be considered monoenergetic.

The energy spectrum at upper left shows a not satisfactory distribution of energy in the straight, primary beam (upper curve) and improved spectrum (below), after separation of energies.

Nowadays there are separate, dedicated accelerators with straight beam for pulse radiolysis and high power accelerators for commercial radiation processing in the Institute.

The first aspect of irradiation of composites is the distribution of dose. The ideal case is the homogenous object irradiated with a broad beam of electrons. The use of proper dosimeter should yield the characteristic depth dose curve. Such curve is usually recorded using a wedge (Fig. 10) with inserted thin film dosimeters. To obtain a proper curve one has to use the wedge from the same material as the dosimetric material. The goal was achieved using the wedge made of polyvinylchloride (PVC). The dosimetric material was also PVC and therefore no corrections from the point of view of radiation physics were needed. All papers publishing depth dose curves from dosimetric measurements on wedges in which material of the wedge and of the dosimeter differ, need careful inspection. The PVC dosimetric film is inserted inbetween precisely machined pieces of the PVC wedge, as matrix polymer. The arrangement shown in the Figure refers to electrons of 8–15 MeV energy (the wedge has the slope of $\alpha = 30^0$). For electrons of energy <5 MeV, the thinner wedge is used, with slope of $\alpha = 10^0$ [10].

The use of PVC as the material of the wedge and the dosimeter don't rise other doubts. For instance there were no objections from the point of view of safety, because the irradiated material conducts electricity and fully deposited electrons cannot build up and cause a multimillion volt discharge like electrons accumulated in polymethylmetacrylate, or any dielectric, transparent thick block, used to show "frozen lightning".

All depth dose curves in the present chapter refer to the broad beam geometry, achieved by scanning of the beam of electrons over the object (Fig. 11). Single

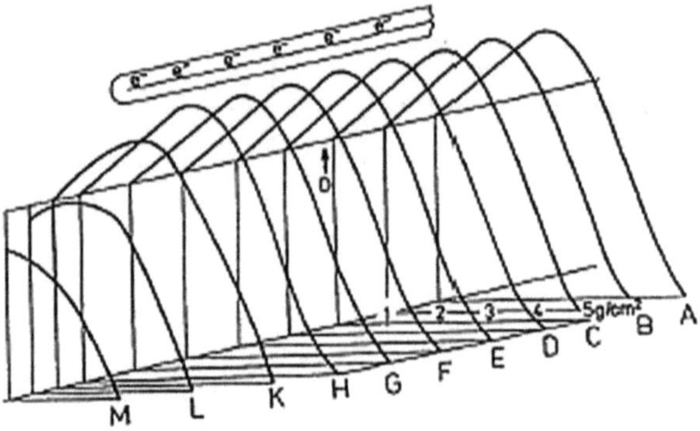


**Fig. 11** Depth dose curves from the broad beam of electrons. At the edge of the scanned range of electrons the shape of the depth dose *curve* changes into *narrow* beam geometry and the maximum of dose is at the entrance of the beam into the material (*curve* M)

beam, seldom used in radiation processing but exclusively in pulse radiolysis, results in depth dose curve without increase of the dose under the surface, resembling curve M (on Fig. 11) at the edge of scanned beam. The depth-dose curve from the straight, unscanned beam of electrons is ill defined and depends on the beam diameter, on the distance from the window to the object and cannot be used for precise calculations and imaging of the distribution of energy in the irradiated object.

Figure 12 shows the effect of tuning the accelerator to the better monoenergetics of nominal 10 MeV: the maximum dose in the depth is shifted back to the beam entrance site, due to the participation of lower energy electrons, but the range is extended due to the participation of higher energy electrons. The most useful part of the depth-dose curve, i.e. the distance in the object from entrance of the beam to the site where the entrance dose is equal the exit dose, is shortened.

The new generation of electron accelerators (Rhodotron) yields electron beam of excellent monoenergeticity, without additional operation, like one described above in the case of linacs. The monoenergenicity is inherent in the idea of Rhodotron, without that the mode of acceleration would not work.

Consideration of homogeneity is important if comparisons are in mind, e.g. if different systems are irradiated, as it is the case in the study of radiation induced dehydrogenation [23, 24].

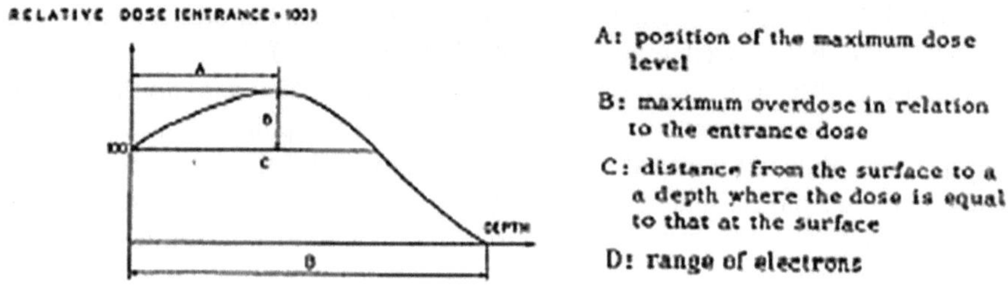


**Fig. 12** Depth dose *curve* for broad beam of electrons—the beam scanned over the irradiated object, e.g. an elastomer transported under the eit windows of an accelerator

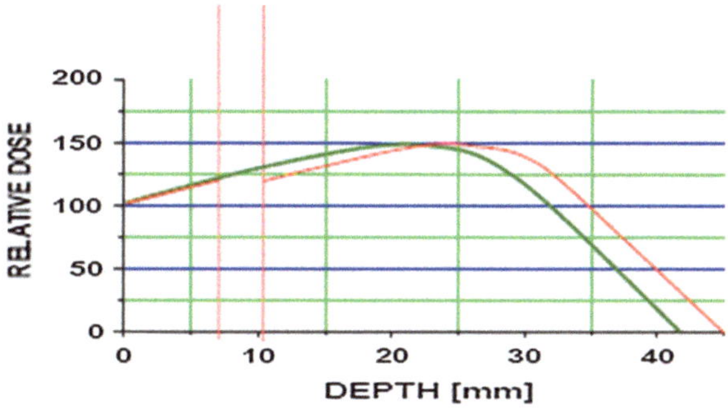


**Fig. 13** The 3 mm air gap in the block of elastomer does not make much change, except shifting virtually the range of electrons. *Green line* the same material without the gap, the *red*—with a gap

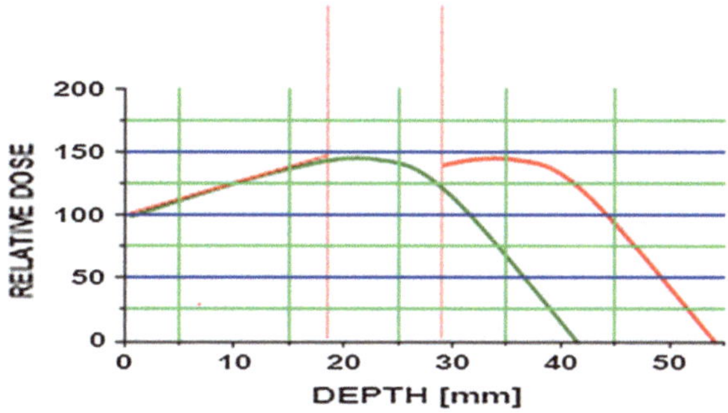


**Fig. 14** The 10 mm air gap in the layer of elastomer is shifting virtually the range of electrons more. Notice the different place of the gap in comparison to previous figure. The *red line* is the real depth dose *curve*

Any deviation from homogeneity, i.e. dealing with composites, introduces more or less disturbance into the depth dose curves. Keeping in mind the definition of the composite as the object with two or more bodies of different absorption characteristics towards interaction with radiation, one can start with very common system of reinforced elastomers. Foams, porous polymers, fibers, powders etc. in which the second component of the system is air have to be specially considered. Air is absorbing roughly by three orders of magnitude less per thickness than solid phase composed of low Z number chemical compounds. Therefore the depth-dose curve looks like on following Figures. Figure 13 shows an object (PVC block) with a 3 mm gap, e.g. two plates of PVC separated by 3 mm of air. Figure 14 shows the curve for the case of 10 mm ap inbetween the plates. The case of the same material, but divided into small particles gives the same basic curve for the same polymer, but of apparent lower density. Gaps between particles are too small to be reflected in details in the curve but the apparent bulk density is lowered. In that case the depth-dose curve is simply extended and the object is behaving like the original polymer "diluted" by air, which does not contribute much to the absorption of energy (but can have enormous effect on the final results of radiation processing from other reasons, for instance due to oxidation of the surface of particles).

More complicate situations as concerns depth dose curves occur, if the second phase in the composite of higher density material. That can be the case in the radiation processing of materials, containing metallic parts. The depth dose curve

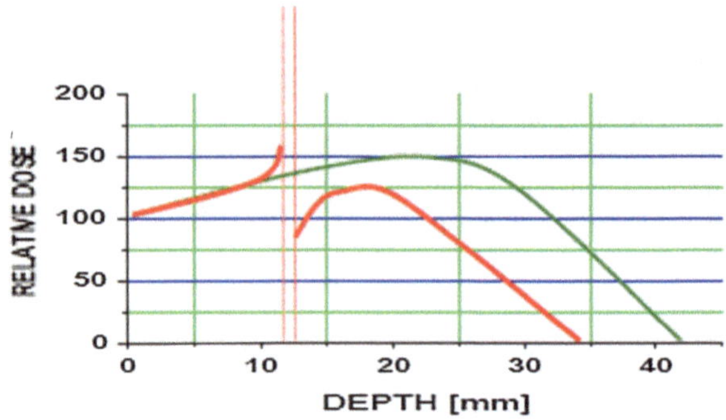


**Fig. 15** One millimeter thick layer of aluminum causes serious distortion of the depth dose curve in the PVC block (interrupted *red curve*). Notice important changes of doses in the vicinity of the elastomer

is squeezed in that case, because the metal absorbs per volume more energy than the polymer (Fig. 15). Whats more, the metal radiates degraded, low energy quanta back in the direction of incoming radiation and forwards into the polymer. Behind the piece of metal, the dose absorbed by the polymer is lower in comparison to basic depth-dose curve. The distance of optimum thickness, i.e. when the entrance and exit dose are equal, is shortened. The increase of dose by introduction of higher density material behind the polymer is sometimes used for improvement of the depth dose curve towards better economics of using the energy.

The presence of a high Z material as the composite introduces more complication, discussed later. It can be the case in the radiation processing of reinforced materials. The presence of even small metallic parts in the material need a careful preparation of the procedure of radiation processing. Sometimes extensive investigation of real dose distribution in irradiated object is needed to reach the decision, or demanding the change of the construction of the irradiated object.

The dose distribution is disturbed not only by high Z present in the composite or irradiated device. The same mechanisms are working if some metals are outside the object, i.e. if the arrangements for irradiation contain metallic parts, what is hardly to avoid. That is the case in a thermobox for electron beam irradiations at temperatures lower or higher than ambient [25]. There is an increase of dose close to the walls of the irradiation chamber, or walls of the metallic container for irradiation.

The consequences of the presence of two phases of different interactions with ionizing radiation are many, but the principal one is unfavorable increase of DUR (dose uniformity ratio). In the case of homogenous polymeric material in liquid state or solid in a block shape or as homogenous porous material, or medical device like a syringe, the DUR will not exceed the value of 2 in most favorable cases. In the case of composites, reaching of such excellent value is usually not possible and preparation of irradiation procedure has to be discussed, how high the value of DUR can be tolerated in particular application, or basic research.

Thermal effects in radiation processing of composites can be complicated. Second important aspect of radiation processing of composites is the different temperature of both phases during high dose rate irradiation. Electron beam creates such condition during adiabatic supply of energy, without possibility of rapid equilibration of temperature, otherwise possible in gamma irradiation. Specific heats capacities of both phases are always different. Differences are especially large if one phase is aqueous. One of the first observation by the present co-author [26] in this respect was connected with radiation processing of rubber latex in the purpose of crosslinking. Water shows especially high specific heat and the latex's rubbery constituents a low one. It is peculiar and hard to accept, that latex spheres, after the dose of 100 kGy, reach the temperature of 73 °C, whereas the aqueous host is only 49 °C hot. In view of expected chemical effects the organic phase is too hot, and the system demands rather split technique irradiation by the electron beam, i.e. two times by 50 kGy, separated by cooling period. The phenomenon does not appear in the case of gamma radiation where the dose rate is low enough to allow the heat exchange between the aqueous and organic phase.

All details connected with the heating of irradiated objects in radiation processing till 1992 are collected in monographic chapter by Zagórski [27]. However, they do not take into consideration composites and the present chapter is trying to fill this gap. It is astonishing that the thermal effect is not always taken into consideration in spite of the fact, that the consequences can be serious. Radiation processing consists of external energy delivery into an object. This operation would appear to be similar to heating, were it not for the fact, that ionizing radiation is able to create deep chemical changes in the exposed material at, or even below, room temperature. The equivalent amount of common heat seldom induces chemical change at such low temperature. Ionizing radiation creates reactive species. They can be formed by heat, but only at elevated temperatures of thousands of degrees Celsius. The creation of new species, both intermediate and stable, change the internal energy of the material. The participation of the resulting chemical energy is low and the main part of the injected energy degrades to heat.

That energy balance applies to most cases of radiation processing of homogenous and composite materials which involve radiation induced crosslinking, grafting, oxidation, controlled degradation. Radiation yields are described by the change of single molecules or effects (e.g. formation of one double bond) per 100 eV absorbed energy. The situation changes in dramatic way if a chemical chain reaction occurs. That is the case when a monomer is added as one of the composite and ionizing radiation acts as an initiator of reaction. The radiation yield jumps to thousand and more in the case of gamma radiation. In the case of electron beam radiation, the yield is not so high, according to $I^{0.5}$ law (where I is intensity of radiation, or dose rate). Even in the case of EB the rate of reaction is usually so high, that the thermal effect is much larger than the heat effect of absorbed radiation. The jump of temperature can be so high, that the system is close to melting or even boiling and direction of reactions runs in unexpected directions. Therefore in many systems the UV initiation is rather applied which controls polymerization without starting of a chain reaction, as in the case of dental fillings.

Under adiabatic irradiation, i.e. with high power electron beam, the target receives the entire dose in time too short to permit a significant exchange of heat with all constituents of the material and with the environment. Thus, all energy absorbed contributes to the increase of temperature of the irradiated system. Consequently, the 10 kGy (10 J/g) dose causes an increase of the temperature by 2.4 K, if the main constituent of the material is water, as in polymeric hydrogels for medical purposes and provided that the chemical reaction, resulting in the change of enthalpy due to, e.g. the formation of crosslinking bonds, decarboxylation etc., is negligible. Water has a high specific heat capacity ($c_p$) and therefore there is a small increase in the temperature upon irradiation. For the same dose, almost all other materials show higher increases of temperature because of much lower specific heats. One can expect dramatic effects in the case of metals which show, as we have seen, increased absorption of radiation energy per volume and at the same time very low specific heat, resulting in a jump of temperature. A reckless planning of the processing can lead to melting of polymer, at the site where it is touching a piece of metal.

The knowledge of specific heat capacity of materials is important in planning the procedure of radiation processing, especially if very different compounds are involved in particular composites. The data of specific heats can be found in the literature, but not always, because demand for that information is limited. Very often the need arises to measure the heat capacity, what can be done simply with the application of DSC technique to determine specific heat capacity of any material occurring in the composite. It is often a mixture of compounds, for which the data in the literature is not likely to be found. The same refers to heavy elements additives which change the specific heat dramatically, like bromine compounds and metalorganic compounds as flame retardants, which lower the specific heat.

Specific heat capacity depends also on the substrate temperature. Simple dividing the energy evolved by the specific heat to give the temperature rise is precise enough only if the dose is low and the specific heat is high. Otherwise, one has to analyze the changes in $c_p$ as a function of temperature [27]. The changes can be high: as water crystallizes into ice, the $c_p$ abruptly drops to 50 % and in liquid nitrogen to 16.2 % of its value for water at 0 °C. Therefore calculations of expected temperature rise in cryogenic irradiations must take into account the change of specific heat with the temperature. Figure 16 provides an estimated error if the change in $c_p$ during irradiation is not taken into account. The error increases at low temperatures and is already pronounced under liquid nitrogen irradiations.

When processing at liquid nitrogen temperatures, one has to remember that the heating from doses of the order of tens of grays is as high as that from doses of tens of kilograys at room temperature. Therefore, irradiation with high doses without significant temperature rise may be realized only by delivering single, low frequency, pulses, resulting in a low average dose rate. Such fractional irradiation is unavoidable when processing materials of low heat capacity.

The phenomenon of different temperatures reached in adiabatic conditions by the main constituent and the composite phase is well pronounced if the latter has micrometer dimensions. With diminishing size of the second phase the heat transport is more and more effective and eventually the thermal equilibration is comparable to the supply of ionizing energy. Calculations of the heat transport are

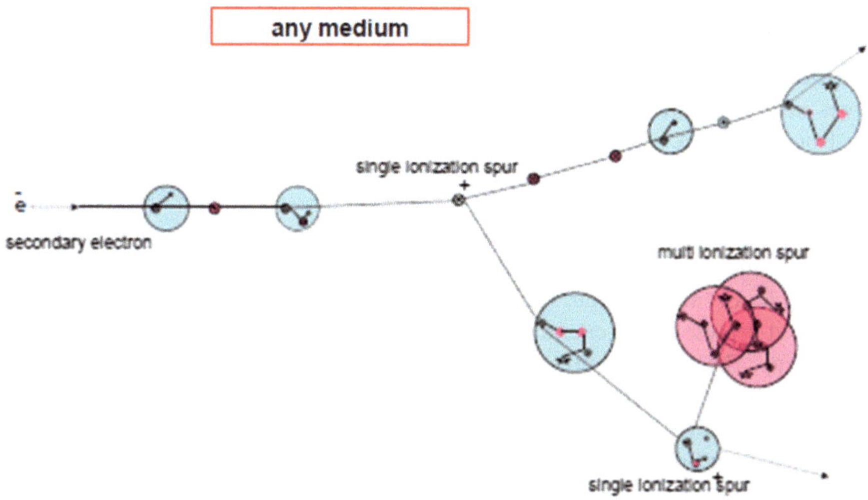


**Fig. 16** First acts of ionizing radiation interaction with the matter. Secondary electrons from any ionizing event are loosing energy in subsequent ionizations. Most ionization spurs are single (80 % of deposited energy, in the case of low LET) but remaining 20 % ionizations are so close one to another that the resulting chemical composition is different (multi-ionization spurs). In aqueous solutions these are $H_2$ and $H_2O_2$, in C,H polymers $H_2$ from the degraded chain

complicated, but simplified estimations show, that composites in which the second phase has indeed dimensions of single nanometers, can be treated as thermally homogenous. Strictly speaking the equilibration of the temperature proceeds as fast as the equilibration of temperature in multi-ionization spurs versus the body of the system.

Large group of irradiated elastomers deals with incorporation of inorganics or carbon into polymeric base. The literature is often referring to the second, added phase as "nanoparticulates" what not always is true and particles, fasers, and folies have their shortest dimension well in the micrometers range, but such classification is understandable because of current fashion in science nowadays. The purpose of looking for advanced plastics, as composites can be called, reaches from simple, cheap applications, like improvement of mechanical properties to more sophisticated cases [e.g. 28].

Sometimes no practical aim is indicated and the purpose of investigation seems to be basic research. [29–33]. Papers referring to practical work on composites have also similar character, contributing to the basic research on composites. Thus the engineering perspective works in the same direction as fundamental ones. Some points from the chemistry of composites, important to radiation processing, already realized, or to be realized may contribute to developments in the field of elastomers. Usual matrix of a composite has seldom elastomeric properties: The most popular are: polypropylene, polyimides, polyurethanes, polyamides, epoxy resins, poly(butylene terephtalate), poly(ethyleneterephtalate).

What is the incorporated material, into or onto the matrix? To name most frequently described inorganics are exfoliated silicate sheet minerals (e.g. montmorrilonite), calcium carbonate, glass fibres (fibreglass), carbon nanotubes, carbon fibres in epoxy resin matrix; other polymers, also in specific geometric shape like cord or short fiber reinforced elastomers, jute fibers, aramid fibres in polyphenylenesulphide or vinylester. Both phases can be chemically very close, e.g. high-density polyethylene fiber/polyethylene matrix composites. Most interesting from the point of view of general radiation chemistry (not the chemistry of elastomers!) is natural biocomposite, i.e. wood, consisting of separated phases of cellulose (carbohydrate) and lignine (aromatic polymer). That example is mentioned because it represents a general rule of protective properties of aliphatic/aromatic composite. The first compound is sensitive to ionizing radiation and degrades easily, the second, as the aromatic compound, is more radiation resistant. As it is the case in radiation chemistry, in the composite, natural wood, there is an intimate contact of both polymers, and therefore the original wood, contrary to pure cellulose is of better resistance towards ionizing radiation. Lignine exhibits a protection effect towards cellulose, by the mechanism of energy transfer. Practical observation to that effect has been made decades ago in the industry of medical supplies. Leaflets included in the box, printed on expensive, pure cellulose paper disintegrated shortly after radiation sterilization, but leaflets printed on ordinary, lignine containing paper, survived the irradiation and the following exposition to air–oxygen.

Similar, interesting composite, prepared from natural resources, but in artificial way is described by Barone [34]. Short-fiber reinforced composites are made from keratin fibers (1 mm long) obtained from poultry feathers and from polyethylene of varying crystallinity. Low crystallinity polyethylenes are reinforced by keratin fibers but high crystallinity polyethylenes are not. The keratin fibers inhibit crystallinity in low crystallinity polyethylenes but enhance crystallinity in high crystallinity polyethylenes. Microscopy shows increased adhesion between the fibers and the polymer for the more amorphous polyethylenes.

It is almost 20 years ago when Kojima et al. [35] added small amounts of montmorillonite to Nylon (polyamid 6) and observed increased modulus, yield strength, heat distortion temperature, as well as improved barrier properties. Their observation started applications and chain of publications on similar systema. Polymer–clay composites attract more and more, especially recently, considerable scientific and industrial interests, because they exhibit significant improvements in physical and mechanical properties over virgin polymers with minimal increase in density [36]. The interest has focused nowadays on rheological properties polymer–clay nanocomposites. From the applied perspective, determining the rheological properties of nanocomposites is vital to optimize processing during the manufacture of engineering components. From the scientific research point of view, nanocomposites provide a nanoscale space to study confined polymers and examine the effect of nanoclays on the rheology of nanocomposites. Since nanoclays significantly affect the rheological properties of nanocomposites, the network formation would be related to the microstructure of nanocomposites and interaction between nanoclays and polymer matrix. Interaction and the mechanism for the formation of the microstructure are not readily understood yet.

Introduction of spectroscopic methods is improving the knowledge of the chemistry of montmorillonite/polymer. Nascimento et al. [37] applied the resonance Raman spectroscopic characterization of doped poly(benzidine) and poly(benzidine)/clay nanocomposite. The organic–inorganic hybrid material was synthesized by oxidative polymerization of benzidinium ions intercalated in the montmorillonites (MMT) nanospace. The formation of poly(benzidine) between the layers was confirmed by X-ray diffraction and scanning electron microscopy. The new material shows electrical conductivity similar to the conductivity of free doped benzidine. The mechanism of he phenomina consist in the intercalation and polymerization process of benzidine (PBZ) in the montmorillonite. The intercalated polymer chains lie flat between the layers. The polymeric chains of the PBZ-MMT nanocomposites present more dication segments than radical cation units.

Loo and Gleason [38] returned to the classical system of nylon 6/montmorillonite, with the method to determine the orientation distribution function of montmorillonite clay in nylon 6 nanocomposite films by a combination of infrared trichroic analysis and transmission electron microscopy image analysis. For the first time the orientation distribution function of exfoliated montmorillonite clay in nylon 6 matrix has been determined. The clay orientation can be described simply by one parameter $\sigma$ using a Gaussian function. The purpose of the project is not clear, especially in view of the affiliation of Authors (Institute for Soldier Nanotechnologies, Massachusetts Institute of Technology, Cambridge, MA, USA).

Effect of clay orientation on the tensile modulus of syndiotactic polypropylene-nanoclay composites, as produced by melt extrusion, was investigated by Galgali et al. [39]. The orientation of the clay tactoids in extruded tape samples was quantified using 2D X-ray diffraction data. In the case of compatibilized hybrids the tensile modulus increased with the extrusion shear rate until a saturation value, whereas for the uncompatibilized hybrids the tensile modulus was nearly independent of the shear rate.

Stretz et al. [40] investigated poly(styrene-*co*-acrylonitrile, SAN)/montmorillonite (MMT) organoclay mixtures as a model system for acrylonitrile-butadiene-styrene (ABS) nanocomposites. The SAN/MMT nanocomposites effectively model the organoclay dispersion seen in more complex ABS/MMT nanocomposites.

There is not always a need to use clay of high montmorillonite content. In many cases the bentonite is sufficient. That composite is reported by Filho et al. [41] in the chapter on thermal stability of nanocomposites based on polypropylene and bentonite. A polycationic bentonite clay was modified with a quaternary organic salt and added to isotactic polypropylene. Compression moulded films were exposed to a thermal environment at 110 °C to evaluate the thermal stability of polypropylene matrix after chemical modification of bentonite. The carbonyl index results, obtained by IR spectroscopy showed that polypropylene with modified clay had higher thermal stability than with the natural clay.

There are many approaches to preparation of clays before combining with the polymer. For instance Zheng [42] and Zhang [43] propose oligomerically modified clays for successful creation of a composite with styrenic, polyethylene and polypropylene matrix with sodium montmorillonite.

The preparation of a nanocomposite is critical, if full advantages of the material have to be achieved. All clays are rather hydrophilic and combination with most hydrophobic elastomers used as matrix is difficult, considering enormous surface area of contact of both phases, mentioned above. One can achieve that using organo-clay modified by organophilic surfactant.

There are complications in real imaging of composites in publications. Imaging is usually limited to pictures of dispersed phase [44]. Some authors managed to record pictures closer to reality. Li et al. [45] have synthesized exfoliated polystyrene/montmorillonite nanocomposite by emulsion polymerization using a zwitterion (aminoundecanoic acid) as the clay modifier. In contrary to intercalated composites, the exfoliation of clay layers in this approach is confirmed by using TEM. The micrograph shows dark thin lines corresponding to the silicate layers on nanometers thickness, but several hundreds nanometers long, dispersed in the polystyrene matrix evenly. The exfoliated nanocomposite shows a greatly improved modulus, higher glass transition temperature and better thermal stability compared to the neat polystyrene and the intercalated polystyrene/montmorillonite composites.

Pehlivan et al. [46] investigated pure and silver exchanged natural zeolite filled polypropylene composite films. The thermal characterization studies showed that the addition of the zeolite increased the crystallinity of the structure, acting as a nucleating agent in the polypropylene crystallization as well as retarding the degradation temperature of polypropylene. Loading with silver had the intention to introduce antifungal, germicidal, bactericidal and antiseptic properties, but results in that respect were not commented.

Already at the beginning of application of montmorillonite as the key constituent of clay/polymer composites, the observation was made, that these materials show poor resistance at outdoor applications in comparison to pure polymer matrix. Morlat-Therias et al. [47] looked for origin of that phenomenon investigating photooxidation of ethylene-propylene-diene(EPDM)/montmorillonite, grafted with maleic anhydride as compatibilising agent. Massive UV illumination caused faster degradation of the 80-100 μm film than the pristine elastomer. The UV absorbtion spectrum of the composite was not interpreted, therefore it is difficult to speak out how the photochemical reaction starts. According to authors, the adsorption of the antioxidant and additives is most likely to be the key factor behind the photodegradation of the exfoliated nano-composites. The tremendous increase of accessible clay surface upon exfoliation enhances the probability of additive adsorption on the solid surface, and, hence would help reduce dramatically its efficiency as antioxidant agent. Consideration of photo destruction of stabilizer (rapid disappearance of maximum in the UV at ca 280 nm in the function of illumination time) was not discussed.

Attractive, but expensive composites are making use of carbon nanotubes.

Computer assisted simulations have been applied also in the discussed field. Chan et al. [48] have performed molecular simulations to study self-assembly of tetratethered nanoparicles with a cubic geometry. Minimal model on nanoscale building blocks (NBB) represented a polyhedral oligomeric silsesquioxane (POSS) molecule with polymeric functionalities. Thus the rich nanostructures formed from self-assembly can make analogies with the morphologies observed in block copolymers and surfactans. Unfortunately, no comparison with experimental findings, if any, have been made. That approach is mentioned for the sake of complecity only, but the meaning for theory and praxis is minor at the time being.

Our discussion is not complete yet, if there is no clear situation whether the system represents miscible blend or immiscible blends [49]; only the second case falls into category of composites, i.e. separate radiations chemistry and energy and material transfers characteristic from the point of view of radiation chemistry. For the same reason we are leaving without comment our papers on radiation chemistry of blends of elastomers with polyolefins [50, 51].

There are no papers yet published on the possibility of curing elastomers in the presence of added monomers. Just for case, one has to mention that in other systems the processing is accompied with the generation of heat. For instance, in the case of composites of wood impregnated with monomer [52] the heat of polymerization is enormous. In that case the temperature of $\gamma$-irradiated, monomer impregnated wood rises well over 100 °C. In another system, Singh and Saunders [53] deal with carbon fiber-acrylated epoxy composites, cured with 10 MeV electron beam from the accelerator. At the high dose rate secured by the accelerator, the radiation yield of the chain reaction is lower than in the case of $\gamma$ radiation. At the same time there is no influence of the gas over the material, because at such high dose rates, oxygen cannot diffuse into the polymer fast enough to inhibit polymerization significantly. That group continued EB curing in next years [54, 55]. Spadaro [56] has introduced new elements into radiation curable composites.

Although at present main experimental effort is directed towards the development of composites as such, and investigation of their specific properties, mechanical, physicochemical and physical, the radiation processing will enter the field on the wider scale, especially as concerns specialized, high value, but of limited amount, materials. It will happen under the acceptance of high cost of ionizing radiation. Some publications, like [57] underestimate the expenses, taking erroneously the energetic expense as the power of the electron beam, but neglecting the cost of producing the beam. In that way the cost of radiation induced crosslinking looks energetically cheaper than chemical initiations. In real energetic expenses, the cost is by two or three orders of magnitude higher in the case of radiation processing. Therefore one can expect limitation of radiation processing of composites to cases of high-tech materials and not oriented to mass products.

In preparation of procedures, two fundamental points are advised to be taken into account: changes in dose distribution in irradiated composites in comparison to homogenous material and thermal effects which can be more far reaching than it is the case with homogenous material. Nevertheless, introduction of radiation processing to composites in which that has not been tried yet, can simplify the production of some materials or improve their quality.

From the basic point of view, the absorbtion of ionizing energy runs in every phase of the composite like in homogenous material, with a variety according to the electron density and correction for the Z value for both phases, discussed above. Unexpected phenomena can occur on the interface between phases and their participation can be quite high in view of enormous surface area in the case of really nano-materials. These phenomena involving energy and material transfer are not fully understood yet. How much can happen on interfaces in composites shows as an example the reference by Patel et al. [57], on gamma radiation induced effects on silica and on silica-polymer interfacial interactions in filled polysiloxane rubber. Investigation was made by ESR and NMR methods. Results suggest presence of trapped paramagnetic species in the silica matrix. These studies and many which follow aim at generating an improved understanding of the polymer-second phase interface. It is hoped that the understanding gained will ultimately lead to the development of composites, where the interface may be tuned to requirements so as to minimize the age related change on properties that are critical for the function of the material.

Whatever can happen on boundary of two different component it is always connected with energy transfer [58].

More and more stressed is the new feature of a composite, i.e. that it belongs to nanotechnology. If the size of dispersed phase is really of nanometers size, demands the look from the perspective of radiation processing. Reducing the size of dispersed phase from millimeter, to micrometer, to nanometer dimensions, the rate of equalization of the temperature in the sample improves, making the system in that respect more similar to homogenous one. However, at the same time the surface area of the interphase increases enormously, as it was shown in the beginning of the section. The importance of interphase reactions, initiated by ionizing radiation increases enormously, revealing new reactions. In that respect the classification to nanotechnology is justified.

Proper explanation of dose distribution in composites demands description of basic experimental approach allowing the optimal condition for applications In gaining basic facts, accelerators with controlled energy distribution spectrum of the beam are preferred. Such accelerator is used in our research. Veracity of depth-dose curves is secured by using monoenergetic electrons. The straight beam of not satisfactory spectrum of energy is bent by $270^0$, separating electrons of lower and higher energy from desired. One-third of power is lost, but the bent and scanned beam can be considered monoenergetic.

Composites with air, in other words porous polymers, are very important in preparation and application of scaffolds for growing cells.

The distance of optimum thickness, i.e. when the entrance and exit dose are equal, is shortened. The consequences of the presence of two phases of different interactions with ionizing radiation are many, but the principal one is unfavorable increase of DUR (dose uniformity ratio). In the case of homogenous polymeric material in liquid state or solid in a block shape or as homogenous porous material, or medical device, the DUR will not exceed the value of 2 in most favorable cases. In the case of composites, reaching of such excellent value is usually not possible and preparation of irradiation procedure has to be discussed, how high the value of DUR can be tolerated in particular application, or basic research.

# 9 EB: Crosslinking of Elastomers, How Does it Compare with Radiation Crosslinking of Other Polymers?

Electron beam crosslinking of polyethylene is a well-established technology, applied commercially for decades. World wide efforts have been directed towards the crosslinking of elastomers already in the second half of the 20th century. In principle, radiation chemistry of crosslinking of any polymer is governed by similar rules. Most important are steric effects that can prevent efficient crosslinking, second next are additives present in irradiated material. The latters cannot be avoided if commercial polymers are radiation processed. Their fate is shown in the function of increasing doses, resulting in the growth of gel content. Changes of radiation yield of hydrogen during crosslinking process as well as mechanical properties of the product are also discussed. Basics of mechanisms in different polymers are interpreted as phenomena of single ionization spurs (80 % of energy deposited) and multi-ionization spurs. Small spurs generate crosslinks of the X type, formed between neighboring macromolecules, whereas multi-ionization spurs, energy rich, cause chain scission. Some fragments of the chains form crosslinks, this time of the Y type, by reacting with their active end with undamaged chains present in the neighborhood. As the representative elastomers, acrylonitrile-butadiene rubbers were chosen, also in the hydrogenated version, described in separate section. These are high technology materials, rather expensive, which are therefore excellent objects of successful commercial radiation processing.

A comparison between the crosslinking of polyethylene and of elastomers starts with an exercise in the history of radiation chemistry. The exact origins of first observations, that polyethylene is crosslinking due to the absorbed energy of ionizing radiation, are not clear. From remarks dispersed in the literature from the 50s, discussed below one can conclude, that the main concern connected with unintentional but unavoidable irradiation of cables close to the active parts of nuclear reactor, was the deterioration, fire and possible explosion of hydrogen, originating in the polyethylene isolation. The reason was, that the content of hydrogen in polyethylene ($CH_3$–$[CH_2–]_n$–$CH_3$) was not very much lower than in the basic hydrocarbon fuel for internal combustion engines ($CH_3$–$[CH_2–]_m$–$CH_3$) where $n \gg m$. The low knowledge of radiation chemistry in middle 40s could not eliminate even an assumption, that a chemical chain reaction can develop, causing a massive production of hydrogen, combined with worsening of electric isolation properties, with the production of elementary carbon.

To the astonishment of reactor chemists, nothing like that, in a black scenario happened; on the contrary, polyethylene was leaving the radiation field, whatever intensive, with improved mechanical and thermal properties. The watershed of 40s/50s brought the explanation what is going on in the polyethylene placed in the ionizing radiation field, i.e. the formation of crosslinks between neighboring chains of the polymer. Arthur Charlesby [59], the physicist, in the fundamental, impressively illustrated paper in the Nucleonics gave the best description of the reaction. Radiation chemist, Malcolm Dole went deeper into the chemistry of crosslinking reaction, assuming that free radicals are transferred along the chain of polyethylene until the place of preferred reaction with another chain be reached. This author has claimed, also in his memoir Refs. [60–62], that his ideas of radiation induced crosslinking of polyethylene were earlier than Charlesby's, but were published in less exposed media, like summaries at small conferences. We can agree that his contributions to the chemistry of crosslinking were equal to Charlesby's but independent and parallel.

Estimating highly early, well documented effort on development of radiation chemistry of polyethylene, done 50 years ago, one has to mention the very important paper published almost parallely to Charlesby's papers in the 50s, but in *the Nature*. Three researchers [63] from the General Electric Laboratories at Shenectady, N.Y. have made fundamental work on radiation chemistry of polymers irradiated by electrons, accelerated to the energies of 800 keV (max). Their efforts can be called fundamental, as they have helped to separate the radiation chemistry of polymers from isotopic sources of high power ionizing radiation, either inconvenient channels of a nuclear reactor or more convenient, but of low dose rate sources of gamma radiation emitted by cobalt-60. The next 50 years have shown, that electric sources of radiation dominate the commercial crosslinking of polymers, whereas large cobalt-60 sources are practical in radiation sterilization of medical supplies, made from polymers, very often from elastomers.

The three researchers form the General Electric Laboratories did not limit their efforts to polyethylene, but extended to 21 most popular polymers, synthetic and natural. They have divided them into undergoing radiation induced crosslinking (14) and those which do not crosslink (7), becoming degraded. Table 1 is reproduced from their paper showing how proper was their investigation and interpretation. This list was reproduced many times throughout textbooks and even now is true, demanding only few supplements, like with polypropylene, which did not exist at that time yet, and belongs to the column with not-crosslinking polymers.

Basic progress in understanding of crosslinking of polyethylene was done in 1959 by Charlesby [see 14] again, who formulated together with Pinner quantitative relations, which are in wide use until now, as Charlesby-Pinner equation. Presentation of results of radiation crosslinking of polymers in this equation coordinates is a rule now (c.f. section on the HNBR).

Polyethylene dominates the column of crosslinkable polymers, and already in the 50s became the object of successful commercialization, especially when irradiated by electron beams. However, there are other polymers on the list, especially elastomers which should become crosslinked on the industrial scale.

Why polyethylene only was the first candidate for commercialization, from the list that was determined at Shenectady in the 1953? The answer is, that there was no other simple and cheap method of crosslinking the polyethylene in contrary to crosslinking of elastomers, called vulcanization of rubber after addition of specific chemicals and application of temperature!

The best example of the latter is natural rubber, which in the 50s had already an almost 100 years long history of successful crosslinking by mixing with sulfur and heat treatment. The crosslinking by chemical additives has been and is continuously refined and possible introduction of radiation induced crosslinking did not find application on the scale comparable with polyethylene. The biggest market for elastomer crosslinking, i.e. the tire industry sponsored the pilot plant [64] for partial crosslinking of green rubber, but the current real engagement of onizing radiation now is not clear. Probably there are problems connected with the application of most expensive form of energy, which is the high power ionization radiation, to extremely high volume production.

However, that obstacle should not be the case with high-tech polymers, produced in modest size batches, like hydrogenated acrylonitrile-butadiene rubber (HNBR), showing excellent properties after radiation crosslinking processing. The thesis of the present section is, that radiation crosslinked HNBR is a high quality product, because the mechanisms of radiation crosslinking are similar to these in radiation crosslinking of polyethylene. The purpose of the present presentation from the point of view of basic radiation chemistry of polymers is to show, that radiation crosslinking runs according to the same rules whether it is polyethylene or more complicated elastomers. The evidently identical role of sizes of spurs is stressed.

Chemistry of radiation crosslinking, 50 years later is more complicated and rather undeveloped in comparison to radiation chemistry of water and aqueous solutions. Even now it is rather backward in comparison to the latter, where one can calculate the result of irradiation of many systems, on hand of proper computer simulations, e.g. by Chemsimul developed in Denmark. Starting radiation yields of radiolysis products of water are independent from the kind of the solute in it and are entered into the system of competing reactions.

Already 25 years ago, on the Second International Meeting on Radiation Processing, [65] in his Conference introductory paper has stated, that “Most radiation chemists are under the impression that the fundamental aspects of radiation chemistry are well understood and are available for ready application by industry. This is not the case”. After the more than 30 years since Silvermans statement, we are certainly better informed, but still far from the mentioned, satisfactory status of knowledge of water radiolysis. For instance, in the case of radiation chemistry of polyethylene, Silverman has stated “that the G-value for crosslinking in the crystalline region is known to be zero”. The same Author has stressed this view again later [66]. Today this statement is again doubtful and has to be revised in view of the actual picture of deposition of energy in the irradiated polymer.

It is astonishing to realize; that an enormous scale of industrial implementation of radiation crosslinking of polyethylene was not accompanied by basic research on chemical mechanisms of radiation induced effects. It is sad to find out, that a substantial part of technologies of radiation processing of polyethylene is founded on trial and error findings. It is a pity, because some present difficulties in successful applications of procedure could be better negotiated, on hand of basic knowledge of phenomena. Every change of starting material for crosslinking, any change of additives, change of the type of accelerator and of the energy spectrum of electrons etc., demands new trial and error procedures.

The starting point for the discussion of mechanisms of reactions in the radiation chemistry of polyethylene is radiation chemistry of water and aqueous solutions. Basics developed in the very beginnings of radiation chemistry are valid to any irradiated matter. These are:

- Ionizing radiation interacts statistically with outer sphere electrons, no matter what compound in the system they belong to (not like in photochemistry where the quanta interact specifically with chromophoric groups).
- Most of energy of absorbed low LET radiations (EB of usual energies of 0.1–10 MeV) appears as single ionization spurs; the remaining 20 % of energy is located in large, multi-ionization spurs due to the low range of final degradation energy electrons. The origin of multi-ionization spurs is explained by Fig. 16, modified from [67] in the reference by [3, 68]. In water, these large spurs are the sites of so called molecular products formation: $H_2$ and $H_2O_2$. Radiation yields of these products are fixed and independent from the solute, like yields of radical products ($e^-_{aq}$, H, OH) from single ionization spurs. Multi-ionization products in water are of low reactivity and enter the reactions with specific solutes only, e.g. with iron(II) ions. Most important reactions in aqueous solutions are between radical products and solutes, which can be quantitatively foreseen from radiation yields and their absolute rate constants. Competition for active reactants is common. Aqueous solutions of polymers behave like other solutes [69]; they can crosslink in the mechanism started by water derived free radicals.
- Products of multi-ionization spurs in water are easily distinguished from products of single ionization spurs, which are very reactive in comparison to hydrogen peroxide and molecular hydrogen. Unfortunately, such distinction is not the case with nonaqueous media.

It is a pity, that "nonaqueous" radiation chemists are missing these points. Mostly neglected are multi-ionization spurs. The existence of multi-ionization spurs is obvious, as the physics of interaction of ionizing radiation with organic matter is very similar to that of water. The rule of the partition of deposited energy in ca 80 % to single ionization spurs and 20 % of multi-ionization is, as usually in condensed low Z matter, preserved. It has been confirmed in the case of solid alanine [3, 68], that the G value of decarboxylation, which is due to multi-ionization spurs, is 0.95 and does not depend on the dose. The case of alanine is easy to verify, because the $CO_2$ is formed only as the product of multi-ionization spurs.

Unfortunately, the fact of different sizes of spurs, so easily accepted in aqueous radiation chemistry, and well understandable in the case of solid alanine, is ignored in the case of neat polymers, i.e. not dissolved in a solvent, which has its own radiation chemistry like water has. It is impossible to meet a paper dealing with radiation chemistry of solid polymer, which takes into account the multi-ionization spurs, in spite of the fact that they can explain results of product analysis. These are usually very different and many products of low molecular weight are found. The deposition of high dose of energy (20–500 eV is usuall in multi-ionization spurs) in a limited volume causes the scission of the chain of any polymer and formation of some debris in addition. There is no other explanation of formation of debris and it is astonishing that authors who have found low molecular weight debris in irradiated polymers, did not look for explanation. Single ionizations which make changes on the chain without breaking it, can cause the scission only after additional secondary reactions, e.g. with the participation of oxygen.

The chain scission caused by multi-ionization spurs is final in the case of no-crosslinking polymers, like polypropylene, as an example. In others, like in polyethylene and especially in the elastomer HNBR [5, 6], a rupture of the main chain results in the formation of terminal, reactive radicals. These free ends of the chain react easily with the neighboring molecule of an intact, original polymer, creating a crosslink of the Y type. The cited paper has shown, that the yield of chain scissions is lower than expected from the 20 % rule. For every 100 acts of crosslinking there are 6–11 acts of degradation. That is caused by partial recovery of scissions as type Y crosslinks [70].

Single ionization spurs, covering ca 80 % of deposited energy are generally accepted as the main origin of radiation induced reactions. In contrast to multi-ionization spurs, which are localized immediately at the point of origin, single ionization spurs and their secondary consequences, are able to transfer along the chain. The proof of that is clear: without energy transfer the fact of protection by additives present in minute concentration in the polymer, could not be explained. The question is what is transferred. The early, 1953, assumption by Dole that travelling species are free radicals, still doubtful 10 years later [71] is still difficult to defend now. It would demand separation of hydrogen in one separate act after the detachment of electron and immediate neutralization of the positive hole. More likely is the travelling of the positive hole, while electrons temporarily waiting in some distant places. Anyway, the energetic disturbance on the chain moves along the chain, until it reaches a favorable place to be localized. In the case of crosslinkable polymer, it is the place where chains meet and in the effect the chemical bond can be formed with the release of $H_2$. The crosslink in the shape of the letter X is formed.

However, in the case of polyethylene the problem is, that it is semicrystalline, i.e. roughly half it occurs in crystallites, which, by the way, are responsible for opacity of that polymer. Although the content of the crystalline phase can be measured, e.g. from the heat of melting, as recorded by DSC and expressed in percentage, the size and shape of crystallites is not so easy to determine. In view of [65] statement that crystalline moieties in PE do not crosslink, one can have doubts what is the picture of phenomena. The crystalline site must participate in the absorption of energy in exactly same way like the amorphous one, in proportion resulting from its participation in the pool of outer sphere electrons, with all consequences. Crystalline moieties must either crosslink, or the deposited energy transfers outside the crystal, where it meets amorphous chains. The last mentioned option is not very likely; the crystalline phase contains many defects and there are many possibilities, that crosslinking occurs, before the energetical disturbance will reach the boundary between crystalline and amorphous phase. In other words, an assumption that there is no crosslinking in the crystalline moiety would suggest, that the ionizations are transferred outside the crystalline moiety, to amorphous part of polyethylene—rather low probability phenomenon.

Situation in the case of elastomers is much simpler, due to the absence of crystalline moiety. One has to see this polymer as one giant system of amorphous, homogenous conglomerate of macromolecules, with many points of mutual contacts of different energy sites. They can be considered as imperfections, attractive to higher energy defects formed on the chain as single ionization spurs. As it is the case, reactive species are moving along the chains to specific sinks of reaction like in the case of polyethylene, i.e. to places of possible crosslink, but also to reactive additives. Elastomers can be stressed before irradiation, creating partial orientation of macromolecules, a pseudo-crystallization, but that possibility of influencing the crosslinking has not been fully explored yet. Multi-ionization spurs in the case of elastomers work as in the case of polyethylene.

The HNBR, taken as an example of radiation crosslinking of elastomers has the formula:

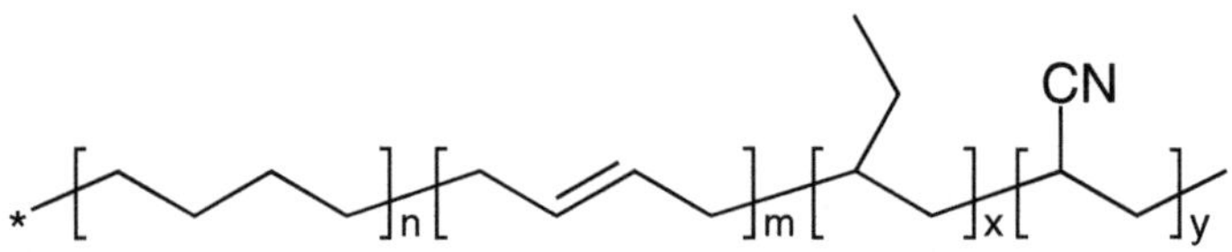


Results of radiation crosslinking show that side groups do not interfere with crosslinking mechanism. It is the main chain, which takes part in the X-crosslinking due to single spurs (movement of active site along the chain), as well in the Y-crosslinking due to degradation caused by multi-ionization spurs, exactly like in the case of polyethylene. That statement is sufficient in discussion of analogies between radiation crosslinking of polyethylene and rubbers. There are small effects of residual double bonds in the HNBR and their higher concentration in the NBR and of participation of nitrile moiety, to be published in a separate paper soon. They are not described here, because they do not influence the similarity of the role of the main chain of the polymer, even if it contains double bonds.

Radiation processing of polymers, both for basic purposes and for commercial purposes, meets several difficulties in comparison to other objects. First is the usual insufficient definition of the composition of the material. If commercial polymers are irradiated, the exact amount and kind of additives is unknown, but it is obvious that all commercial polymers contain several nonpolymeric compounds, allowing any starting processing, especially thermal. Unfortunately, some researchers preparing home compositions do not care to publish precise information. For example, the important reference by [72] informs, that one part of phenolic antioxidant per 110 or 140 parts of composition is applied. Exactly the same addition is of a lubricant, not defined even roughly as the antioxidant mentioned above. Low concentrations of additives should not suggest their low importance, because they do not take significant part in the primary process. However, they can play a key role in secondary reactions after the ionization, due to the energy transfer. We are using, always if possible, the virgin polymer obtained at the producer, before any additives are introduced. The concentration of nonpolymeric material is thus limited to traces of catalyst and solvents.

Represented ideas were checked experimentally in irradiations, executed with our linear electron accelerator type LAE 13/9, with 270 deg bent beam, which secures satisfactory, low stray of energy of electrons around the the nominal average value [10]. Elimination of electrons of much lower energy from the assumed average value prevents unexpected deposition of extra charge in radiation processed polymer. The thickness of irradiated layer is adjusted usually to the rule that the entrance and the exit doses in the material are equal. If the electron beam is reasonably monoenergetical, the main amount of charge is not deposited in the material. If however, the beam contains low energy electrons they can cause discharges, very dangerous in crosslinking of any polymers. The discharge can cause fire and at least is burning holes in the material.

Radiation crosslinking brings problems of heating, because doses exceed by two orders of magnitude, e.g. doses applied to sterilization of polymeric medical devices. To avoid thermal effects in irradiated materials, the split dose irradiation was always applied, keeping the adiabatic jump of temperature never higher than by 30 °C. The maximum dose allowed in one pass of the material under the window of electrons depends on the specific heat of the irradiated material. The rapid acquisition of the dose allows avoiding extensive oxygen effects, which has proved to be negligible in the case of investigated elastomers. Therefore a common source of undesired phenomena of oxidations of polymers irradiated in gamma sources, without proper isolation from the air access, is avoided.

A good analogy between the radiation crosslinking of polyethylene and the HNBR is comparison of shapes of curves of radiation yields of hydrogen in both cases. Hydrogen determination in irradiated polymers has been made in a system, which covers several orders of magnitude of dose, necessary to find out the progress of crosslinking and the role of additives. The basic irradiation vessel is typical 3 ml volume ampoule. Vials with material to be irradiated are closed with septa and the material irradiated in the narrow beam of electrons. Only the lower part of the vial, containing the polymer is irradiated and the strayed, weak ionizing radiation does not influence the rubber septa, what has been proved by irradiation of material which did not contain hydrogen. The vials were positioned in the electron beam with the help of laser beam.

In some experiments, irradiations of vials with investigated material were made with scanned beam and samples placed horizontally on the conveyer, under the electron window. In that case the top of the vial with the septa was contained in a thick "hat" made of lead. Again no traces of hydrogen have been found in irradiated empty vials.

All modes of irradiation were controlled with different dosimetric systems. Irradiated vessel was irradiated with split doses causing the warming by 30 K from the starting temperature. That period of time was sufficient to attain stable concentration of hydrogen in the gas phase in the vial. After cooling down to the temperature of air conditioned laboratory with the gas chromatograph, the measured amount of gas was taken by needle and syringe into the column of the gas chromatograph, Shimadzu type 14 B.

Many polymers have been analyzed for the radiation yield of hydrogen; comparisons of hydrogen production from irradiated polyethylene (PE) with hydrogenated acrylonitrile-butadiene (HNBR) elastomer indicates mechanisms of radiolysis. Diagrams show the initial zone of doses, where the production of hydrogen is lower. Curves of crosslinking or crosslinking connected parameters also show a diminished effect of radiation at the beginning, i.e. at starting doses.

The low additives polyethylene shows constant production of hydrogen versus the dose in all zones of the dose. The HNBR however, shows diminished production of hydrogen at the beginning of irradiation. It is the same range of doses where less efficient crosslinking is observed as well. Evidently, the energy of absorbed radiation is consumed by additives. The radiation yield of hydrogen in that zone is roughly 5 times lower than with larger doses. This value can be considered as the yield of multi-ionization spurs: the production of hydrogen is that of debris at the chain scission. The chain scission occurs in random site, not connected with additives. This location cannot move to additives. Single ionization sites can move and are neutralized by additives without hydrogen production, e.g. by dissipation of energy on aromatic groups.

Considering the consumption of additives by energy of radiation, comparison of crosslinking of the HNBR with crosslinking of polyolefins, in Charlesby-Pinner coordinates, one can speak about identical behavior.

It is impossible to cover, in a short chapter all approaches to elastomers, which have been exercised during years of study of radiation crosslinking of polymers. Other topics omitted here:

- Results of diffuse reflection spectrophotometry (DRS) measurements [73, 74] of irradiated rubbers, measured versus unirradiated samples shows unsaturations formed during irradiation and possible products of oxidation at the surface of elastomer. The method allows optical UV-Viz spectroscopy even with samples containing opaque additives (see the end of the chapter).
- Sufficient transparency of some elastomers (no crystallites in the structure!) makes possible any optical measurements, including absorption spectrophotometry, also time resolved in pulse radiolysis approach. As transients in irradiated rubbers disappear too fast to be observed even with nanosecond pulses and nanosecond time scale observation of absorption afterward, the lowering of temperature of experiment was necessary. Deep freezing allows to immobilize transients and measure them with low temperature spectrophotometry

equipment. At increasing temperature spectra of transients change their position and eventually disappear.

- As in all cases of radiation chemistry of polymers, the EPR (electron paramagnetic resonance) method contributes to understanding of mechanisms of reactions. EPR investigations on hydrogenated and unhydrogenated nitrile butadiene rubbers contribute much to the understanding of mechanisms of radiation induced crosslinking. Also in the case of HNBR, that method shows limited participation of the CN group, which does not interfere with the crosslinking process.

In conclusion, already half a century ago, the crosslinking of polymers, induced by ionizing radiation has been discovered and the list of polymers, which do not crosslink, and those, which do, formulated. The list is still valid now, but it is hard to understand why polyethylene crosslinking only has been commercialized very early. One reason is, that from the beginnings, there were no other competitive methods of crosslinking available. Irradiation became an approach of choice.

On the other hand, six elastomers in the early list of crosslinking polymers, were conveniently crosslinked by chemicals, especially the natural rubber, since the 19th century. There was no apparent need to introduce new methods. It is recently shown on the example of high-tech elastomer, that radiation induced crosslinking proceeds exactly in a controllable mode like in the case of polyethylene. The model of single- and multi-ionization spurs works in the same way in both, very different groups of polymers. High price of these elastomers will justify an application of expensive radiation processing.

Parallel radiation research on different types of polymers contributes to better understanding of mechanisms of chemical reactions, especially using the fact of low crystallinity of elastomers.

Polymers becoming cross-linked:

Polyacrylic esters, Polyesters, Nylon, Polyethylene, Chlorinated polyethylene, Chlorosulphonated polyethylene, Natural rubber, GRS (Government Rubber-Styrene—styrene-butadiene rubber), Butadiene-acrylonitrile copolymers, Neoprene-W, Neoprene-GN, Polydimethylsiloxanes, Styrene-acrylonitrile copolymers.

Polymers becoming degraded: Polymethyl methacrylate, Polyvinyl chloride, Polyvinylidene chloride, Polytetrafluoroethylene, Polychlorotrifluoroethylene, Cellulose, Polyisobutylene, (Wording as in the original Ref. [63]).

In conclusion, concerning modification of elastomers by ionizing radiation. Absorbtion of ionizing radiation energy by polymers results in four, parallely running chemical processes: crosslinking, degradation, formation of unsaturation, including multiple bonds and oxidation. If crosslinking prevails over degradation, the modification of properties of polymer goes in useful direction. Intentional, radiation induced crosslinking is in many respects better than chemical, because it is realized at ambient temperature and is readily controlled by adjustment of optimal dose. From the practical point of view, especially important is recognition of inhomogeneous deposition of energy and the role of multi-ionization spurs on processes of formation of the net of crosslinking bonds in elastomers. Introduction of analytical methods used in radiation chemistry of polymers revealed new facts. In particular, gas chromatographic determination of radiation yield of hydrogen in irradiated HNBR rubber has shown, that only the half of crosslinks, as determined by gel fraction, is due to hydrogen abstraction from two neighboring chains and the rest is due to entanglements. Hydrogen is not determined in conventional, chemical crosslinking reactions, because this gas is emitted only in the non typical, radiation induced crosslinking. Another non-typical analytical method, the diffuse reflected light spectrophotometry (DRS) is applied to irradiated elastomers, showing the radiation induced chemical reactions.

# 10 Selected Cases of Radiation Crosslinking of Composite Elastomers

Crosslinking of elastomers started in 1844 (Ch. Goodyear, patent in 1844) consisting in using sulphur in the composite and has been refined technologically to the level not demanding completely new approaches like introduction of ionizing radiation energy. That was not the case with synthetic polymers invented much later. These, especially polyethylene were the field of successful introduction of crosslinking by ionizing radiation. Therefore the number of papers and patents for crosslinking of polyethylene outnumbers these devoted to elastomers.

Radiation chemistry of polymers in general, including elastomers was started by Arthur Charlesby. His equation (Charlesby-Pinner) [14] is still applied in description of radiation induced crosslinking, paper [59, 75–77] starts with rubber already in the title and publication [78] appeared in the journal devoted to rubber. Book [79] by Charlesby completes four most important, early publications on radiation chemistry of polymers, including elastomers.

In our Laboratory application of radiation induced crosslinking has started with natural rubber latex. The material was supplied in the shape of aqueous emulsion, stabilized with ammonia. In the 60s the dominating source was cobalt 60 in the absence of accelerators. The beginnings encountered unexplained phenomena: The crosslinking without mixing proceede in the gamma chamber excellently. Introduction of intensive mixing of latex in an open vessel placed in the gamma chamber caused lowering of crosslinking yield and was worsening of the rubber quality. Analytical polarographical investigation has shown that original latex stored in closed vessels did not contain oxygen [80], consumed completely by bacteria present in the material. Intensive mixing, lowering the viscosity of thixotropic system was introducing air and oxygen was entering rapidly in radiation induced reactions, causing completely differet sequence of reactions, spoiling crosslinking and leading to the degradation of polymer. In experiments on oxygenfree latex emulsion the addition of certain chlorine containing compounds was improving the radiation yield of crosslinking.

Pilot plant production of dipped medical products from radiation processed rubber latex resulted in trade mark Radiatex, patented and published [81–85]. However, the product was not perfectly biocompatible and production was withheld. The shelflife was also not satisfactory. Informal information from Japan was, that balloons produced from irradiated latex was not durable enough, this time to the satisfaction of organizers of festoons. Baloons were rapidly decomposed and were not endangering birds.

Attempts to crosslink cis-polybutadiene [86] were successful but not introduced into production because of high expense of irradiation with gamma sources. Perhaps with new cheaper, accelerator sources of radiation the commercial production will be reconsidered.

The commercial success of radiation processing of polyethylene was still tempting to repeat the success with elastomers. The main object was tires and publication by Hunt and Allinger [64] seemed to announce construction of production line for vulcanization of tires. In spite of the premature information, known promoter of radiation processing of polymers, Silverman [65] noticed in 1992 [66, 87] that irradiated green rubber and not the tire will be the object of radiation processing. Some doubts on the situation of participation of radiation in the production of tires were expressed by Cleland et al. [88] from Ion Beam Applications in San Diego, USA, quoting as an single position the publication from 1979, [64]!

The situation in the tire industry is not clear—there are no truthworty publications. On the other hand, on the IRaP 2002 conference (Irradiation and Polymers), Katsumura [89, 90] has informed that in Japan ca 300 accelerators are busy in factories and financial aspects of results are bigger than from the production of electrical energy in nuclear power stations. Most probably radiation processing is connectd with components of tires, but complete tires are not irradiated.

Literature contains inconsistent information, e.g. [88], where radiation processing in Japan is compared to the effort in the USA and Canada together. Category of electron accelerators of most convenient energy of electrons 9–13 MeV in the case of tires, the number is not 300 but nil in both group of countries. Informations about application of radiation processing in tires industries can be safely rejected, especially in view of the fact that processing in that case demand doses two to three orders of magnitude higher than in the case of radiation sterilization of medical products produced from polymers. That group of applications of radiation is transparent as concerns the size of production, prices involved, competition of other methods of sterilization, etc. The same applies to radiation crosslinking of polyethylene for hot water.

The survey of recent applications of radiation processing of polymers Chmielewski et al. [89, 90] devotes little space to elastomers and few paragraphs to elastomers. As concerns rubber tires, there are only 3 citations [91–93]. "Six companies produced about 170 mln tires in Japan, according to estimates of 1997. Five major companies have installed 23 electron accelerators for prevulcanization of carcass ply to increase green strength. One accelerator (500 keV, current 75–150 mA) can treat 30,000–50,000 piles [probably it should mean plies] per day, [93]. Production of automobile radial tires using EB crosslinking of rubber sheet is the second largest components in both USA and Japan in the industry value of 13.5 bilion US$ in the USA and 8.4 in Japan. The share of EB processing of passenger tires in Japan is 95 % or more [93]".

The lack of technical details of large scale applications in tire industry does not mean the lack of important scientific publications on the subject, especially as concerns composites of rubber. Interesting combination of traditional approach with new technology represents the reference by Aoshuang et al. [94], comparing mechanical properties of the blend of natural rubber with butadiene rubber in proportion 60/40, crosslinked initially by sulphur in the presence of chemical accelerators. Pieces 2 mm thick, under pressure at temp. 150 °C where irradiated 10 min (1.5 parts w. of S) or 15 min (0.5 parts w. S) $\gamma$ at 20 kGy/h to the dose of 50–90 kGy. Mechanical properties and resistance to aging were better than with sulphur alone.

Another attempts to connect old methods to new ones was done by Basfar [95], in radiation laboratory of Silverman in Maryland, USA, the purpose was improvement of resistence of butadiene—styrene rubbers (SBR) crosslinked by sulphur and ionizing radiation towards ozone. Blends contained carbon black, sulphur, zinc oxide, stearinic acid and antiozonant (Antiozite), antioxidant (Agerite White, Agerite Resin D and Santocure). Massive doses of gamma radiation were applied from cobalt, to monstruous doses of 1 MGy. Basic research method was spectrophotometry. Authors think that resistance to ozone of theirs rubbers containing S and irradiated was protective action of remaining groups C=C, left due to the finishing of crosslink process by radiation. That explanation is questionable.

During next years much effort to the question of radiation induced crosslinking was paid by research laboratories in countries producing natural rubber. Majority of papers is devoted to that raw material, but very often blends of it with synthetic rubbers were investigated. From rubber institute in Sri Lanka, in collaboration with Takasaki Centre in Japan, the Ref. [96] originated, dealing with gamma vulcanization of latex with addition of trimethacrylate of trimethylolopropane and phenoxyethyl acrylate (PEA) as crosslinking coagent. The second mentioned additive proofed to be more efficient but still not enough to lower radiation doses needed for crosslinking.

More interesting is the paper by Ratnam et al. [97] from Malesia, in which lower energy of electrons was used (3 MeV). It means that thinner objects have to be used than in the case of usuall 10 MeV. The irradiated elastomer was epoxydated natural rubber (ENR 50), with addition of trimethyl-propane triacrylate as coagent of crosslinking. Another object was ENR50 with Irganox, and the last one, ENR50 with lead sulphate as a stabilizer. Obtained gel fractions (determined in THF) in the function of dose up to 200 kGy were in logical order: ENR without additives crosslinks max to 80 % gel fraction only after 100 kGy, blend with coagent reaches 95 % of gel fraction already after 20 kGy. Mixture with Irganox does not gel up to 20 kGy, because first Irganox is consumed, and 85 % of gel fraction appears after 160 kGy. The curve for the mixure with lead sulphate is similar to the curve for ENR without additives and 90 % of the material insoluble in the THF is reached after 160 kGy. That picture can be viewed as typical, at the same time it is indicating the importance of known identity of additives, which are sometimes not known in commercial samples.

Same authors describe [98, 99] radiation processing of ENR in 50:50 mixture with polyvinyl chloride. The stabilizer of PVC was basic lead sulphate, and the TMPTA was added as coagent of crosslinking. Dose of 200 kGy caused 90 % crosslinking (as measured by solubility in THF). In the absence of Irganox no starting threshold was observed, usually present in the presence of stabilizers. The curve of crosslinking indicated almost linear increase up to 70 % after the dose of 80 kGy and later a slower to 85 % after 200 kGy.

Limiting the report to most important papers only, one has to mention the reference by Chattopadhyaya [100] on structural effects after irradiation by 2 MeV electrons to the dose of 500 kGy, by the technique of split dose irradiation to avoid excessive warming of films. The objects were thermoplastic elastomers with polyethylene and copolymer ethylene—vinyl acetate. As coagent of crosslinking ditrimethylolopropane tetraacrylate (DTMPTA, Ebecryl 140) was used in concentration of 1–5 % by weight.

In France, electrons of 2 MeV energy [101] were used to investigate viscoelastic properties of different compositions of ethylene-propylene-diene rubbers (EPDM). In comparison, gamma radiation (cobalt 60) in the presence of air caused, after 100 kGy, a far reaching degradation of chains. Temperature of irradiation was pushed to 100 °C, because the purpose of investigation was the application of elastomers as isolators of electric cables in nuclear power stations.

Radiation processing was proposed in modification of thermoplastic polystyrene elastomers and their mixtures with polyolefins [102].

An interesting application of gamma radiation processing of elastomeric composites and blends was presented by Magda and Zeid [103]. The object was NBR (acrylo-nitrylo-butadiene rubber) with EPDM (monomer etyleno-propylenowy) with addition of HAF—high abrasion furnace carbon black. The research was performed in the purpose of improving properties of the material [104]. The traditional vulcanization proceeds in the presence of sulphur or a peroxide and heating to 180 °C. Additional, not easy to be controlled chemical reactions have sometimes not desired effects. Radiation induced crosslinking proceed at ambient temperature under easy to be controlled parameters of dose rate and total delivered dose. Densities of crosslinking can be similar to classic ones, but the type of crosslinking is different and consists in C–C bonds, securing better mechanical properties at higher temperatures, better reply to stretch, better resistance to wear and tear—abrasion and better resistance towards damage by ozone exposure.

Rubbers are not electric current conducting materials, because atoms in the chain are bound covalently. There is no place for delocalization of valence electrons and there is no conductivity path. Such road can be created by addition conducting component as carbon black, carbon threads or even metallic particles. Usually used rubber additive of quasi-graphite structure secures mobility over hexagonal layer. Material of different content of carbon black was gamma irradiated to different doses of radiation. The conductivity of the blend without carbon black did not change to the dose of 200 kGy ($10^{-9}$ $ohm^{-1}$ $cm^{-1}$). At the maximum content of 100 phr of carbon black was increasing to $10^{-5}$ after the dose of 150 kGy and was remaining on the same level to the dose of 250 kGy. Electric current flow is determined by crossing of two barriers. First obstacle is the width between two aggregates which should be narrow enough to allow the jump of an electron. The second barrier is the layer of chemically adsorbed oxygen between phases, acting as an isolator. There is an assumption that radiation may cause the crosslinking between carbon black and the elastomer. Exact mechanisms are waiting for an explorer.

Another case is radiation processing of rubber blends with a compatibilizer. Multicomponent rubber blends are more and more often investigated from the point of view of radiation chemistry, because the behaviour of a composite can be different from the sum of reations in the particular component irradiated in a 100 % single state. Senna et al. [105] report in a voluminous paper results of irradiation of the blend containing epoxidized natural rubber (ENR), also polypropylene (PP) and a reactive compatibilizer. Polypropylene is known in radiation chemistry as rather a degrading polymer. Positive resultants were achieved due to the use of additional compatibilizer, polypropylene grafted with maleineic anhydride (PP-g-MAH, PP graft copolymer with maleic anhydride) or 1,6-heksandiol-diacrylate (HDDA, 1,6-hexandiol-diacrylate). Positive result is explained probably as due to formation interphase bond between phases of PP and ENR during preparation of the blend in molten state. Compatibilization creates coaction between nonmiscible surfaces on the molecular level or between two noncompatible polymer domains. It demands supply of energy from the outside of the systems and radiation processing helps in the realization. Here radiation chemistry meets elastomers with polyolefins. Polypropylene is one of most important polymers in production of medical devices, but degrades under ionizing radiation used for sterilization. Degradation proceeds in the postirradiation after effect due to the presence of peroxy radicals starting chain reaction during the shelflife of the product.

Polypropylene with maleic acid anhydride grafted on the elastomer polyethylene acetate (POE-g-MAH), also with monomere HDDA 1,6-(hexandiol diacrylate) reduces separation of phases between polypropylene and epoxydized natural rubber. Irradiated blend does not show negative properties of irradiated polypropylene, especially autooxidation of the material is reduced. It can be done by an interesting method to avoid oxidation: samples are irradiated in air after moulding between Mylar films [106].

Research and development in the field of radiation chemistry is not easy. First difficulty is the exact composition needed if the interpretations of results, recognizing the mechanisms of reactions, have to be done with scientific precision. Commercial elastomers contain very often additives not revealt as to chemical formula and concentration. General knowledge of physico-chemical phenomena is very often not sufficient. Rare are publications like by George and Thomas [107] discussing energy transfer in highly viscous state of elastomers. There is sometimes lack of data on unsaturation, mobility of segments, free volume between molecules. Little is known a possibility of diffusion of small molecules through elastomers. It would be a very important information for radiation chemists, opening possibilities of radiation induced grafting and formation of co-polymers, by soaking the elastomer with proper monomer before irradiation. Elastomers can be modifies by addition of the monomer or another reagent from the gas phase or from solution. The review of possibilities in the case of urethane elastomers and thermoplastic SBS give Jonquieres et al. [108].

As it was explained in the section on radiation chemistry of composites they have to be treated with special attention as a combination of well separated different phases. That is not the case with blends which are homogenous from the point of view of radiation chemistry. An example is the case of radiation processing of blends of nitrylo-butadiene-rubbers with other polymers.

Radiation chemistry of HNBR elastomer has been described above. There are several attempts to irradiate blends of NBR and HNBR elastomers with other polymers. Das et al. [109] describe blends onf nylon 6 with HNBR, 50/50, with doses up to 80 kGy, only sometimes to 200 kGy, using low energy accelerator delivering electrons of 1.8 MeV energy, useful for films only and not to bulk material. The material is two phase, HNBR rubber is dissipated in continuous phase of nylon. Results of the paper are inconclusive.

Selected investigations connected to radiation crosslinking of elastomers ask for looking into comparison with radiation chemistry of better known phenomena in polyethylene.

Elements of radiation chemistry common for well known modification of polyethylene, but also for elastomers, are closely connected with the background of radiation chemistry of solid phase. General elements of chemistry and physical chemistry of elastomers, more complicated than in the case of polyethylene influence the course of reactions initiated by ionizing radiation. Resulting technologies of radiation processing of elastomers must be more complicated. Nevertheless, from the basic research point of view the conclusion can be drawn, that the transfer of the positive hole $h^+$ in irradiated HNBR is very similar to that in the case of polyethylene.

The main difference is that in the case of HNBR the process of crosslinking is accompanied by the disruption of chains. In that case the relation of number of chain scissions to chain crosslinkings is 0.41

Zhao et al. [110] at gel dose of 38 kGy. In the previous section radiolysis of the HNBR has been described, here radiolysis of it with other polymers is mentioned.

Vallat et al. [111, 112] in CNRS UPR (France) used electron beam (2.2 MeV, dose rate 2.5 kGy/s, doses 10, 20, 30, 70, 250, 500 kGy) of HNBR/polyethylene blends: HNBR—Therban A 3907 (100 %), HNBR/PE: 80/20, 60/40, 40/60, 20/80, PE (100 %) as 0.5–0.7 mm films separated by Mylar blends, at temperature 160 °C and 7 MPa pressure. No crosslinking agents were added. HNBR had 39 % of bound acrylonitryl, $\leq$0.9 % double bonds, $T_g = -25$ °C. Soluble fraction was, in %, resp. to doses listed above 6, 8, 9, 13, 22, 15.

An interesting phenomenon which occurs during irradiation of polymers and elastomers in particular is formation of hydrogen. It is not released in gaseous form during chemical crosslinking, because it is bound by crosslinking agent. It must be detached, otherwise crosslinking tetra- and trifunctional bonds could not be formed. Radiation induced release of hydrogen reaches the value of 2–3 molecules of $H_2$ per 100 eV of absorbed energy. Radiation yields of hydrogen help to draw important basic conclusions. Our investigations show, that in some cases the yield of hydrogen is too low to explain the crosslinking. The 50 % of expected hydrogen matching the crosslinking could be explained entangling of elastomer. The phenomenon needs further explanation by topology of space net.

Type Y crosslinks dominate and therefore yield of degradation of the chain is lower than expected 20 %. For every 100 acts of crosslinking there are in average 6–11 acts of degradation. Both phenomena are caused by single ionization spurs responsible for 80 % of radiation induced processes. In contrast to multi-ionization spurs taking 20 % of deposited low LET energy, single ionization spurs can move along the chain. Radiation yields of crosslinking of HNBR are 2.5–2.8 per 100 eV and do not depend on double bond content and concentration of oxygen. However, radiation yields of hydrogen, determined in our laboratory are only 1.17–1.30 per 100 eV, therefore the half of crosslinking causing insolubility is due to entanglements, belonging to the field of supramolecular chemistry.

The critical approach to reported papers asks to draw attention to improper application of experimental techniques. In the study on radiation chemistry of NBR rubbers, Hill et al. [111, 112] have described the application of the NMR for the investigation structural changes in irradiated three samples, containing resp. (18, 30 i 45 %) acrylonitrile. Due to the low sensitivity of NMR on radiatin induced changes, they have applied mammoth dose of radiation of 1.5–9 MGy, i.e. almost two orders of magnitude higher than needed for crosslinking. The same applies to EPR method working well at low doses and at very high doses giving misleading conclusions. Very high doses act on already crosslinked material, completely different to the starting one. Observed results refer to anoher starting material. Therefore Hills results are not considered.

Many publications describing radiation processing of elastomers are comparing new approach with traditional one. It is important, because radiation processing is more expensive and strong arguments have to be involved to justify the introduction of new technology. The oldest method of vulcanization, i.e. crosslinking in the new terminology, consists in preparation of the blend with sulphur, or more recently with peroxides. Both technologies demand supply of activation energy as heat. Heating of the blend to the temperature of 150–180 °C starts the desired reactions of crosslinking, but at the same time causes running of undesired side reactions usually not under control. That argument is rised by Wang et al. [113].

There are other comparisons, like by Hafezi et al. [114] who deal with physicochemical properties of blends NBR-PCV crosslinked by sulphur and in comparison crosslinked by electron beam from the most sofisticated electron accelerator—Rhodotron. The object was the blend NBR-PCV (60/40) taken as 100 phr, carbon black N330, 40 phr, zinc oxide 3 phr and stearinic acid 1 phr. Doses were reaching up to 150 kGy, but 75 kGy were optimal. The blend for traditional vulcanization contained 1–2 phr of sulphur, MBTS (disulphide of dibenzotiazyl) 1–2 phr and wulkacide CBS (N-cykloheksy-2-benzotiazyl sulfonamide) 0.3–0.7 phr. Blend was wulkanized at the temperature of 160 °C, under pressure. Authors are observing the radiation processing as giving better results, as securing better control.

The just quoted paper is interesting also because of using a new method of determination of crosslinking, i.e. by the instrument called MRCDS (magnetic resonance crosslink density spectrometer). Innovative work on the new approach started a decade ago and the instrument became commercial only recently. It allows to determine the density of crosslinking in $10^{-5}$mol $cm^{-3}$ units, but a strict comparison is missing, except information of the producer, the company IIC Co. Ltd in Germany. Blends of NBR (23.5 % w of styrene) contained 1,1,1-trimetylopropane triacrylate (TMPTA) up to optimal 8 phr. Density of crosslinking was going higher up to a certain concentration of TMPTA, and after that was diminishing in the range of doses of 20–200 kGy. Polymer films were irradiated by electron beam 2 MeV, by split dose (by 10 kGy) method to avoid overheating.

Another comparison of thermal properties of rubbers crosslinked by different methods was done by Wang et al. [113] styrene-butadiene rubber. Thermal stability was investigated at heating rate of 5 K/min. Two steps of decomposition were observed, first one caused by degradation of additives to rubber, the second one was explained by decomposition of the SBR itself. Density of crosslinking is an important parameter: the higher density, the better is thermal stability of rubber due to higher activation energy needed for thermal degradation of the wulcanizate.

Investigation of FTIR spectra of radiation crosslinked rubbers gives important information about concentration of conjugated bonds C–C, aromatic C–C, and C=C. Concentration of saturated groups $-CH_2-$, formed during degradation of the SBR is diminishing with increase of density of crosslinking. As it was shown in the section on radiation chemistry of HNBR rubber the concentration of $-CH_2-$ is diminishing as hydrogen atoms are removed with the formation of new C–C bonds in macromolecules of SBR. Authors of mentioned papers did not determine the radiation yields of $H_2$.

There is continuous interest in radiation crosslinking of natural latex. The product is better than conventionally wulcanized rubber, using sulphur. Mitra et al. [115] indicate, that conventionally produced latex-rubber contains toxic nitrosoamines, excluding medical applications. Radiation crosslinked rubber-latex is superior.

Studies on the radiation yield of detachment of hydrogen should be continued, because they yield many conclusions of both basic and technological character. For instance, curves of hydrogen production vs dose measured up to 280 kGy are linear only after 20 kGy. Initial doses of radiation energy are used for the destruction of non-rubber substances in raw material. Therefore crosslinking starts only after passing that initial doses.

Another method of research not sufficiently exploited yet in elastomer chemistry is electron paramagnetic research (EPR, also called ESR for electron spin resonance). That method shows the presence of unpaired electron, formed during absorption of ionizing radiation and changing position and subsequentent locations. Results are useful in elucidation of mechanisms of radiation induced phenomena and eventually in the technological approach. Our research presented on conferences at the time being was done with the help of Bruker instrument. Samples are gamma or electron beam irradiated at controlled temperatures and transferred into the spectrometers cave, also of controlled temperature. If the latter is low, the warming allows to study the decay of free macroradicals. Spectra are processed in the computer and if needed simulated with proper programme for identification. The HNBR elastomer was investigated after irradiation at different temperatures and spectra observed during increase of the temperature. The signal of unpaired electron was still visible at 280 K but has disappeared completely at room temperature. It understandable because HNBR like most elastomers do not contain crystalline phase. The shock cooling of the sample did not provoke formation of crystalline phase which stabilizes unpaired electrons. Spectra at low temperature are superposition of same spectra but in different proportions with changing temperature. Starting with simple polyethylene, low and high field spectra observed after irradiation indicate the presence of $CH_2\,{}^{\bullet}CH\,CH_2$, alkyl radical forming 6 lines at the distance of ca 3.0 mT, after low temperature radiolysis. The second component of the spectrum obtained at 77 K is the signal with strong central line and two small piks at the distance of 9.0 mT. Nitrile rubbers were not analyzed by the EPR but per analogy with other third order radicals one can identify the described spectrum as a radical product $CH_2\,{}^{\bullet}C(CN)\,CH_2$. At the temperature of 250 K the allyl radical ${}^{\bullet}CH\,CH{=}CH$ appears, formed by secondary alkyl radical. Distances between seven lines are ca. 1.1 mT. The intensity of the signal is higher for fully hydronenated elastomer, what indicates that it is not the only secondary reaction. The most stable paramagnetic species shows a quartet of t lines of over subtle splitting 1.2 mT. It is a secondary product, because its spectrum is not observed at low temperatures. One can conclude that two radicals, i.e. allil and the radical showing a quartet are formed from the mentioned third order radical with nitrile group.

$$\underset{\textbf{(a)}}{CH_2\,{}^{\bullet}CH\,CH_2} \rightarrow \underset{\textbf{(c)}}{{}^{\bullet}CH\,CH{=}CH}$$

$$\underset{\textbf{(b)}}{CH_2\,{}^{\bullet}C(CN)CH_2\,CH_2\,CH_2} \rightarrow \underset{\textbf{(d)}}{CH_2\,{}^{\bullet}CH(CN)} + CH_2{=}CH\,CH_2$$

One can assume that nitrile group participates in radical reactions and is one of factors causing disappearance of radical products in room temperature. It is stressed that there is no presence of unpaired electron centered on nitrogen, which shows characteristic signals, not to be missed.

EPR spectroscopy of polymers is more complicate than in the case of low molecular weight simple compounds. However, already the conclusion about no formation of crystallisation of elastomer after shock cooling and elimination of nitrogen assumed as centrum of the chain of reactions in irradiated nitrile rubber are worth of the EPR research. That method is not able to speak out about very fast reactions, e.g. the travel of positive hole over the chain of a polymer, which Malcolm Dole assumed in 1959 in the case of irradiated polyethylene. Even the TREPR (time resolved electron paramagnetic rezonans) is not able to solve the problem. Other experiments described in other places of the chapter show that the movement of positive hole along the chain of polyolefins in general and elastomers, including the HNBR in particular, to places energetically preferable are the most probable picture of phenomena in radiation induced crosslinking.

UV-Vis investigations of elastomers are seldom applied in spite of the fact that they contribute important information about radiation induced effects. Many polymers are transparent enough to permit direct registration of absorption spectrum before and after irradiation. Majority of elastomers is poorly or even not transparent, demanding special approach in absorption spectroscopy. We have chosen diffuse reflection spectroscopy (DRS), using spectrophotometer Perkin Elmer Lambda 9 with reflection attachment. A typical measurement consists in irradiation of samples of an elastomer with a dose from 20 to 300 kGy. [73, 74].

As an example, HNBR was investigated by DRS, using UV 200 do 450 nm. Spectra of Therban A4307 irradiated in oxygen has shown intensive band in 370–375 nm shifting upwards with incresing doses to apparent OD of 0.75. Distinct piks were at 315, 265 i 225 nm.

Spectra of Therban A4307 in argon have shown a band around 370, but of lower intensity (max 0.65). Additional phenomenon was observed of removing original chromophores present in the material.

Therban C4367 was investigated in oxygen and in argon. One can ascribe observed bands to new identities in irradiated elastomer:

| | $\lambda$ {nm} | $\varepsilon$ |
|---|---|---|
| –CH=CH–CH=CH– | 236 | 25,000 |
| $–CH_2(CH{=}CH)_3CH_2–$ | 274 | 41,800 |
| $–CH_2(CH{=}CH)_4CH_2–$ | 310 | 58,900 |
| $–CH_2\dot{C}HCH_2–$ | 215 | 1,800 |
| $–\dot{C}HCH{=}CH–$ | 258 | 7,300 |
| $–\dot{C}HCH{=}CH–CH{=}CH–$ | 285 | 29,000 |
| –O–O– | 210 | |
| –CO– | 260–280 | |

Bands in lower UV are connected with oxidation of the elastomer. It is clear, that concentration of double bonds is vital in the oxidation of an elastomer. Lower wavelength bands are connected with oxidation of the elastomer, what is connected with formation of peroxides and keton groups.

DRS measurement concentrates at phenomena on the surface, therefore specially planned and executed experiments can help in investigation of oxidation phenomena. For that purpose the investigated film have to be conditioned for a long time in argon atmosphere and in the same conditions electron beam irradiated. The parallel investigation consists in conditioning in oxygen atmosphere and irradiation in the same environment. Such experiments demand an accelerator with single beam port, covering it together with the sample with a tent, because creation of described conditions on the conveyor with scanned electron beam is not possible.

Nonhydrogenated NBR was also investigated. It shows intensive absorption band at 320–330 nm, more intensive than in hydrogenated elastomers. Intensive absorption at 235 nm increases with dose. Broad bands are at 250–300 nm. Intensive bands can be ascribed to multiple double bonds, which are present in higher concentration in comparison to less hydrogenated elastomers.

Unfortunately there are only few papers which compare new approaches—and radiation processing belongs here—with conventional, old technologies. An example is blends of NBR/EPDM with carbon black. An example of application of radiation processing to elastomers is investigation of the influence of ionizing radiation on the NBR (acryl-nitrylo-butadiene rubber) with the EPDM (monomer etylene-propylene) monomer, with addition of carbon black (HAF—high abrasion furnace carbon black). The purpose of the project was improvement of properties of the product [103]. Traditional vulcanization consists in addition of sulphur and/or peroxide and heating of the blend to the temperature of 150–180 °C. Crosslinking takes place, accompanied with uncontrolled side reactions. Radiation induced crosslinking proceeds at room temperature without presence of sulphur or peroxides. Controlled parameters are applied dose and dose rate. Crosslinking densities are similar to those obtained by conventional means. However, the main difference is that the type of crosslinking is different and consists in –C–C– bonds securing better strength to stretch at higher temperatures, better resistance to abrasion and to destructive action of ozone.

Rubbers are in principle nonconducting materials, because atoms in the chain are bonded covalentionally. In saturated carbon compounds there is no place for delocalization of valence electrons and there is no conductivity path. Incorporated additives like carbon black, carbon fibres or metals secure electrical conductivity. Described blend is a composite of quasi-graphite structure securing mobility of electrons on hexagonal layers.

The irradiations were done in cobalt 60 source. Conductivity of the blend without carbon black was low even after the dose of 200 kGy ($10^{-9}$ ohm$^{-1}$cm$^{-1}$), but with the presence of maximum of 100 phr carbon black, was rising to $10^{-5}$ after the dose of 150 kGy and remained on that level till the dose of 250 kGy. The flow of electric current through the composite is possible after negotiation of two barriers. The first one is the width between aggregates, the second is the layer of chemically adsorbed oxygen on the phase boundaries, acting as an isolator. Most probably the second barrier disapears during the irradiation, due to the crosslinking between carbon black and rubber matrix. C.f. also [104].

How radiation processing is related to rubber blends with a compatibilizer? Multicomponent rubber blends are often investigated by radiation processing technique, which shows different behaviour of components in 100 % state. Senna et al. [105] report results on blends containing beside epoxidized natural rubber (ENR) also polypropylene (PP) and reactive compatibilizer. Polypropylene is known in radiation chemistry as a degrading and not crosslinking polymer, but positive results were achieved due to the application of additive compatibilizer, i.e. polypropylene grafted with maleinic anhydride (PP-g-MAH, PP graft copolymer with maleic anhydride) or 1,6-hexandiol-diacrylate (HDDA, 1,6-hexandiol-diacrylate). Positive result is explained by interphase bond between PP i ENR occurring during preparation of the blend in molten state. Compatibilization is a technique of creation of bonds between unmiscible surfaces on the molecular level or between different polymer domenes. That process demands extra energy which is supplied by ionizing radiation. Compatibilization prevents also undesirable action of air. Another compatibilizators are polyethylene acetate (POE-g-MAH) and PP- g-MAH with monomer HDDA (1,6-heksanodiol diacrylate) which reduce separation of phases between polypropylene and epoxidized natural rubber. Irradiated blend does not show negative features of irradiated polypropylene. Autooxidation of the material is reduced.

Among research method listed in the chapter one has to stress the importance of determination of hydrogen released during irradiation of elastomers and comparison of radiation yields with production of hydrogen from other polymers. Several types of release of hydrogen has been selected. Determination of hydrogen is done gaschromatographically in sophisticated operation of irradiation combined with transfer of the gas phase from irradiation ampoule to gas chromatograph.

The most simple case of hydrogen study is that of polyethylene, where the radiation yield is 4.0–4.6 per 100 eV of absorbed energy, and the similar value of radiation yield of crosslinking coincides well. Both reactions are closely connected and they illustrate the mechanizm of radiation induced crosslinking. Radiation yield value is that of single ionization spur, which travels along the chain to the site of close encounter of two polyethylene chains. That energetically preferred site allows the arriving positive hole to react with the resting place of the electron—the second product of ionization. In that place hydrogen is released and the bond between two chains is fordem. This is the modern version of mechanism of crosslinking proposed in 1959 by Malcolm Dole [62]. The situation is not as simple as in the case of elastomers, due to their more complicated formula, steric hindrances etc. Our results on radiation yield of the HNBR elastomer show the radiation yield of hydrogen only 2.5–2.8. Radiation yield of hydrogen is only 1.17–1.30, half the yield of crosslinking. Proposed explanation consists in formation of entanglements, discussed in terms of supramolecular chemistry. Determination of hydrogen yield has given also further conclusions. Curves of hydrogen radiation yields in irradiated rubbers, versus doses from 5 to 280 kGy are running linear only from the dose of 20 kGy. First portions of radiation energy are used for the destruction of non-rubber compounds present in the material.

| | HNBR | acrylonitryle | Hydrogenation | Atmosphere |
|---|---|---|---|---|
| H43 | Therban A4307 | 43 % mas AN | >99.5 % mol | Oxygen |
| H43 | Therban A4307 | 43 % mas AN | >99.5 % | Argon |
| S43 | Therban C4367 | 43 % mas AN | 94.5 % | Oxygen |
| S43 | Therban C4367 | 43 % mas AN | 94.5 % | Argon |

There is no place in the present chapter to discuss in full thermal investigations of elastomers. However, one paper has to be mentioned [113], because it is devoted to comparison between conventional product and that crosslinked by radiation. The same method was used for both product thus eliminating the instrument error. The styrene-butadiene rubber was investigated for thermal stability with the rate of 5 K/min. Two steps of decomposition were observed—first connected with the decomposition of additives, the second with the decomposition of the SBR itself. Density of crosslinking is an important factor influencing thermal stability of the rubber. The higher the density of crosslinking the better the thermal stability, due to to the increase of activation energy of thermal degradation of rubber.

Another comparison was done by Hafezi et al. [114] on physicochemical properties the NBR-PVC blend, EB irradiated rubber.

Excellent evaluation of changes occurring during heating of radiation cross-linked rubbers gives FTIR examination, which shows the concentration of conjugated C–C bonds, aromatic C–C, aromatic C=C. Concentration of saturated groups $-CH_2-$, formed during degradation of the SBR lowers during SBR degradation with increase of crosslinking density. FTIR spectroscopy show that active hydrogen atoms are detached from $-CH_2-$ groups during radiation crosslinking what creates New C–C bonds in macromolecules, with parallel release of $H_2$. Our measurements of hydrogen formation in irradiated HNBR can confirm that. Unfortunately, Authors of the paper did not determine hydrogen.

Discussing comparisons one can menton continuing interest in natural rubber crosslinking. Conventional crosslinking with sulphur is not for all applications safe, because product contains toxic nitrosamine [116]. Radiation crosslinked elastomer is better for medical applications.

Role of DRS (diffussed reflection spectrophotometry) in the investigation of radiation chemistry of elastomers.

Investigation of spectra of irradiated elastomers gives important information on mechanisms involved in chemical changes. However, majority of elastomers is not transparent enough to make conventional measurements, simply inserting the sample instead of the cell into a spectrophotometer. The technique of DRS, possible to be applied in any spectrophotometer equipped with proper apparative attachment, allows to detect in the UV/Viz any changes in absorption spectra, due to radiation induced changes (71). Investigations of radiation chemistry of polypropylene have shown that observation of products of oxidation is by two orders of magnitude more sensitive than traditional FTIR method in the case of keto-groups.

Absorption spectra of the HNBR elastomer Therban A4307 irradiated in oxygen show very intensive lines in the zone of 370–375 nm, shifted in the direction higher wavelengths with the increase of dose from almost 0 to 0.75 (Kubelka–Munk OD). There are also distinct lines at 315, 265 i 225 nm.

Spectra of Therban A4307 (in argone) show a band also in the 70 nm region, but less intensive, max 0.65. Remarkably, lower zone of wavelengths lacks absorbtion bands, and even negative values against nonirradiated samples are observed, increasing with the dose, what indicates the removal chromofors.

Therban C4367 (irradiated in oxygen), similarly to all samples hydrogenated >99.5 % shows the 370 nm band, less intensive (max. 0.5). Instead, there are very intensive bands in the lower zone, overcovering mutually.

Therban C4367 in argone shows a spectrum in the 370 nm zone similar to that of hydrogenated Therban irradiated in argon. There is also an intensive band at 240 nm.

We ascribe following bands to exact groups on the chain with according molar absorbtion coefficients:

| Fragment | $\lambda$ {nm} | $\varepsilon$ |
|---|---|---|
| –CH=CH–CH=CH– | 236 | 25,000 |
| $-CH_2(CH{=}CH)_3CH_2-$ | 274 | 41,800 |
| $-CH_2(CH{=}CH)_4CH_2-$ | 310 | 58,900 |
| $-CH_2\dot{C}HCH_2-$ | 215 | 1,800 |
| $-\dot{C}HCH{=}CH-$ | 258 | 7,300 |
| $-\dot{C}HCH{=}CH-CH{=}CH-$ | 285 | 29,000 |
| –O–O– | 210 | |
| –CO– | 260–280 | |

In conclusion, intensity of bands at 370 nm only slightly depends on irradiation conditions (oxygen, argon), but are distingly lower from nonhydrogenated elastomers. They are connected, probably with formation of double bonds. Bands in lower regions are connected with oxidation of the polymer and formation of peroxides and probably of ketones. Important is the number of double bonds; they stabilize, probably, the oxidation precess.

The 240 nm band in nonhydrogenated elastomer irradiated in argone may be ascribed to free radical form connected with double bond, which rapidly attaches oxygen formin peroxide group. Negative values in 260 nm band in hydrogenated elastomer in argone may be due to transformation of double bonds into another configuration.

Absorbance of elastomers in the function of the radiation dose was investigated for TS34 and HS34 (Therban resp. C3446 i 3467), H34 (Therban A3407). The nonhydrogenated NBR shows intensive band at 320–330 nm, more intensive in comparison to hydrogenated elastomers. Intensive bands at 235 is increasing with dose what may be connected with the increase of number of bonds of the CH=CH–CH=CH– type.

Figures 17 and 18 are examples of DRS optical spectra in irradiated elastomers.

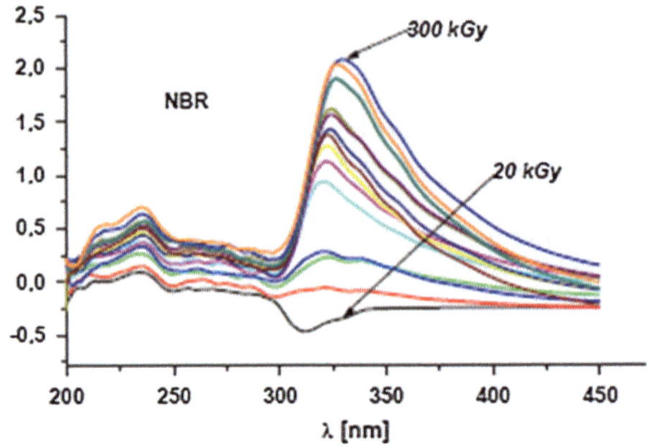


**Fig. 17** An example of DRS experiment with irradiation in air of NBR (nitrylbutadien rubber). The spectrum is measured against unirradiated sample. The negative absorption effect is seen, due to the destruction of the UV-chromofore constituent of rubber

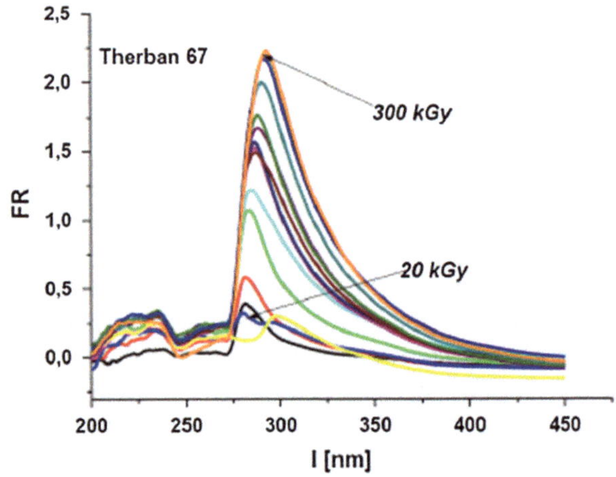


**Fig. 18** Absorption spectra as Kubelka–Munk function for different doses of radiation from 20 to 300 kGy irradiated HNBR Recent investigations in the field of radiation chemistry of elastomers show [116] that applications of DRS and GC to crosslinking of elastomers may show synergistic effects of two method of applied crosslinking

## 11 Examples of Patents Connected with Radiation Processing of Elastomers

Specific references are patents, which do not need conform to scientific rules.

Examples of patents connected with the topic of the chapter can be found in *Sage Publications, Journal of Elastomers and Plastics*, connected with the industry. For instance, USA Patent 20030190484, for Shin-Etsu Chemical Co., Ltd., Radiation curing silicone rubber composition and adhesive silicone elastomer film:

Components of irradiated silicon rubber are listed. Polysiloxane is formed into the film, irradiated and used in electronics.

Another USA Patent 20060055089, is for The Procter & Gamble Company, under the title "Zoned radiation cross-linked elastomeric material" for medical purposes. As in the former patent, scientific background is difficult to recognize.

There are patents for the crosslinking of polyurethane polymers with UV, e.g. for Bajer Material Science LLC.

**Acknowledgments** Project is supported by the Polish Ministry of Science and Higher Education grant No N N209083838 (Synergistic systems of crosslinking elastomers).

# References

1. Zagórski, Z.P.: Radiation chemistry of spurs in polymers. In: Advances in Radiation Chemistry of Polymers, pp. 21–31. IAEA-TECDOC-1420, Vienna (2004)
2. Zagórski, Z.P.: Dosimetry as an Integral Part of Radiation Processing, pp. 257–264. IAEA-TECDOC-1070, Vienna, (1999)
3. Zagórski, Z.P.: Solid state radiation chemistry—features important in basic research and applications. Radiat. Phys. Chem. **56**, 559–565 (1999)
4. Głuszewski, W., Zagórski, Z.P.: Radiation effects in polypropylene/polystyrene blends as the model of aromatic protection effects. Nukleonika **53**(Suppl.1), s21–s24 (2008)
5. Bik, J., Głuszewski, W., Rzymski, W.M., Zagórski, Z.P.: EB radiation crosslinking of elastomers. Radiat. Phys. Chem. **67**, 421–423 (2003)
6. Bik, J.M., Głuszewski, W., Rzymski*, W.M., Zagórski*, Z.P.: Electron beam crosslinking of hydrogenated acrylonitrile-butadiene rubber. Kautschuk, Gummi und Kunststoffe **57**, 651–655 (2004)
7. Mirza, J., Schön, N., Thörmer, J.: ®Therban – Eigenschaftsbild und Vorzűge eines verschleiβfesten Elastomeren mit hoher Hitze-, Ozon- und Őlbeständigkeit. Kautsch. Gummi Kunstst. **39** (7/86), 615 (1986)
8. Casper, R., Rohde, E.: Zunehmende Bedeutung der Spezialkautschuke in Westeuropa. Kautsch. Gummi Kunstst. 41 (6/88), 541 (1988)
9. Thörmer, J., Mirza, J., Szentivanyi, Z., Obrecht, W.: Therban – Einfluβ des Vernetzungssystems auf das Verarbeitungsverhalten und das Eigenschaftsprofil von hydriertem Nitrilkautschuk (HNBR). Kautsch. Gummi Kunstst. **41** (12/88), 1208 (1988)
10. Zagórski, Z.P.: Dependence of depth-dose curves on the energy spectrum of 5 to 13 MeV electron beams. Radiat. Phys. Chem. **22**, 409–418 (1983)
11. Zagórski, Z.P.: Thermal and electrostatic aspects of radiation processing of polymers. In: Singh, A., Silverman, J. (eds.) Radiation processing of polymers, pp. 271–287. Hanser Publishers, Munich (1992)
12. Zagórski, Z.P.: Modification, degradation and stabilization of polymers in view of the classification of radiation spurs. Radiat. Phys. Chem. **63**, 9–19 (2002)
13. Zagórski, Z.P.: Radiation chemistry of spurs in polymers. In: Advances in Radiation Chemistry of Polymers. Proceedings of a Technical Meeting Held in Notre Dame, pp. 21–31. Indiana, USA, 13–17 Sept 2003, IAEA-TECDOC-1420, Vienna (2004)
14. Charlesby, A., Pinner, S.H.: Analysis of the solubility behaviour of irradiated polyethylene and other polymers. Proc. Royal Soc. Lond. **A249**, 367–386, (1959)
15. Zhao, W., Yu, L., Zhong, X., Zhang, Y., Sun, J.: Radiation vulcanization of hydrogenated nitrile butadiene rubber (HNBR). J. Appl. Polymer Sci. **54**, 1199–1205 (1994)
16. Fedors, R.F., Landel, R.R.: Relationship between maximum extensibility of network and the degree of crosslinking and primary molecular weight. J. Appl. Polym. Sci. **19**, 2709 (1975)
17. Grobler, J.H.A., McGill, W.J.: Effect of network heterogeneity on tensile and tear strengths of radiation, peroxide, efficient and conventional cured polyisoprene. J. Polym. Sci. Pol. Phys. **32**, 287–295 (1994)

18. Perraud, S., Vallat, M.-F., Dawid, M.-O., Kuczynski, J.: Network characteristics of hydrogenated nitrile butadiene rubber networks obtained by radiation crosslinking by electron beam. Polym. Degrad. Stabil. **95**, 1495–1501 (2010)
19. Mc Graw-Hill Encyclopedia of Science & Technology, 7th edn. New York (1992)
20. Ullmans Encyclopedia of Industrial Chemistry, English Language Edition, VCH,Weinheim, Germany 1985–1988
21. Marwanta, E., Mizumo, T., Nakamura, N., Ohno, H.: Improved ionic conductivity of nitrile rubber/ionic liquid composites. Polymer **46**, 3795–3800 (2005)
22. Nastase, C., Nastase, F., Dumiru, A., Ionescu, M., Stamatin, I.: Thin film composites of nanocarbons-polyaniline obtained by plasma polymerization technique. Compos Part A Appl Sci Technol **36**, 481–485 (2005)
23. Zagórski, Z.P.: Radiation induced dehydrogenation of organics: from amino acids, to synthetic polymers, to bacterial spores. Indian J. Rad. Res. **3**, 89–93 (2006)
24. Zagórski, Z.P., Głuszewski, W.: Irreversible radiolytic dehydrogenation of polymers—the key to recognition of mechanisms. INCT Annual Report, pp. 40–42 (2003)
25. Zagórski, Z.P.: Design and applications of a constant temperature box for high-energy electron-beam processing at temperatures −200 °C to 700 °C. Int. J. Appl. Radiat. Isot. **36**, 243–245 (1985)
26. Zagórski, Z.P.: Thermal effects in radiation processing (in Polish) Przem. Chem. (Warsaw) **51**, 640–645 (1972)
27. Zagórski, Z.P.: Chapter: Thermal and electrostatic aspects of radiation processing of polymers. In: Singh, A., Silverman, J. (eds.) Radiation processing of polymers, pp. 271–287. Hanser Publishers, Munich (1992)
28. Broza, G., Kwiatkowska, M., Rosłaniec, Z., Schulte, K.: Processing and assessment of poly(butylene terephtalate) nanocomposites reinforced with oxidised single wall carbon nanotubes. Polymer **46**, 5860–5867 (2005)
29. Yanagishita, H., Arai, J., Sandoh, T., Negishi, H., Kitamoto, D., Ikegami, T., Haraya, K., Idemoto, Y., Koura, N.: Preparation of polyamide composite membranes grafted by electron beam irradiation. J. Membrane Sci. **232**, 93–98 (2004)
30. Sirisinha, K., Kawko, K.: Crosslinkable polypropylene composites made by the introduction of silane moieties. J. Appl. Polym. Sci. **97**, 1476–1483 (2005)
31. Mizutani, Y., Nago, S., Sasai, M.: Polypropylene composites with fine particles of poly(styrene-divylbenzene). J. Appl. Polym. Sci. **77**, 1614–1620 (2000)
32. Zilli, D., Chiliotte, C., Escobar, M.M., Bekeris, V., Rubiolo, G.R., Cukierman, A.L., Goyanes, S.: Magnetic properties of multi-walled carbon nanotube-epoxy composites. Polymer **46**, 6090–6095 (2005)
33. Pattanayak, A., Jana, S.C.: Thermoplastic polyurethane nanocomposites of reactive silicate clays: effects of soft segments on properties. Polymer **46**, 5183–5193 (2005)
34. Barone, J.R.: Polyethylene/keratin fiber composites with varying polyethylene crystallinity. Compos. Part A. Appl. Sci. Manuf. **36**, 1518–1524 (2005)
35. Kojima, Y., Usuki, A., Kawasumi, M., Okada, A., Kurauchi, T., Kamigaito, O.: Synthesis of nylon 6—clay hybrid by montmorillonite intercalated with e-caprolactam. J. Polym. Sci. Part A: Polym. Chem. **31**, 983–986 (1993)
36. Pinnavaia, T.J., Beall, G.W.: Polymer-Clay Nanocomposites. Wiley, New York (2000). (Chapter 6)
37. Nascimento, G.M., Constantino, V.R.L., Temperini, M.L.A.: Spectroscopic characterization of doped poly(benzidine) and its nanocomposite with cationic clay. J. Phys. Chem. B **108**, 5564–5571 (2004)
38. Loo, L.S., Gleason, K.K.: Investigation of polymer and nanoclay orientation distribution in nylon 6/montmorillonite nanocomposite. Polymer **45**, 5933–5939 (2004)
39. Galgali, G., Agarwal, S., Lele, A.: Effect of clay orientation on the tensile modulus of polypropylene-nanoclay composites. Polymer **45**, 6059–6069 (2004)

40. Stretz, H.A., Paul, D.R., Cassidy, P.E.: Poly(styrene-co-acrylonitryle)/montmorillonite organoclay mixtures: A model system for ABS nanocomposites. Polymer **46**, 3818–3830 (2005)
41. Filho, F.G.R., Melo, T.J.A., Rabello, M.S., Silva, S.M.L.: Thermal stability of nanocomposites based on polypropylene and bentonite. Polym. Degrad. Stabil. **89**, 289–297 (2005)
42. Zheng, X., Jijang, D., Wang, D., Wilkie, Ch.A.: Flammability of styrenic polymer clay nanocomposites based on a methyl methacrylate oligomerically-modified clay. Polym. Degrad. Stabil. **91**, 289–297 (2005)
43. Zhang, J., Jijang, D., Wilkie, Ch.A.: Thermal and flame properties of polyethylene and polypropylene nanocomposites based on an oligomerically-modified clay. Polym. Degrad. Stabil. (2005) in printnie ma wogole
44. Szleifer, I., Yerushalmi-Rosen, R.: Polymers and carbon nanotubes-dimensionality, interactions and nanotechnology. Polymer **46**, 7803–7818 (2005)
45. Li, H., Yu, Y., Yang, Y.: Synthesis of exfoliated polystyrene/montmorillonite nanocomposite by emulsion polymerization using a zwitterions as the clay modifier. Europ. Polym. J. **41**, 2016–2022 (2005)
46. Pehlivan, H., Balkõse, D., Ülkü, S.: Tihminlioğlu, characterization of pure and silver exchanged natural zeolite filled polypropylene composite films. Compos. Sci. Technol. **65**, 2049–2058 (2005)
47. Morlat-Therias, S., Mailhot, B., Gardette, J.-L., Da Silva, C., Haidar, B., Vidal, A.: Photooxidation of ethylene-propylene-diene/montmorillonite nanocomposites. Polym. Deg. rad. Stabil. **90**, 78–85 (2005)
48. Chan, E.R., Zhang, X., Lee, C.-Y., Neurock, M., Glotyer, S.C.: Simulations of tetrathethered organic/inorganic nanoco-polymer assemblies. Macromolecules **38**, 6168–6180 (2005)
49. Van Gisbergen, J., Overbergh, N.: Radiation effects on polymer blends. In: Singh, A., Silverman, J. (eds.) Radiation Processing of Polymers, pp. 51–69. Hanser Publishers, Munich (1992)
50. Przybytniak, G.K., Zagórski, Z.P., Żuchowska, D.: Free radicals in electron beam irradiated blend of polyethylene and butadiene-styrene block copolymer. Radiat. Phys. Chem. **55**, 655–658 (1999)
51. Żuchowska, D., Zagórski, Z.P., Przybytniak, G.K., Rafalski, A.: Influence of butadiene/styrene copolymers on the modification of polypropylene in electron beam irradiation. Int. J. Polym. Mater. **52**, 335–344 (2003)
52. Czvikovsky, T.: Radiation processing of wood-plastic composites. In: Singh, A., Silverman, J. (eds.) Radiation Processing of Polymers, pp. 121–148. Hanser Publishers, Munich (1992)
53. Singh, A., Saunders, Ch.B.: Radiation processing of carbon fiber-acrylated epoxy composites. In: Singh, A., Silverman, J.(eds.) Radiation Processing of Polymers, pp. 187–203, Hanser Verlag, Munich (1992)
54. Singh, A., Saunders, C.B., Barnard, J.W., Lopata, V.J., Kremers, W., McDougall, T.E., Ching, M., Tateishi, M.: Electron processing of fibre-reinforced advanced composites. Radiat. Phys. Chem. **48**, 153–170 (1996)
55. Lopata, V.J., Saunders, C.B., Singh, A., Janke, C.J., Wrenn, G.E., Havens, S.J.: Electron beam-curable epoxy resins for the manufacture of high performance composites. Radiat. Phys. Chem. **56**, 405–415 (1999)
56. Spadaro, G., Calderaro, E., Tomarchio, E., Dispenza, C.: Novel epoxy formulations for high energy radiation curable composites. Radiat. Phys. Chem. **72**, 465–473 (2005)
57. Patel, M., Morrell, P.R., Murphy, J.J., Skinner, A., Maxwell, R.S.: Gamma radiation induced effects on silica and on silica-polymer interfacial interactions in filled polysiloxane rubber. Polym. Degrad. Stabil. **91**, 406–413 (2006)

58. Żuchowska, D., Zagórski, Z.P., Przybytniak, G.K., Rafalski, A.: Influence of butadiene/styrene copolymers on the modification of propylene in electron beam irradiation. Int. J. Polym. Mater. **52**, 335–344 (2003)
59. Charlesby, A.: How radiation affects long-chain polymers. Nucleonics **12**(6), 18–25 (1954)
60. Dole, M.: History of the radiation cross-linking of polyethylene. J. Macromol. Sci-Chem. A **1597**, 1403–1409 (1981)
61. Dole, M.: My research in the field of the radiation chemistry of high polymers, in early developments in radiation chemistry, pp. 81–90. Royal Society of Chemistry, Cambridge (1989)
62. Williams, T.F.W., Dole, M.: Irradiation of polyethylene. J. Phys. Chem. **81**, 2919 (1956)
63. Lawton, E.J., Bueche, A.M., Balwit, J.S.: Irradiation of polymers by high-energy electrons. Nature **172**, 76–77 (1953)
64. Hunt, J.D., Alliger, G.: Rubber—application of radiation to tire manufacture. Radiat. Phys. Chem. **14**, 39–53 (1979)
65. Silverman, J.: Current status of radiation processing. Radiat. Phys. Chem. **14**, 17–21 (1979)
66. Silverman, J.: Radiation-induced and chemical crosslinking: A brief comparison. In: Radiation Processing of Polymers, pp. 15–22, Hanser Verlag, Munich, (1992)
67. Henglein, A.: Einfűhrung in die Strahlenchemie, p. 63. Verlag Chemie, Weinheim (1969)
68. Zagórski, Z.P.: Pulse radiolysis of solid and rigid systems. In: Mayer, J., Warsaw, PWN. (eds.) Properties and Reactions of Radiation Induced Transients, Selected Topics (1999), str. 219–233
69. Zagórski, Z.P.: Aqueous gelatine gels as the medium of pulse radiolysis. Radiat. Phys. Chem. **34**, 839–847 (1989)
70. Zagórski, Z.P.: EB-crosslinking of elastomers, how does it compare with radiation crosslinking of other polymers? Radiat. Phys. Chem. **71**, 261–267 (2004)
71. Cracco, F., Arvia, A.J., Dole, M.: ESR studies of free radical decay in irradiated polyethylene. J. Phys. Chem. **37**, 2449–2457 (1962)
72. Sethi, M., Gupta, N.K., Srivastava, A.K.: Dynamic mechanical analysis of polyethylene and ethylene vinylacetate copolymer blends irradiated by electron beam. J. Appl. Polym. Sci. **86**, 2429–2434 (2002)
73. Zagórski, Z.P.: Diffuse reflection spectrophotometry (DRS) for recognition of products of radiolysis in polymers. Int. J. Polym. Mater. **52**, 323–333 (2003)
74. Zagórski, Z.P., Rafalski, A.: Radiation chemistry of polymers as recognized by diffuse reflectance spectrophotometry (DRS), INCT Annual Report 2000 (publ. 2001), pp. 35–37
75. Charlesby, A.: Effect of molecular weight distribution on gel formation by high energy radiation. J. Polym. Sci. **14**, 547–653 (1954)
76. Charlesby, A: The crosslinking of rubber by pile radiation. Atomics **5**, 12–21 and 27 (1954)
77. Charlesby, A.: Effect of ionizing radiation on long chain olefins and acetylenes. Radiat. Res. **2**, 96–107 (1955)
78. Charlesby, A., Groves, D.: Crosslinking and radiation effects in some natural and syntheticrubbers. Rubber Chem. Technol. **30**, 27–41, (1957)
79. Charlsby A., Atomic radiation and polymers, Pergamon Press, Oxford 1960
80. Zagórski, Z.P.: Consumption of oxygen in two- and three-phase systems under radiolytic conditions. (Pol.) Przemysł Chem. **48**, 746–749 (1969)
81. Ambroż, H., Zieliński, W.: Sensitization of radiation cross-linking in the form of polyisoprene natural rubber latex. (in Polish). Parts I, II i III, Polimery **3,4,5**, 112–114, 145–151, 194–201 (1968)
82. Ambroż, H., Zieliński W., Golecka E.: Radiatex – Nature of rubber latex, vulcanized by radiation, for use in medicine. (in Polish) Problemy Techn. Med. **1**, 53 (1970)
83. Ambroż, H., Zieliński, W.: Radiation vulcanization of nature of latex. (in Polish) Polimery **17**, 559 (1972)
84. Ambroż, H.: Some aspects of -radiolysis of polyisoprene in the form of natural rubber latex. J. Polym. Sci. Part C **42**, 1339–1345 (1973)

85. Ambroż, H., Zieliński, W., Jaworska, E.: PURITEX—natural rubber latex for medical articles. Polim. Med. **3**, 181–190 (1973)
86. Jankowski, B., Kroh, J.: Crosslinking of cis-polybutadiene by 60Co $\gamma$ rays. Appl. Polym. Sci. **9**, 1363–1366 (1965)
87. Silverman, J.: Radiation-induced and chemical crosslinking: A brief comparison. In: Radiation Processing of Polymers, pp. 15–25. Hanser Verlag, Munich, (1992)
88. Cleland, M.R., Parks, L.A., Cheng, S.: Application for radiation processing of materials.Nucl. Instr. Meth. Phys. Res. B **208**, 66–73, (2003)
89. Chmielewski, A.G., Haji-Saeid, M., Ahmed, S.: Progress in radiation processing of polymers. Nucl. Instr. Meth. Phys. Res. B **236**, 44–54 (2005)
90. Katsumura Y.: An overview of the industrial applications of radiation in Japan. Workshop, presentation W-158 on the 5th International Symposium on Ionizing Radiation and Polymers (IRaP 2002), Canada, Sept 2002, full next not published
91. Machi, S.: Emerging Applications of Radiation Technology, TECDOC-1386, p. 5. IAEA, Vienna (2004)
92. Makkuchi, K.: In: Role of Radiation Processing in Material Science Applications, KAST, Riyadh, Saudi Arabia, 1999, after Chmielewski et al. [86], full text not published
93. Mizusawa K., Baba T.: In: Recent Developments in Electron Accelerator, Technology and Applications, IAEA, Internal Report Consultation Meeting, Quebec, Canada, 18–20 September 2002, after Chmielewski et al. [86], full text not published
94. Aoshuang, Y., Zhengtao, G., Li, L., Ying, Z., Peng, Z.: The mechanical properties of radiation-vulcanized NR/BR blending system. Radiat. Phys. Chem. **63**, 497–500 (2002)
95. Basfar, A.A., Silverman, J.: Improved ozone resistance of styrene-butadiene rubber cured by a combination of sulfur and ionizing radiation. J.Polym. Degr. Stabil. **46**, 1–8, (1994)
96. Jayasuriya, M.M., Makuuchi, K., Yoshi, F.: Radiation vulcanization of natural rubber latexusing TMPTMA and PEA. Eur. Polym. J. **37**, 93–98, (2001)
97. Ratnam, C.T., Nasir, M., Baharin, A., Zaman K.: Electron beam irradiation of epoxidized natural rubber. Nucl. Instrum. Meth. B **171**, 455–464 (2000)
98. Ratnam, C.T., Nasir, M., Baharin, A., Zaman, K.: Evidence of irradiation induced crosslinking in miscible blends of poly(vinyl chloride)/epoxidized natural rubber in presence of trimethylolpropane triacrylate. J. Appl. Polym. Sci. **81**, 1914–1925 (2001)
99. Ratnam, C.T., Nasir, M., Baharin, A., Zaman, K.: Electron-beam irradiation of poly(vinyl chloride)/epoxidized natural rubber blends in presence of trimethylolpropane triacrylate. J. Appl. Polym. Sci. **81**, 1926–1935 (2001)
100. Chattopadhyay, S., Chaki, T.K., Bhowmick, A.K.: Structural characterization of electron-beam crosslinked thermoplastic elastomeric films from blends of polyethylene and ethylene-vinyl acetate copolymers. J. Appl. Polym. Sci. **81**, 1936–1950 (2001)
101. Davenas, J., Stevenson, I., Celette, N., Vigier, G., David L.: Influence of the molecularmodifications on the properties of EPDM elastomers under irradiation. Nucl. Instrum. Meth. B**208**, 461–465, (2003)
102. Żuchowska, D., Zagórski, Z.P., Przybytniak, G.K., Rafalski, A.: Influence of butadiene/styrene copolymers on the modification of polypropylene in electron beam irradiation. Int. J. Polymeric Mater. **52**, 335–344, (2003)
103. Magda, M.: Abou Zeid, Radiation effect on properties of carbon black filled NBR/EPDM rubber blends. Europ. Polym. J. **43**, 4415–4422 (2007)
104. Senna, M.M.H., Abdel-Monem, Y.K.: Effect of electron beam irradiation and reactive compatibilizers on some properties of polypropylene and epoxidized natural rubber polymer blends. J. Elastomers Plast. **42**, 275–295 (2010)
105. Senna, M.M., Abdel Fatah, A.A., Manem, A.: Spectroscopic analysis and mechanical properties of elektron beam irradiate popylpropylene/epoxidized natura rubber (PP/ENR) polymer blends. Nucl. Instr. Meth. B **266**, 2599–2609 (2008)
106. Perraud, S., Vallat, M.-F., Kuczynski, J.: Radiation crosslinking of poly(ethylene-co-octene) with electron beam radiation. Macromol. Mater. Eng. **288**, 117–123 (2003)

107. George, S.C., Thomas, S.: Transport phenomena through polymer systems. Prog. Polym. Sci. **26**, 985–1017, (2001)
108. Jonquieres, A., Clément, R., Lonchon, P.: Permeability of block copolymers to vapors and liquids. Prog. Polym. Sci. **27**, 1803–1877, (2002)
109. Das, P.K., Ganguly, A., Banerji, M.: Electron-beam curing of hydrogenated acrylonitrile–butadiene rubber. J. Appl. Polym. Sci. **97**, 648–651, (2005)
110. Zhao, W., Yu, L., Zhong, X., Zhang, Y., Sun J.: Radiation vulcanization of hydrogenated acrylonitrile butadiene rubber (HNBR). J. Appl. Polym. Sci. **54**, 1199–1205, (1994)
111. Vallat, M.-F., Perraud, S., Kuczynski, J.: Curing of polymer blends by electron beam, Poster P-155, 5th International Symposium on Ionizing Radiation and Polymers, IRaP 2002, Canada Sept 2002, full text not published
112. Hill, D.J.T., O'Donnell, J.H., Perera, M.C.S., Pomery, P.J.: An investigation of radiation induced structural changes in nitrile rubber. J. Polym. Sci. A: Polym. Chem. **34**, 2439-2454 (1996)
113. Wang, Q., Wang, F., Cheng, K.: Effect of crosslink density on some properties of electron beam-irradiated styrene-butadiene rubber. Rad. Phys. Chem. **78**, 10001–11005 (2009)
114. Hafezi, M., Nouri Khorasani, S., Ziali, F., Zim, H.R.: Comparison of physicochemical properties of NBR-PV blend cured by sulfur and electron beam. J. Elastomers Plast. **39**, 151–163 (2007)
115. Mitra, S., Chattopadhyay, S., Bharadwaj, Y.K., Sabharwal, S., Bhowmick, A.K.: Effect of electron beam-crossinked rels on the rheological properties of raw natural rubber. Rad. Phys. Chem. **77**, 630–642 (2008)
116. Gluszewski W., Zagórski Z.P., Rajkiewicz M., Synergistic effects in the processes of crosslinking of elastomers, Summary published in book of abstracts for the Conference 10th IRAP 2012, submitted for publication

# Electro-Elastic Continuum Models for Electrostrictive Elastomers

**R. Vertechy, G. Berselli, V. Parenti Castelli and M. Bergamasco**

**Abstract** A continuum finite-deformation model is described for the study of the isothermal electro-elastic deformations of electrostrictive elastomers. The model comprises general balance equations of motion, electrostatics and electro-mechanical energy, and together with phenomenological invariant-based constitutive relations. The model is presented in both Eulerian (spatial) and Lagrangian (material) description. Specialization of the considered model is also presented for "Dielectric Elastomers", which are a specific class of electrostrictive elastomers having dielectric properties independent of deformation.

## 1 Introduction

Electrostrictive elastomers are a special kind of "electroactive polymeric material" whose electrical and structural behavior is highly non-linear and strongly coupled. In essence, they are rubber-like dielectric solids that experience large finite deformations in response to applied electric fields while, at the same time, alter the existing electrostatic fields in response to the deformations undergone.

---

R. Vertechy (✉) · M. Bergamasco
Scuola Superiore Sant'Anna, Pisa, Italy
e-mail: r.vertechy@sssup.it

M. Bergamasco
e-mail: m.bergamasco@sssup.it

G. Berselli
University of Modena and Reggio, Emilia, Italy
e-mail: giovanni.berselli@unimore.it

V. P. Castelli
University of Bologna, Bologna, Italy
e-mail: vincenzo.parenti@unibo.it

P. M. Visakh et al. (eds.), *Advances in Elastomers I*,
Advanced Structured Materials 11, DOI: 10.1007/978-3-642-20925-3_14,

Practical electrostrictive elastomers are polyurethanes [1–3], ferroelectric polymers [4, 5], graft elastomers [6], silicone and polyacrylate elastomers [7–11], styrene-butadiene and styrene-isoprene-styrene elastomers [11], liquid-crystal elastomers [12], interpenetrating-polymer-network elastomers and nano-structured rubber [13–15], as well as a number of other elastomeric composites employing either conducting, semiconducting or ferroelectric particles as fillers [11, 16–20]. Typical stresses and strains that can be induced in electrostrictive elastomers via electrical activation are [6, 7, 21]: 2 MPa and 11 % for polyurethanes; 54 Mpa and 7 % for ferroelectric elastomers; 4 MPa and 4 % for graft elastomers; 3 MPa and 120 % for silicone elastomers; 7 MPa and 380 % for polyacrylate elastomers; 1 MPa and 4 % for liquid-crystal elastomers. Typical rates of such electrically-induced strains are larger than 1,000 %/s for almost all electrostrictive elastomers [21].

Thanks to the peculiar electromechanical coupling, along with the intrinsic compliance, lightness, malleability, easy-manufacturability and low-cost, electrostrictive elastomers are perfectly suited for the development of novel solid-state mechatronic transducers which are more resilient, lightweight, integrated, economic and disposable than traditional devices obtainable via conventional material technologies. Practical transduction devices that can be developed by using electrostrictive elastomers are [22–27]: compliant muscle-like actuators; compact and portable Braille displays; distributed force/displacement sensors; solid-state electrical energy harvester/generators; large solid-state loud-speakers. Potential applications are in the field of machine interfaces for human assistance and entertainment, safe and low-cost robotics, as well as disposable and consumable mechatronics.

To give some example, a linear actuator prototype based on an electrostrictive elastomer is shown in Fig. 1. As depicted in Fig. 2, the device is a lozenge-shaped planar actuator featuring two electroactive sheets that are connected to the opposing sides of a four-bar mechanism having identical rigid links and tension-tape hinges. Each electroactive sheet is made of three compliant electrodes made of a carbon conductive grease, which are separated by two acrylic elastomer films

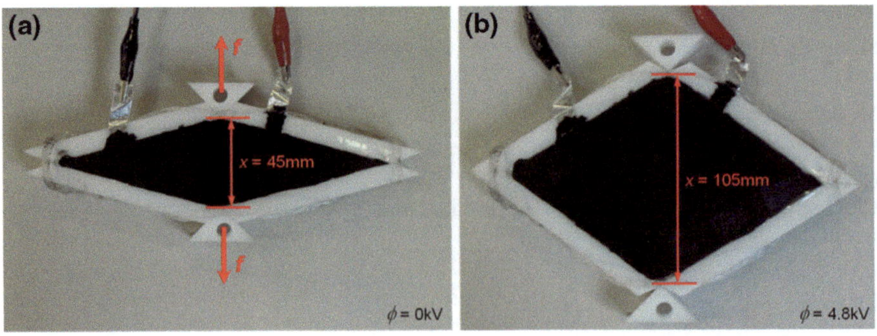


**Fig. 1** Lozenge actuator prototype: **a** actuator inactive and **b** actuator active

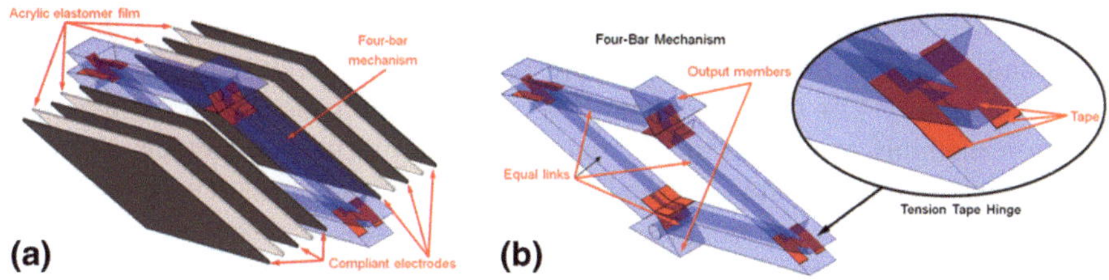


**Fig. 2** Lozenge actuator CAD drawings: **a** actuator exploded view and **b** four-bar mechanism

that are bi-axially pre-stretched and connected along their boundary to the links of the four-bar mechanism. The placement of an electric potential difference (hereafter also called voltage) between the inner and the outer electrodes of the stack induces the acrylic elastomer films to shrink in thickness and to expand in area (see Fig. 1). In this device, the four-bar mechanism constrains the area expansion of the acrylic elastomer films to be uniform and enables the transmission to the film boundary of the useful mechanical work which can be performed by any external force acting on the mechanism links. Further details concerning the design and the performance of this linear actuator can be found in [28].

Typical force-length characteristic curves of this lozenge actuator are shown in Fig. 3 for different activation voltages $\phi$, where the length $x$ is the distance between the axes of the two joints of the four-bar mechanism that lie on one of the lozenge diagonals and the force $f$ is the external equilibrating force acting on the same joint axes with direction equaling that of the same lozenge diagonal (see Fig. 1a). As shown in Fig. 3, the electrically-induced deformations occurring in electrostrictive elastomers can be quite large and rather non-linear. Owing to this complex multi-physical behavior, the design and the control of practical devices based on electrostrictive elastomers are extremely challenging problems. Intuitive approaches are usually defied, which calls for the availability of model-based engineering simulation tools.

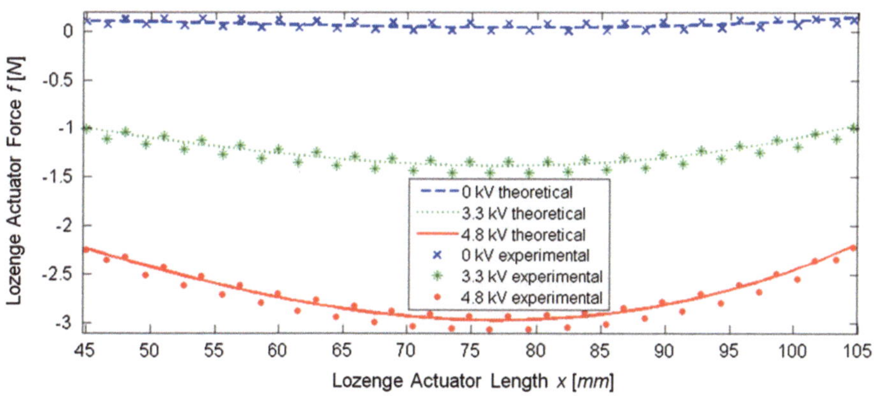


**Fig. 3** Force-length curves of the lozenge actuator

The characterization of the electromechanical interaction between a dielectric body and an electric field has been a topic of intensive research since the end of the nineteen century [29–67]. Based on diverse interaction models, postulates, and solution methods, a number of continuum electro-mechanical models with dissimilar expressions for the electrically-induced force and stress-tensor have been proposed by different authors [29–67], which have been a subject of controversy for over a century [36, 50, 52, 53, 59, 64, 66, 68–71]. In particular: some authors directly postulated a particular expression for the electrically-induced body force and/or stress tensor, and simply added them to the continuum balance law of linear mechanical momentum (for instance in [34, 38, 40, 41]); other authors postulated elementary models, such as charges and dipoles, which characterize the macroscopic behavior of the polarized material (for instance in [40, 45, 49, 51]), thus deriving an expression of the electrically-induced body force and/or stress tensor from them; other authors postulated that the interaction at the microscopic scale is characterized by electrons moving in ether, thus obtaining the electrically-induced force and/or stress tensor by a statistical averaging process (for instance in [33, 47, 54]); other authors postulated a particular form of the global energy balance from which the coupled balance laws of mass, momenta (directly including the electrically-induced forces and/or stress tensor) and electricity are obtained by imposing certain invariance properties (for instance in [45, 52, 53, 55]); other authors postulated a given form of the electric enthalpy from which the coupled balance laws of linear momentum (directly including the electrically-induced forces and/or stress tensor) and electricity are obtained via some variational principle (for instance [38, 56, 63, 66, 67]).

In this context, this chapter presents a finite-deformation electro-elastic continuum model for the study of the isothermal electrically-induced deformations of electrostrictive elastomers that are conservative and isotropic. The model relies on the standard balance laws of mass, momenta (which do not explicitly include any electrically-induced force and stress tensor term) and electrostatics, and accounts for the electromechanical coupling via an appropriately chosen phenomenological constitutive relation which is consistent with the principles of thermodynamics and based on the theory of invariants. Specialization of the constitutive theory is also presented for the so called "Dielectric Elastomers" [25], which are a special kind of electrostrictive elastomers whose dielectric properties are deformation independent. The model is formulated both in the Eulerian (spatial) description, which enables for a more immediate comprehension of the electro-elastic phenomena that occur in electrostrictive elastomers, and in the Lagrangian (material) description, which is more suited for the numerical simulation of virtual prototypes of electrostrictive-elastomer-based devices via engineering tools alike finite element analysis software. For further reference, the finite element implementation of the considered model is discussed in [72], whereas its extension to the study of the thermo-electro-elastic deformations of electrostrictive elastomers is presented in [73].

## 2 Kinematics of Continuous Media

Consider the closed and electrically isolated system $\mathcal{B}$ depicted in Fig. 4, which comprises dielectric and conducting bodies as well as free space. Both dielectric and conducting bodies can move and deform under the action of externally applied loads of both mechanical and electrical origin. During such body motions/deformations, the physical region occupied by $\mathcal{B}$ does not remain fixed, but it moves/deforms accordingly. However, while these body motions/deformations occur: (1) no mass can enter or leave the boundary of $\mathcal{B}$; (2) energy (in the form of work or heat) can cross the boundary of $\mathcal{B}$; (3) no interaction occurs between the electrical charges that lay within $\mathcal{B}$ and those outside [i.e. the boundary of $\mathcal{B}$ either is electrically shielded from its exterior or has an infinite extent].

In this perspective, define with $\Im_l$ the known reference (laboratory) frame with respect to which the motions/deformations of $\mathcal{B}$ are measured. For any arbitrary time instant $t \geq 0$, identify the moving/deforming region occupied by an arbitrary inner subsystem of $\mathcal{B}$ with the current (deformed) volume $\Omega(t)$ and its boundary surface $\partial\Omega(t)$; $\bar{\Omega} = \Omega(t = \bar{t} \equiv 0)$ and $\partial\bar{\Omega} = \partial\Omega(t = \bar{t} \equiv 0)$ being the referential (undeformed) volume and boundary surface respectively.

Consider now a general material point $P$ (i.e. a particle) belonging to the arbitrary subsystem of $\mathcal{B}$, and indicate with $X$ and $x(t)$ the position vectors expressing the location (relative to the origin $O$ of $\Im_l$) occupied by $P$ when the system $\mathcal{B}$ is in its undeformed (i.e. $t = \bar{t} \equiv 0$) and deformed configuration respectively. Then, for any $\boldsymbol{X} \in \bar{\Omega}$ and any $t \geq 0$, the motion of $\mathcal{B}$ can be described by

$$\boldsymbol{x} = \boldsymbol{\chi}(\boldsymbol{X}, t), \tag{1}$$

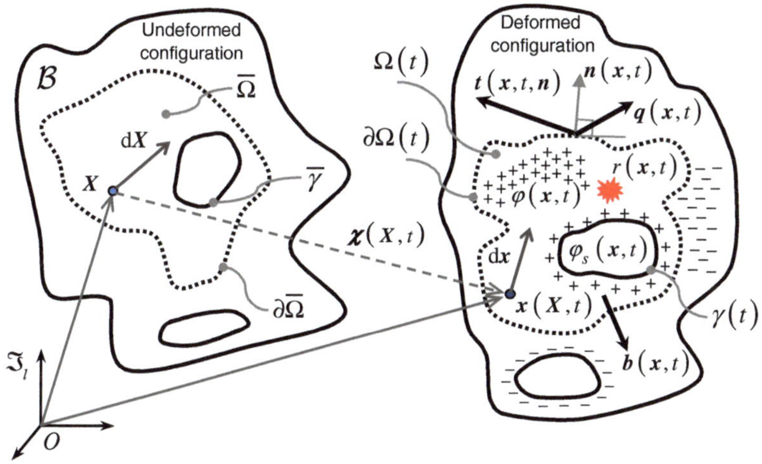


**Fig. 4** Electrostrictive elastomer system: undeformed and deformed configuration

Where the map $\chi(X, t)$ is a suitably regular and single-valued vector field (i.e. it is invertible and possesses continuous time derivatives with respect to both position and time) that carries all points $P$ from places $X$ (within the referential configuration $\bar{\Omega}$) to places $x$ (in the current configuration $\Omega(t)$ at time $t$). By definition, the map $\chi$ needs to satisfy the two conditions $X = \chi(X, 0)$ and $X = \chi^{-1}(x, t)$.

Time differentiation of the map described by Eq. (1) provides the velocity field of the arbitrary subsystem of $\mathcal{B}$ expressed in either the Lagrangian form, $V(X, t)$ (i.e. the material description with respect to the coordinate $X$ in the referential configuration), or the Eulerian form, $v(x, t)$ (i.e. the spatial description with respect to the coordinate $x$ in the current configuration),

$$V(X, t) = \frac{\partial \chi(X, t)}{\partial t}, \tag{2}$$

$$v(x, t) = V(\chi^{-1}(x, t), t). \tag{3}$$

The material gradient (i.e. the derivative with respect to the referential coordinate $X$) and the spatial gradient (i.e. the derivative with respect to the spatial coordinate $x$) respectively yield the deformation gradient, $\mathbf{F}(X, t)$, and its inverse, $\mathbf{F}^{-1}(x, t)$,

$$\mathbf{F} = \mathrm{Grad}x = \frac{\partial \chi(X, t)}{\partial X}, \tag{4}$$

$$\mathbf{F}^{-1} = \mathrm{grad}X = \frac{\partial \chi^{-1}(x, t)}{\partial x}, \tag{5}$$

which essentially provide the map

$$\mathrm{d}x = \mathbf{F}\mathrm{d}X, \tag{6}$$

between the undeformed line element, $\mathrm{d}X$, and the deformed line element $\mathrm{d}x$. Equation (6) describes the change in length and orientation (i.e. a rigid body rotation) of a physical line element when system $\mathcal{B}$ passes from the undeformed to the deformed configuration. Whenever the rigid body rotation is not of interest, one may conveniently resort to the quadratic forms

$$|\mathrm{d}x|^2 = \mathrm{d}X \cdot \mathbf{C}\mathrm{d}X, \tag{7}$$

$$\mathbf{C} = \mathbf{F}^T\mathbf{F}, \tag{8}$$

$$|\mathrm{d}X|^2 = \mathrm{d}x \cdot \mathbf{B}^{-1}\mathrm{d}x, \tag{9}$$

$$\mathbf{B} = \mathbf{F}\mathbf{F}^T, \tag{10}$$

where $\mathbf{C}$ and $\mathbf{B}$ are symmetric tensors usually known as the Green deformation tensor and the Finger deformation tensor respectively.

The determinant of the deformation gradient gives the Jacobian, $J(\boldsymbol{X}, t)$,

$$J = \det \mathbf{F} > 0, \tag{11}$$

which is also known as the volume ratio because of the relationship

$$\mathrm{dv} = J\mathrm{dV}, \tag{12}$$

between the infinitesimal undeformed volume element, dV, and the infinitesimal deformed volume element, dv.

By considering

$$\mathrm{dV} = \mathrm{d}\boldsymbol{X} \cdot \boldsymbol{N}\mathrm{dS}, \tag{13}$$

$$\mathrm{dv} = \mathrm{d}\boldsymbol{x} \cdot \boldsymbol{n}\mathrm{ds}, \tag{14}$$

where dS and ds are the infinitesimal undeformed and deformed surface elements, with $\boldsymbol{N}$ and $\boldsymbol{n}$ being the respective unit vector normals, use of Eqs. (6) and (12) gives the Nanson's formula

$$\boldsymbol{n}\mathrm{ds} = J\mathbf{F}^{-T}\boldsymbol{N}\mathrm{dS}. \tag{15}$$

Regarding the differential kinematics, time-differentiation of Eq. (4) and use of the Schwarz's theorem together with the definition given in Eq. (2) provides

$$\mathrm{Grad}\bar{\boldsymbol{v}} = \dot{\mathbf{F}}, \tag{16}$$

which, because of the chain rule differentiation, makes it possible to write

$$\mathrm{grad}\boldsymbol{v} = \mathrm{Grad}\bar{\boldsymbol{v}}\mathbf{F}^{-1} = \dot{\mathbf{F}}\mathbf{F}^{-1}. \tag{17}$$

Moreover, the derivative of the Jacobian with respect to the deformation gradient yields

$$\frac{\partial J}{\partial \mathbf{F}} = J\mathbf{F}^{-T}, \tag{18}$$

which, by the chain rule differentiation, makes it possible to write

$$\dot{J} = J\mathbf{F}^{-T} : \dot{\mathbf{F}} = J\mathbf{1} : \mathrm{grad}\boldsymbol{v} = J\mathrm{div}\boldsymbol{v}, \tag{19}$$

where $\mathbf{1}$ is the $3 \times 3$ identity matrix, while the operator (:) indicates the double contraction of two tensors [as defined in Eq. (185)].

Finally, by time-differentiating the identity $\mathbf{F}^{-1}\mathbf{F} = \mathbf{1}$, the derivative of the inverse of the deformation gradient follows as

$$\frac{\mathrm{d}\mathbf{F}^{-1}}{\mathrm{d}t} = -\mathbf{F}^{-1}\dot{\mathbf{F}}\mathbf{F}^{-1} = -\mathbf{F}^{-1}\mathrm{grad}\boldsymbol{v}. \tag{20}$$

# 3 Conservation of Mass

Under the assumption that the arbitrary inner subsystem of $\mathcal{B}$ (which has been depicted in Fig. 4) is a closed system, no mass can enter or leave the subsystem boundary $\partial\Omega(t)$. That is, the overall mass $m$ of the volume $\Omega(t)$ must remain constant at any time instant $t \geq 0$. In mathematical terms, this requirement reads as

$$\dot{m} = \frac{\mathrm{d}}{\mathrm{d}t} \int\limits_{\Omega(t)} \rho \mathrm{dv} = 0, \tag{21}$$

where $\rho(x,t)$ is the spatial mass density. Equation (21) represents the Continuity Mass Equation (CME) expressed in global form and in the current configuration.

## *3.1 Continuity Mass Equation in Eulerian Description*

Using the Reynold's transport theorem [Eq. (207)] and since $\Omega(t)$ is an arbitrary volume, Eq. (21) can also be written as

$$\frac{\partial \rho}{\partial t} + \mathrm{div}(\boldsymbol{v}\rho) = \dot{\rho} + \rho \mathrm{div}(\boldsymbol{v}) = 0, \tag{22}$$

corresponding to the local (rate) form of the CME in the Eulerian (spatial) description.

### 3.1.1 Continuity Mass Equation in Lagrangian Description

Resorting to Eqs. (12, 21) can be rewritten in the referential configuration as

$$\dot{m} = \frac{\mathrm{d}}{\mathrm{d}t} \int\limits_{\bar{\Omega}} \bar{\rho} \mathrm{dV} = \int\limits_{\bar{\Omega}} \dot{\bar{\rho}} \mathrm{dV} = 0, \tag{23}$$

where $\bar{\rho}(X)$ is defined as the reference mass density

$$\bar{\rho} = J\rho. \tag{24}$$

By considering that $\bar{\Omega}$ is an arbitrary volume, Eq. (23) immediately yields

$$\dot{\bar{\rho}} = 0, \tag{25}$$

representing the local (rate) form of the CME in the Lagrangian (material) description.

# 4 Balance of Mechanical Momentum

In the framework of Newtonian mechanics, the motion and deformation of the bodies contained in any arbitrary closed subsystem of $\mathcal{B}$ (which has been depicted in Fig. 4) require the rate of change of the associated mechanical momentum to be equivalent to the ensemble of forces acting on the enclosed bodies themselves. In mathematical terms, given the definitions of linear momentum, $\boldsymbol{L}(t)$,

$$\boldsymbol{L}(t) = \int\limits_{\Omega(t)} \rho \boldsymbol{v} \mathrm{dv}, \tag{26}$$

and of angular momentum, $\boldsymbol{A}_0(t)$, with respect to a general point identified by the position vector $\boldsymbol{x}_0$

$$\boldsymbol{A}_0(t) = \int\limits_{\Omega(t)} \boldsymbol{r} \times \boldsymbol{v} \rho \mathrm{dv}, \tag{27}$$

with $\boldsymbol{r} = \boldsymbol{x} - \boldsymbol{x}_0$, the equivalence requirement stated above amounts to the two following momentum balance principles

$$\frac{\mathrm{d}}{\mathrm{d}t}\boldsymbol{L} = \int\limits_{\partial\Omega(t)} \boldsymbol{t} \mathrm{ds} + \int\limits_{\Omega(t)} \boldsymbol{f} \mathrm{dv}, \tag{28}$$

$$\frac{\mathrm{d}}{\mathrm{d}t}\boldsymbol{A}_0 = \int\limits_{\partial\Omega(t)} \boldsymbol{r} \times \boldsymbol{t} \mathrm{ds} + \int\limits_{\Omega(t)} \boldsymbol{r} \times \boldsymbol{f} \mathrm{dv}, \tag{29}$$

where $\boldsymbol{f}(\boldsymbol{x}, t)$ is the (spatial) body force density measured per unit of the current volume $\Omega(t)$, while $\boldsymbol{t}(\boldsymbol{x}, t, \boldsymbol{n})$ is the Cauchy (or true) traction vector (a force measured per unit surface area defined in the deformed configuration) acting on the current surface $\partial\Omega(t)$ and having unit normal $\boldsymbol{n}(\boldsymbol{x}, t)$. In particular, Eqs. (28) and (29) respectively represent the Balance of Linear Momentum (BLM) and the Balance of Moment of Momentum (BMM), both expressed in global forms and in the current configuration.

## *4.1 Cauchy's First Equation of Motion in Eulerian Description*

Consider the Cauchy's Stress Theorem (in the spatial description)

$$\boldsymbol{t}(\boldsymbol{x}, t, \boldsymbol{n}) = \boldsymbol{\sigma} \cdot \boldsymbol{n}, \tag{30}$$

where $\boldsymbol{\sigma}(\boldsymbol{x}, t)$ denotes a spatial tensor field called the Cauchy (or true) stress tensor and with the scalar product between tensor $\boldsymbol{\sigma}$ and vector $\boldsymbol{n}$ being defined as in Eq. (183). Then, by employing the Reynold's transport theorem [Eq. (207)] and Eq. (22) to its left-hand side, and the Cauchy's Stress Theorem [Eq. (30)] and the divergence theorem [Eq. (204)] to the first term of its right-hand side, Eq. (26) becomes

$$\int_{\Omega(t)} (\text{div}\boldsymbol{\sigma} + \boldsymbol{f} - \rho\dot{\boldsymbol{v}})\text{dv} = \mathbf{0}, \tag{31}$$

which represents the global form of the Cauchy's First Equation of Motion (CFEM) in the current configuration. By considering that $\Omega(t)$ is an arbitrary volume, Eq. (31) implies

$$\text{div}\boldsymbol{\sigma} + \boldsymbol{f} - \rho\dot{\boldsymbol{v}} = \mathbf{0}, \tag{32}$$

that corresponds to the local (rate) form of CFEM in the Eulerian description.

## 4.2 Cauchy's First Equation of Motion in Lagrangian Description

Resorting to the volume ratio relationship (Eq. 12), Eq. (28) can be rewritten in the referential configuration as

$$\frac{\text{d}}{\text{d}t}\int_{\bar{\Omega}} \bar{\rho}\boldsymbol{V}\text{dV} = \int_{\partial\bar{\Omega}} \boldsymbol{T}\text{dS} + \int_{\bar{\Omega}} \bar{\boldsymbol{f}}\text{dV}, \tag{33}$$

where $\bar{\boldsymbol{f}}(\boldsymbol{X}, t)$ is defined as the reference body force

$$\bar{\boldsymbol{f}} = J\boldsymbol{f}, \tag{34}$$

where as $\boldsymbol{T}(\boldsymbol{X}, t, \boldsymbol{N})$ represents the first Piola-Kirchhoff (or nominal) traction vector (a force measured per unit area defined in the undeformed configuration) satisfying the relation

$$\boldsymbol{t}\text{ds} = \boldsymbol{T}\text{dS}. \tag{35}$$

That is, by definition, the direction of the first Piola-Kirchhoff traction vector, $\boldsymbol{T}$, coincides to that of the Cauchy traction vector, $\boldsymbol{t}$.

Then, resorting to Eq. (25) on the left-hand side, and to both the Cauchy's Stress Theorem (in the material description)

$$\boldsymbol{T}(\boldsymbol{X}, t, \boldsymbol{N}) = \boldsymbol{\Pi} \cdot \boldsymbol{N} \tag{36}$$

where $\mathbf{\Pi}(\boldsymbol{X},\boldsymbol{t})$ denotes a material tensor field called the Nominal stress tensor, and the divergence theorem [Eq. (204)] on the first term of the right-hand side, Eq. (33) can also be written as

$$\int_{\bar{\Omega}} \left(\mathrm{Div}\mathbf{\Pi} + \bar{\boldsymbol{f}} - \bar{\rho}\dot{\boldsymbol{V}}\right)\mathrm{dV} = \mathbf{0}, \tag{37}$$

which represents the global form of the CFEM in the referential configuration. Since $\bar{\Omega}$ is an arbitrary volume, Eq. (37) entails

$$\mathrm{Div}\mathbf{\Pi} + \bar{\boldsymbol{f}} - \bar{\rho}\dot{\boldsymbol{V}}, \tag{38}$$

which corresponds to the local (rate) form of CFEM in the Lagrangian description.

## 4.3 Cauchy's Second Equation of Motion in Eulerian Description

By the Reynold's transport theorem [Eq. (207)], Eq. (22) and the equivalence $\dot{\boldsymbol{r}} = \dot{\boldsymbol{x}} - \dot{\boldsymbol{x}}_0 = \boldsymbol{v}$ (since $\dot{\boldsymbol{x}}_0 = \mathbf{0}$), the rate of the angular momentum $\boldsymbol{A}_0(t)$ which is required on the left-hand side of Eq. (29) reduces to

$$\frac{\mathrm{d}}{\mathrm{d}t}\boldsymbol{A}_0 = \int_{\Omega(t)} (\dot{\boldsymbol{r}} \times \boldsymbol{v} + \boldsymbol{r} \times \dot{\boldsymbol{v}})\rho\mathrm{dv} = \int_{\Omega(t)} \boldsymbol{r} \times \dot{\boldsymbol{v}}\rho\mathrm{dv}. \tag{39}$$

Besides, by Eq. (30) and the divergence theorem [Eq. (204)], the first term on the right-hand side of Eq. (29) becomes

$$\int_{\partial\Omega(t)} \boldsymbol{r} \times \boldsymbol{t}\mathrm{ds} = \int_{\partial\Omega(t)} \boldsymbol{r} \times (\boldsymbol{\sigma} \cdot \boldsymbol{n})\mathrm{ds} = \int_{\Omega(t)} (\boldsymbol{r} \times \mathrm{div}\boldsymbol{\sigma} + \mathcal{E} : \boldsymbol{\sigma})\mathrm{dv}, \tag{40}$$

where $\mathcal{E}$ indicates the third-order Levi-Civita (or permutation) tensor

$$\mathcal{E} = \left[\varepsilon_{ijk}\right], \tag{41}$$

with $\varepsilon_{ijk}$ being the permutation symbol defined according to Eq. (191).

Then, use of Eqs. (39) and (40), together with Eq. (32), enables to simplify Eq. (29) as

$$\int_{\Omega(t)} \mathcal{E} : \boldsymbol{\sigma}\mathrm{dv} = \mathbf{0}, \tag{42}$$

representing the global form of the Cauchy's Second Equation of Motion (CSEM) in the current configuration. By considering that $\Omega(t)$ is an arbitrary volume, Eq. (42) gives

$$\mathcal{E} : \boldsymbol{\sigma} = \mathbf{0}, \tag{43}$$

or identically

$$\boldsymbol{\sigma} = \boldsymbol{\sigma}^T, \tag{44}$$

which implies the symmetry of the Cauchy stress tensor $\boldsymbol{\sigma}$.

## 4.4 Cauchy's Second Equation of Motion in Lagrangian Description

According to the definitions introduced in the previous sections, Eq. (29) can be converted in the referential configuration as

$$\frac{\mathrm{d}}{\mathrm{d}t}\int_{\bar{\Omega}} \boldsymbol{r} \times \boldsymbol{V}\bar{\rho}\mathrm{dV} = \int_{\bar{\Omega}} \boldsymbol{r} \times \dot{\boldsymbol{V}}\bar{\rho}\mathrm{dV} = \int_{\partial\bar{\Omega}} \boldsymbol{r} \times \boldsymbol{T}\boldsymbol{d}\boldsymbol{S} + \int_{\bar{\Omega}} \boldsymbol{r} \times \bar{\boldsymbol{f}}\boldsymbol{d}\boldsymbol{V}. \tag{45}$$

Besides, owing to Eq. (36) and the divergence theorem [Eq. (204)], and by the chain-rule differentiation with $\mathrm{Grad}\boldsymbol{r} = \mathrm{Grad}(\boldsymbol{x} - \boldsymbol{x_0}) = \mathbf{F}$, the first term on the right-hand side of Eq. (45) can be rewritten as

$$\int_{\partial\Omega(t)} r \times t\mathrm{ds} = \int_{\partial\Omega(t)} r \times (\sigma \cdot n)\mathrm{ds} = \int_{\Omega(t)} (r \times \mathrm{div}\sigma + \mathcal{E} : \sigma)\mathrm{dv} \tag{46}$$

As a consequence, with the use of Eqs. (46) and (38), Eq. (45) reduces to

$$\int_{\bar{\Omega}} \mathcal{E} : (\mathbf{F\Pi})\mathrm{dV} = \mathbf{0}, \tag{47}$$

which, under the assumption that $\bar{\Omega}$ is an arbitrary volume, gives

$$\mathcal{E} : (\mathbf{F\Pi}) = \mathbf{0}, \tag{48}$$

or identically

$$\mathbf{F\Pi} = \mathbf{\Pi}^T\mathbf{F}^T, \tag{49}$$

highlighting that differently than the Cauchy stress tensor, $\boldsymbol{\sigma}$, the Nominal stress tensor, $\mathbf{\Pi}$, is generally not symmetric. This is indeed confirmed by the Piola transformation

$$\mathbf{\Pi} = J\mathbf{F}^{-1}\boldsymbol{\sigma}, \tag{50}$$

which can be derived by combining Eqs. (35), (30), and (36), together with the Nanson's formula (Eq. 15).

# 5 Electrostatic Interactions

In the framework of Maxwellian electromagnetism [29], assuming that any existing magnetic field is stationary, the electrical phenomena acting between the dielectric and the conducting bodies enclosed in the system depicted in Fig. 4 can be described by the two following laws: the simplified version of Faraday's law

$$\int_{S(t)} \mathrm{rot}\boldsymbol{E} \cdot \boldsymbol{n}\mathrm{ds} = 0, \tag{51}$$

where $\boldsymbol{E}(x,t)$ is the electric field acting on any arbitrary spatial open surface $S(t)$ that belongs to $\mathcal{B}$, and the Gauss's law

$$\int_{\partial\Omega(t)-\gamma(t)} \boldsymbol{D} \cdot \boldsymbol{n}\mathrm{ds} = \int_{\Omega(t)-\gamma(t)} \varphi \mathrm{dv} + \int_{\gamma(t)} \varphi_s \mathrm{dv}, \tag{52}$$

with

$$\boldsymbol{D} = \boldsymbol{D}_0 + \boldsymbol{P}, \tag{53}$$

$$\boldsymbol{D}_0 = \varepsilon_0 \boldsymbol{E}, \tag{54}$$

where $\varepsilon_0 = 8.85\, e^{-12} F/m$ is the constant vacuum permittivity, $\boldsymbol{P}(\boldsymbol{x},t)$ and $\boldsymbol{D}(\boldsymbol{x},t)$ are the electric polarization and the electric displacement permeating any arbitrary volume $\Omega(t)$ that belongs to $\mathcal{B}$, whereas $\varphi(\boldsymbol{x},t)$ and $\varphi_s(\boldsymbol{x},t)$ are the free-charge densities per unit of spatial volume and per unit of spatial surface respectively. Specifically, $\varphi$ resides within the deformed volume $\Omega(t)$ (usually in the form of injected electrons and/or ions), whereas $\varphi_s$ i.e., on some deformed discontinuity surface $\gamma(t)$ (usually a conducting surface electrode) comprised in $\Omega(t)$.

## 5.1 Balance Law of Electrostatic in Eulerian Description

By considering that $S(t)$ is a general open surface, Eq. (51) immediately yields

$$\mathrm{rot}\boldsymbol{E} = 0. \tag{55}$$

That is, the Faraday's law directly implies the existence of an electric potential field, $\phi(\boldsymbol{x},t)$, such that

$$\boldsymbol{E} = -\mathrm{grad}(\phi). \tag{56}$$

Besides, using the divergence theorem [Eq. (205)] to the left-hand side of Eq. (52), the Gauss's law can be rewritten as

$$\int_{\Omega(t)-\gamma(t)} \mathrm{div}\boldsymbol{D}\mathrm{dv} + \int_{\gamma(t)} [\![\boldsymbol{D}]\!] \cdot \boldsymbol{n}\mathrm{ds} = \int_{\Omega(t)-\gamma(t)} \varphi\mathrm{dv} + \int_{\gamma(t)} \varphi_s\mathrm{ds}, \quad (57)$$

where $[\![\boldsymbol{D}]\!] \equiv (\boldsymbol{D}^+ - \boldsymbol{D}^-)$ indicates the jump of the quantity $\boldsymbol{D}$ from the positive (+) side to the negative (−) side of the discontinuity surface $\gamma(t)$. By considering that $\Omega(t)$ and $\gamma(t)$ are arbitrary volumes and surfaces, Eq. (57) immediately implies

$$\mathrm{div}\boldsymbol{D} = \varphi, \quad (58)$$

$$[\![\boldsymbol{D}]\!] \cdot \boldsymbol{n} = \varphi_s. \quad (59)$$

Equations (58) and (59) correspond to the local form and the associated boundary condition of the Gauss's law expressed in the Eulerian description.

More concisely, by considering Eqs. (53), (54), (56), (58) and (59), the electric phenomena occurring within any arbitrary subsystem of $\mathcal{B}$ (such as the one depicted in Fig. 4) can be described by the following balance law (expressed in local form and in the Eulerian description)

$$\varepsilon_0 \mathrm{div}(\mathrm{grad}\phi) = \mathrm{div}\boldsymbol{P} - \varphi, \quad (60)$$

along with the associated boundary condition

$$[\![\boldsymbol{P} - \varepsilon_0 \mathrm{grad}\phi]\!] \cdot \boldsymbol{n} = \varphi_s. \quad (61)$$

## 5.2 Balance Law of Electrostatic in Lagrangian Description

Considering the Faraday's law, by the Stoke's theorem [Eq. (202)] and Eq. (6), the following holds for Eq. (51)

$$\int_{S(t)} \mathrm{rot}\boldsymbol{E} \cdot \boldsymbol{n}\mathrm{ds} = \int_{L(t)} \boldsymbol{E} \cdot \mathrm{d}\boldsymbol{x} = \int_{\bar{L}} \bar{\boldsymbol{E}} \cdot \mathrm{d}\boldsymbol{X} = \int_{\bar{S}} \mathrm{Rot}\bar{\boldsymbol{E}} \cdot \boldsymbol{N}\mathrm{dS} = 0, \quad (62)$$

where $L(t)$ is the closed spatial curve bounding the open spatial surface $S(t)$, $\bar{L}$ and $\bar{S}$ are the material counterparts of $L(t)$ and $S(t)$, while $\bar{\boldsymbol{E}}(\boldsymbol{X}, t)$ is defined as the reference electric field

$$\bar{\boldsymbol{E}} = \mathbf{F}^T \boldsymbol{E}, \quad (63)$$

From Eq. (56) and through the chain-rule differentiation, Eq. (63) yields

$$\bar{\boldsymbol{E}} = -\mathrm{Grad}\bar{\phi}, \quad (64)$$

where $\bar{\phi}(\boldsymbol{X}, t) = \phi(\boldsymbol{x}, t)$ is the electric potential expressed with respect to the referential configuration.

Considering the Gauss's law, by Eqs. (12) and (15), Eq. (52) can be rewritten as

$$\int_{\partial\bar{\Omega}-\bar{\gamma}} \bar{\boldsymbol{D}}\cdot\boldsymbol{N}\mathrm{dS} = \int_{\bar{\Omega}-\bar{\gamma}} \bar{\varphi}\mathrm{dV} + \int_{\bar{\gamma}} \bar{\varphi}_s\mathrm{dS}, \tag{65}$$

where $\bar{\gamma}$ is the material counterpart of the discontinuity surface $\gamma(t)$, whereas $\bar{\boldsymbol{D}}(\boldsymbol{X},t)$, $\bar{\boldsymbol{P}}(\boldsymbol{X},t)$, $\bar{\varphi}(\boldsymbol{X},t)$ and $\bar{\varphi}_s(\boldsymbol{X},t)$ are the reference electric displacement, the reference electric polarization and the reference free-charge densities respectively defined as

$$\bar{\boldsymbol{D}} = \bar{\boldsymbol{D}}_0 + \bar{\boldsymbol{P}} = J\mathbf{F}^{-1}\mathbf{D}, \tag{66}$$

$$\bar{\boldsymbol{P}} = J\mathbf{F}^{-1}\boldsymbol{P}, \tag{67}$$

$$\bar{\boldsymbol{D}}_0 = J\mathbf{F}^{-1}\boldsymbol{D}_0, \tag{68}$$

$$\bar{\varphi} = J\varphi, \tag{69}$$

$$\bar{\varphi}_s = \varphi_s\mathrm{ds}/\mathrm{dS}. \tag{70}$$

Then, use of the the divergence theorem [Eq. (205)] to the left-hand side of Eq. (65) and since $\bar{\Omega}$ and $\bar{\gamma}$ are arbitrary volumes and surfaces, the Lagrangian counterparts of Eqs. (58) and (59) follow as

$$\mathrm{Div}\bar{\boldsymbol{D}} = \bar{\varphi}, \tag{71}$$

$$[\![\bar{\boldsymbol{D}}]\!]\cdot\boldsymbol{N} = \bar{\varphi}_s. \tag{72}$$

Moreover, from Eqs. (68), (54) and (63), the Lagrangian counterpart of Eq. (54) is

$$\bar{\boldsymbol{D}}_0 = \varepsilon_0 J\mathbf{C}^{-1}\bar{\boldsymbol{E}}, \tag{73}$$

which highlights that the electrostatic analysis of a moving and deforming system of dielectric and conducting bodies contained in any arbitrary vacuum volume $\Omega(t)$ is equivalent to the study of a fixed system of dielectric and conducting bodies embedded in a media with fixed volume $\bar{\Omega}$ and characterized by an anisotropic and inhomogeneous dielectric tensor equaling $\varepsilon_0 J\mathbf{C}^{-1}$ [54].

Summarizing, by considering Eqs. (71), (72) (66), (73) and (64), the electric phenomena occurring within any arbitrary subsystem of $\mathcal{B}$ (such as the one depicted in Fig. 4) can be described by the following balance law (expressed in local form and in the Lagrangian description)

$$\varepsilon_0\mathrm{Div}\left(J\mathbf{C}^{-1}\mathrm{Grad}\bar{\phi}\right) = \mathrm{Div}\bar{\boldsymbol{P}} - \bar{\varphi}, \tag{74}$$

along with the associated boundary condition:

$$[\![\bar{\boldsymbol{P}} - \varepsilon_0 J \mathbf{C}^{-1} \mathrm{Grad}\bar{\phi}]\!] \cdot \boldsymbol{N} = \bar{\varphi}_s. \tag{75}$$

## 6 Conservation of Total Energy

For any arbitrary closed subsystem of $\mathcal{B}$ (such as the one depicted in Fig. 4), energy can cross the system boundary $\partial\Omega(t)$ in the form of heat, electrical work and mechanical work. Differently than $\mathcal{B}$, the subsystem is not electrically isolated and thus interactions may exist between the electrical charges that i.e., within the volume $\Omega(t)$ and those outside, which certainly need to be accounted for in the balance of subsystem energy. However, since the charges outside the boundary $\partial\Omega(t)$ can always be replaced, without modifying in any way the electric potential at any interior point of $\Omega(t)$, by an equivalent single and double layer of charges that are distributed on $\partial\Omega(t)$ with surface density equaling [34, 35]

$$\hat{\varphi}_s = -\boldsymbol{D} \cdot \boldsymbol{n} = -\bar{\boldsymbol{D}} \cdot \boldsymbol{N} d\mathrm{S}/ds, \tag{76}$$

then the energy balance of any arbitrary subsystem of $\mathcal{B}$ can be performed by considering an equivalent isolated subsystem as shown in Fig. 5. That is, in mathematical terms, the balance of electro-thermo-mechanical energy (i.e. the first law of thermodynamics) for any arbitrary subsystem of $\mathcal{B}$ which is bounded by the surface $\partial\Omega(t)$ reads as

$$\frac{\mathrm{d}}{\mathrm{d}t}\mathcal{K} + \frac{\mathrm{d}}{\mathrm{d}t}\mathcal{W} + \frac{\mathrm{d}}{\mathrm{d}t}\mathcal{I} = \mathcal{P}_{me} + \mathcal{P}_{el} + \mathcal{Q}, \tag{77}$$

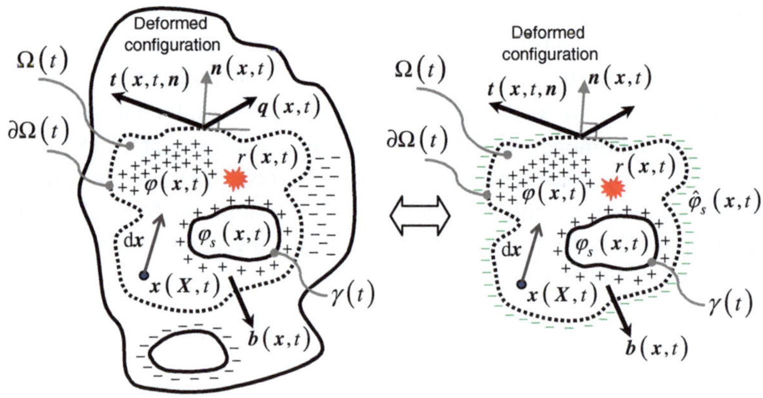


**Fig. 5** Electrostrictive elastomer system: arbitrary electrically-non-isolated interior subsystem and its electrically-isolated equivalent

where $\mathcal{K}(t)$, $\mathcal{W}(t)$ and $\mathcal{I}(t)$ are the kinetic, electrostatic and internal energies associated to the physical space contained within the boundary $\partial\Omega(t)$

$$\mathcal{K} = \int\limits_{\Omega(t)} \frac{1}{2}\rho \boldsymbol{v}^2 \mathrm{dv}, \tag{78}$$

$$\mathcal{W} = \int\limits_{\Omega(t)-\gamma(t)} \left(\frac{1}{2}\varepsilon_0 \boldsymbol{E}^2 + \boldsymbol{E}\cdot\boldsymbol{P}\right)\mathrm{dv}, \tag{79}$$

$$\mathcal{I} = \int\limits_{\Omega(t)-\gamma(t)} \rho U \mathrm{dv}, \tag{80}$$

$U(\boldsymbol{x},t)$ being the internal energy density associated to the bodies contained in the volume $\Omega(t)$; whereas $\mathcal{P}_{me}(t)$, $\mathcal{P}_{el}(t)$ and $\mathcal{Q}(t)$ are the external mechanical, electrical and thermal powers entering in the system from the outside of its boundary $\partial\Omega(t)$

$$\mathcal{P}_{me} = \int\limits_{\partial\Omega(t)} \boldsymbol{t}\cdot\boldsymbol{v}\mathrm{ds} + \int\limits_{\Omega(t)} \boldsymbol{f}\cdot\boldsymbol{v}\mathrm{dv}, \tag{81}$$

$$\mathcal{P}_{el} = \int\limits_{\Omega(t)-\gamma(t)} \phi\frac{\mathrm{d}}{\mathrm{d}t}(\varphi\mathrm{dv}) + \int\limits_{\gamma(t)} \phi\frac{\mathrm{d}}{\mathrm{d}t}(\varphi_s\mathrm{ds}) + \int\limits_{\partial\Omega(t)} \phi\frac{\mathrm{d}}{\mathrm{d}t}(\hat{\varphi}_s\mathrm{ds}), \tag{82}$$

$$\mathcal{Q} = -\int\limits_{\partial\Omega(t)} \boldsymbol{Q}\cdot\boldsymbol{n}\mathrm{ds} + \int\limits_{\Omega(t)} R\mathrm{dv}, \tag{83}$$

where $\boldsymbol{Q}(\boldsymbol{x},t)$ is the Cauchy heat flux vector entering the system (defined per unit area of the deformed surface $\partial\Omega(t)$). And $R(\boldsymbol{x},t)$ is the heat source density (i.e. a reservoir of heat) defined per unit volume of the deformed $\Omega(t)$.

### *6.1 Conservation of Total Energy in Eulerian Description*

By the Reynold's transport theorem [Eq. (207)], Eqs. (19) and (22), the three terms on the left-hand side of Eq. (77) follow as

$$\frac{\mathrm{d}}{\mathrm{d}t}\mathcal{K} = \int\limits_{\Omega(t)} \rho\dot{\boldsymbol{v}}\cdot\boldsymbol{v}\mathrm{dv}, \tag{84}$$

$$\frac{d}{dt}\mathcal{W} = \int\limits_{\Omega(t)-\gamma(t)} \left(\varepsilon_0 \boldsymbol{E}\cdot\dot{\boldsymbol{E}} + \boldsymbol{P}\cdot\dot{\boldsymbol{E}} + \boldsymbol{E}\cdot\dot{\boldsymbol{P}}\right)\mathrm{dv} + \int\limits_{\Omega(t)-\gamma(t)} \left(\frac{1}{2}\varepsilon_0 \boldsymbol{E}^2 + \boldsymbol{E}\cdot\boldsymbol{P}\right)\mathrm{div}\boldsymbol{v}\mathrm{dv}, \tag{85}$$

$$\frac{\mathrm{d}}{\mathrm{d}t}\mathcal{I} = \int\limits_{\Omega(t)-\gamma(t)} \rho\dot{U}\mathrm{dv}. \tag{86}$$

As for the first term on right-hand side of Eq. (77), by using Eq. (30) and the divergence theorem [Eq. (204)], and by additionally resorting to Eqs. (198), (81) can also be rewritten as

$$\mathcal{P}_{me} = \int\limits_{\Omega(t)} \mathrm{div}(\sigma\boldsymbol{v})\mathrm{dv} + \int\limits_{\Omega(t)} \boldsymbol{f}\cdot\boldsymbol{v}\mathrm{dv} = \int\limits_{\Omega(t)} \boldsymbol{\sigma}^T : \mathrm{grad}\boldsymbol{v}\mathrm{dv} + \int\limits_{\Omega(t)} (\mathrm{div}\boldsymbol{\sigma} + \boldsymbol{f})\cdot\boldsymbol{v}\mathrm{dv}, \tag{87}$$

which, by considering Eqs. (77) and (32), yields

$$\mathcal{P}_{me} - \frac{\mathrm{d}}{\mathrm{d}t}\mathcal{K} = \int\limits_{\Omega(t)} \boldsymbol{\sigma}^T : \mathrm{grad}\boldsymbol{v}\mathrm{dv}. \tag{88}$$

As for the second term on right-hand side of Eq. (77), using Eqs. (59), (76), (12) and (15), Eq. (82) can be rewritten as

$$\mathcal{P}_{el} = \int\limits_{\bar{\Omega}-\bar{\gamma}} \phi\frac{\mathrm{d}}{\mathrm{d}t}(\varphi J\mathrm{dV}) + \int\limits_{\bar{\gamma}} \phi\frac{\mathrm{d}}{\mathrm{d}t}\left([\![\boldsymbol{D}]\!]\cdot J\mathbf{F}^{-T}\boldsymbol{N}\mathrm{dS}\right) - \int\limits_{\partial\bar{\Omega}} \phi\frac{\mathrm{d}}{\mathrm{d}t}\left(\boldsymbol{D}\cdot J\mathbf{F}^{-T}\boldsymbol{N}\mathrm{dS}\right), \tag{89}$$

which, upon differentiation and use of the kinematic Eqs. (19) and (20), becomes

$$\begin{aligned}\mathcal{P}_{el} = &\int\limits_{\Omega(t)-\gamma(t)} \phi[\dot{\varphi} + \varphi\mathrm{div}\boldsymbol{v}]\mathrm{dv} + \int\limits_{\gamma(t)} \left[\phi\left([\![\dot{\boldsymbol{D}}]\!] + \mathrm{div}\boldsymbol{v}[\![\boldsymbol{D}]\!] - \mathrm{grad}\boldsymbol{v}[\![\boldsymbol{D}]\!]\right)\right]\cdot\boldsymbol{n}\mathrm{ds}\\ &- \int\limits_{\partial\Omega(t)} \left[\phi\left(\dot{\boldsymbol{D}} + \mathrm{div}\boldsymbol{v}\boldsymbol{D} - \mathrm{grad}\boldsymbol{v}\boldsymbol{D}\right)\right]\cdot\boldsymbol{n}\mathrm{ds}.\end{aligned} \tag{90}$$

Then, application of the divergence theorem [Eq. (205)] to the last two terms on the right-hand side of Eq. (90) yields

$$\mathcal{P}_{el} = \int\limits_{\Omega(t)-\gamma(t)} \phi[\dot{\varphi} + \varphi\mathrm{div}\boldsymbol{v}]\mathrm{dv} - \int\limits_{\Omega(t)-\gamma(t)} \mathrm{div}\left[\phi\left(\dot{\boldsymbol{D}} + \mathrm{div}\boldsymbol{v}\boldsymbol{D} - \mathrm{grad}\boldsymbol{v}\boldsymbol{D}\right)\right]\mathrm{dv}, \tag{91}$$

which, with the use of Eq. (199) and owing to Eq. (56), can be more conveniently rewritten as

$$\mathcal{P}_{el} = \int\limits_{\Omega(t)-\gamma(t)} \phi(\dot{\varphi} + \varphi \mathrm{div}\boldsymbol{v})\mathrm{dv} - \int\limits_{\Omega(t)-\gamma(t)} \phi \mathrm{div}\left(\dot{\boldsymbol{D}} + \mathrm{div}\boldsymbol{v}\boldsymbol{D} - \mathrm{grad}\boldsymbol{v}\boldsymbol{D}\right)\mathrm{dv} + \int\limits_{\Omega(t)-\gamma(t)} \boldsymbol{E} \cdot \left(\dot{\boldsymbol{D}} + \mathrm{div}\boldsymbol{v}\boldsymbol{D} - \mathrm{grad}\boldsymbol{v}\boldsymbol{D}\right)\mathrm{dv}. \tag{92}$$

Further, since

$$\mathrm{div}\dot{\boldsymbol{D}} = \dot{\varphi} + (\mathrm{grad}\boldsymbol{D})^T : \mathrm{grad}\boldsymbol{v}, \tag{93}$$

$$\mathrm{div}(\mathrm{div}\boldsymbol{v}\boldsymbol{D}) = \mathrm{grad}(\mathrm{div}\boldsymbol{v}) \cdot \boldsymbol{D} + \varphi \mathrm{div}\boldsymbol{v}, \tag{94}$$

$$\mathrm{div}(\mathrm{grad}\boldsymbol{v}\boldsymbol{D}) = \mathrm{grad}(\mathrm{div}\boldsymbol{v}) \cdot \boldsymbol{D} + (\mathrm{grad}\boldsymbol{D})^T : \mathrm{grad}\boldsymbol{v}, \tag{95}$$

Eq. (92) simplifies into

$$\mathcal{P}_{el} = \int\limits_{\Omega(t)-\gamma(t)} \boldsymbol{E} \cdot \left(\dot{\boldsymbol{D}} + \mathrm{div}\boldsymbol{v}\boldsymbol{D} - \mathrm{grad}\boldsymbol{v}\boldsymbol{D}\right)\mathrm{dv}, \tag{96}$$

which, by considering Eq. (85) together with the identities $\dot{\boldsymbol{D}} = \varepsilon_0\dot{\boldsymbol{E}} + \dot{\boldsymbol{P}}$, $\mathrm{div}\boldsymbol{v} = \mathbf{1} : \mathrm{grad}\boldsymbol{v}$ and $\boldsymbol{E} \cdot (\mathrm{grad}\boldsymbol{v}\boldsymbol{D}) = (\boldsymbol{E} \otimes \boldsymbol{D}) : \mathrm{grad}\boldsymbol{v}$ [$\boldsymbol{E} \otimes \boldsymbol{D}$ indicating the tensor product between vectors $\boldsymbol{E}$ and $\boldsymbol{D}$ as defined in Eq. (181)], yields

$$\mathcal{P}_{el} - \frac{\mathrm{d}}{\mathrm{d}t}\mathcal{W} = \int\limits_{\Omega(t)-\gamma(t)} \left[-\boldsymbol{P} \cdot \dot{\boldsymbol{E}} + \left(\frac{1}{2}\varepsilon_0 \boldsymbol{E}^2 \mathbf{1} - \boldsymbol{E} \otimes \boldsymbol{D}\right) : \mathrm{grad}\boldsymbol{v}\right]\mathrm{dv}. \tag{97}$$

As for the last term on right-hand side of Eq. (77), by the divergence theorem [Eq. (203)], Eq. (83) can be rewritten as

$$\mathcal{Q} = \int\limits_{\Omega(t)} (R - \mathrm{div}\boldsymbol{Q})\mathrm{dv}. \tag{98}$$

In summary, by considering Eqs. (86), (88), (97) and (98), and by assuming that no discontinuity across any surface $\gamma(t)$ exists in the variables $\rho$, $\boldsymbol{\sigma}$, and $\boldsymbol{Q}$, Eq. (77) reduces to

$$\int\limits_{\Omega(t)-\gamma(t)} \left[-\boldsymbol{P} \cdot \dot{\boldsymbol{E}} - \rho\dot{U} + R - \mathrm{div}\boldsymbol{Q} + \left(\boldsymbol{\sigma}^T - \boldsymbol{E} \otimes \boldsymbol{D} + \frac{1}{2}\varepsilon_0 \boldsymbol{E}^2 \mathbf{1}\right) : \mathrm{grad}\boldsymbol{v}\right]\mathrm{dv} = 0 \tag{99}$$

which, since $\Omega(t)$ and $\gamma(t)$ are arbitrary volumes and surfaces, yields

$$-\boldsymbol{P}\cdot\dot{\boldsymbol{E}}-\rho\dot{U}+R-\mathrm{div}\boldsymbol{Q}+\left(\boldsymbol{\sigma}^{T}-\boldsymbol{E}\otimes\boldsymbol{D}+\frac{1}{2}\varepsilon_{0}\boldsymbol{E}^{2}\mathbf{1}\right):\mathrm{grad}\boldsymbol{v}=0,\tag{100}$$

representing the local form of the First Law of Thermodynamics (FLT) for electrostrictive elastomers expressed in Eulerian description.

## 6.2 Conservation of Total Energy in Lagrangian Description

According to the notation introduced in the previous sections, the Lagrangian counterparts of Eqs. (78–80) are

$$\mathcal{K}=\int_{\bar{\Omega}}\frac{1}{2}\bar{\rho}\boldsymbol{V}^{2}\mathrm{dV},\tag{101}$$

$$\mathcal{W}=\int_{\bar{\Omega}-\bar{\gamma}}\left(\frac{1}{2}\bar{\boldsymbol{E}}\cdot\bar{\boldsymbol{D}}_{0}+\bar{\boldsymbol{E}}\cdot\bar{\boldsymbol{P}}\right)\mathrm{dV},\tag{102}$$

$$\mathcal{I}=\int_{\bar{\Omega}-\bar{\gamma}}\bar{\rho}\bar{U}\mathrm{dV},\tag{103}$$

where $\bar{U}(\boldsymbol{X},t)=U(\boldsymbol{x},t)$ is the internal energy density expressed with respect to the referential configuration. Time derivation of Eqs. (101–103) directly yields

$$\frac{\mathrm{d}}{\mathrm{d}t}\mathcal{K}=\int_{\bar{\Omega}}\bar{\rho}\dot{\boldsymbol{V}}\cdot\boldsymbol{V}\mathrm{dV},\tag{104}$$

$$\frac{\mathrm{d}}{\mathrm{d}t}\mathcal{W}=\int_{\bar{\Omega}-\bar{\gamma}}\frac{1}{2}\left(\dot{\bar{\boldsymbol{E}}}\cdot\bar{\boldsymbol{D}}_{0}+\bar{\boldsymbol{E}}\cdot\dot{\bar{\boldsymbol{D}}}_{0}+\bar{\boldsymbol{P}}\cdot\dot{\bar{\boldsymbol{E}}}+\bar{\boldsymbol{E}}\cdot\dot{\bar{\boldsymbol{P}}}\right)\mathrm{dV},\tag{105}$$

$$\frac{\mathrm{d}}{\mathrm{d}t}\mathcal{I}=\int_{\bar{\Omega}-\bar{\gamma}}\bar{\rho}\dot{\bar{U}}\mathrm{dV}.\tag{106}$$

Besides, the Lagrangian counterparts of Eqs. (81)–(83) are

$$\mathcal{P}_{me}=\int_{\partial\bar{\Omega}}\boldsymbol{T}\cdot\boldsymbol{V}\mathrm{dS}+\int_{\bar{\Omega}}\bar{\boldsymbol{f}}\cdot\boldsymbol{V}\mathrm{dV},\tag{107}$$

$$\mathcal{P}_{el} = \int\limits_{\bar{\Omega}-\bar{\gamma}} \bar{\phi}\frac{\mathrm{d}}{\mathrm{d}t}(\bar{\varphi}\mathrm{dV}) + \int\limits_{\bar{\gamma}} \bar{\phi}\frac{\mathrm{d}}{\mathrm{d}t}(\bar{\varphi}_s\mathrm{dS}) - \int\limits_{\partial\bar{\Omega}} \bar{\phi}\frac{\mathrm{d}}{\mathrm{d}t}(\bar{\boldsymbol{D}}\cdot\boldsymbol{N}\mathrm{dS})$$
$$= \int\limits_{\bar{\Omega}-\bar{\gamma}} \bar{\phi}\dot{\bar{\varphi}}\mathrm{dV} + \int\limits_{\bar{\gamma}} \bar{\phi}\dot{\bar{\varphi}}_s\mathrm{dS} - \int\limits_{\partial\bar{\Omega}} \bar{\phi}\dot{\bar{\boldsymbol{D}}}\cdot\boldsymbol{N}\mathrm{dS}, \tag{108}$$

$$\mathcal{Q} = -\int\limits_{\partial\bar{\Omega}} \bar{\boldsymbol{Q}}\cdot\boldsymbol{N}\mathrm{d}S + \int\limits_{\bar{\Omega}} \bar{R}\mathrm{d}V, \tag{109}$$

where the Piola-Kirchhoff heat flux, $\bar{\boldsymbol{Q}}(\boldsymbol{X},t)$, which is defined per unit area of the reference surface $\partial\bar{\Omega}$, and the heat source, $\bar{R}(\boldsymbol{X},t)$, which is defined per unit reference volume, read as

$$\bar{\boldsymbol{Q}} = J\mathbf{F}^{-1}\boldsymbol{Q}, \tag{110}$$

$$\bar{R} = JR. \tag{111}$$

Starting with $\mathcal{P}_{me}(t)$, use of Eq. (36), of the divergence theorem [Eq. (204)] and of Eq. (200), enables to write

$$\mathcal{P}_{me} = \int\limits_{\bar{\Omega}} \mathrm{Div}(\boldsymbol{\Pi}\boldsymbol{V})\mathrm{dV} + \int\limits_{\bar{\Omega}} \bar{\boldsymbol{f}}\cdot\boldsymbol{V}\mathrm{dV}$$
$$= \int\limits_{\bar{\Omega}} \boldsymbol{\Pi}^T\!:\!\mathrm{Grad}(\boldsymbol{V})\mathrm{dV} + \int\limits_{\bar{\Omega}} (\mathrm{Div}\boldsymbol{\Pi}+\boldsymbol{f})\cdot\boldsymbol{V}\mathrm{dV}, \tag{112}$$

which, by considering Eqs. (38) and (16), yields

$$\mathcal{P}_{me} - \frac{\mathrm{d}}{\mathrm{d}t}\mathcal{K} = \int\limits_{\bar{\Omega}} \boldsymbol{\Pi}^T : \dot{\mathbf{F}}\mathrm{dV}. \tag{113}$$

Skipping to $\mathcal{P}_{el}(t)$, considering Eqs. (71) and (72), Eq. (108) can be rewritten as

$$\mathcal{P}_{el}(t) = \int\limits_{\bar{\Omega}-\bar{\gamma}} \bar{\phi}\mathrm{Div}\dot{\bar{\boldsymbol{D}}}\mathrm{dV} + \int\limits_{\bar{\gamma}} \bar{\phi}[\![\dot{\bar{\boldsymbol{D}}}]\!]\cdot\boldsymbol{N}\mathrm{dS} - \int\limits_{\partial\bar{\Omega}} \bar{\phi}\dot{\bar{\boldsymbol{D}}}\cdot\boldsymbol{N}\mathrm{dS}, \tag{114}$$

which, by applying the divergence theorem [Eq. (205)] to the second term on the right-hand side, and because of Eqs. (201) and (64), gives

$$\mathcal{P}_{el}(t) = \int\limits_{\bar{\Omega}-\bar{\gamma}} \bar{\phi}\mathrm{Div}\dot{\bar{\boldsymbol{D}}}\mathrm{dV} - \int\limits_{\bar{\Omega}-\bar{\gamma}} \mathrm{Div}\left(\bar{\phi}\dot{\bar{\boldsymbol{D}}}\right)\mathrm{dV} = \int\limits_{\bar{\Omega}-\bar{\gamma}} \bar{\boldsymbol{E}}\cdot\dot{\bar{\boldsymbol{D}}}\mathrm{dV}. \tag{115}$$

From Eqs. (105) and (115), the Lagrangian counterpart of Eq. (97) is

$$\mathcal{P}_{el}(t) - \frac{\mathrm{d}}{\mathrm{d}t}\mathcal{W}(t) = \int\limits_{\bar{\Omega}-\bar{\gamma}} \left[ -\bar{\boldsymbol{P}} \cdot \dot{\bar{\boldsymbol{E}}} + \frac{1}{2}\left( \bar{\boldsymbol{E}} \cdot \dot{\bar{\boldsymbol{D}}}_0 - \dot{\bar{\boldsymbol{E}}} \cdot \bar{\boldsymbol{D}}_0 \right) \right] \mathrm{dV}, \tag{116}$$

which, since

$$\bar{\boldsymbol{E}} \cdot \dot{\bar{\boldsymbol{D}}}_0 = \bar{\boldsymbol{E}} \cdot \frac{\mathrm{d}}{\mathrm{d}t}\left( \varepsilon_0 J \mathbf{C}^{-1} \bar{\boldsymbol{E}} \right) = \dot{\bar{\boldsymbol{E}}} \cdot \bar{\boldsymbol{D}}_0 - \varepsilon_0 J (2\boldsymbol{E} \otimes \boldsymbol{E} - \boldsymbol{E} \cdot \boldsymbol{E}\mathbf{1}) \mathbf{F}^{-T} : \dot{\mathbf{F}}, \tag{117}$$

simply reduces to

$$\mathcal{P}_{el}(t) - \frac{\mathrm{d}}{\mathrm{d}t}\mathcal{W}(t) = \int\limits_{\bar{\Omega}-\bar{\gamma}} \left[ -\bar{\boldsymbol{P}} \cdot \dot{\bar{\boldsymbol{E}}} - \Pi_M^T : \dot{\mathbf{F}} \right] \mathrm{dV}, \tag{118}$$

where

$$\mathbf{\Pi}_M = J \mathbf{F}^{-1} \boldsymbol{\sigma}_M, \tag{119}$$

$$\boldsymbol{\sigma}_M = \boldsymbol{\sigma}_M^T = \varepsilon_0 \left( \boldsymbol{E} \otimes \boldsymbol{E} - \frac{1}{2} \boldsymbol{E} \cdot \boldsymbol{E}\mathbf{1} \right), \tag{120}$$

$\boldsymbol{\sigma}_M$ being the well known Maxwell stress tensor (i.e. a true stress tensor) for the vacuum [29] expressed in the spatial description, whereas $\mathbf{\Pi}_M$ being its material counterpart (i.e. a nominal stress tensor).

Regarding $\mathcal{Q}(t)$, use of the divergence theorem [Eq. (203)] enables to rewrite Eq. (109) as

$$\mathcal{Q} = \int\limits_{\bar{\Omega}} (\bar{R} - \mathrm{Div}\bar{\boldsymbol{Q}}) \mathrm{dV}. \tag{121}$$

In summary, by considering Eqs. (106), (113), (118) and (121), and by assuming that no discontinuity across any surface $\bar{\gamma}$ exists in the variables $\bar{\rho}$, $\mathbf{\Pi}$, and $\bar{\boldsymbol{Q}}$, Eq. (78) also reads as

$$\int\limits_{\bar{\Omega}-\bar{\gamma}} \left[ -\bar{\boldsymbol{P}} \cdot \dot{\bar{\boldsymbol{E}}} - \bar{\rho}\dot{\bar{U}} + \bar{R} - \mathrm{Div}\bar{\boldsymbol{Q}} + \left( \Pi^T - \Pi_M^T \right) : \dot{\mathbf{F}} \right] \mathrm{dV} = 0, \tag{122}$$

which, since $\bar{\Omega}$ and $\bar{\gamma}$ are arbitrary volumes and surfaces, yields

$$-\bar{\boldsymbol{P}} \cdot \dot{\bar{\boldsymbol{E}}} - \bar{\rho}\dot{\bar{U}} + \bar{R} - \mathrm{Div}\bar{\boldsymbol{Q}} + \left( \Pi^T - \Pi_M^T \right) : \dot{\mathbf{F}} = 0, \tag{123}$$

representing the local form of the FLT for electrostrictive elastomers expressed in Lagrangian description.

# 7 Entropy Inequality Principle

In addition to Eq. (77), complete description of the energy transfer occurring across any arbitrary subsystem of $\mathcal{B}$ (such as the one depicted in Fig. 4) also requires the total entropy, which is produced during all the admissible thermo-electro-mechanical processes, to be non-negative. In mathematical terms, this can be stated by the inequality

$$\frac{\mathrm{d}}{\mathrm{d}t}\mathcal{S} - \mathcal{H} \geq 0, \tag{124}$$

where $\mathcal{S}(t)$ and $\mathcal{H}(t)$ are the total entropy of the subsystem occupying the volume $\Omega(t)$ and the rate of entropy input entering from the subsystem boundary $\partial\Omega(t)$ respectively, and defined as

$$\mathcal{S}(t) = \int\limits_{\Omega(t)} \rho\eta \mathrm{dv}, \tag{125}$$

$$\mathcal{H}(t) = -\int\limits_{\partial\Omega(t)} \frac{\boldsymbol{Q}}{T} \cdot \boldsymbol{n} \mathrm{ds} + \int\limits_{\Omega(t)} \frac{R}{T} \mathrm{dv}, \tag{126}$$

$\eta(\boldsymbol{x}, t)$ and $T(\boldsymbol{x}, t)$ (with $T(\boldsymbol{x}, t) \geq 0$) respectively being the entropy density (per unit mass) and the absolute temperature both expressed in the current configuration.

## 7.1 *Entropy Inequality in Eulerian Description*

By applying the Reynold's transport theorem [Eq. (207)] and Eq. (22) to the time derivative of Eq. (125), and by employing the divergence theorem [Eq. (203)] to the first term on the right-hand side of Eqs. (126), (124) reads as

$$\int\limits_{\Omega(t)} \left[\rho\dot{\eta} + \mathrm{div}\left(\frac{\boldsymbol{Q}}{T}\right) - \frac{R}{T}\right] \mathrm{dv} \geq 0, \tag{127}$$

which, since $\Omega(t)$ is an arbitrary volume, yields

$$\rho\dot{\eta} + \mathrm{div}\left(\frac{\boldsymbol{Q}}{T}\right) - \frac{R}{T} \geq 0, \tag{128}$$

corresponding to the local form of the Clausius-Duhem inequality in the Eulerian description.

Resorting to Eqs. (100), (128) becomes

$$\rho T\dot{\eta} - \rho\dot{U} - \boldsymbol{P}\cdot\dot{\boldsymbol{E}} + \left(\boldsymbol{\sigma}^T - \boldsymbol{E}\otimes\boldsymbol{D} + \frac{1}{2}\varepsilon_0\boldsymbol{E}^2\mathbf{1}\right) : \mathrm{grad}\boldsymbol{v} - \frac{\boldsymbol{Q}}{T}\cdot\mathrm{grad}T \geq 0, \quad (129)$$

which, by considering the following Legendre transformation (i.e. a procedure used to replace one or more variables with their conjugate counterparts [74])

$$\Psi = U - T\eta, \quad (130)$$

where $\Psi(\boldsymbol{x}, t)$ is a free-energy function, can also be written as

$$\rho\eta\dot{T} + \rho\dot{\Psi} + \boldsymbol{P}\cdot\dot{\boldsymbol{E}} - \left(\boldsymbol{\sigma}^T - \boldsymbol{E}\otimes\boldsymbol{D} + \frac{1}{2}\varepsilon_0\boldsymbol{E}^2\mathbf{1}\right) : \mathrm{grad}\boldsymbol{v} + \frac{\boldsymbol{Q}}{T}\cdot\mathrm{grad}T \leq 0. \quad (131)$$

Further, by considering that heat does not flow against a temperature gradient, i.e.

$$\boldsymbol{Q}\cdot\mathrm{grad}T \leq 0, \quad (132)$$

Eq. (131) can be reduced into the stronger form

$$\rho\eta\dot{T} + \rho\dot{\Psi} + \boldsymbol{P}\cdot\dot{\boldsymbol{E}} - \left(\boldsymbol{\sigma}^T - \boldsymbol{E}\otimes\boldsymbol{D} + \frac{1}{2}\varepsilon_0\boldsymbol{E}^2\mathbf{1}\right) : \mathrm{grad}\boldsymbol{v} \leq 0, \quad (133)$$

which represents the local form of the Clausius-Planck Inequality (CPI) of thermodynamics for electrostictive elastomers expressed in the Eulerian formulation.

## 7.2 Entropy Inequality in Lagrangian Description

According to the notation introduced in previous sections, the Lagrangian counterparts of Eqs. (125) and (126) are

$$\mathcal{S}(t) = \int_{\bar{\Omega}} \bar{\rho}\bar{\eta}\mathrm{dV}, \quad (134)$$

$$\mathcal{H}(t) = -\int_{\partial\bar{\Omega}} \frac{\bar{\boldsymbol{Q}}}{\bar{T}}\cdot\boldsymbol{N}\mathrm{dS} + \int_{V\bar{\Omega}} \frac{\bar{R}}{\bar{T}}\mathrm{dV}, \quad (135)$$

$\bar{\eta}(\boldsymbol{X}, t) = \eta(\boldsymbol{x}, t)$ and $\bar{T}(\boldsymbol{X}, t) = T(\boldsymbol{x}, t)$ being the entropy density and the absolute temperature expressed with respect to the referential configuration.

By differentiating Eq. (134) and by employing the divergence theorem [Eq. (203)] to the first term on the right-hand side of Eqs. (135), (124) can also be rewritten as

$$\int_{\bar{\Omega}} \left( \bar{\rho}\dot{\bar{\eta}} + \mathrm{Div}\left(\frac{\bar{\boldsymbol{Q}}}{\bar{T}}\right) - \frac{\bar{R}}{\bar{T}} \right) \mathrm{dV} \geq 0, \tag{136}$$

which, since $\bar{\Omega}$ is an arbitrary volume, yields

$$\bar{\rho}\dot{\bar{\eta}} + \mathrm{Div}\left(\frac{\bar{\boldsymbol{Q}}}{\bar{T}}\right) - \frac{\bar{R}}{\bar{T}} \geq 0, \tag{137}$$

corresponding to the local form of the Clausius-Duhem inequality in the Lagrangian description.

Resorting to Eqs. (123), (137) becomes

$$\bar{T}\bar{\rho}\dot{\bar{\eta}} - \bar{\rho}\dot{\bar{U}} - \bar{\boldsymbol{P}} \cdot \dot{\bar{\boldsymbol{E}}} + \left(\Pi^T - \Pi_M^T\right):\dot{\mathbf{F}} - \frac{\bar{\boldsymbol{Q}}}{\bar{T}}\mathrm{Grad}\bar{T} \geq 0, \tag{138}$$

which, by considering the following Legendre transformation

$$\bar{\Psi} = \bar{U} - \bar{T}\bar{\eta}, \tag{139}$$

where $\bar{\Psi}(\boldsymbol{X}, t) = \Psi(\boldsymbol{x}, t)$ is the same free-energy function defined in Eq. (130) but expressed with respect to the reference configuration, can also be rewritten as

$$\bar{\rho}\bar{\eta}\dot{\bar{T}} + \bar{\rho}\dot{\bar{\Psi}} + \bar{\boldsymbol{P}} \cdot \dot{\bar{\boldsymbol{E}}} - \left(\boldsymbol{\Pi}^T - \boldsymbol{\Pi}_M^T\right) : \dot{\mathbf{F}} + \frac{\bar{\boldsymbol{Q}}}{\bar{T}}\mathrm{Grad}\bar{T} \leq 0. \tag{140}$$

Furthermore, considering the Lagrangian counterpart of Eq. (132)

$$\bar{\boldsymbol{Q}} \cdot \mathrm{Grad}\bar{T} \leq 0, \tag{141}$$

Eq. (140) reduces to the stronger form

$$\bar{\rho}\bar{\eta}\dot{\bar{T}} + \bar{\rho}\dot{\bar{\Psi}} + \bar{\boldsymbol{P}} \cdot \dot{\bar{\boldsymbol{E}}} - \left(\boldsymbol{\Pi}^T - \boldsymbol{\Pi}_M^T\right) : \dot{\mathbf{F}} \leq 0, \tag{142}$$

representing the local form of the CPI of thermodynamics for electrostrictive elastomers expressed in the Lagrangian description.

## 8 Constitutive Equations

According to the relations described in the previous sections, complete knowledge of the thermo-electro-mechanical state of the closed and electrically isolated system depicted in Fig. 4 requires the determination of the variables $\rho$, $\boldsymbol{v}$ (or $\mathbf{F}$), $\boldsymbol{\sigma}$, $\varphi$, $\varphi_s$, $\boldsymbol{E}$, $\boldsymbol{P}$, $\boldsymbol{Q}$, $R$, $T$, $\Psi$ and $\eta$ [or equivalently by their Lagrangian counterparts $\bar{\rho}$, $V$ (or $\mathbf{F}$), $\boldsymbol{\Pi}$, $\bar{\varphi}$, $\bar{\varphi}_s$, $\bar{\boldsymbol{E}}$, $\bar{\boldsymbol{P}}$, $\bar{\boldsymbol{Q}}$, $\bar{R}$, $\bar{T}$, $\bar{\Psi}$ and $\bar{\eta}$], whose evolution is subjected to the balance Eqs. (22), (32), (44), (56), (58), (59), (100) and constrained by the inequalities (132) and (133) [or equivalently governed by the balance Eqs. (25), (38), (49), (71), (72), (123) and constrained by the inequalities (141) and (142)].

Despite the free-charge densities, $\varphi$ and $\varphi_s$, and the heat source, $R$, are usually externally imposed (and thus known) quantities, the balance equations are not sufficient by themselves to make the problem determined. Therefore, other relationships need to be introduced.

In the derivation of the balance laws, no specification was provided regarding the nature of the substance constituting the dielectric and conducting deformable bodies comprised within the closed system $\mathcal{B}$ depicted in Fig. 4. Balance laws are indeed valid for all types of substances (e.g. gas, fluids, or solids having a variety of different properties). Of course, complete determination of the thermo-electro-mechanical problem requires the phenomenological properties of specific materials to be considered and adequately expressed in clean mathematical forms, usually known as constitutive relations. Contrarily to balance equations, constitutive relations are only aimed at modeling the important features of material responses such as stress-strain behavior, material electrical polarizability, as well as cross-effects alike electrostriction.

Constitutive relations can be physically-based or phenomenological invariant-based. The former are mechanistically motivated relationships which come from a microstructurally-based statistical mechanics treatment of the interaction between matter and field. The latter are mathematically motivated relationships which come from a continuum mechanics treatment of the matter-field interaction.

Notwithstanding the approach used, constitutive equations require certain constants to be determined from a (preferably small) number of experiments. In the case of mechanistically motivated theories, these constants are physically-based parameters which are independent of the thermo-electro-mechanical state of the material and usually provide the constitutive model with strong predictive capabilities. Instead, in the case of mathematically motivated theories, these constants lack a direct physical connection to the underlying mechanism of matter-field interaction and, therefore, provide the constitutive model with weaker predictive capabilities and require a stability check after the constitutive equations have been fitted to the experimental data.

Though the use of physically-based approaches should be best advised because of their inherent stability and because of the enhanced predictive capabilities, since theories of this kind have not yet been developed for electrostrictive elastomers, a phenomenological invariant-based constitutive theory is described in the following [54, 57, 60, 62]. This theory should suffice in most cases whenever the resulting constitutive equations are made fit experimental data which are representative of the effective working state of the material in the practical system under investigation.

## 8.1 Requirements of Phenomenological Constitutive Theories

Systematic development of phenomenological constitutive equations requires the necessary satisfaction of certain physical and mathematical requirements, also called axioms, which are either verified experimentally or considered as obvious. No matter the physics under investigation, the basic axioms for the construction of consistent constitutive equations are thoroughly described in [75]. For electrostrictive elastomers with no memory effect (i.e. with no dependence on the specific history of the occurring thermo-electro-mechanical processes), the fundamental axioms to be satisfied are [54, 76]:

- *Axiom of Admissibility*: Constitutive equations must be consistent with balance laws and entropy inequalities.
- *Axiom of Causality*: Deformation, temperature and electric field of any material point belonging to an electroelastic body are self-evident and observable in any thermo-electro-mechanical behavior of the system and can be considered as independent variables. Correspondingly, the density, the free-energy density, the entropy, the stress tensor, the electric polarization and the heat flux vector are considered as dependent variables.
- *Axiom of Neighborhood*: Constitutive relations are subjected to continuity requirements.
- *Axiom of Equipresence*: Constitutive relations should be considered as dependent on the same constitutive variables, until the contrary is deduced.
- *Axiom of Objectivity* (Principle of Material Frame Indifference): Constitutive relations need to be form-invariant under arbitrary rigid motions of the spatial frame of reference as well as under a constant shift of the origin of time. That is, the thermo-electro-mechanical properties of materials cannot depend on the motion of the observer.
- *Axiom of Time Reversal*: Entropy production must be nonnegative under time reversal.
- *Axiom of Material Invariance*: Constitutive relations must be form-invariant with respect to rigid body motions superimposed on the referential configuration as well as to microscopic time reversals representing specific material symmetry conditions.

In the following, these axioms are used to develop an appropriate constitutive theory for the isothermal electro-elastic behavior of conservative and isotropic electrostrictive elastomers. According to this restriction, the study of the system depicted in Fig. 4 simplifies since the energy balance equations [either Eq. (100) or Eq. (123)] can be neglected together with the constitutive equations for the heat flux vector (either $\boldsymbol{Q}$ or $\bar{\boldsymbol{Q}}$).

## 8.2 Constitutive Equations for the Isothermal Behavior of Conservative and Isotropic Electro-Elastic Solids

Owing to the axioms of causality, neighborhood and equipresence, the free-energy function, $\Psi$ (and $\bar{\Psi}$), the entropy density, $\eta$ (and $\bar{\eta}$), the stress tensor, $\boldsymbol{\sigma}$ (and $\boldsymbol{\Pi}$), and the electric polarization, $\boldsymbol{P}$ (and $\bar{\boldsymbol{P}}$), are taken as continuous functions of $\mathbf{F}$, $\boldsymbol{E}$ (or $\bar{\boldsymbol{E}}$) and $T$. In particular

$$\Psi = \Psi(\mathbf{F}, \boldsymbol{E}, T) = \bar{\Psi}(\mathbf{F}, \bar{\boldsymbol{E}}, T). \tag{143}$$

Then, according to Eq. (143), the CPI given by Eqs. (133) and (142) can respectively be rewritten as

$$\rho\left(\eta + \frac{\partial \Psi}{\partial T}\right)\dot{T} + \left(\boldsymbol{P} + \rho\frac{\partial \Psi}{\partial \boldsymbol{E}}\right) \cdot \dot{\boldsymbol{E}} + \left[\rho\frac{\partial \Psi}{\partial \mathbf{F}}\mathbf{F}^T - \left(\boldsymbol{\sigma}^T - \boldsymbol{E} \otimes \boldsymbol{D} + \frac{1}{2}\varepsilon_0 \boldsymbol{E}^2 \mathbf{1}\right)\right] : \mathrm{grad}\boldsymbol{v} \leq 0, \tag{144}$$

$$\bar{\rho}\left(\bar{\eta} + \frac{\partial \bar{\Psi}}{\partial T}\right)\dot{\bar{T}} + \left(\bar{\boldsymbol{P}} + \bar{\rho}\frac{\partial \bar{\Psi}}{\partial \bar{\boldsymbol{E}}}\right) \cdot \dot{\bar{\boldsymbol{E}}} + \left(\bar{\rho}\frac{\partial \bar{\Psi}}{\partial \mathbf{F}} - \Pi^T + \Pi_M^T\right) : \dot{\mathbf{F}} \leq 0, \tag{145}$$

which, for all admissible thermo-electro-mechanical processes (i.e. for the axioms of admissibility and time-reversal), yield the following identities in the Eulerian description

$$\eta(\mathbf{F}, \boldsymbol{E}, T) = -\frac{\partial \Psi}{\partial T}, \tag{146}$$

$$\boldsymbol{P}(\mathbf{F}, \boldsymbol{E}, T) = -\rho\frac{\partial \Psi}{\partial \boldsymbol{E}}, \tag{147}$$

$$\sigma^T(\mathbf{F}, \boldsymbol{E}, T) = \rho\frac{\partial \Psi}{\partial \mathbf{F}}\mathbf{F}^T + \boldsymbol{E} \otimes \boldsymbol{D} - \frac{1}{2}\varepsilon_0 \boldsymbol{E}^2 \mathbf{1}, \tag{148}$$

or equivalently

$$\sigma^T(\mathbf{F}, \boldsymbol{E}, T) = 2\rho\frac{\partial \Psi}{\partial \mathbf{B}}\mathbf{B} + \boldsymbol{E} \otimes \boldsymbol{D} - \frac{1}{2}\varepsilon_0 \boldsymbol{E}^2 \mathbf{1}, \tag{149}$$

and the following identities in the Lagrangian description

$$\bar{\eta} = -\frac{\partial \bar{\Psi}}{\partial T}, \tag{150}$$

$$\bar{\boldsymbol{P}} = -\bar{\rho}\frac{\partial \bar{\Psi}}{\partial \bar{\boldsymbol{E}}}, \tag{151}$$

$$\mathbf{\Pi}^T = \bar{\rho}\frac{\partial\bar{\Psi}}{\partial\mathbf{F}} + \mathbf{\Pi}_M^T, \tag{152}$$

or equivalently

$$\mathbf{\Pi}^T = 2\bar{\rho}\mathbf{F}\frac{\partial\bar{\Psi}}{\partial C} + \mathbf{\Pi}_M^T. \tag{153}$$

Equations (146–153) highlight that the complete solution of the isothermal electro-elastic problem only requires the definition of the free-energy function, $\Psi$ (or identically $\bar{\Psi}$).

In this regard, for the axiom of objectivity [75, 76], the free-energy function of the considered materials needs to satisfy

$$\Psi(\mathbf{F}, \boldsymbol{E}, T) = \Psi(\mathbb{R}\mathbf{F}, \mathbb{R}\boldsymbol{E}, T), \tag{154}$$

for every proper orthogonal tensor $\mathbb{R}$ (i.e. $\mathbb{R}^T\mathbb{R} = 1$ and $\det\mathbb{R} = +1$) representing an arbitrary rotation superimposed on the current configuration. Note that the objectivity requirement given by Eq. (154) is immediately guaranteed whenever the free-energy is an explicit function of the Green deformation tensor $\mathbf{C}$ and the reference electric field $\bar{E}$.

Besides, for isotropic electrostrictive elastomers, the axiom of material invariance also requires

$$\bar{\Psi}(\mathbf{F}, \bar{\boldsymbol{E}}, T) = \bar{\Psi}\left(\mathbf{F}\mathbb{R}^T, \mathbb{R}\bar{\boldsymbol{E}}, T\right), \tag{155}$$

for every proper orthogonal tensor $\mathbb{R}$ (i.e. $\mathbb{R}^T\mathbb{R} = \mathbf{1}$ and $\det\mathbb{R} = +1$) representing an arbitrary rotation superimposed on the reference configuration. Note that the isotropy requirement expressed by Eq. (155) is immediately guaranteed whenever the free-energy is an explicit function of the Finger deformation tensor $\mathbf{B}$ and the spatial electric field $\boldsymbol{E}$; in which case, owing to the representation theorem for invariants [77], $\Psi$ (and $\bar{\Psi}$) admits the following irreducible representation [78–82]

$$\Psi = \Psi(I_1, I_2, I_3, I_4, I_5, I_6, T) = \bar{\Psi}(\bar{I}_1, \bar{I}_2, \bar{I}_3, \bar{I}_4, \bar{I}_5, \bar{I}_6, T) \tag{156}$$

in terms of the minimal set of invariants

$$I_1 = \mathrm{tr}\mathbf{B} = \bar{I}_1 = \mathrm{tr}\mathbf{C} \tag{157}$$

$$I_2 = \frac{1}{2}\left[(\mathrm{tr}\mathbf{B})^2 - \mathrm{tr}\left(\mathbf{B}^2\right)\right] = \bar{I}_2 = \frac{1}{2}\left[(\mathrm{tr}\mathbf{C})^2 - \mathrm{tr}\left(\mathbf{C}^2\right)\right] \tag{158}$$

$$I_3 = \det\mathbf{B} = \bar{I}_3 = \det\mathbf{C} = J^2 \tag{159}$$

$$I_4 = \boldsymbol{E}\cdot\boldsymbol{E} = \bar{I}_4 = \bar{\boldsymbol{E}}\cdot\left(\mathbf{C}^{-1}\bar{\boldsymbol{E}}\right) \tag{160}$$

$$I_5 = \boldsymbol{E}\cdot(\mathbf{B}\boldsymbol{E}) = \bar{I}_5 = \bar{\boldsymbol{E}}\cdot\bar{\boldsymbol{E}} \tag{161}$$

$$I_6 = \boldsymbol{E} \cdot \left(\mathbf{B}^2 \boldsymbol{E}\right) = \bar{I}_6 = \bar{\boldsymbol{E}} \cdot (\mathbf{C}\bar{\boldsymbol{E}}) \tag{162}$$

Of course this irreducible representation is also objective since Eq. (156), with Eqs. (157)–(162), immediately satisfies Eq. (154).

Then, by assuming a free-energy function in the form of Eqs. (156–162), and with the definition

$$a_i = \partial\Psi/\partial I_i = \partial\bar{\Psi}/\partial\bar{I}_i, \text{ for } i = 1,\ldots,6 \tag{163}$$

the constitutive laws for the electric polarization and the stress tensor reduce to

$$\boldsymbol{P} = -2\rho\left(a_4\boldsymbol{E} + a_5\mathbf{B}\boldsymbol{E} + a_6\mathbf{B}^2\boldsymbol{E}\right) \tag{164}$$

$$\begin{aligned} \boldsymbol{\sigma}^T = & \; 2\rho\left[(a_1 + a_2 I_1)\mathbf{B} - a_2\mathbf{B}^2 + a_3 I_3 \mathbf{1}\right] \\ & + \left[(\varepsilon_0 - 2\rho a_4)\boldsymbol{E} \otimes \boldsymbol{E} - \frac{1}{2}\varepsilon_0 I_4 \mathbf{1} + 2\rho a_6(\mathbf{B}\boldsymbol{E}) \otimes (\mathbf{B}\boldsymbol{E})\right] \end{aligned} \tag{165}$$

for the Eulerian description, and

$$\bar{\boldsymbol{P}} = -2\bar{\rho}\left(a_4\mathbf{C}^{-1}\bar{\boldsymbol{E}} + a_5\bar{\boldsymbol{E}} + a_6\mathbf{C}\bar{\boldsymbol{E}}\right), \tag{166}$$

$$\begin{aligned} \Pi^T = & \; 2\bar{\rho}\mathbf{F}\left[(a_1 + a_2 I_1)\mathbf{1} - a_2\mathbf{C} + a_3 I_3\mathbf{C}^{-1} + \left(\frac{\varepsilon_0}{2\bar{\rho}}\sqrt{I_3} - a_4\right)\left(\mathbf{C}^{-1}\bar{\boldsymbol{E}}\right)\right. \\ & \left. \otimes\left(\mathbf{C}^{-1}\bar{\boldsymbol{E}}\right) - \frac{\varepsilon_0}{4\bar{\rho}}\sqrt{I_3}I_4\mathbf{C}^{-1} + a_6\bar{\boldsymbol{E}} \otimes \bar{\boldsymbol{E}}\right] \end{aligned} \tag{167}$$

for the Lagrangian description. Of course, Eqs. (164) and (166) identically satisfy Eq. (67), whereas Eqs. (165) and (167) identically satisfy Eq. (50).

## 8.3 Reduced Constitutive Equations for Dielectric Elastomers

Among the class of electrostrictive rubber-like solids, Dielectric Elastomers are a special type of elastic insulating material whose spatial electric polarization, $\boldsymbol{P}$, only depends on the spatial electric field, $\boldsymbol{E}$(and not on the deformation gradient $\mathbf{F}$). Consequently, in mathematical terms, Dielectric Elastomers can be characterized by a free-energy function with the form

$$\Psi = \Psi_{hyp}(I_1, I_2, I_3, T) - \frac{(\varepsilon_a - \varepsilon_0)}{2\bar{\rho}} I_4\sqrt{I_3}, \tag{168}$$

where $\Psi_{hyp}(I_1, I_2, I_3, T)$ is some hyperelastic strain-energy function (i.e. a free-energy only depending on deformation), while the material parameter $\varepsilon_a$ is the absolute permittivity of the dielectric elastomer. Note that the absolute permittivity may depend on the isothermal temperature of the material, i.e. $\varepsilon_a = \varepsilon_a(T)$ (see for instance in [73]).

Then, substitution of Eq. (168) into Eqs. (164–167), provides the following constitutive equations for general Dielectric Elastomers in isothermal conditions

$$\boldsymbol{P} = (\varepsilon_a - \varepsilon_0)\boldsymbol{E}, \tag{169}$$

$$\boldsymbol{\sigma}^T = 2\rho\left[\left(a_1' + a_2' I_1\right)\mathbf{B} - a_2'\mathbf{B}^2 + a_3' I_3 \mathbf{1}\right] + \left[\varepsilon_a \boldsymbol{E}\otimes\boldsymbol{E} - \frac{1}{2}\varepsilon_a \boldsymbol{E}^2\mathbf{1}\right], \tag{170}$$

for the Eulerian description, whereas

$$\bar{\boldsymbol{P}} = J(\varepsilon_a - \varepsilon_0)\mathbf{C}^{-1}\bar{\boldsymbol{E}}, \tag{171}$$

$$\begin{aligned}\Pi^T &= 2\bar{\rho}\mathbf{F}\left[\left(a_1' + a_2' I_1\right)1 - a_2'\mathbf{C} + a_3' I_3 \mathbf{C}^{-1} + \frac{\varepsilon_a}{2\bar{\rho}}\sqrt{I_3}\left(\mathbf{C}^{-1}\bar{\boldsymbol{E}}\right)\otimes\left(\mathbf{C}^{-1}\bar{\boldsymbol{E}}\right)\right.\\ &\left.- \frac{\varepsilon_a}{4\bar{\rho}}\sqrt{I_3}\bar{\boldsymbol{E}}\cdot\left(\mathbf{C}^{-1}\bar{\boldsymbol{E}}\right)\mathbf{C}^{-1}\right]\end{aligned} \tag{172}$$

for the Lagrangian description. In Eqs. (170)–(172), the quantities $a_i'$ read as

$$a_i' = \partial\Psi_{hyp}/\partial I_i, \quad \text{for } i = 1, \ldots, 3. \tag{173}$$

Furthermore, for typical Dielectric Elastomers that are incompressible (i.e. $I_3 = J = 1$) and with deviatoric deformation response which is well described by a Yeoh's hyperelastic model [83], the strain-energy function can be assumed as

$$\Psi_{hyp}(I_1, I_2, I_3, T) = \left[c_1(I_1 - 3) + c_2(I_1 - 3)^2 + c_3(I_1 - 3)^3 - p\left(\sqrt{I_3} - 1\right)\right]\Big/\bar{\rho}, \tag{174}$$

where $c_1$, $c_2$ and $c_3$ are material parameters only depending on the temperature of the material (see for instance in [73]), while $p$ is an hydrostatic pressure which can only be determined from the equilibrium equations and the associated boundary conditions. Accordingly, use of Eq. (174) into Eqs. (170) and (172), respectively provides the following constitutive equations for the Cauchy's and the Nominal stress tensors of incompressible Dielectric Elastomers in isothermal conditions

$$\boldsymbol{\sigma}^T = -p\mathbf{1} + 2\left[c_1 + 2c_2(I_1 - 3) + 3c_3(I_1 - 3)^2\right]\mathbf{B} + \varepsilon_a\boldsymbol{E}\otimes\boldsymbol{E} - \frac{1}{2}\varepsilon_a\boldsymbol{E}^2\mathbf{1}, \tag{175}$$

$$\begin{aligned}\boldsymbol{\Pi}^T &= \mathbf{F}\left[-p\mathbf{C}^{-1} + 2\left[c_1 + 2c_2(I_1 - 3) + 3c_3(I_1 - 3)^2\right]\mathbf{1} + \varepsilon_a\left(\mathbf{C}^{-1}\bar{\boldsymbol{E}}\right)\right.\\ &\left.\otimes\left(\mathbf{C}^{-1}\bar{\boldsymbol{E}}\right) - \frac{\varepsilon_a}{2}\bar{\boldsymbol{E}}\cdot\left(\mathbf{C}^{-1}\bar{\boldsymbol{E}}\right)\mathbf{C}^{-1}\right],\end{aligned} \tag{176}$$

which, by masking the hydrostatic electrically-induced terms into the unknown pressure $p$, identically read as

$$\sigma^T = -p\mathbf{1} + 2\left[c_1 + 2c_2(I_1 - 3) + 3c_3(I_1 - 3)^2\right]\mathbf{B} + \varepsilon_a \boldsymbol{E} \otimes \boldsymbol{E}, \tag{177}$$

$$\Pi^T = \mathbf{F}\left[-p\mathbf{C}^{-1} + 2\left[c_1 + 2c_2(I_1 - 3) + 3c_3(I_1 - 3)^2\right]\mathbf{1} + \varepsilon_a\left(\mathbf{C}^{-1}\bar{\boldsymbol{E}}\right) \otimes \left(\mathbf{C}^{-1}\bar{\boldsymbol{E}}\right)\right]. \tag{178}$$

## 9 Conclusions

This chapter presented a fully-coupled electromechanical continuum model for the study of the isothermal electro-elastic large deformations of general electrostrictive elastomers which are conservative and isotropic. The model comprises the standard equilibrium equations of both electrostatics and finite elasticity, along with their associated boundary conditions, and a specific electro-hyperelastic constitutive equation which accounts for all the coupling effects that arise from the interaction between polarizable elastic bodies and electrostatic fields. As a special case of this electromechanical model, a specific electro-elastic continuum model for Dielectric Elastomer materials has been also deduced. To enable for both immediate understanding on the underlying electro-mechanical coupling and easy implementation in numerical multi-physics simulation environments, the model has been formulated both in the Eulerian (spatial) description and in the Lagrangian (material) description.

**Acknowledgments** Rocco Vertechy acknowledges the financial support from the EC, in the framework of the project PolyWEC - New mechanisms and concepts for exploiting electroactive Polymers for Wave Energy Conversion (FP7–ENERGY.2012.10.2.1, grant: 309139).

## Appendix A. Mathematical Operators

This appendix defines the mathematical operators that have been employed throughout this chapter.

Consider the 3 × 1 vectors, $\boldsymbol{a}$ and $\boldsymbol{b}$, and the 3 × 3 matrices, $\mathbf{A}$ and $\mathbf{B}$,

$$\boldsymbol{a} = [a_i] = \begin{bmatrix} a_1 \\ a_2 \\ a_3 \end{bmatrix}, \boldsymbol{b} = [b_i] = \begin{bmatrix} b_1 \\ b_2 \\ b_3 \end{bmatrix}, \mathbf{A} = [A_{ij}] = \begin{bmatrix} A_{11} & A_{12} & A_{13} \\ A_{21} & A_{22} & A_{23} \\ A_{31} & A_{32} & A_{33} \end{bmatrix}, \mathbf{B} = [B_{ij}] = \begin{bmatrix} B_{11} & B_{12} & B_{13} \\ B_{21} & B_{22} & B_{23} \\ B_{31} & B_{32} & B_{33} \end{bmatrix} \tag{179}$$

The scalar product of vectors $\boldsymbol{a}$ and $\boldsymbol{b}$, yielding a scalar quantity, is defined as

$$\boldsymbol{a} \cdot \boldsymbol{b} = \sum_{l=1}^{3} a_l b_l. \tag{180}$$

The tensor product (or the dyad) of vectors $\boldsymbol{a}$ and $\boldsymbol{b}$, yielding a 3 × 3 matrix, is defined as

$$\boldsymbol{a} \otimes \boldsymbol{b} = \left[(\boldsymbol{a} \otimes \boldsymbol{b})_{ij}\right] = \left[a_i b_j\right]. \tag{181}$$

The product between a matrix **A** and a vector $\boldsymbol{b}$, yielding a 3 × 1 vector, is defined as

$$A\boldsymbol{b} = \left[(\mathbf{A}\boldsymbol{b})_i\right] = \left[\sum_{l=1}^{3} a_{il} b_l\right]. \tag{182}$$

The scalar product between a matrix **A** and a vector $\boldsymbol{b}$, yielding a 3 × 1 vector, is defined as

$$\mathbf{A} \cdot \boldsymbol{b} = \left[(\mathbf{A} \cdot \boldsymbol{b})_i\right] = \left[\sum_{l=1}^{3} a_{li} b_l\right]. \tag{183}$$

Note that Eqs. (182) and (183) differ in the matrix index used for the summation.

The product between matrices **A** and **B**, yielding a 3 × 3 matrix, is defined as

$$\mathbf{AB} = \left[(\mathbf{AB})_{ij}\right] = \left[\sum_{l=1}^{3} A_{il} B_{lj}\right]. \tag{184}$$

The double contraction between matrices **A** and **B**, yielding a scalar, is defined as

$$\mathbf{A} : \mathbf{B} = \sum_{m=1}^{3} \sum_{l=1}^{3} A_{lm} B_{lm}. \tag{185}$$

Given a third-order tensor $\mathcal{A} = \left[\mathcal{A}_{ijk}\right]$, the double contraction of $\mathcal{A}$ and a matrix **B**, yielding a 3 × 1 vector, is defined as

$$\mathcal{A} : \mathbf{B} = \left[(\mathcal{A} : \mathbf{B})_i\right] = \sum_{l=1}^{3} \sum_{m=1}^{3} \mathcal{A}_{ilm} B_{lm}. \tag{186}$$

Given the spatial position coordinate $\boldsymbol{x} = \left[x_1 \quad x_2 \quad x_3\right]^T$, the spatial gradient (grad) of a scalar quantity $\phi = \phi(x, t)$, $t$ being the time variable, is defined as

$$\operatorname{grad}\phi = \left[(\operatorname{grad}\phi)_i\right] = \left[\partial\phi / \partial x_i\right]; \tag{187}$$

whereas the spatial gradient (grad), divergence (div) and rotation (rot) of a vector $\boldsymbol{a} = \boldsymbol{a}(\boldsymbol{x}, t)$, yielding $3 \times 3$ matrix, a scalar and a $3 \times 1$ vector respectively, are defined as

$$\mathrm{grad}\boldsymbol{a} = \left[(\mathrm{grad}\boldsymbol{a})_{ij}\right] = \left[\partial a_i / \partial x_j\right], \tag{188}$$

$$\mathrm{div}\boldsymbol{a} = \sum_{l=1}^{3} \partial a_l / \partial x_l, \tag{189}$$

$$\mathrm{rot}\boldsymbol{a} = \left[(\mathrm{rot}\boldsymbol{a})_i\right] = \left[\sum_{l=1}^{3}\sum_{m=1}^{3} \varepsilon_{ilm} \partial a_m / \partial x_l\right], \tag{190}$$

where $\varepsilon_{ijk}$ is the permutation symbol such that

$$\varepsilon_{ijk} = \begin{cases} 1, & \text{for even permutation of } (i,j,k), \text{ (i.e. 123, 231, 312)} \\ -1, & \text{for odd permutation of } (i,j,k), \text{ (i.e. 132, 213, 321)} \\ 0, & \text{if there is a repeated index} \end{cases}. \tag{191}$$

Besides, the spatial divergence (div) of a matrix $\mathbf{A} = \mathbf{A}(\boldsymbol{x}, t)$ is defined as

$$\mathrm{div}\mathbf{A} = \left[(\mathrm{div}\mathbf{A})_i\right] = \sum_{l=1}^{3} \partial A_{li} / \partial x_l. \tag{192}$$

Given the material position coordinate $\boldsymbol{X} = \begin{bmatrix} X_1 & X_2 & X_3 \end{bmatrix}^T$, the material gradient (Grad) of a scalar quantity $\phi = \phi(\boldsymbol{X}, t)$ is defined as

$$\mathrm{Grad}\phi = \left[(\mathrm{Grad}\phi)_i\right] = \left[\partial \phi / \partial X_i\right]; \tag{193}$$

whereas the material gradient (Grad), divergence (Div) and rotation (Rot) of a vector $\boldsymbol{a} = \boldsymbol{a}(\boldsymbol{X}, t)$ are respectively defined as

$$\mathrm{Grad}\boldsymbol{a} = \left[(\mathrm{Grad}\boldsymbol{a})_{ij}\right] = \left[\partial a_i / \partial X_j\right], \tag{194}$$

$$\mathrm{Div}\boldsymbol{a} = \sum_{l=1}^{3} \partial a_l / \partial X_l, \tag{195}$$

$$\mathrm{Rot}\boldsymbol{a} = \left[(\mathrm{Rot}\boldsymbol{a})_i\right] = \left[\sum_{l=1}^{3}\sum_{m=1}^{3} \varepsilon_{ilm} \partial a_m / \partial X_l\right]. \tag{196}$$

Besides, the material divergence (Div) of a matrix $\mathbf{A} = \mathbf{A}(\boldsymbol{X}, t)$ is defined as

$$\mathrm{Div}\mathbf{A} = \left[(\mathrm{div}\mathbf{A})_i\right] = \sum_{l=1}^{3} \partial A_{li} / \partial X_l. \tag{197}$$

With regard to the divergence of the products between vector $\boldsymbol{a}$ and either matrix $\mathbf{A}$ or scalar $\phi$

$$\mathrm{div}(\mathbf{A}\boldsymbol{a}) = \mathrm{div}\mathbf{A} \cdot \boldsymbol{a} + \mathbf{A}^T : \mathrm{grad}\boldsymbol{a}, \tag{198}$$

$$\mathrm{div}(\phi\boldsymbol{a}) = \phi\mathrm{div}\boldsymbol{a} + \mathrm{grad}\phi \cdot \boldsymbol{a}, \tag{199}$$

$$\mathrm{Div}(\mathbf{A}\boldsymbol{a}) = \mathrm{Div}\mathbf{A} \cdot \boldsymbol{a} + \mathbf{A}^T : \mathrm{Grad}\boldsymbol{a}, \tag{200}$$

$$\mathrm{Div}(\phi\boldsymbol{a}) = \phi\mathrm{Div}\boldsymbol{a} + \mathrm{Grad}\phi \cdot \boldsymbol{a}. \tag{201}$$

# Appendix B. Fundamental Mathematical Theorems

This appendix summarizes the fundamental mathematical theorems that have been employed throughout this chapter.

For any open surface $S(t)$ with bounding closed curve $L(t)$, the Stokes' theorem states

$$\int_{S(t)} \mathrm{rot}\boldsymbol{a} \cdot \boldsymbol{n}\mathrm{ds} = \int_{L(t)} \boldsymbol{a} \cdot \mathrm{d}\boldsymbol{x}. \tag{202}$$

where ds is the infinitesimal surface belonging to $S(t)$ and with unit normal $\boldsymbol{n}$, whereas $\mathrm{d}\boldsymbol{x}$ is the infinitesimal line element belonging to $L(t)$.

For any volume $V(t)$ with bounding closed surface $\partial V(t)$, the Gauss' divergence theorem states

$$\int_{V(t)} \mathrm{div}\boldsymbol{a}\mathrm{dv} = \int_{\partial V} \boldsymbol{a} \cdot \boldsymbol{n}\mathrm{ds}, \tag{203}$$

$$\int_{V(t)} \mathrm{div}\mathbf{A}\mathrm{dv} = \int_{\partial V} \mathbf{A} \cdot \boldsymbol{n}\mathrm{ds}, \tag{204}$$

where dv is the infinitesimal volume belonging to $V(t)$, whereas ds is the infinitesimal surface belonging to $\partial V(t)$ and with unit normal $\boldsymbol{n}$. In the presence of a discontinuity surface $\gamma(t)$, within volume $V(t)$, across which some vector $\boldsymbol{a}$ and tensor $\mathbf{A}$ admit non-continuous values, the Gauss' divergence theorem states

$$\int_{V(t)-\gamma(t)} \mathrm{div}\boldsymbol{a}\mathrm{dv} + \int_{\partial\gamma(t)} [\![\boldsymbol{a}]\!] \cdot \boldsymbol{n}\mathrm{ds} = \int_{\partial V-\gamma(t)} \boldsymbol{a} \cdot \boldsymbol{n}\mathrm{ds}, \tag{205}$$

$$\int_{V(t)-\gamma(t)} \mathrm{div}\mathbf{A}\mathrm{dv} + \int_{\gamma(t)} [\![\mathbf{A}]\!] \cdot \boldsymbol{n}\mathrm{ds} = \int_{\partial V-\gamma(t)} \mathbf{A} \cdot \boldsymbol{n}\mathrm{ds}, \tag{206}$$

where $[\![\boldsymbol{a}]\!] \equiv \boldsymbol{a}^{+} - \boldsymbol{a}^{-}$ and $[[\mathbf{A}]] \equiv \mathbf{A}^{+} - \mathbf{A}^{-}$ indicate the jumps of vector $\boldsymbol{a}$ and matrix $\mathbf{A}$ from the positive (+) side to the negative (−) side of the discontinuity.

For any given scalar quantity $\phi$ $(\boldsymbol{x},t)$, the Reynolds' transport theorem states

$$\frac{\mathrm{d}}{\mathrm{d}t}\int_{V(t)} \phi \mathrm{d}v = \int_{V(t)} \left[\dot{\phi} + \phi \mathrm{div}\boldsymbol{v}\right]\mathrm{d}v = \int_{V(t)} \left[\frac{\partial \phi}{\partial t} + \mathrm{div}(\phi \boldsymbol{v})\right]\mathrm{d}v. \tag{207}$$

## References


1. Zhenyi, M., Scheinbeim, J.I., Lee, J.W., Newman, B.A.: High field electrostrictive response of polymers. J. Polym. Sci. Part B: Polym. Phys. **32**(16), 2721–2731 (1994)
2. Zhang, Q.M., Su, J., Kim, C.H., Ting, R., Capps, R.: An experimental investigation of electromechanical responses in a polyurethane elastomer. J. Appl. Phys. **81**(6), 2770–2776 (1997)
3. Guillot, F.M., Balizer, E.: Electrostrictive effect in polyurethanes. J. Appl. Polym. Sci. **89**(2), 399–404 (2003)
4. Zhang, Q.M., Bharti, V., Zhao, X.: Giant electrostriction and Relaxor Ferroelectric behavior in electron-irradiated Poly(vinylidene fluoride-trifluoroethylene) copolymer. Science **280**(5372), 2101–2104 (1998)
5. Bauer, F., Fousson, E., Zhang, Q.M.: Recent advances in highly electrostrictive P(VDF–TrFE–CFE) terpolymers. IEEE Trans. Dielectr. Electr. Insul. **13**(5), 1149–1154 (2006)
6. Su, J., Ounaies, Z., Harrison, J.S., Bar-Cohen, Y., Leary, S.: Electromechanically active polymer blends for actuation. Proc. Smart Struct. Mater. 2000: Electroact. Polym. Actuators Devices (EAPAD) **3987**, 65–72 (2000)
7. Pelrine, R., Kornbluh, R., Joseph, J.: Electrostriction of polymer dielectrics with compliant electrodes as a means of actuation. Sens. Actuators A **64**, 77–85 (1998)
8. Pelrine, R., Kornbluh, R., Pei, Q., Joseph, J.: High-speed electrically actuated elastomers with strain greater than 100 %. Science **287**(5454), 836–839 (2000)
9. Kofod, G., Sommer-Larsen, P., Kornbluh, R., Pelrine, R.: Actuation response of Polyacrylate dielectric elastomers. J. Intell. Mater. Syst. Struct. **14**, 787–793 (2003)
10. Mathew, G., Rhee, J.M., Nah, C., Leo, D.J.: Effects of silicone rubber on properties of dielectric acrylate elastomer actuator. Polym. Eng. Sci. **46**(10), 1455–1460 (2006)
11. Ludeelerd, P., Niamlang, S., Kunaruksapong, R., Sirivat, A.: Effect of elastomer matrix type on electromechanical response of conductive polypyrrole/elastomer blends. J. Phys. Chem. Solids **71**, 1243–1250 (2010)
12. Lehmann, W., Skupin, H., Tolksdorf, C., Gebhard, E., Zentel, R., Krüger, P., Lösche, M., Kremer, F.: Giant lateral electrostriction in ferroelectric liquid-crystalline elastomers. Nature **410**, 447–450 (2001)
13. Ha, S.M., Yuan, W., Pei, Q., Pelrine, R., Stanford, S.: Interpenetrating networks of elastomers exhibiting 300 % electrically-induced area strain. Smart Mater. Struct. **16**, S280–S287 (2007)
14. Shankar, R., Ghosh, T.K., Spontak, R.J.: Electromechanical response of nanostructured polymer systems with no mechanical pre-strain. Macromol. Rapid Commun. **28**, 1142–1147 (2007)
15. Zhang, H., During, L., Kovacs, G., Yuan, W., Niua, X., Pei, Q.: Interpenetrating polymer networks based on acrylic elastomers and plasticizers with improved actuation temperature range. Polym. Int. **59**, 384–390 (2010)
16. Liu, B., Shaw, M.T.: Electrorehology of filled silicone elastomers. J. Rheol. **45**(6), 641–657 (2001)

17. Carpi, F., De Rossi, D.: Improvement of electromechanical actuating performances of a silicone dielectric elastomer by dispersion of titanium dioxide powder. IEEE Trans. Dielectr. Electr. Insul. **12**(4), 835–843 (2005)
18. Puvanatvattana, T., Chotpattananont, D., Hiamtup, P., Niamlang, S., Sirivat, A., Jamieson, A.M.: Eletric field induced stress moduli in polythiophene/polyisoprene elastomer blends. React. Funct. Polym. **66**, 1575–1588 (2006)
19. Kunanuruksapong, R., Sirivat, A.: Poly(p-phenylene) and acrylic elastomer blends for electroactive application. Mater. Sci. Eng. A **454–455**, 454–460 (2007)
20. Yuse, K., Guyomar, D., Kanda, M., Seveyrat, L., Guiffard, B.: Development of large-strain and low-powered electro-active polymers (EAPs) using conductive fillers. Sens. Actuators A **165**, 147–154 (2011)
21. Madden, J.D.W., Vandesteeg, N.A., Anquetil, P.A., Madden, P.G.A., Takshi, A., Pytel, R.Z., Lafontaine, S.R., Wieringa, P.A., Hunter, I.W.: Artificial muscle technology: physical principles and naval prospects. IEEE J. Oceanic Eng. **29**(3), 706–728 (2004)
22. Nalwa, H.S.: Ferroelectric Polymers: Chemistry: Physics, and Applications, ISBN: 0824794680, CRC Press, New York, (1995)
23. Bar-Cohen, Y.: Electroactive Polymer (EAP) Actuators as Artificial Muscles: Reality, Potential and Challenges, vol. PM136, ISBN: 0-8194-5297-1. SPIE Press, Washington, (2004)
24. Kim, K.J., Tadokoro, S.: *Electroactive Polymers for Robotic Applications Artificial Muscles and Sensors*, ISBN13: 9781846283710, Springer, Berlin, (2007)
25. Carpi, F., De Rossi,D., Kornbluh, R., Pelrine, R., Sommer-Larsen, P.: *Dielectric Elastomers as Electromechanical Transducers: Fundamentals, Materials, Devices, Models and Applications of an Emerging Electroactive Polymer Technology*, ISBN13: 9780080474885, Elsevier, Oxford, (2008)
26. O'Halloran, A., O'Malley, F., McHugh, P.: A review on dielectric elastomer actuators, technology, applications, and challenges. J. Appl. Phys. **104**, 071101 (2008)
27. Dubowsky, S., Hafez, M., Lichter, M., Weiss, P., Wingert, A.: Dielectric elastomer actuated systems and methods. US Patent No. 7411331 (2008)
28. Vertechy, R., Berselli, G., Parenti Castelli, V., Vassura, G.: Optimal design of lozenge-shaped dielectric elastomer linear actuators: mathematical procedure and experimental validation. J. Intell. Mater. Syst. Struct. **21**, 503–515 (2010)
29. Maxwell, J.C.: *A treatise on Electricity and Magnetism*, ISBN13: 9781108014045, Dover Publications Inc., New York, (1891)
30. Larmor, J.: A dynamical theory of the electric and luminiferous medium. Part III, Relat. Mater. Media Phylosophical Trans. Roy. Soc. London: A **190**, 205–493 (1897)
31. Einstein, A., Laub, J.: Uber die Elektromagnmetischen Felde auf Ruhende Korpen Ausgeubten Ponderomotorishen Krafte. Ann. Phys. **331**(8), 541–550 (1908)
32. Minkowski, H.: Die Glundgleichungen fur die elektromagnetischen Vorgange in bewegten Korper. Math. Ann. **68**(4), 472–525 (1909)
33. Lorentz, H.A., *The Theory of Electrons and Its Application to the Phenomena of Light and Radiat Heat*, 2nd edn. ISBN13: 9780486495583, Dover Publications Inc., New York, (1916)
34. Livens, G.H.: *The Theory of Electricity*, 2nd edn. ISBN 13: 9781178439892, Cambridge University Press, London, (1926)
35. Stratton, J.A.: *Electromagnetic Theory*, ISBN13: 9780070621503, McGraw-Hill Book Co., New York, (1941)
36. Smith-White, W.B.: On the mechanical forces in dielectrics. Philos. Mag. Ser. 7 **40**(303), 466–479 (1949)
37. Panofsky, W.K.H., Phillips, M.: *Classical Electricity and Magnetism*, 2nd edn. ISBN13: 9780486439242, Dover Publication Inc., New York, (1955)
38. Toupin, R.A.: The elastic dielectrics. J. Rational Mech. Anal. **5**, 849–915 (1956)

39. Landau, L.D., Lifshitz, E.M.: *Electrodynamics of Continuous Media*, ISBN13: 9780750626347, Addison-Wesley, New York, (1960)
40. Fano, R.M., Chu, L.J., Adler, R.B.: *Electromagnetic Fields, Energy, and Forces*, ISBN13: 9780262561709, Wiley, New York, (1960)
41. Toupin, R.A.: A dynamical theory of elastic dielectrics. Int. J. Eng. Sci. **1**, 101–126 (1963). Pergamon Press, NY
42. Becker, R.: Electromagnetic Fields and Interactions. English edition: Sauter, F. (Electromagnetic theory and relativity, Vol. 1. (Trans: Knudson, A.W.) ISBN13: 9780216874169, Blackie, London, (1964)
43. Chu, L.J., Haus, H.A., Penfield, P.: The force density in polarizable and magnetizable fluids. Proc. IEEE **54**(7), 920–935 (1966)
44. Grindlay, J.: Elastic dielectric. Phys. Rev. **143**(2), 637–645 (1966)
45. Penfield, P., Haus, H.A.: Electrodynamics of moving media, ISBN13: 9780262160193. The MIT Press, Cambridge (1967)
46. Lax, M., Nelson, D.F.: Linear and nonlinear electrodynamics in elastic anisotropic dielectrics. Phys. Rev. B **4**, 3694–3731 (1971)
47. De Groot, S.R., Suttorp, L.G.: *Foundation of Electrodynamics*, ISBN13: 9780720402483, North-Holland Publication Co., Amsterdam, (1972)
48. Van de Ven, A.A.F.: Interaction of electromagnetic and elastic fields in solids. Ph.D. Thesis, Technical University of Eindhoven, The Netherlands, (1975)
49. Pao, Y.H., Hutter, K.: Electrodynamics for moving elastic solids and viscous fluids. Proc. IEEE **63**(7), 1011–1021 (1975)
50. Robinson, F.N.H.: Electromagnetic stress and momentum in matter. Phys. Rep. **16**(6), 313–354 (1975)
51. Tiersten, H.F., Tsai, C.F.: On the interaction of the electromagnetic field with heat conducting deformable insulator. J. Math. Phys. **13**(3), 361–378 (1972)
52. Pao, Y.H.: Electromagnetic Forces in Deformable Media, *Mechanics Today*, vol. 4. Pergamon Press Inc, New York (1978)
53. Hutter, K., Van de Ven, A.A.F.: Field matter interactions in thermoelastic solids. Lecture Notes in Physics, Vol. 710, ISBN: 9783540372394, Springer, Berlin, (1978)
54. Eringen, A.C., Maugin, G.A.: *Electrodynamics of Continua I: Foundations and solid media*, ISBN: 0387969365, Springer, Berlin, (1990)
55. Green, A.E., Naghdi, P.M.: A unified procedure for construction of theories of deformable media. II. Generalized Continua. Proc. Roy. Soc. London A **448**(1934), 357–377 (1995)
56. Yang, J.S., Batra, R.C.: Mixed variational principles in non-linear electroelasticity. Int. J. Non-Linear Mech. **30**(5), 719–725 (1995)
57. Rajagopal, K.R., Wineman, A.: A constitutive equation for non-lincar electro-active solids. Acta. Mech. **135**(3–4), 219–228 (1999)
58. Kovetz, A.: *Electromagnetic Theory*, ISBN13: 9780198506034, Oxford University Press, Oxford, (2000)
59. Bobbio, S.: *Electrodynamics of Materials: Forces, Stresses, and Energies in Solids and Fluids*, ISBN13: 9780121082604, Academic, San Diego, (2000)
60. Dorfmann, A., Ogden, R.W.: Nonlinear electroelasticity. Acta Mech. **174**, 167–183 (2005)
61. McMeeking, R.M., Landis, C.M.: Electrostatic forces and stored energy for deformable dielectric materials. J. Appl. Mech. **72**, 581–590 (2005)
62. Bustamante, R., Ogden, R.W.: Universal relations for nonlinear electroelastic solids. Acta Mech. **182**, 125–140 (2006)
63. Ericksen, J.L.: Theory of elastic dielectric revisited. Arch. Ration. Mech. Anal. **183**(2), 299–313 (2007)
64. Ericksen, J.L.: On formulating and assessing continuum theories of electromagnetic fields in elastic materials. J. Elast. **87**(2–3), 95–108 (2007)

65. Suo, Z., Zhao, X., Greene, W.H.: A nonlinear field theory of deformable dielectrics. J. Mech. Phys. Solids **56**(2), 467–486 (2008)
66. Bustamante, R., Dorfmann, A., Ogden, R.W.: Nonlinear electroelastostatics: a variational framework. Zeitschrift für Angewandte Mathematik und Physik (ZAMP) **60**(1), 154–177 (2009)
67. Trimarco, C.: On the lagrangian electrostatics of elastic solids. Acta Mech. **204**(3–4), 193–201 (2009)
68. Cade, R.: On mechanical action in dielectrics. Proc. Phys. Soc.: Section A **64**, 665 (1951)
69. Smith-White, W.B.: Reply to on mechanical action in dielectrics. Proc. Phys. Soc.: Section A **65**, 289 (1952)
70. Cade, R.: On mechanical action in dielectrics. Proc. Phys. Soc.: Section A **65**, 287 (1952)
71. Hakim, S.S.: Mechanical forces in dielectrics. Proc. IEEE: Part C: Monographs **109**(15), 158–161 (1962)
72. Vertechy, R., Frisoli, A., Bergamasco, M., Carpi, F., Frediani, G., De Rossi, D., Modeling and experimental validation of buckling dielectric elastomer actuators. Smart. Mater. Struct. ISSN 0964-1726, **21**(9), 94005 (2012). doi: 10.1088/0964-1726/21/9/094005
73. Vertechy, R., Berselli, G., Parenti Castelli, V., Bergamasco, M., Continuum thermo-electro-mechanical model for electrostrictive elastomers. J Intell. Mater. Syst. Struct. (2012) doi: 10.1177/1045389X12455855, first published on August 28
74. Abraham, R., Marsden, J.E.: *Foundations of Mechanics*, 2nd revised edition, ISBN13: 9780201408409, Westview Press, Cambridge, (1978)
75. Eringen, A.C.: *Mechanics of Continua*, ISBN13: 9780882756639, Krieger Publication Co., New York, (1980)
76. Eringen, A.C., Maugin, G.A.: *Electrodynamics of Continua II*: *Fluids and complex media*, ISBN: 0387970053, Springer, Berlin, (1990)
77. Truesdell, C., Noll, W.: The Non-Linear Field Theories of Mechanics, 2nd edn. Springer, Berlin (1992)
78. Spencer, A.J.M.: Theory of invariants. In: Eringen, A.C. (ed.) Continuum Physics, vol. 1, pp. 239–353. Academic, New York (1971)
79. Wang, C.C.: On representation for isotropic functions I & II. Arch. Ration. Mech. Anal. **33**, 249–287 (1969)
80. Smith, G.F.: On a fundamental error in two papers of C.C. Wang 'on representation for isotropic functions I & II'. Arch. Ration. Mech. Anal. **36**, 161–165 (1970)
81. Wang, C.C.: A new representation theorem for isotropic functions I & II. Arch. Ration. Mech. Anal. **36**, 166–223 (1970)
82. Smith, G.F.: On isotropic functions of symmetric tensors. Int. J. Eng. Sci. **19**, 899–916 (1971)
83. Yeoh, O.H.: Characterization of elastic properties of carbon-black filled rubber vulcanizates. Rubber Chem. Technol. **63**, 792–805 (1990)

# Editors Biography

**Visakh P. M.** (M.Sc., M.Phil.) is a Research Fellow at the School of Chemical Science Mahatma Gandhi University, Kottayam, Kerala, India. He edited 5 books with Sabu Thomas from Wiley & Springer and more than 8 books in press (from Wiley, Springer, American Chemical Society and Royal Society of Chemistry and Elsevier). He has been invited as a visiting student in Czech Republic (2012, 2013), Italy (2009, 2012), Argentina (2010), Sweden (2010, 2011, 2012), Switzerland (2010), Spain (2011, 2012), Slovenia (2011), France (2011), Belgium (2012) and Austria (2012) for his research work and he published more than 5 publications and more than 10 book chapters. He has attended and presented more than 25 conferences. Now he is working in Charles University, Czech Republic as pre-post doc. His research interests include: polymer nanocomposites, bio-nanocomposites, rubber based nanocomposites, fire retardant polymers and liquid crystalline polymers and silicon sensors.

**Sabu Thomas** (Ph.D.) is a Professor of Polymer Science and Engineering at the School of Chemical Sciences, as well as the Director of Centre for Nanoscience and Nanotechnology, Mahatma Gandhi University, India. He received his Ph.D. in 1987 in Polymer Engineering from the Indian Institute of Technology (IIT), Kharagpur, India. He is a fellow of the Royal Society of Chemistry, London and a member of the American Chemical Society. He has been ranked no.5 in India with regard to the number of publications (most productive scientists). He also received the coveted Sukumar Maithy Award for the best polymer researcher in the country for the year 2008. The research group of Prof. Thomas has received numerous awards and honors for excellent work in polymer science and engineering.

P. M. Visakh et al. (eds.), *Advances in Elastomers I*,
Advanced Structured Materials 11, DOI: 10.1007/978-3-642-20925-3,

**Arup K. Chandra** (M.Tech. & Ph.D.). Rubber Technology from IIT (Indian Institute of Technology), Kharagpur is presently working as Global Head (RM and Compound Development) in M/s. Apollo Tyres Ltd. He worked in different capacities in Technical Department and R&D Centre of M/s. J. K. Industries Ltd. from 1983 to 1997. He is one of the founder members of Indian Rubber Institute (IRI)—Rajasthan Branch and Postgraduate Diploma in Polymer Science and Technology course in Mohanlal Sukhadia University, Udaipur.

He remains the Honorary Secretary of IRI, Rajasthan Branch till he left J. K. Industries. He is member and fellow members of different professional bodies both in India and Abroad. He is a regular visiting Professor, question setter and examiner of polymer Science and Technology Course of different Universities and PGDIRI, Conducted by IIT—Kharagpur. He has contributed more than 120 papers (including conference papers) in different Journal of National/International Repute. He received extensive training/exposure at different leading Research Centre/Institute/Universities and travelled across the Globe (USA, Canada, The Netherlands, Belgium, Germany, Spain, Italy, Korea, Kenya, Malaysia, Singapore, Thailand, South Africa, Hong Kong and China to mention a few).

**Aji. P. Mathew** (Ph.D.) is an Associate Professor in the field of Bionanocomposites at Luleå University of Technology, Luleå, Sweden. She recieved her Ph.D. in Polymer Chemistry, in 2001 from Mahathma Gandhi University, Kottayam, Kerala, India. She worked as Postdoc. at CNRS, Grenoble, France and Norwegian University of Science and Technology (NTNU), before taking up the faculty position at Luleå University of Technology. Prof. Mathew is active in the field of biobased nanoreinforcments, nanocomposites, membranes and medical implants and products and published about 50 papers in these research areas.

FSC
www.fsc.org
MIX
Papier aus verantwortungsvollen Quellen
Paper from responsible sources
FSC® C105338

If you have any concerns about our products,
you can contact us on
**ProductSafety@springernature.com**

In case Publisher is established outside the EU,
the EU authorized representative is:
**Springer Nature Customer Service Center GmbH**
**Europaplatz 3, 69115 Heidelberg, Germany**

Printed by Libri Plureos GmbH
in Hamburg, Germany